PRODUCING SPEECH:
CONTEMPORARY
ISSUES
for Katherine Safford Harris

AIP Series in
Modern Acoustics and Signal Processing

ROBERT T. BEYER, Series Editor-in-Chief
Physics Department, Brown University

EDITORIAL BOARD

BOOKS IN SERIES

Computational Ocean Acoustics, by Finn B. Jensen, William A. Kuperman, Michael B. Porter, and Henrik Schmidt

Oceanography and Acoustics: Prediction and Propagation Models, edited by Alan R. Robinson and Ding Lee

Handbook of Condenser Microphones, edited by George S. K. Wong and Tony F. W. Embleton

Producing Speech: Contemporary Issues for Katherine Safford Harris, edited by Fredericka Bell-Berti and Lawrence J. Raphael

PRODUCING SPEECH: CONTEMPORARY ISSUES

for Katherine Safford Harris

Fredericka Bell-Berti
St. John's University
Jamaica, New York

Haskins Laboratories
New Haven, Connecticut

Lawrence J. Raphael
Herbert H. Lehman College, CUNY
Bronx, New York

The Graduate School, CUNY
New York, New York

Haskins Laboratories
New Haven, Connecticut

American Institute of Physics   **New York**

©1995 by American Institute of Physics
All rights reserved.
Printed in the United States of America.

AIP Press
American Institute of Physics
500 Sunnyside Boulevard
Woodbury, NY 11797-2999

Library of Congress Cataloging-in-Publication Data
Producing speech : contemporary issues : for Katherine Safford Harris/
edited by Fredericka Bell-Berti, Lawrence J. Raphael.
    p. cm.
  Includes indexes.
  ISBN 1-56396-286-1
  1. Speech--Physiological aspects. 2. Harris, Katherine S.
I. Harris, Katherine S. II. Bell-Berti, Fredericka, 1945–
III. Raphael, Lawrence J.
QP399.P76 1995   95-12369
616.7'8--dc20   CIP

10   9   8   7   6   5   4   3   2   1

Contents

Section 2

Producing Speech: The Source

Section 3

Producing Speech: The Utterance

Contributors

Arthur S. Abramson, Haskins Laboratories, 270 Crown Street, New Haven, CT 06511

Peter J. Alfonso, Department of Speech and Hearing Science, University of Illinois, Urbana-Champagne, 901 South Sixth Street, Champagne, IL 61820

Ronald S. Baken, Department of Speech and Language Pathology and Audiology, Teachers' College, Columbia University, Box 180, 525 West 120 Street, New York, NY 10027

Fredericka Bell-Berti, Haskins Laboratories, 270 Crown Street, New Haven, CT 06511

Suzanne E. Boyce, Speech Communication Group, Research Laboratory of Electronics, M.I.T., 44 Cummington Street, Boston, MA 02215

Catherine P. Browman, Haskins Laboratories, 270 Crown Street, New Haven, CT 06511

Melanie McNutt Campbell, Haskins Laboratories, 270 Crown Street, New Haven, CT 06511

Claude Chevrie-Muller, Laboratoire de la Recherche sur Le Langage, L'Hopital de la Salpetriere, Bâtiment de la Pharmacie, 47, Blvd. de l'Hopital, Paris, 75651 CEDEX 13, France

René Collier, Institute for Perception Research, Hearing and Speech Group, P. O. Box 513, 5600 MB, Eindhoven, The Netherlands

Mary T. Cord, Audiology and Speech Center, Walter Reed Army Medical Center, Washington, DC 20307

Sarita Eisenberg, Communication Disorders Department, William Paterson College, 300 Pompton Road, Wayne, NJ 07470

Alice Faber, Haskins Laboratories, 270 Crown Street, New Haven, CT 06511

Carol A. Fowler, Haskins Laboratories, 270 Crown Street. New Haven, CT 06511

Frances J. Freeman, University of Texas at Dallas, Callier Center for Communication Disorders, 1966 Inwood Road, Dallas, TX 75235

Carole E. Gelfer, Communication Disorders Department, William Paterson College, 300 Pompton Road, Wayne, NJ 07470

Louis J. Goldstein, Haskins Laboratories, 270 Crown Street, New Haven, CT 06511

Vincent L. Gracco, Haskins Laboratories, 270 Crown Street, New Haven, CT 06511

Hajime Hirose, Department of Otolaryngology, School of Medicine, Kitasato University, 1-15-1 Kitasato, Saagmihara-shi, Kanagawa 228, Japan

David M. Hogue, University of New Orleans, New Orleans, LA 70148

Kiyoshi Honda, ATR Auditory and Visual Perception Research Laboratory, 2-2 Hikaridai, Seika-cho, Soraku-gun, Kyoto 619-02, Japan

Bruce Kay, Department of Cognitive and Linguistic Sciences, Brown University, Box 1978, Providence, RI 02912

J. A. S. Kelso, Center for Complex Systems, Florida Atlantic University, Boca Raton, FL 33431

Ray D. Kent, Department of Communicative Disorders, University of Wisconsin, 1975 Willow Drive, Madison, WI 53706

Jeff Kinsella-Shaw, Haskins Laboratories, 270 Crown Street, New Haven, CT 06511

H. Betty Kollia, Haskins Laboratories, 270 Crown Street, New Haven, CT 06511

Rena A. Krakow, Department of Speech-Language-Hearing, Temple University, 265-65, Philadelphia, PA 19122

Donna R. Levandowski, Audiology and Speech Center, Walter Reed Army Medical Center, Washington, DC 20307

Leigh Lisker, Haskins Laboratories, 270 Crown Street, New Haven, CT 06511

Anders Löfqvist, Haskins Laboratories, 270 Crown Street, New Haven, CT 06511

Christy Ludlow, National Institutes of Health, Speech and Voice Unit, NIDCD, Bldg. 10, Room 5D38, Bethesda, MD 20892

Kiyoshi Makiyama, Department of Otolaryngology, Nihon University Surugadai Hospital, Tokyo, Japan

Michael D. McClean, Audiology and Speech Center, Walter Reed Army Medical Center, Washington, DC 20307

Nancy S. McGarr, Department of Speech, Communication Sciences, and Theatre, St. John's University, 8000 Utopia Parkway, Jamaica, NY 11439

Richard McGowan, Haskins Laboratories, 270 Crown Street, New Haven, CT 06511

Dale E. Metz, Department of Speech Pathology and Audiology, State University of New York, 208-C Sturges Hall, Geneseo, NY 14454

Doron Milstein, Ph.D. Program in Speech and Hearing Sciences, The Graduate School, CUNY, 33 West 42nd Street, New York, NY 10036

Tova Most, School of Communication Disorders and School of Education, University of Tel Aviv, Tel Aviv, Israel

Kevin Munhall, Department of Psychology, Queen's University, Kingston, Ontario K7L 3N6, Canada

Sieb G. Nooteboom, Trans 10, Research Institute for Language and Speech (OTS), 3512JK Utrecht, The Netherlands

Robert J. Porter, Department of Psychology, University of New Orleans, New Orleans, LA 70148

Lawrence J. Raphael, Department of Speech and Theatre, Herbert H. Lehman College, CUNY, Bedford Park Boulevard, West Bronx, NY 10468

B. Rockwell, The Wilbur James Gould Voice Research Center, Denver Center for the Performing Arts, Recording and Research Center, 1245 Champa Street, Denver, CO 80236

Philip Rubin, Haskins Laboratories, 270 Crown Street, New Haven, CT 06511

Elliot Saltzman, Haskins Laboratories, 270 Crown Street, New Haven, CT 06511

Angelien Sanderman, Institute for Perception Research, Hearing and Speech Group, P. O. Box 513, 5600 MB, Eindhoven, The Netherlands

Masayuki Sawashima, Yokohama Seamen's Insurance Hospital, 137 Kamadai-cho, Hodogaya-ku, Yokohama 240, Japan

Ronald C. Scherer, The Wilbur James Gould Voice Research Center, Denver Center for the Performing Arts, Recording and Research Center, 1245 Champa Street, Denver, CO 80236

Nicholas Schiavetti, Department of Speech Pathology and Audiology, State University of New York, 208-C Sturges Hall, Geneseo, NY 14454

Naoko Shimazaki, Department of Otolaryngology, Nihon University Surugadai Hospital, Tokyo, Japan

Paula A. Square, Graduate Department of Speech Pathology, Faculty of Medicine, University of Toronto, Tanz Neuroscience Building, 6 Queen's Park Crescent West, Toronto, Ontario M5S 1A8, Canada

Maureen Stone, Department of Electrical & Computer Engineering, Johns Hopkins University, Barton Hall, Baltimore, MD 21218

Ingo Titze, National Center for Voice and Speech, University of Iowa, 330 SCH, Iowa City, IA 52242

Kristin Tjaden, Department of Communicative Disorders, University of Wisconsin, 1975 Willow Drive, Madison, WI 53706

Emily A. Tobey, Department of Communication Disorders, Louisiana State University Medical Center, 1900 Gravier, New Orleans, LA 70112

Yishai Tobin, Department of Foreign Literatures and Linguistics, Ben-Gurion University of the Negev, Beersheva, Israel

Betty Tuller, Center for Complex Systems, Florida Atlantic University, Boca Raton, FL 33431

V. J. Vail, The Wilbur James Gould Voice Research Center, Denver Center for the Performing Arts, Recording and Research Center, 1245 Champa Street, Denver, CO 80236

Q. Emily Wang, Haskins Laboratories, 270 Crown Street, New Haven, CT 06511

Gary Weismer, Department of Communicative Disorders, University of Wisconsin, 1975 Willow Drive, Madison, WI 53706

Series Preface

"...Soun is noght but air y-broke"
— *Geoffrey Chaucer*
end of the 14th century

Traditionally, acoustics has formed one of the fundamental branches of physics. In the twentieth century, the field has broadened considerably and become increasingly interdisciplinary. At the present time, specialists in modern acoustics can be encountered not only in Physics Departments, but also in Electrical and Mechanical Engineering Departments, as well as in Departments of Mathematics, Oceanography, and even Psychology. They work in areas spanning from musical instruments to architecture to problems related to speech perception. Today, six hundred years after Chaucer made his brilliant remark, we recognize that sound and acoustics is a discipline extremely broad in scope, literally covering waves and vibrations in all media at all frequencies and at all intensities.

This series of scientific literature, entitled *Modern Acoustics and Signal Processing (MASP)*, covers all areas of today's acoustics as an interdisciplinary field. It offers scientific monographs, graduate level textbooks, and reference materials in such areas as: architectural acoustics; structural sound and vibration; musical acoustics; noise; bioacoustics; physiological and psychological acoustics; speech; ocean acoustics; underwater sound; and acoustical signal processing.

Acoustics is primarily a matter of communication. Whether it be speech or music, listening spaces or hearing, signaling in sonar or in ultrasonography, we seek to maximize our ability to convey information and, at the same time, to minimize the effects of noise. Signaling has itself given birth to the field of signal processing, the analysis of all received acoustic information or, indeed, all information in any electronic form. With the extreme importance of acoustics for both modern science and industry in mind, AIP Press is initiating this series as a new and promising publishing venture. We hope that this venture will be beneficial to the entire international acoustical community, as represented by the **Acoustical Society of America**, a founding member of the American Institute of Physics, and other related societies and professional interest groups.

It is our hope that scientists and graduate students will find the books in this series useful in their research, teaching, and studies.

James Russell Lowell once wrote: "In creating, the only hard thing's to begin." This is such a beginning.

Robert T. Beyer
Series Editor-in-Chief

Preface

During the 1950's, Haskins Laboratories was the principal center of research into the nature of speech perception. When Katherine Safford Harris began her research activity at the Laboratories, her first contributions were to the paramount interests of the day: Acoustic cues to speech perception and modelling the speech perception process. It was the development of the Haskins model, a motor theory of speech perception, that provided the impetus for much of her subsequent research into speech physiology. Motor theory, in its attempt to establish a link between the processes of speech perception and speech production, demanded a clearer understanding of the productive process than was available at the time. The task of mounting the research effort necessary for such understanding fell to her.

It should be understood that she had to start from scratch: Facilities for investigating speech production, either in terms of gross articulation or physiology, were nonexistent at Haskins. Nor was there a cadre of researchers trained to perform such research. Such a state of affairs is, more than three decades later, difficult to imagine. Under Kathy's leadership, Haskins has developed an impressive array of instruments and personnel. The enterprise she established has generated a vast body of influential research literature, one that is unmatched by any other institution. Her presence has attracted many colleagues to Haskins and students to the Ph.D. program in Speech and Hearing Sciences at the Graduate School of the City University of New York, where she is a Distinguished Professor. Her presence in both institutions has attracted some of the brightest and most promising graduate students and some of the most established and productive scholars. They have come from all parts of the United States, from Canada, from Europe, Asia, and Africa. Some have come for (often repeated) short visits. Others have come and stayed for years. Some are still here. Many of them have contributed to the present volume. All are in debt to Kathy. Indeed, even those who have never done research at Haskins are in debt because of the importance of the research she has fostered.

When we decided that it was time to begin this *Festschrift,* we invited scientists from many universities and laboratories in many cities and countries. We invited scientists who have been taught by Kathy Harris, who have worked with her, who have met her at scholarly society meetings—all of them scientists whose work reflects Kathy's interest in all aspects of speech production. The response was overwhelming—of the 36 persons or groups invited, all accepted. In the end, all but two were able to complete manuscripts in time for inclusion. Of importance to note, in addition, was the enthusiasm of the contributors: some said "It's about time!" and others said "Do you really think my work is good enough to honor Katherine Harris?"

Katherine Safford Harris has had a profound influence not only on our understanding of speech, and especially speech production, but also on the ways in which we study speech. Her influence extends beyond narrowly defined areas of academic study, reaching at the least to departments of speech-language pathology, psychology, neurology, linguistics, and engineering, and beyond even national geographical boundaries. This volume represents only the smallest part of that influence.

Fredericka Bell-Berti
New Haven, Connecticut

Lawrence J. Raphael
New Haven, Connecticut

Acknowledgments

Any project like this necessarily involves the help of many individuals and the support of many institutions. At the end of each chapter, the contributors have provided their individual acknowledgments; what remains, then, is for us to recognize our indebtedness publicly, and offer our thanks to those who have contributed so much to this project.

First, we must thank the Haskins Laboratories, and the Laboratories' President, Carol A. Fowler, for their support and encouragement for this project—this book would have been impossible without the Laboratories' facilities and support. We must also acknowledge our gratitude to the National Institutes of Health, and particularly the National Institute of Deafness and Other Communication Disorders, whose support of the research program at Haskins Laboratories has made possible many of the papers in this volume, as well as the volume itself.

We must also acknowledge the contributions of Maria Taylor and the editorial staff at the American Institute of Physics. When we were concerned, upon realizing that the *Festschrift* would have to be longer than the originally proposed length (because of the *abbondanza* of papers), they did not bat an eye! And from the day that Charles Doering, Senior Editor at AIP, took over the project, and Kim Kleinstiver became the Editorial Assistant on the project, all questions and worries have been resolved almost before they occurred.

We saved the best for last: The enthusiasm as well as the skill of Yvonne Manning-Jones have not only made this volume the joy it is to look at, but have also improved its stylistic quality and substantially reduced its errors. Without her skill and judgment (known by now to all the contributors) this book would have taken months longer and would not be nearly as esthetically pleasing. The standard answer to our questions to each other about how to handle a problem was "Ask Yvonne." She always knew—how to format text, how to format problematic references, how to create an index—and was always pleasant in repeating a set of directions for the twelfth time. We can never thank her enough.

Introduction

Almost from the outset of her career, the research that Katherine Harris performed and promoted looked in the directions of both normal and abnormal speech production. That it is essential to study both of these aspects of speech follows from her view of the extreme importance of speech as human behavior. Accordingly, research into both normal and abnormal speech behavior must also be extremely important. Furthermore, she has often explained how the study of speech disorders can provide a window through which we can observe normal behaviors and learn much about the control systems of speech production. Because of the critical value studies of speech disorders have for both the ultimate remediation of the disorders and a more complete understanding of speech production in general, we have included chapters on normal speech production and chapters on speech disorders within four of the sections of this *Festschrift*.

The papers in this volume are organized into five sections, each with the same general heading: Producing Speech. The subheadings for the five sections are: The Segment, The Source, The Utterance, Feedback, The Segment (Reprise). Section 1, The Segment, contains chapters describing research and theories on the nature of the segmental structure and organization of speech, both in normal speakers and in adults and children with articulatory deficits. Section 2, The Source, contains chapters describing research and theories on respiratory and phonatory functions, directed towards understanding normal respiratory and phonatory behaviors as well as how those behaviors are disrupted/distorted by several communication disorders. Section 3, The Utterance, contains chapters that examine suprasegmental patterns in speech in fluent and disfluent speech. Section 4, Feedback, contains chapters on the role of feedback in speech production and the effects of the absence of auditory feedback on speech production. Finally, in Section 5, The Segment (Reprise) we return to studies that address theoretical questions about the segmental organization of speech.

Producing Speech: The Segment

1   Limited Lookahead in Speech Production

Sieb G. Nooteboom
Research Institute for Language and Speech (OTS)

Many years ago, at a meeting of the Acoustical Society of America, Katherine Harris asked me whether declination of pitch in speech could provide information on the amount of preprogramming in speech production. I do not recall my answer, but it must have been very unsatisfactory. Below I make a second attempt, considering also some other sources of empirical evidence on preprogramming in speech.

1. INTRODUCTION

Wolfgang Amadeus Mozart, reporting about a new composition taking form in his mind, said: "The work grows; I keep expanding it, conceiving it more and more clearly until I have the entire composition finished in my head though it may be long. Then my mind seizes it as a glance of my eye a beautiful picture or a handsome youth. It does not come to me successively, with various parts worked out in detail, as they will later on, but in its entirety that my imagination lets me hear it" (Hadamard, 1945; cf. Penrose, 1991).

Obviously, composing takes time, but "hearing" the end result as "inner music" for Mozart was immediate, all the music being mentally present simultaneously to be written or performed. Analogously, we can imagine a

3

gifted orator who "composes" a long speech in his mind, where the whole speech sits waiting, as a massive phonetic plan, to be either further refined by mental editing or to be spoken. People like Mozart or our imaginary orator (not fully imaginary: I have known at least one man who seemed to prepare his speeches as Mozart did his compositions), are, however, rare. Most composers and most orators have either to use script as an external memory during preparation or to improvise while performing. One can, of course, also learn a whole composition or long speech by heart before performing, but this rarely seems to lead to the experience described by Mozart. Learning by heart involves all kinds of memory tricks that help in retrieving all details in the right order from long term memory. It does not normally lead to a whole composition or speech being simultaneously present to the conscious mind.

Yet, during normal spontaneous speech production, there must be something in the mind before it can be produced. Interestingly, it does not help very much to look into your own mind while speaking, and ask "What's there? What is my phonetic plan? How many words does it contain? Which are the details of speech taken care of by my phonetic plan, and which are the ones only added by the output mechanisms?" During actual speech production the mind does not seem to have the ability to keep track of its own workings to such an extent that these questions can be answered. The same is true for speech that is imagined but not spoken. Although most people I asked understand the experience of inner speech, no one could give me any detailed answers as to, for example, the amount of inner speech that is simultaneously present to the conscious mind.

This brings us to observing overt speech, in the hope that it will provide a window on the mental processes underlying speech production. In doing so, we use our mental capacities to study our own mental processes as if they belonged to some alien species. Let us look at the following quotation from Chapter 1 of the *Speech Science Primer* by Borden and Harris (1980, p. 11):

"It seems probable that chunks of the message are briefly stored in a buffer (temporary storage) ready for output. The chunks are perhaps of sentence length or phrase length. Evidence for this storage comes from slips of the tongue. The fact that people make mistakes such as 'He cut the knife with the salami,' Victoria Fromkin's example, indicates the existence of such a buffer in order for the speaker to have substituted what should have been the last word for the fourth-to-last word." The quotation is revealing. "It seems probable..." adequately suggests that it is not easy to say anything definite about the contents of the temporary storage supposedly preceding overt speech. The suggestion that the "chunks" might be of sentence length or phrase length, although perhaps somewhat naive in not taking into account the possibility of incremental processing during phrase or sentence production, is, precisely in its naiveté, suggestive of how slippery this area of research is. The quotation also shows how phenomena in overt speech might be used as hints about the underlying mental processes. The error of speech quoted from Fromkin would suggest that the phonetic plan contains at least four words, in this case "salami with the knife," unintendedly realized as "knife with the salami." But note that we cannot be sure that at the level of programming where the error occurs the function words "with the" were already selected. It is at least imaginable that there exists a level of mental programming at which content words are selected together with all the semantic, syntactic, and morphological information necessary for later insertion of function words and affixes. If so, the error would be suggestive of only two words being simultaneously present in the mental program at that level, or of one word "lookahead" as it is defined by Levelt (1989). Function words and affixes would then be comparable to the "various parts worked out in detail" that do not enter into Mozart's timeless mental picture of a symphony.

Errors of speech are not the only source of hints about the size of the mental program for speaking. Acoustic/phonetic details like vowel duration or the course of pitch, depending on aspects of the message yet to come, may also tell us something about the size of lookahead during speaking. Below I will discuss some relevant observations from four different areas: starting frequency and declination of pitch, accent-lending rises and falls in Dutch intonation, anticipatory shortening, and errors of speech. At some points I will take issue with statements or arguments in Chapter 10 of Levelt's admirable book "Speaking: From intention to articulation"(1989). Throughout this book Levelt assumes, as many others have done, that the mental production of speech is a multi-stage process, including, for example, separate stages for conceptualizing, grammatical encoding, phonological encoding, and articulating, and that it is incremental, meaning that the next stage of processing does not have to wait until the output of the preceding stage has been completed. Rather, processing at each successive stage starts on the basis of partial, incomplete output from the preceding stage. Therefore, processing at different stages overlap in time. Much less lookahead is necessary if we assume incremental rather than serial processing. In fact, Levelt throughout his book assumes that the minimally necessary lookahead at the level of the "phonetic plan," being the output of phonological encoding and the input of articulation, is only one word, meaning that the phonetic plan does not have to contain more than two words at a time, the word under production and the next word. Occasionally more lookahead may occur, for example, during reading aloud, or perhaps in very gifted speakers with an unusual mental attention span, but this is not necessary for the production of fluent speech. More lookahead may, according to Levelt, make speech more esthetically pleasing, though. The reader may notice that it would not be easy to show that Levelt's position is incorrect. There is very little systematic acoustic/phonetic evidence relating to spontaneous speech. Any evidence of more than one-word lookahead can be explained as being occasional, and not indicative of what is minimally necessary for the production of fluent speech. What I set out to do below, then, is not to argue that on the level of the phonetic plan more lookahead is minimally necessary for the production of acceptable spontaneous speech, but to review some empirical evidence as to the amount of lookahead actually present during normal speech production. It will also become apparent that at some points, notably with respect to spontaneous speech, empirical evidence is insufficient. I will assume that on the level of the phonetic plan, which results from phonological encoding, all words are spelled out from early-to-late as sequences of segments, together with temporal structures and pitch patterns. The output of grammatical encoding, containing a sequence of lexical items with grammatical structure, is not necessarily fully specified as a sequence of all morphemes to be produced.

2. DECLINATION

During the course of coherent stretches of speech, often called intonational phrases, the pitch gradually drifts down. This phenomenon is known as declination (Cohen & 't Hart, 1967). It has also been observed that declination gets less steep with increasing length of the intonational phrases concerned (Cooper & Sorensen, 1981; De Pijper, 1983; 't Hart, Collier, & Cohen, 1990; Ohala, 1978). As mentioned by Levelt (1989, p. 400), this has been taken as evidence for preprogramming over the length of an intonational phrase. Levelt points out (p. 400), however, that, first, declination is much more clearly present in reading than in spontaneous speech, and second, that the causal relation may be inverse: "If a speaker, for whatever reason, makes his pitch decline rapidly, he will sooner feel the urge to rest. This may induce him to take an early break option. Consequently, the running intonational phrase will be a short one." There is , however, something Levelt does not mention. Both in reading aloud and in spontaneous speech, pitch tends to move towards the same speaker-specific value at the ends of intonational phrases, even though its value at the beginnings of intonational phrases may vary considerably, being higher in longer phrases (Cooper & Sorensen, 1981; 't Hart et al., 1990). This suggests that speakers adapt their starting frequency to the length of the intonational phrase to be spoken, or, conversely, that speakers adapt the length of the intonational phrase to both the starting frequency and the slope of declination.

Willems (1983) measured declination slopes in 35 read-out British-English sentences and in 35 spontaneous British-English utterances varying in duration from 0.6 to 6.3 seconds. He compared measured slopes with slopes predicted from the following formulas:

For $t < 5$: $D = -11 / t + 1.5$
For $t > 5$: $D = -8.5 / t$

in which D is the slope (in semitones) to be calculated and t is duration in seconds. The formulas predict a slope of –5 semitones/second for the shortest utterance and of –1.35 semitones/second for the longest utterance. End frequencies were set at speaker-specific fixed values. Average differences between predicted and measured slopes were −0.51 semitones/second for reading aloud and −0.3 semitones/second for spontaneous utterances. Standard deviations were 0.79 semitones/second for reading aloud and 2.05 semitones/second for spontaneous speech.

't Hart, Collier, and Cohen (1990, p. 134) conclude from Willems' data "...that speakers apply a certain amount of preplanning: if the duration of the utterance they are about to produce is known in advance, they can choose a start frequency and a slope suitable to finish at their individual end frequency. Estimating the duration can, understandably, be done fairly accurately if the speaker is reading from text. The fact that such a preprogramming is less successful in individual cases of spontaneous speech does not entirely rule out its occurrence." I would add that, given that on the average the effect shows up in a corpus of only 35 spontaneous utterances, there is at least the suggestion that such preplanning is not exceptional. Unfortunately, the corpus is too limited and the variance for spontaneous utterances too big to say anything definite on the size of the supposed lookahead. In principle, this could be remedied in the future by carrying out similar studies with more extensive databases. Let us assume for the moment that the correlation between starting frequency and length of intonational phrases for individual speakers would be significant up to a length of intonational phrases of three or four words.

What then about Levelt's inverse causal relation? Logically we could argue that, if a speaker for whatever reason starts high, he would then be inclined by whatever mechanism not to let his pitch drift down very rapidly, and thus he would not be forced by his pitch getting too low to use an early optional breakpoint, and would often go for a later optional breakpoint. This would explain the correlation between starting frequency, slope of declination, and sentence length without the assumption of considerable lookahead. There is, however, a problem with this line of argumentation. Many intonational phrases are also sentences, or at least complete utterances, the shortest being only one word in length. If we believe in Levelt's inverse causal relation, we have also to assume that, in case a speaker for whatever reason starts on a relatively high pitch with a relatively slow declination, this causes him to make a longer sentence. So now we have to believe that the number of words in an utterance or spoken sentence is determined by the relative height of pitch and the relative slope of declination at the moment speaking starts, and not by the conceptual intentions of the speaker. I would not be surprised if even Levelt would find this hard to believe. Levelt's inverse causal relation is hard to disprove, but rather implausible.

In summary, both starting frequency and slope of declination in intonational phrases correlate with the length of the phrase. Assuming, according to Levelt's line of argumentation, that phrasal length is caused by high starting frequency and slope of declination, forces us also to assume that the number of words in utterances consisting of one intonational phrase is caused by the way declination is programmed. This seems highly improbable. Instead, in principle starting frequency and slope of declination can be used as indicators of the amount of lookahead at the level of the phonetic plan. For reading aloud, available evidence suggests that lookahead corresponds to a stretch of speech lasting a number of seconds, and containing more than a few words. There are to my knowledge no studies available from which the amount of actual lookahead can be estimated with any precision. For spontaneous speech, data on the correlation mentioned are too limited and variable to give any estimate of the lookahead involved. In principle, this could be remedied by further research on larger data bases. In such research one should focus on complete utterances, consisting of one intonational phrase.

3. THE HAT PATTERN

In Dutch one of the possible pitch configurations consists of an accent-lending rise followed by an accent-lending fall. This configuration is called the "hat pattern" (Cohen & 't Hart, 1967; 't Hart et al., 1990). In normal emotionally and attitudinally neutral utterances, an accent-lending rise has to be followed by an accent-lending fall: what goes up must come down. So the very moment a speaker makes an accent-lending rise, he must, in order to produce a correct hat pattern, have sufficient lookahead to know that there will be an accented word within the same intonational phrase to be marked with an accent-lending fall. The distances between such rises and falls in spontaneous speech may, therefore, be indicative of lookahead in the speech program, as pointed out by Levelt (p. 405): "Lookahead is a condition for producing the hat pattern." He also points out (p. 405) that "the hat pattern is a more likely pitch contour when two accents are to be made in close succession," thus suggesting that in actual practice lookahead is fairly limited. Of course, what we need here are real data. Some relevant data can be found in Collier (1972) who among other things measured the distances between accent-lending rises and falls in hat patterns in a corpus of 750 spontaneous Dutch utterances. Unfortunately for the present purpose, distances were calculated in syllables, not in words. Below, distances are recalculated in words on the basis of the average word length of 1.8 syllables in Dutch spontaneous speech. Sixty-five percent of Collier's utterances contained a hat pattern. In 39% of all hat patterns rise and fall fell compulsorily on the same syllable, and thus on the same word, because there were no later accented words in the utterance that could attract a fall. Of the remaining cases, where speakers had an option to produce the hat pattern over more than one word, 46% still had rise and fall on the same word, with no necessary lookahead. The next and last accent in such cases was produced by a new rise-fall configuration. This may indicate that speakers tend to avoid the necessity of considerable lookahead. Of all cases where speakers did take the option of a hat pattern over more than one word, 32% had rise and fall on two consecutive words, meaning that there was one-word lookahead, 12% had a rise and fall span of three words, indicating two-word lookahead, and the remaining 10% covered estimated spans of more than four words, suggesting more than three-word lookahead. Thus, in 22% of these cases, corresponding to 8.5% of the utterances in the entire corpus of 750 utterances, lookahead had to be more than one word.

These numbers can be taken to support Levelt's idea that generally lookahead is not necessarily more than one word, "generally" meaning here in about 92% of utterances. On the other hand one can argue that it is only possible to say anything about the lookahead involved in those cases where speakers had an option of producing a hat pattern over more words. Of those cases, 22% show a minimum lookahead of more than one word, suggesting that a lookahead of more words, although still not attested in the majority of cases, is far from exceptional. Of course, we cannot be sure that if speakers decline to take the option of an extended hat pattern, they do so because of lack of necessary lookahead. They may have other reasons, for example, of a melodic nature. If that were generally the case, we could only say something about the lookahead in those cases where people are so kind to take the option of making an extended hat pattern. Of those cases 60% show a minimum lookahead of one word, 22% of two words, and 18% of more than two words. So in all, 40% of those cases where we really can say something about the minimum lookahead required, show a lookahead of more than a single word. But there seems to be no way in which we can make out whether this is representative of the amount of lookahead at all other moments during speech production.

Assuming that Collier's corpus is representative of Dutch spontaneous speech in general, we may conclude that in at least one out of every twelve Dutch spontaneous utterances lookahead is more than one word. For 11 out of 12 utterances there is no such evidence, one way or the other.

4. ANTICIPATORY SHORTENING

Anticipatory shortening is a well known and systematically occurring phenomenon in speech: the segments of a syllable, particularly a stressed syllable, become shorter as more syllables follow within the same word (Lindblom, 1968; Nooteboom, 1972). Anticipatory shortening is not limited to the level of words. The segments of a stressed word also shorten as the number of syllables or words coming later in the same intonational phrase increases (De Rooij, 1979; Nakatani, O'Connor, & Aston, 1981). This suggests that the number of following words that still contribute to the shortening of the first syllable might be indicative of the amount of lookahead. Levelt, discussing this phenomenon, here again resorts to the inverse causal relation: "...one might conjecture that the speaker 'blindly' increases the duration of successive stressed syllables till he reaches the end of the intonational phrase, and that he then resets the duration parameter to the initial value for the next phrase"

(Levelt, 1989, p. 390). Levelt then continues: "Another possible explanation involves nuclear stress. Phrase-final stresses naturally 'grow' toward the end of the sentence, owing to the mechanisms of the Nuclear-Stress Rule. And more heavily stressed syllables tend to be longer." In both explanations it is implied that durations of stressed syllables spoken in isolation are shortest, and that they get longer as more material is added in front of the syllable concerned: if a speaker, by lack of sufficient lookahead, does not know how many words are following, his realization of any stressed word should be such that it is suitable for being the last or one-but-last word in an utterance. This is contrary to what we know about the temporal structure of speech: a word spoken in isolation and the same word at the end of an utterance have approximately the same duration. (See, for example, relevant data in De Rooij, 1979.) The word duration gets systematically shorter as more material follows in the utterance, it does not get systematically longer, compared to the duration in isolation, as more material precedes it. The anticipatory shortening effect can be illustrated with some data from De Rooij (1979), who measured durations of sequences of a vowel plus following plosive silent interval in the Dutch stressed monosyllabic words [pe:t] and [Xa:t], as a function of increasing the number of following words in the sentence. Durations, for the present purpose averaged over 12 different cases, were as follows:

TABLE 1. *Average durations of vowel plus silent interval in Dutch stressed monosyllable words, for 0 – 5 following words in the same utterance. N = 12. (After De Rooij (1979).)*

0 following words:	242 ms
1 following word:	216 ms
2 following words:	186 ms
3 following words:	176 ms
4 following words:	167 ms
5 following words:	167 ms

That a duration of 167 ms would not be suitable when no words follow in the utterance, can easily be confirmed in a simple classroom demonstration: an utterance-initial word, artificially isolated from the utterance by gating, sounds unacceptably short. An utterance-final word presented in isolation sounds all right. This demonstration also works after intonational cues have been made identical by means of LPC analysis, manipulation of pitch, and resynthesis. De Rooij's data suggest that his speakers had a lookahead of three or four words.

This may not be surprising: these data were obtained from read sentences, where lookahead was prepared for the speakers in print. To my knowledge no such or similar data exist for spontaneous speech. If in future research similar effects are found in spontaneous speech, these effects can be interpreted in terms of lookahead during speech production.

In summary, anticipatory shortening is really what the term says: it is not perseveratory lengthening, as suggested by Levelt. Therefore, anticipatory shortening can be used as an indication of the amount of lookahead in speech production. In read sentences, lookahead appears to be in the order of three or four words. Data on spontaneous speech are still lacking.

5. ANTICIPATORY ERRORS OF SPEECH

When someone says: "sil...filter cigarette" (example taken from Hockett, 1967), we assume that the [s] of "cigarette" is anticipated inappropriately in the pronunciation of the word "filter," replacing the [f] of "filter." This can only be explained by assuming a lookahead of at least one word. By the same token a slip like "knife with the salami" instead of "salami with the knife" seems to suggest a lookahead of at least three words. From Lashley (1951) onwards, many students of speech errors have argued in these or similar terms that anticipatory errors show "that speakers must have access to a representation that spans more than the next word of the utterance" (Shattuck-Hufnagel, 1979). Of course, as errors of speech themselves by their very nature are the exception rather than the rule, they do not tell us what the amount of lookahead normally is. They can only provide hints about what the lookahead sometimes can be.

In order to discuss errors of speech in terms of lookahead, we have to distinguish between 'origin' and 'target.' 'Origin' is the position where a particular entity belongs in the error-free version of the utterance. 'Target' is the position where this entity ends up in the speech error. So in the lapse quoted from Hockett, the origin is the first phoneme of "cigarette," and the target is the first phoneme of "filter." We assume that the distance between target and origin in the intended utterance provides a clue as to the amount of lookahead at the time the error was produced. As we cannot be certain that speech errors involving phonemes as misplaced entities and speech errors involving morphemes or words as entities are generated at the same stage of speech programming, the two classes of errors are best kept separate.

Many years ago, I made some counts of distances between targets and origins in a collection of Dutch and German errors of speech (cf. Nooteboom and Cohen, 1975). For phonological errors distances were expressed in syllables. This gave the following numbers:

TABLE 2. *Distances in syllables between origin and target in phonological anticipatory errors of speech in Dutch and German. The syllable containing the origin is, and the syllable containing the target is not, counted (N = 1057).*

1 syll. lookahead:	34%
2 syll. lookahead:	29%
3 syll. lookahead:	16%
4 syll. lookahead:	10%
5 syll. lookahead:	3%
> 5 syll. lookahead:	8%

Recalculating these numbers on the basis of an average word length of 1.8 syllables (an estimate obtained from the spontaneous utterances in the same collection of speech errors), gives the following estimates of lookahead in numbers of words:

TABLE 3. *Distances in words in phonological anticipatory errors of speech in Dutch and German. The word containing the origin is, and the word containing the target is not, counted (N =1057).*

0 words lookahead:	c. 15%
1 word lookahead:	c. 50%
2 words lookahead:	c. 23%
> 2 words lookahead:	c. 12%

In order to check whether these estimates are at all realistic, I counted lookahead spans expressed in words in the selection of errors published in Fromkin (1973), assuming that the selection was random with respect to material spans involved, and skipping all nonanticipatory errors and lexical errors and also some errors I found hard to interpret. I also excluded a list of within-word errors because these were obviously selected according to the material span involved. There remained 231 anticipatory phonological errors, of which 10 (4%) showed zero lookahead, 129 (56%) one-word lookahead, 62 (27%) two-word lookahead, 22 (9%) three-word lookahead, and 9 (4%) of four-word lookahead. Given the limited size of this sample, these numbers come reassuringly close to the above estimates. Some examples of Fromkin's speech errors are:

TABLE 4. *Examples of phonological errors of speech, taken from Fromkin (1973).*

0 words lookahead:	significantly	significlantly
1 word lookahead:	roman numeral	noman numeral
2 words lookahead:	a Canadian from Toronto	a Tanadian...
3 words lookahead:	the hiring of minority faculty	the firing of...
4 words lookahead:	Paris is the most beautiful city	Baris...

The estimates obtained from the Dutch/German corpus indicate that in 35% of anticipatory phonological speech errors lookahead is more than a single word. This is a considerable proportion, suggesting that lookahead of more than one word may not be all that exceptional.

Not only phonemes move around in speech errors. Morphemes and whole words also get misplaced. The following numbers were obtained in counting words between targets and origins for anticipatory speech errors involving lexical items (morphemes and words) as entities changing position:

TABLE 5. *Distances in words between origin and target, in anticipatory lexical errors of speech in Dutch and German. The origin is, the target is not, counted (N = 147).*

0 words lookahead:	7%
1 word lookahead:	34%
2 words lookahead:	24%
3 words lookahead:	22%
4 words lookahead:	10%
> 4 words lookahead:	3%

Fifty-nine percent of these errors involve a necessary lookahead of more than one word! That is much more than the 35% we estimated for phonological speech errors. Obviously speakers look farther ahead when selecting and ordering lexical items than when spelling out the selected lexical items as strings of ordered phonemes. This seems to provide an answer to a question raised by Shattuck-Hufnagel (1979, p. 329): "Does the size of the span change during the planning process; e.g., is it longer when syntactic structure is being computed, shorter when phonological details are being worked out?" But whether the span changes or not depends on how we count the entities in the span. I will shortly come back to this.

One way of interpreting the difference between the two classes of speech errors is to assume that they reflect two different stages of mental programming. One stage, generating the surface structure, is concerned with selecting and ordering lexical items, and one stage deals with spelling out phonological forms and setting up phonetic plans for speaking these items in coherent stretches of speech. This interpretation is in line with Levelt (1989), who discusses speech errors involving exchanges of words and morphemes in his chapter on the generation of surface structure, and discusses phonological errors in his chapter on phonetic plans for words. Levelt also points out that misplaced lexical items attract the pitch accent, case marking, and inflectional forms that go with their new position:

(a) "the knife with the SALAMI" instead of "the salami with the KNIFE"
(b) "Bis er es bei Dir abholt" instead of "Bis Du es bei ihm abholt"
(c) "Dat is nieuwer dan een dure" instead of "Dat is duurder dan een nieuwe"

In (a) "salami" gets the pitch accent that "knife" should have had in the intended utterance. In (b) the German pronouns for second and first person not only swap positions, but then receive the grammatically correct case markings going with their new positions. In (c) the Dutch content morphemes "nieuw" and "duur" exchange position, and then the comparative suffix changes correctly from "-der" to "-er," adapting itself to the incorrectly placed content morpheme. There are many such examples in the literature, showing that errors of this type take place during grammatical encoding rather than during phonological encoding. Errors of type (b) and (c) show that function words and inflectional morphemes are not yet spelled out phonologically at the moment the speech error is generated. Presumably they are present as abstract syntactic and/or semantic functions or labels, that are about to be attached to appropriately or inappropriately selected content morphemes, and only thereafter receive their phonological form. For the present purpose this means that it is hard to know how to interpret the quantitative data given earlier in terms of lookahead. Do we count formless abstract function words as words, and formless abstract inflections as morphemes? Or do we only count content morphemes in estimating the lookahead? If we do the latter, the distribution of amounts of lookahead for lexical errors becomes much more similar to the one for phonological errors, showing a majority of cases with only one item lookahead. Such items are then, of course, much more grammatically complex than the phonological forms in the phonetic plan. However, there does not seem to be a principled way to decide what should and what should not count as ordered entities at the stage of grammatical encoding.

In summary, phonological speech errors may be used as a source of information about the amount of lookahead at the stage of phonological encoding, generating a phonetic plan. As in an estimated 35% of such errors lookahead is at least two words, we may conclude that lookahead of more than a single word is not exceptional. Lexical speech errors, on the other hand, reflect planning at the stage of grammatical encoding. In terms of the utterance to be produced, speakers look farther ahead at this stage than at the stage of phonological encoding, three- or four-word lookahead not being exceptional. It is difficult, however, to decide on the nature and number of ordered entities actually involved at this stage of programming.

6. DISCUSSION

We have no direct access to the size of the phonetic plan underlying speech production. Quite literally, we do not know what we do when we speak. Estimates of the extent of preprogramming during speech production can only come from indirect evidence, such as acoustic/phonetic aspects of speech depending on what is yet to come, and recorded slips of the tongue. We have seen that lexical speech errors, such as "knife with the salami," cannot be taken to reflect spans of attention on the level of the phonetic plan. They rather betray lookahead during grammatical encoding, often comprising three of four words, and sometimes more. It seems to me that lookahead during grammatical encoding in spontaneous speech production could in principle also be investigated by looking at material spans over which selection dependencies are maintained, for example, dependencies between early auxiliary verbs and later past participles in Dutch. My main concern here, however, is not with grammatical but with phonological encoding.

The brief review of some available empirical evidence given above suggests that, although one-word lookahead may be sufficient for the production of fluent speech, a lookahead of more than a single word is far from exceptional in spontaneous speech production. Often lookahead is two words and occasionally lookahead may be three or four words. The strongest evidence to this effect stems from phonological speech errors. There is no way to know, of course, whether these estimates are biased: It may be that the probability of speech errors increases with the amount of lookahead. If so, the frequencies of occurrence of particular material spans over which errors occur would not reflect frequencies of occurrence of amounts of lookahead in error-free speech production. On the other hand, the material spans counted contain what is minimally necessary to explain the errors concerned. There is no way of knowing whether actual lookahead is generally more than this. Because of such uncertainties it is worthwhile to look at empirical evidence from different sources.

As we have seen, anticipatory shortening of stressed syllables in sentence production can be used as evidence for the amount of lookahead during speech production generally. Available evidence suggests a lookahead of three or four words during reading aloud. Unfortunately, there are no available data on spontaneous speech. This is, of course, not accidental. Speech sound durations are strongly affected by a great many factors that also show strong interactions (Klatt, 1976; Nooteboom, 1991; van Santen, 1992). In order to get a precise estimate of how many upcoming words still have an effect on the shortening of an utterance-initial stressed word or syllable, one needs rather precise control over the speech material. Spontaneous speech yet to be produced, by its very nature does not easily allow us such control. But perhaps in the future this lack of control can be compensated for by using large speech data bases and quantitative techniques such as proposed and tested on read texts by Van Santen (1992). Further data on anticipatory shortening as a measure of lookahead would be particularly interesting because it seems to be virtually always present and would thus potentially provide a relatively rich source of empirical evidence.

We have also seen that the material spans covering rise-fall configurations or hat patterns in Dutch speech melodies betray the amount of lookahead minimally present at the moment the rise is produced. Of course, this situation is hardly representative of what happens at other moments in speech production: Hat patterns over more than a single word are not always allowed and are never obligatory. In many utterances speakers do not have the option to produce a hat pattern over more words, for example, because there is only one accented word in the utterance. In most other utterances there is only one point where the speaker has an option to produce a hat pattern over more words. Of all cases where speakers have this option, they decline taking it in 46% of the cases, possibly because they lack the necessary lookahead at that moment, or for other unknown reasons. It would be interesting to know whether the accented words at which speakers decline to take the option of an extended hat pattern do or do not show anticipatory shortening as a function of the number of upcoming words in the utterance. As it is, the data on hat patterns at least show that a lookahead of more than a single word is not something very special, and that lookahead of three or more words does occur during normal spontaneous speech production.

We started our search for empirical evidence for lookahead with the starting frequency and slope of declination. In principle this could be a fairly reliable indicator of planning ahead in speech production. Although pitch is influenced by the segmental structure of utterances, the influence is minor as compared to what happens to speech sound durations. We do not need very precise control over the speech material in order to test hypotheses as to the correlation between the length of intonational phrases and the starting frequency and slope of declination. A large data base of spontaneous utterances, preferably with many isolated utterances from a single speaker, or from each of a few speakers, should allow us to assess how many upcoming words affect the course of pitch at the onset of the utterance. Data on reading aloud show that there is preprogramming involved. Data on spontaneous speech suggest that there, also, preprogramming is not excluded. Statistical studies specifically directed at the amount of preprogramming are still lacking, though, both for reading aloud and for spontaneous speech. My prediction is that such studies will show that a lookahead of three or four words in spontaneous speech is far from exceptional.

REFERENCES

Borden, G. J., & Harris, K. S. (1980). Speech science primer. *Physiology, acoustics, and perception of speech* (1st ed.). Baltimore: Williams and Wilkins.

Cohen, A., & 't Hart, J. (1967). The anatomy of intonation. *Lingua, 19,* 177-192.

Collier, R. (1972). *From pitch to intonation.* Unpublished doctoral dissertation, Louvain.

Cooper, W. E., & Sorensen, J. M. (1981). *Fundamental frequency in sentence production.* Berlin/Heidelberg/NewYork: Springer Verlag.

De Pijper, J. R. (1983). *Modelling British-English intonation.* Unpublished doctoral dissertation, Utrecht.

De Rooij, J. J. (1979). *Speech punctuation. An acoustic and perceptual study of some aspects of speech prosody in Dutch.* Unpublished doctoral dissertation, Utrecht.

Fromkin, V. A. (Ed.) (1973). *Speech errors as linguistic evidence.* The Hague: Mouton.

Hadamard, J. (1945). *The psychology of invention in the mathematical field.* Princeton: Princeton University Press.

Hockett, C. F. (1967). Where the tongue slips, there slip I. *To Honor Roman Jakobson.* The Hague: Mouton.

Hockett, C. F. (1967). Where the tongue slips, there slip I. In V. A. Fromkin (Ed.), *Speech errors as linguistic evidence* (pp. 93-119). The Hague: Mouton.

Klatt, D. H. (1976). Linguistic uses of segmental duration in English: Acoustic and perceptual evidence. *Journal of the Acoustical Society of America, 59,* 1208-1221.

Lashley, K. S. (1951). The problem of serial order in behaviour. In L. A. Jefferess (Ed.), *Cerebral mechanisms in behavior. The Hixon symposium.* New York: Wiley.

Levelt, W. J. M. (1989). *Speaking: From intention to articulation.* Cambridge/London: MIT Press.

Lindblom, B. E. F. (1968). Temporal Organization of Syllable Production. *Speech Transmission Laboratory, Quarterly Progress Report No 2-3/1968* (pp. 1-5). Stockholm: Royal Institute of Technology.

Nakatani, L. H., O'Connor, J. D., & Aston, C. H. (1981). Prosodic aspects of American English speech rhythm. *Phonetica, 38,* 84-106.

Nooteboom, S. G. (1972). *Production and perception of vowel duration.* Unpublished doctoral dissertation, Utrecht.

Nooteboom, S. G. (1991). Some observations on the temporal organisation and rhythm of speech. *Proceedings of the XIIth International Congress of Phonetic Sciences* (Vol. 1, pp. 228-237). Aix-en-Provence, August 19-24, 1991.

Nooteboom, S. G., & Cohen, A. (1975). Anticipation in speech production and its implications for perception. In A. Cohen & S. G. Nooteboom (Eds.), *Structure and process in speech perception* (pp. 124-142). Berlin/Heidelberg/New York: Springer-Verlag.

Ohala, J. J. (1978). Production of tone. In V. A. Fromkin (Ed.), *Tone: A linguistic survey.* New York: Academic Press.

Penrose, R. (1989) (1991). *The emperors new mind.* London: Vintage (First published in 1989, Oxford: Oxford University Press).

Shattuck-Hufnagel, S. (1979). Speech errors as evidence for a serial-ordering mechanism in sentence production. In W. E. Cooper & E. C. T. Walker (Eds.), *Sentence processing: Psycholinguistic studies presented to Merrill Garrett* (pp. 295-342). Hillsdale, NJ: Lawrence Erlbaum Associates, Publishers.

't Hart, J., Collier, R., & Cohen, A. (1990) *A perceptual study of intonation. An experimental-phonetic approach to speech melody.* Cambridge: Cambridge University Press.

van Santen, J. P. H. (1992). Contextual effects on vowel duration. *Speech Communication, 11,* 513-546.

Willems, N. J. (1983). Towards an objective course in English intonation: standardized precepts. In M. van den Broecke, V. van Heuven, & W. Zonneveld (Eds.), *Sound structures, studies for Antonie Cohen* (pp. 281-296). Dordrecht/Cinnaminson: Foris Publications.

2 Gestural Syllable Position Effects in American English

Catherine P. Browman
Haskins Laboratories

Louis Goldstein
Haskins Laboratories
Yale University

INTRODUCTION

In a 1984 paper, Harris and Bell-Berti observed that "it remains in the future, then, for us to develop a theory of syllabification and coarticulation using evidence gathered from the articulatory domain with a net of mesh that has a smaller gauge than that which has produced our present views" (p. 94). The framework of Articulatory Phonology, which we have been developing over the last several years (Browman & Goldstein, 1986, 1989, 1992), provides one approach to meeting the challenge that they laid out.

In Articulatory Phonology, the basic phonological units are gestures, and lexical units are composed of dynamically defined articulatory gestures and their organization. Each gesture corresponds to the formation of a constriction within one of the relatively independent articulatory subsystems of the vocal tract, and can be modeled as a task-dynamic system or control regime (Saltzman & Munhall, 1989). For oral gestures, constriction tasks are defined along pairs of coordinates called tract variables, each of which specifies one of the two spatial dimensions of a constriction involving the lips, the tongue tip, or the tongue body (for English). In order to characterize the constrictions, movements are analyzed into discrete underlying gestures, and the values of the coefficients for a dynamic equation for each tract variable, or dimension of the constriction, are determined. Velic and glottal opening-and-closing gestures are considered to be unidimensional, and therefore consist of a single tract variable each. Gestures are associated with each other in terms of their phasing; the resultant gestural group is referred to as a gestural constellation.

One phonological property for which gestural constellations provide a particularly interesting account is syllable structure. Often, in nonarticulatory phonologies (e.g., Kahn, 1976) syllable structure is expressed as a hierarchical structure imposed on a linear (or multilinear) sequence of phonological units. One of the reasons that such structures are posited is to account for variation in phonological units as a function of their position within this structure. For example, Kahn attempted to account for allophonic differences in consonants (particularly /t/) as a function of being in syllable initial vs. noninitial position. However, in such approaches there is no intrinsic connection between the configurational structure (the syllable hierarchy) and the kinds of variation that it is described as conditioning.

In the gestural approach, there are configurational properties intrinsic to gestural constellations that identify syllable-initial vs. syllable-final consonants without necessarily invoking any hierarchical structure. Thus, affiliated consonants both preceding and following a vowel ('syllable-initial' and 'syllable-final' consonants) are phased with respect to the vowel, but in different ways. Moreover, because the configurational properties (the phase relations) also determine the patterns of overlap among gestures, differences in configuration are automatically associated with differences in overlap, and in turn, with the articulatory and acoustic consequences of such overlap. Thus, there is an intrinsic, lawful relation between configurational patterns that correspond to different syllable positions, and (at least some of) the kinds of articulatory and acoustic variation that can be observed for a given phonological unit. Put another way, within this view syllable structure is a characteristic pattern of coordination among gestures; certain types of variation are automatic consequences of this pattern of coordination, and are, therefore, necessary correlates of syllable structure. Moreover, this view entails that, although the exact nature of the observed correlates will vary as a function of the nature of the gestures being organized, the patterns of coordination among disparate gestures may be identical. We will see an example of such identity of coordination in the next section.

TIMING OF GESTURES FOR NASALS AND LATERALS

In American English, both nasals and laterals consist of two gestures. Nasals, of course, involve a velic opening gesture in addition to an oral closure gesture.

Apparently, laterals also consist of two gestures. Sproat and Fujimura (1993) showed that in American English both light and dark [1] (i.e., both syllable-initial and syllable-final [1]) consist of two movements, one involving the tongue tip and one involving the tongue dorsum. In the language of Articulatory Phonology, this would mean that [1] consists of two gestures. This double aspect of [1] can be seen in Figure 1, which shows a mid-sagittal view for a sustained [1] for six different speakers of American English, acquired using Magnetic Resonance Imaging. From this perspective, any lateral side-channel could be purely a secondary consequence of the stretching of the tongue caused by the tongue tip raising toward the upper teeth and tongue dorsum retraction. That these two gestures, without specification of lateral channels, can indeed produce a percept of [1] was demonstrated using a modified version of the Haskins vocal tract model to produce the shape shown in Figure 2, to which there are no side channels added (see Rubin, Baer, & Mermelstein, 1981, for a description of the original model). This shape produced an acoustic signal that sounded like an [1] in informal perceptual testing.

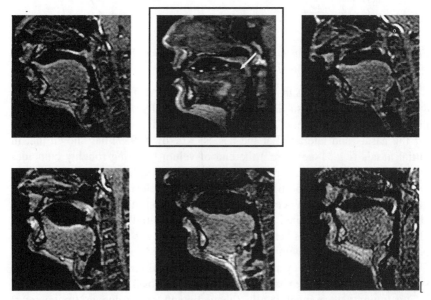

FIGURE 1. MRI images of [1] (held for several seconds) for 6 speakers of American English. The speakers lived in different regions between the ages of 0 and 17: left to right, top row, Connecticut; Chicago (0-13)/Cincinnati (13-17); Washington state; left to right, bottom row: Texas/Oklahoma; San Francisco area (0-7)/primarily Australia (7-12)/Minnesota (12-17); Philadelphia. The image in the middle of the top row that is outlined with a box is from the speaker for whom X-ray information on [1] is reported in Figures 3 through 5. The white ellipses on the surface of the tongue approximate the placement of the gold pellets in the microbeam experiments. Movement data for the frontmost pellet and for a pellet in the position indicated by the arrow are shown in Figure 3.

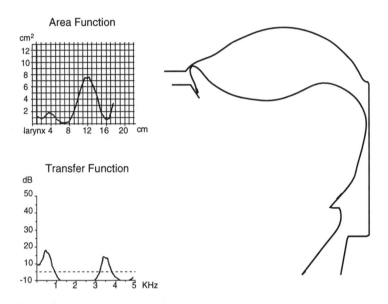

FIGURE 2. Bi-gestural [1] in vocal tract model (with no side channels). Midsagittal view on right, area function and transfer function on left.

Both nasals and laterals in American English show differential timing between their component gestures syllable-initially and -finally. Moreover, as discussed in Browman and Goldstein (1992), the temporal patterns appear to be the same for the nasals and laterals. For the nasals, Krakow (1989) has shown that in initial nasals (e.g., "see more") the end of velum lowering roughly coincides with the END of the lip closing movement, whereas in final nasals (e.g., "seem ore") it coincides with the BEGINNING of the lip closing movement. That is, the velic opening gesture occurs much earlier with respect to its associated oral closure when syllable-final than when syllable-initial. Sproat and Fujimura (1993) showed analogous behavior for laterals, with the tongue dorsum gesture occurring much earlier with respect to its associated tongue tip closure when final than when initial (they attribute the differential timing to the affinity for the syllable margin of gestures having extreme constrictions).

This timing pattern involving the tongue tip and tongue dorsum (tongue rear) for American English laterals is illustrated in the movement of the relevant tongue pellets for "leap" and "peel." These utterances are shown in Figures 3(a) and 3(b), respectively, which display representative tokens of movement data collected using the X-ray microbeam system in Madison, Wisconsin. (Four productions each of "leap" and "peel" in the carrier phase "Give ___ buttons," where the words "leap" and "peel" were accented, were collected from a female speaker of midwestern American English, approximately 40 years old. The speaker read the "leap" phrase and the "peel" phrase in alternation.)

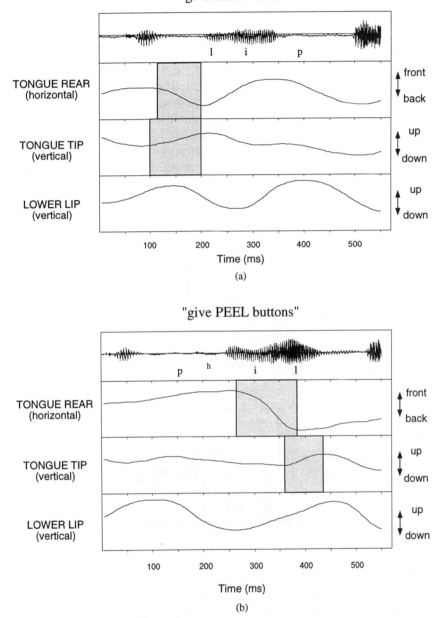

FIGURE 3. X-ray microbeam pellet data showing difference in relative timing of tongue tip raising and tongue dorsum (tongue rear) retraction for initial and final [1]. Tongue tip pellet is the frontmost pellet in the outlined image in Figure 1; tongue rear pellet is in the position indicated by the arrow in that image. (a) "leap" (b) "peel."

The shaded regions in Figure 3 highlight the movements presumably being controlled for the laterals. We would expect these regions to correspond roughly to the intervals of activation of the two gestures, as would be represented in the gestural score employed in Articulatory Phonology. Note there are two gestures present both syllable-initially and syllable-finally, that is, in both "leap" and "peel." The two gestures are roughly synchronous in "leap," but the dorsum gesture leads substantially in "peel." In fact, as can be seen in Figure 4 for all the tokens, in "leap" the end of the tongue dorsum retraction is nearly synchronous with the END of tongue tip movement (Figure 4(a)), while in "peel" the end of the tongue dorsum retraction is nearly synchronous with the BEGINNING of the tongue tip movement (Figure 4(b)). These patterns are analogous to the timing patterns discussed above for the nasals.

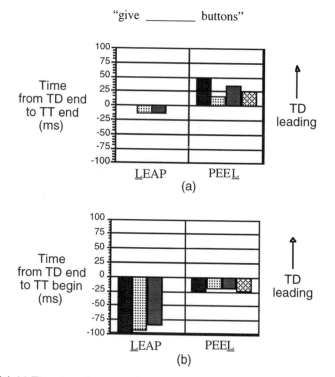

FIGURE 4. (a) Time from the end of the tongue-dorsum retraction (TD) movement to the end of the tongue-tip raising (TT) movement for "leap" and "peel." Each bar represents one token; there are three tokens of "leap" (the first of which has a time value of 0 ms) and four of "peel." (b) Time from the end of the tongue-dorsum retraction (TD) movement to the beginning of the tongue-tip raising (TT) movement for "leap" and "peel." Each bar represents one token; there are three tokens of "leap" and four of "peel." Movement beginning and end times for both (a) and (b) were automatically detected when the movement velocity for a given pellet/dimension increased above (beginning) or decreased below (end) a threshold of 30 mm/s.

Thus, for both the laterals and the nasals, it is the wider constriction that precedes the narrower constriction syllable-finally, as pointed out also by Sproat and Fujimura (1993). That is, for the nasal [m], the velum-lowering gesture has a wider constriction than the labial closing gesture, and this wider gesture precedes the narrower when syllable-final. Similarly, for the laterals, the wider tongue-dorsum constriction precedes the narrower tongue-tip closure syllable-finally. Gesturally there appears to be a single generalization here, applicable at least to both nasals and laterals (and perhaps to other gestures), namely that syllable-final linked gestures are phased so that the gesture with the wider constriction degree comes earlier, whereas syllable-initially these gestures are phased roughly synchronously. This single configurational principle results in different lawful acoustic consequences in the two examples—consequences which have been described as a nasalized vs. non-nasalized vowel, or a light vs. dark /1/.

The gestural generalization, of course, contrasts with the feature- and segment-based characterizations of syllable-position effects for nasals and laterals. American English syllable-initial and -final nasals are typically considered to differ in a featural approach only in that, in final position, the feature [+nasal] is extended to co-occur with the preceding vowel. That is, the same feature value is used for initial and final nasals (cf. Cohn, 1993). Laterals, however, are often considered to have different feature values initially and finally, so that, for example, American English initial and final laterals are often presumed to be [−back] and [+back], respectively (see discussion in Sproat & Fujimura, 1993, who also argue against this view). The syllable position similarity between nasals and laterals is totally missed in this kind of feature-based approach.

It is interesting to note that the tongue tip closure in American English final laterals can occur even in the silence following an utterance (Recasens & Farnetani, 1992). That is, the tongue tip articulation can occur even though vocalization has ceased. Such a situation might conceivably serve as an explanation for the errors in language acquisition in which a word ending with [1] is apparently mispronounced by the child as having a final [w] instead of a final [1] (e.g., in Dutch, /kas'tel/ 'castle' pronounced by a child of 2 years, 1 month as [tas'te:u], reported in Fikkert, 1992). In such a situation, presumably the inaudibility of the tongue tip closure for [1] would contribute causally to the interpretation of the final consonant as an approximant containing a dorsal gesture, i.e., [w]. One can speculate that similar inaudible articulations might be causally involved in those historical sound changes that involve the loss of a final nasal consonant, with concomitant nasalization of the preceding vowel. Presumably in such a case the velic opening and oral closure gestures would be timed as per the above discussion, with the velic opening gesture occurring during the vowel when syllable-final. That would mean, of course, that no additional process would need to be invoked to nasalize the vowel. If, as for [1], the closure gesture for the nasal could occur so late as to occur in utterance-final silence, then presumably the deletion of the closure gesture for the nasal would be a listener-based sound change (Ohala, 1981) caused by the inaudibility of the closure gesture when utterance-final.

FINAL REDUCTION

The perceptual deletion of the closure gesture could also be the result of its articulatory reduction. That is, regardless of what other gestures are involved, the position of the tongue tip is reduced in final position for [1] as well as for [t] and [n] (this appears to be a general positional effect, as we shall see later, although it is more prominent for articulations involving the tongue tip than for those involving other articulators). The cause of this reduction is not known, although it might be caused by a general reduction of speaking effort over the time course of the unit. As Browman and Goldstein (1992) have argued, this reduction can be stated as a single generalization about gestures, while in traditional phonological descriptions, a number of different types of rules (some selecting allophones, some specifying quantitative values) involving different features would be invoked.

In the microbeam data for "leap" and "peel" discussed above, reduction of the tongue-tip closure in syllable-final [1] can be observed. (Reduction for final-position [1] was found earlier by Giles & Moll, 1975.) Figure 5(a) shows the reduction of the tongue-tip component of syllable-final [1], by comparing the vertical positions of a gold pellet placed about 7 mm behind the tongue tip in each of the four repetitions of "leap" and "peel," measured at the maximum tongue-tip height for the [1]. In all cases, the tongue tip for the syllable-final [1] is lower than that for syllable-initial [1]. It is not the case that the tongue-tip gesture for syllable-final [1] is absent, since in most cases there is tongue-tip movement present, as is shown by the fact that tongue-tip height is generally above the baseline (which is the value of tongue-tip height that occurs when no tongue-tip gesture is present). That is, in general, these cases appear to be instances of syllable-final reduction of the tongue-tip movement. Figure 5(b) shows the same information when a low vowel, instead of a high vowel, occurs.

The reduction of the tongue-tip movement for syllable-final [1] is consistent with traditional transcriptions of variability in syllable-final [1], for example, those indicating the possibility of its "vocalization" or deletion in certain contexts, for some speakers (Bailey, 1985). However, this reduction is not specific to [1]: the tongue tip shows a reduction in the maximum height achieved finally in other gestural contexts as well—for example, in words like "pot" and "pawn," for which the traditional descriptions differ both from [1] and from each other (Bailey, 1985).

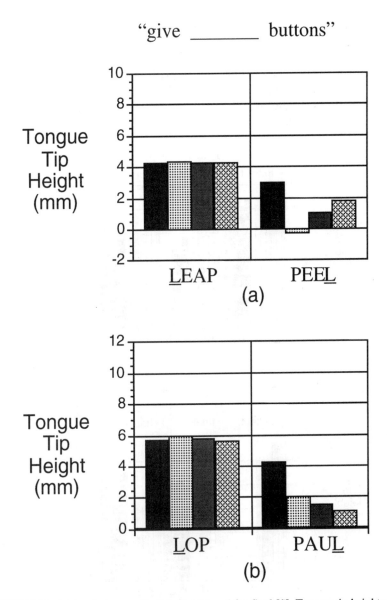

FIGURE 5. Reduction in tongue-tip height achieved for final [1]. Tongue-tip height was measured at the temporal center of a 1 mm noise band at the peak of the movement. Each bar represents one token. (a) "leap" vs. "peel." The tongue-tip height displayed for each token is relative to the baseline tongue-tip height (0 mm), which in this subfigure is the average maximum vertical tongue-tip position occurring during the final labial gesture in tokens of "leap" for this speaker. (b) "lop" vs. "Paul." Here the baseline (0 mm) is the average maximum vertical tongue-tip position occurring during the final labial gesture in tokens of "lop" rather than in tokens of "leap."

Again using the X-ray microbeam system in Madison, Wisconsin, we collected C1VC2 words in carrier phrases varying in accent and in the gesture following the target word. The speaker was a college-aged male speaker of California English. Figure 6 shows the results for five consecutive repetitions of "tot," "top," "pot," "Don," and "pawn," with the targeted word always accented and occurring in each of two contexts "(My BEAR doesn't huddle, but) my___ huddles" and "(My BEAR doesn't puddle, but) my ___ puddles."

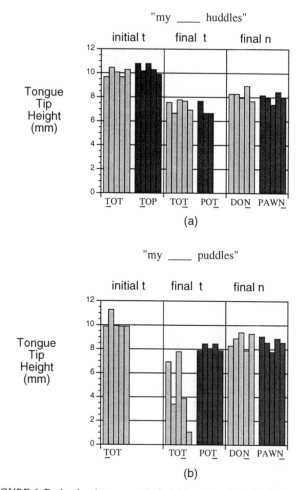

FIGURE 6. Reduction in tongue-tip height achieved for final [t] and [n]. Time points for tongue-tip height measurement were chosen as described for Figure 5. The baseline (0 mm) is the mean maximum vertical tongue-tip position occurring during the final labial gesture in tokens of "top" for this speaker in this accent condition. Each bar represents one token; all the utterances have five tokens except "pot huddles," which has three. (a) context "huddles," (b) context "puddles."

(The parenthetical portions of the sentences were shown to the speaker to elicit the desired accent placement, but were not actually produced by the speaker. Also, the speaker produced the same vowel in all the utterances, as judged by informal listening, which is not surprising since there is no [O] in the dialect of English spoken in California, as in many other dialects; i.e., the vowel in "pawn" is pronounced like the vowel in the second syllable of "upon.") The figure shows the final reduction in tongue-tip height of both [t] and [n], by comparing the maximum heights of a gold pellet (placed 10 mm behind the tongue tip) for the final tokens to the maximum heights of the same tongue tip pellet in initial [t].

Looking first at Figure 6(a), in which the word immediately following does NOT begin with an oral closure gesture ("huddles"), it is clear that all the final tokens of [t] and [n] have reduced tongue-tip positions when compared to the initial [t] on the left. That the tongue-tip gestures are reduced but not deleted is indicated by the fact that all the positions are far above the baseline value of 0 mm, which represents the position of the tongue tip during a final labial stop. Similarly, in Figure 6(b), in which the word immediately following DOES begin with an oral closure gesture ("puddles"), the final tokens are reduced. Several of the tokens in "tot puddles" do behave somewhat differently, however, with two of the tokens being reduced even further, and a third token being effectively deleted. While there is something additional going on in the case of "tot puddles" (additional reduction and possibly also deletion), this does not change the fact that, in general, final tongue-tip gestures are reduced. Similar results were obtained with the other accent patterns that were also collected—where either the word before or the word after the targeted word was accented, e.g., "(YOUR __ doesn't huddle, but) MY __ huddles," and "(It may not DANCE, but) my __ HUDDLES." A subset of these results (for "tot" but not for "pot") is shown in the middle rows of Figures 7(a) and 7(b).

While tongue-tip movements are reduced the most, the reduction effect also occurs in gestures that involve other oral articulators. Figure 7 shows the results for some of the other target words employed in this experiment—"pop" and "caulk" (pronounced [kAk]), as well as for "tot." As can be seen in the top row of Figures 7(a) and 7(b) for labials, and the bottom row of Figures 7(a) and 7(b) for dorsals, gestures involving the lips and tongue dorsum are also reduced finally. The position effect is significant at the 0.01 level for each of [p], [t], and [k], using analysis of variance (with 3 factors: accent placement, following word starting with oral closure gesture or not, and position within the word/syllable, which was treated as a repeated measures factor). For coronals ($p < 0.01$) and marginally for dorsals ($p < 0.021$), the magnitude of final reduction also varies according to the accent placement, presumably because there is more final reduction when the accent is placed on the word preceding the word being measured. In general, then, oral closure gestures decrease in magnitude when in final position.

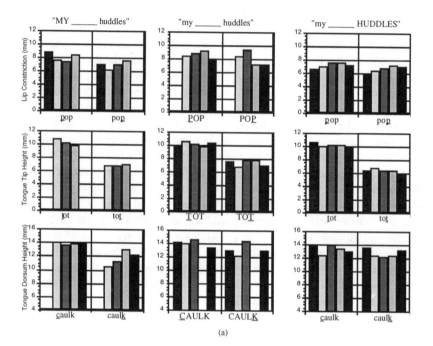

(a)

FIGURE 7. Reduction in position achieved of relevant articulators for final voiceless stops, with three different places of articulation and three different accent patterns. Top row shows reduction in lip constriction for the labials in [pɑp]; the middle row reduction in the height of the tongue tip for the coronals in [tɑt]; and the bottom row reduction in the height of the tongue dorsum for the dorsals in [kɑk]. The three columns show reduction for three different patterns of accent placement with respect to the measured word: left, accent preceding; middle, accent on the measured word; right, accent following. Time points at which the measurements were made were all determined using the method described in the caption for Figure 5. The values displayed in the top 2 rows were computed as follows: Lip constriction was measured by finding the vertical distance

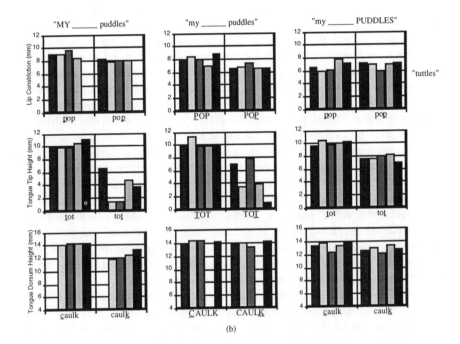

(b)

between the pellets on the upper and lower lips (Lip Aperture) and then, to invert the curve for visual compatibility with other rows, subtracting that distance from 30 mm, which is a rough estimate of the Lip Aperture during initial tongue-tip gestures for this speaker. Thus, the greater the value of lip constriction, the smaller the distance between the upper and lower lips. For tongue-tip height (middle row), the baseline (0 mm) is the mean maximum vertical tongue-tip position occurring during the final labial gesture in tokens of "top" for the appropriate accent condition. For tongue dorsum height, no special baseline is employed. (a) context "huddles," (b) context "puddles" (or "tuttles", following [pap]).

CONCLUSION

In this chapter we have attempted to demonstrate how, within a gestural perspective, the notion of syllable position corresponds to different modes of configuration of physical events (the gestures). Three implications of this view are worth summarizing. First, it is often possible to capture the phonological facts that make reference to syllable position without reifying any hierarchical syllable structure, because syllable position inheres in the patterns of gestural coordination. Second, the gestural approach allows a single generalization to be stated for some superficially unrelated cases involving nasals and laterals. These phenomena, which have been described as involving variation of phonological units as a function of syllable position (e.g., allophonic variation), can be better understood, we argue, as the lawful physical consequences of the different patterns of gestural overlap that in effect define the different syllable positions. Third, as also posited by Sproat and Fujimura (1993), there may be a single gestural generalization for English, namely that, syllable-finally, gestures involving wider constrictions precede those with narrower constrictions. While this chapter considered only relatively simple gestural constellations (involving two gestures), the regularity uncovered suggests that it might be rewarding to investigate the more complex codas of American English phonology (constellations involving multiple gestures, e.g., in "howled") to see what additional generalizations emerge, and to see what light can be shed on the distributional patterns of such structures when viewed as gestural constellations rather than being parsed into a string of segment-sized units.

It seems also to be the case that there is a very general reduction in the size of gestures in final position. Such a result is perhaps not surprising, given the general weakness of the syllable-final position in the languages of the world (many languages have been described as having no syllable-final consonants at all), but it is striking that such reduction occurs in English, a language with a sizable inventory of final consonants. Further investigation of this reduction effect should determine its precise domain (is it really syllable-final, or word-final, phrase-final?) and how the reduction is distributed over the component articulators that constitute a gesture's coordinative structure (do all the articulators cooperate to produce a less extreme gesture, or is the reduction caused by the activity of a single articulator, possibly the jaw?). Answers to such questions could provide insight both into the causes of the reduction effect, and into the ways it might be expected to show up in the phonology of languages.

ACKNOWLEDGMENTS

This work was supported by NSF grant DBS-9112198 and NIH grants HD-01994 and DC-00121 to Haskins Laboratories. Our thanks to the Yale School of Medicine, General Electric Company, and the Esther A. and Joseph Klingenstein

Fund for providing the capacity to obtain the Magnetic Resonance Images reported on herein. Thanks also to Alice Faber for comments on an earlier version.

REFERENCES

Bailey, C.-J. N. (1985). *English phonetic transcription.* Arlington, TX: Summer Institute of Linguistics.

Browman, C. P., & Goldstein, L. (1986). Towards an articulatory phonology. *Phonology Yearbook, 3,* 219-252.

Browman, C. P., & Goldstein, L. (1989). Articulatory gestures as phonological units. *Phonology 6,* 201-251.

Browman, C. P., & Goldstein, L. (1992). Articulatory phonology: An overview. *Phonetica, 49,* 155-180.

Cohn, A. C. (1993). Nasalization in English: Phonology or phonetics. *Phonology, 10,* 43-81.

Fikkert, P. (1992). A prosodic account of truncation in child language. Presentation at 7th International Phonology Meeting, Krems, Austria, July 1992.

Giles, S. B., & Moll, K. L. (1975). Cinefluorographic study of selected allophones of English /l/. *Phonetica, 31,* 206-227.

Harris, K. S., & Bell-Berti, F. (1984). On consonants and syllable boundaries. In L. J. Raphael, C. B. Raphael, & M. R. Valdovinos (Eds.), *On consonants, and syllable boundaries* (pp. 89-95). Plenum Publishing Corporation.

Kahn, D. (1976). *Syllable-based generalizations in English phonology.* Bloomington, IN: Indiana University Linguistics Club.

Krakow, R. A. (1989). *The articulatory organization of syllables: a kinematic analysis of labial and velar gestures.* Unpublished doctoral dissertation. Yale University.

Ohala, J. J. (1981). The listener as a source of sound change. In C. S. Masek, R. A. Hendrick, & M. F. Miller (Eds.), *The listener as a source of sound change* (pp. 178-203). Chicago: Chicago Linguistic Society.

Recasens, D., & Farnetani, E. (1992). Spatiotemporal properties of different allophones of /l/. Phonological implications. Presentation at 7th International Phonology Meeting, Krems, Austria, July, 1992 .

Rubin, P. E., Baer, T., & Mermelstein, P. (1981). An articulatory synthesizer for perceptual research. *Journal of the Acoustical Society of America, 70,* 321-328.

Saltzman, E. L., & Munhall, K. G. (1989). A dynamical approach to gestural patterning in speech production. *Ecological Psychology, 1,* 333-382.

Sproat, R., & Fujimura, O. (1993). Allophonic variation in English /l/ and its implications for phonetic implementation. *Journal of Phonetics, 21,* 291-311.

3 Speech Production Theory and Articulatory Behavior in Motor Speech Disorders

Gary Weismer, Kristin Tjaden, and Ray D. Kent
University of Wisconsin-Madison

INTRODUCTION

Some Considerations of Timing in Speech Production Theories

Theories of speech production have always required some component responsible for the temporal flow of articulatory, and hence acoustic, events. Some years ago, Fowler (1980) separated this component into two potential classes, so-called 'extrinsic' vs. 'intrinsic' timing mechanisms. Extrinsic mechanisms were usually envisioned as involving some sort of clock, possibly not specific to the speech production mechanism, that 'metered out' a sequence of discretely specified items (usually columnar arrays of feature bundles specifying phonemes: see Daniloff & Hammarberg, 1973; Kent & Minifie, 1977; Lindblom, 1963; Perkell, 1980). The capabilities and constraints of this metering have never been discussed in great detail, although concepts such as inherent duration of phonemes (Klatt, 1976), compressibility and expandability of the metered elements (Fujimura, 1987), and the relative stability of elements-to-be-timed (Kent & Minifie, 1977; Fujimura, 1986) have been given some consideration.

Extrinsic timing mechanisms require, by their very label, the *imposition* of temporal form on the string of discrete, timeless elements mentioned above. This

35

view has been challenged by various proponents of the task dynamic approach to speech production (Fowler, 1980; Kelso, Saltzman, & Tuller, 1986), who argue that the timing of articulatory gestures emerges from the dynamics of the system. The timing of articulatory gestures is, then, intrinsic to the gestures being produced (see Fowler, 1980, for an in-depth exposition of the issues) and is not 'controlled' by some unknown agent. Intrinsic timing theories are not only attractive because the details of the temporal form of articulatory gestures can be worked out and modeled in an explicit analytical form (Saltzman & Kelso, 1987), but also because they open the door to a conception of gestures as phonological units (Browman & Goldstein, 1986, 1991). If the distinctive function of phonetic contrasts could be accounted for by a gestural phonology that in turn could be shown to account for the kinematic details of speech production, the need for a level of abstract, timeless entities (phonemes) could be eliminated from the structure of speech production theories.

Clearly, the challenge of incorporating timing into a theory of speech production has generated a substantial amount of empirical and theoretical effort and progress. For the most part, the theoretical development has been aimed at an account of timing in speech produced by adults with normal speech mechanisms. Because the form of speech timing is apparently a function of aging (at both ends of the life span: see Kent, 1983; Weismer, 1991) and the neurological integrity of the speech mechanism (Kent & Adams, 1989; Weismer & Martin, 1992), a theory capable of handling the variety of speech timing events among various speakers would be desirable. This is not to say that timing theories cannot be tested effectively against data produced by normal, adult speakers; rather, the various theoretical perspectives should be considered carefully as likely candidates to account for speech timing in a variety of populations.

In the present chapter, we will focus our attention on timing issues in motor speech disorders. Although some persons with neurological disease are known to produce speech with different timing characteristics than those observed in the normal, adult population, there has been little effort to consider how the choice or development of speech production theory might be affected by these differences.[1] Weismer, Tjaden, and Kent (in press) reviewed speech production phenomena in persons with motor speech disorders, and concluded that some

---

[1]It could be argued, of course, that a theory of speech production that is based on data from 'prototypical' speakers (adults with normal neurological and structural profiles) should not be responsible for data obtained from nonprototypical speakers. It may be more fruitful, for example, to allow theory to develop independently for separate speaker groups, and then to identify the domains of agreement and disagreement between the separate theories. This exercise might identify the core axioms that should be included in a general explanatory account of speech production as well as axioms specific to particular groups (e.g., speakers with neurological damage to the cerebellum). For reasons discussed in Weismer, Tjaden, and Kent (in press), we prefer the other approach, wherein the first step is to identify the theory most likely to explain data from a variety of speaker groups.

version of gesture theory (e.g., Browman & Goldstein, 1986) was more likely to account for the facts of neurologically-disordered speech than a translation theory (e.g., Daniloff & Hammarberg, 1973).

Gesture Theory and Motor Speech Disorders: Predictions

In gesture theory, the basic form of a gesture is assumed to be fairly constant across phonetic contexts and speaking rates. The gesture slides with respect to other surrounding gestures in response to variations in context and rate, but is not substantially changed by these factors (e.g., Boyce, Krakow, Bell-Berti, & Gelfer, 1990; Munhall & Löfqvist, 1992). The phasing of a set of these gestures, for the purposes of producing a word, has yet to be explained (see Saltzman & Munhall, 1989) but can be expected to be fairly stable across repetitions of the same word by a normal speaker. In the case of gestures associated with vocalic nuclei, the acoustic manifestation of this stability would be the superimposition of formant trajectories across word repetitions, as seen in the left panel of Figure 1. There the F_2 trajectories are shown for 20 repetitions of the word 'toy' from the sentence 'See four toy cows' produced by a normal, 65-year-old adult male. These trajectories, and all others shown in this chapter, were generated and edited using the formant tracker and spectrographic display of CSpeech (Milenkovic, 1992). For graphical purposes all trajectories have been set at the same starting point, which is the first glottal pulse of the vocalic segment. The superimposition of these trajectories suggests the phasing stability mentioned above; stated otherwise, the relative amount of across-repetition sliding between gestures for the consonant and diphthong in this word appears to have been minimal, a reasonable expectation for a normal, adult speaker.

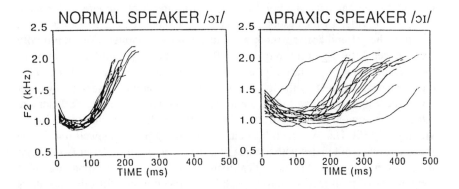

FIGURE 1. Multiple F_2 trajectories for the diphthong /ɔɪ/ in 'toy' produced by a neurologically normal, 65-year old man (left) and a 56-year-old man with apraxia of speech (right). Time zero for all trajectories is set to the formant frequency measured at the initial glottal pulse of the vocalic nucleus.

For selected speakers with motor speech disorders, tight superimposition of formant trajectories is not the rule. Rather, trajectories show variation in both the temporal and spatial (i.e., frequency) dimensions, as shown in the right panel of Figure 1 for 19 repetitions of 'toy' by a 56-year-old speaker with apraxia of speech. Liss and Weismer (1992) argued that some of this repetition-to-repetition variation in trajectory characteristics could be explained, in part, by excessive sliding between adjacent consonantal and vocalic gestures. Based on a qualitative analysis of a limited number of trajectories produced by two apraxic speakers, Liss and Weismer (1992) suggested that the form of the gestures was not necessarily different from normal, but rather the unusually variable phasing across repetitions (i.e., the excessive gestural sliding) was the primary cause of the *appearance* of temporal and spatial variability in the acoustic results. Based on this qualitative analysis and notions from gesture theory, Liss and Weismer (1992) made several predictions concerning the effect of excessive gesture sliding from repetition-to-repetition on temporal properties of the formant trajectories.

The model for these predictions is presented in Figure 2 (see Boyce et al., 1990, Figure 5, for a similar schematic). The falling and rising curve shows the basic form of the F_2 trajectory for the diphthong /ɔɪ/ in 'toy,' which we take as an acoustic manifestation of the vocalic gesture.[2] Superimposed on this curve are two boxes, which are meant to indicate the onset (left edge of box), duration, and offset (right edge of box) of the closure interval for two different productions of /t/ (in the current example the /t/ closure duration is held constant). The closure duration box is sliding with respect to the vocalic gesture, and the primary acoustic effect is to uncover more or less of the onset of the F_2 trajectory, as indicated by the arrows; the same effect would obtain if the vocalic gesture were sliding with respect to the closure gesture, or if both gestures were sliding relative to one another. The point here is that an index of the *effect* of gesture sliding is provided by the onset frequency of the F_2 trajectory (arrows, Figure 2).

When the /t/ and /ɔɪ/ gestures are substantially separated (i.e., minimally 'coproduced'), the F_2 onset frequency reflects a relatively 'pure' articulation of the /t/, one that is relatively unaffected by surrounding context. This circumstance, depicted by the left box in Figure 2, should be associated with an F_2 onset frequency close to the theoretical locus value (Lehiste & Peterson, 1961) for lingua-alveolar obstruents, which for adult males should be in the vicinity of 1800 Hz. Weismer, Tjaden, and Kent (in press) showed that in the speech of two adult males with apraxia of speech, the mean F_2 onset frequencies in 'toy' were closer to this theoretical locus value than the corresponding onset frequencies of a normal, geriatric speaker. Similar findings for other speakers

---

[2]We do not mean to imply that the F_2 trajectory is *the* reflection of the vocalic gesture. Clearly other parts of the acoustic signal (such as other formant trajectories) reflect parts of the vocal tract gesture that are not captured by F_2.

and contexts have been reported by Weismer and Liss (1991) and Liss and Weismer (1992). In motor speech disorders, the sliding between gestures has been said to result in the phenomenon called 'segmentalization' (Kent, 1983; Kent & Rosenbek, 1983; Liss & Weismer, 1992; Weismer & Liss, 1991), wherein successive gestures tend to be 'pulled apart.' If variability of F_2 onset frequency is a reasonable index of the variability of gesture sliding that occurs from repetition to repetition, then the trajectories shown in Figure 1, as well as those previously published by Liss and Weismer (1992, 1994) and Weismer and Liss (1991), suggest that the degree of gesture sliding is greater in persons with motor speech disorders than it is in normal speakers. Note the greater spread in the F_2 onset frequencies for the apraxic speaker, as well as the previously mentioned tendency of these values to be closer to the lingua-alveolar locus of 1800 Hz. Thus some speakers with motor speech disorders not only tend to segmentalize articulatory sequences, but also phase successive gestures with less stability than normal speakers. The impression of 'scanning speech' (Kent & Rosenbek, 1983) in various motor speech disorders is a likely perceptual correlate of segmentalized speech.

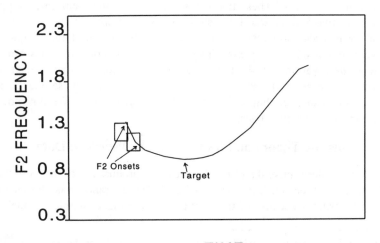

TIME

FIGURE 2. Schematic illustration demonstrating the logic of one of the main predictions of this study. The F_2 formant trajectory for /ɔɪ/, taken here as an acoustic manifestation of the vocalic articulatory gesture, is shown with two hypothetical phasings of the preceding closure interval for /t/. When the closure interval is phased early with respect to the vocalic gesture (left, upper box) more of the formant trajectory is 'uncovered' and the F_2 onset measure (arrows) is closer to the theoretical locus for alveolar obstruents. Gesture theory predicts that the F_2 onset frequency should predict the interval between the F_2 onset and the target of the vocalic nucleus (arrow).

With these considerations in mind, we can use the schematic in Figure 2 to formulate several predictions. First, if successive articulatory gestures are more segmentalized in neurogenically-disordered speech than in normal speech, there should be a difference between the speakers in the F_2 onset frequencies, with apraxic values being closer to the theoretical locus for the obstruent. Second, if in fact the sliding of adjacent gestures is greater in persons with motor speech disorders than in normal speakers, the variance of the F_2 onset frequency should be greater in the former group of speakers. Third, if the gesture itself is relatively stable but the phasing is unusually variable, the index of gestural sliding (i.e., the F_2 onset frequency) should predict the time to the vocalic target frequency (see 'target,' Figure 2). This prediction follows because as the vocalic gesture moves away from the consonantal gesture it moves as a whole (in line with the tenets of gesture theory), displacing the target frequency further from the onset of the trajectory. Thus, the closer the F_2 onset frequency is to the theoretical consonant locus, the greater the duration to the vocalic target. This is the primary prediction formulated by Liss and Weismer (1992), and tested on a limited corpus in Weismer, Tjaden, and Kent (in press).

Weismer et al. (in press) showed that F_2 onsets were closer to the lingua-alveolar locus and somewhat more variable for the apraxic speakers as compared to the normal speakers.[3] These findings seemed to provide preliminary support for the framework suggested by gesture theory. One apraxic speaker and the normal speaker also produced transition times to the vocalic 'target frequency' that were predicted by the F_2 onset frequencies, although only a modest amount of the shared variance (14-31%) was accounted for by these functions. Nevertheless, these results suggested further examination of these phenomena, using the same speakers but a different articulatory sequence. We turn now to the data from this later analysis.

Gesture Theory and Motor Speech Disorders: Data

Table 1 shows descriptive data from the speakers' productions of the word "see" in the sentence "see four toy cows eat hay." This utterance was chosen from existing recordings because of the multiple productions available for analysis.

---

[3]The mean F_2 onset frequencies reported in Weismer et al. for the apraxic speakers actually underestimate the difference from the normal speaker's value. In the original set of measurements (Table 1 in Weismer et al., in press), the F_2 value at the first glottal pulse following release of the /t/ in 'toy' was taken as the F_2 onset. Because both apraxic speakers produced substantially longer VOTs than the normal speakers, this meant that the measured F_2 onsets for the apraxic speakers were much further into the vocalic gesture than they were for the normal speaker. When we remeasured the F_2 onsets for the apraxic speakers by taking the values at equivalent times for all three speakers (i.e., at the mean VOT for the normal speaker, 56 ms), the differences between the apraxic and normal speakers became much more pronounced. This analysis was not reported in Weismer et al.

These multiple productions occurred in both 'neutral' (N) and 'contrastive stress' (CS) speaking conditions (Weismer et al., in press). The shape of the typical (normal) F_2 trajectory for the vocalic nucleus of 'see' is shown in Figure 3, together with the measurements made for the current analysis. For this data, set F_2 target frequency was defined in two ways, including 1) the initial occurrence of the highest frequency in the trajectory, and 2) the frequency at the temporal midpoint of the formant trajectory. The F_2 offset frequency, at the /i/-/f/ interface, was also measured.

The actual trajectories from the contrastive stress condition are shown in Figure 4. For the purposes of this display and the statistical analyses reported below, the trajectories from the normal speaker have been separated into those receiving contrastive stress (N=6, Figure 4, upper right) and those for which contrastive stress occurred on another word in the utterance (N=24, Figure 4, upper left). We separated the trajectories in this way for the normal speaker, but not for the apraxic speakers, because preliminary analyses showed large differences between stressed and nonstressed trajectories only for the normal speaker. These differences are readily apparent in the top part of Figure 4. Figure 5 shows the full set of trajectories from the neutral condition for all three speakers.

TABLE 1. *Descriptive data (Means and Standard Derivations) obtained for two apraxic and one normal speaker in both neutral (N) and contrastive stress (CS) speaking conditions. For the normal subject, CS refers to those productions where the word "see" was not the word contrastively stressed and CSsee refers to those productions where the word "see" did receive the primary stress. The terms 'ΔTAR' and 'ΔOFF' refer to the time in milliseconds (ms) from F_2 onset frequency to F_2 target frequency and the time from F_2 target frequency to F_2 offset frequency, respectively.*

Subject	F_2 Onset	F_2 Target	ΔTAR	F_2 Offset	ΔOFF
A4/N	1802(65)	2129(49)	214(49)	1732(139)	235(66)
A4/CS	1883(55)	2275(73)	281(107)	1895 (181)	220(75)
A1/N	1784(69)	2096(56)	143(54)	1368(185)	181(33)
AI/CS	1801(80)	2138(41)	139(35)	1307(135)	210(76)
N4/N	1862(67)	2189(53)	68(12)	2012(71)	40(13)
N4/CS	1887(48)	2205(52)	71(12)	1889(119)	48(9)
N4/CSsee	1750(74)	2469(40)	174(30)	2135(218)	50(23)

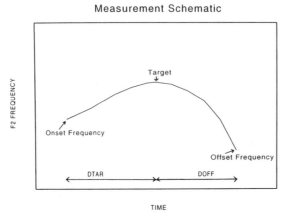

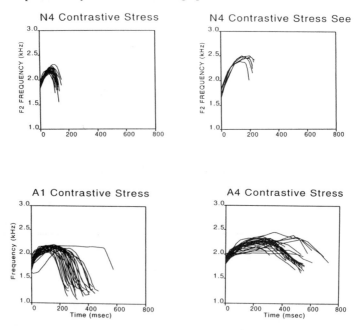

FIGURE 3. F$_2$ formant trajectory for the /i/ vocalic nucleus in the word 'see.' Measures taken in the present study are indicated in the graph.

FIGURE 4. Multiple F$_2$ formant trajectories for the /i/ vocalic nucleus produced in the contrastive stress (CS) condition. Top left: normal speaker, trajectories for all 'see's' except those where 'see' was stressed. Top right: normal speaker, trajectories for stressed 'see's'; bottom left: apraxic speaker (A1), all trajectories; bottom right: apraxic speaker (A4), all trajectories.

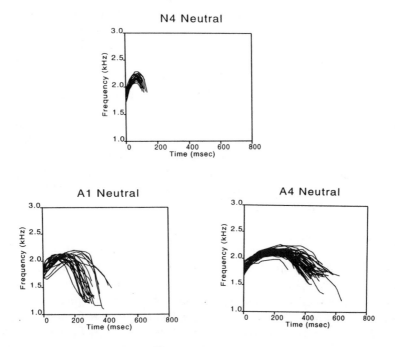

FIGURE 5. Multiple F_2 formant trajectories for the /i/ vocalic nucleus produced in the neutral (N) condition. Top: normal speaker; left, bottom: apraxic speaker (A1); right, bottom: apraxic speaker (A4).

The descriptive data can be summarized as follows. First, excluding the data from the normal speaker's stressed 'see's,' three of the four mean F_2 onset frequencies produced by apraxic speakers were lower than those produced by the normal speaker; the exception was the mean F_2 onset associated with A4's contrastively stressed 'see's.' These lower F_2 onsets are in the correct direction (closer to 1800 Hz) for the interpretation of greater segmentalization among the apraxic speakers. This interpretation also seems to apply to the data for F_2 offset frequency, where the relatively low values for the apraxic speakers (with the exception of A4-CS) suggests that more of the vocalic gesture has been 'uncovered' because of the reduced overlap between the /i/ and /f/ gestures. The relative closeness of the apraxic F_2 offsets (especially A1) to the labial locus supports this interpretation. As in the 'toy' data reported by Weismer et al. (in press), the transition durations produced by the apraxic speakers are clearly longer and more variable than those produced by the normal speaker. If the difference between the F_2 target and F_2 onset (or offset) is divided by the transition durations, it is also obvious that the transition slopes are more shallow for the apraxic speakers.

As stated above, the stressed 'see' tokens of the normal subject had substantially different trajectories than those of the nonstressed syllables. Specifically, the stressed trajectories had lower F_2 onset frequencies, higher F_2 target and F_2 offset frequencies as compared to the nonstressed trajectories.

The results of the regression analyses are given in Table 2; scatterplots for selected significant functions are shown in Figure 6. When the F_2 target was considered as the first occurrence of the highest frequency along the trajectory, three of the eight possible apraxic functions and two of the four possible normal functions were statistically significant. The significant functions occurred in cases where either F_2 onset or F_2 offset was used as the predictor variable, and accounted for between 13 and 52% of the shared variance. When the trajectory midpoint was used as the vowel target, only two functions were statistically significant (both for A1, in the N condition), and these accounted only for a very small portion of the variance.

TABLE 2. *Results of regression analyses. Nonsignificant results are indicated by NS; the symbol (*) denotes significant results (p < 0.05) with the associated r-squared value in parentheses. The top panel includes measurements and analysis performed defining F_2 target frequency as the initial occurrence of the highest frequency in the trajectory. The bottom panel shows the same analysis performed defining F_2 target frequency as the temporal midpoint of the formant trajectory. The first column under each condition heading contains results from the analysis predicting ΔTAR from F_2 onset frequency; the second column under each condition heading reports results from the analysis predicting ΔOFF from F_2 offset frequency.*

	Neutral Onset	Contrastive Stress Offset	Onset	Offset
A4	NS	NS	NS	*(52%)
A1	*(27%)	NS	*(21%)	NS
N4	NS	*(33%)	*(13%)	NS

	Neutral Onset	Contrastive Stress Offset	Onset	Offset
A4	NS	NS	NS	NS
A1	*(14%)	*(15%)	NS	NS
N4	NS	NS	NS	NS

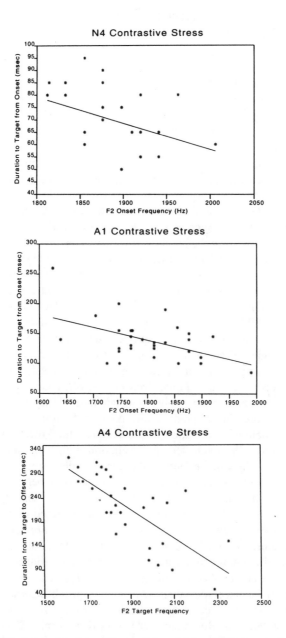

FIGURE 6. Selected scatterplots of F_2 onset (x axis) by ΔTAR (y axis: interval between F_2 onset and vowel target: to two plots) or F_2 offset and ΔOFF (interval between vowel target and F_2 offset: bottom plot). Linear regressions were statistically significant for the displayed plots.

DISCUSSION

Taken together, the data from Weismer et al. (in press) and those presented here suggest that, for some speakers, the duration of the interval between the vocalic onset and target, or between the target and vocalic offset, is partly predictable from the F_2 onset or offset frequency. It is useful to consider briefly 1) why the significant functions account for only a relatively modest amount of the shared variance between F_2 onset frequency and ΔTAR (the time from F_2 onset to the target), and 2) why certain functions fail to account for a significant portion of the shared variance.

There are several possible explanations for the relatively modest degree of prediction efficiency evidenced by the significant functions. First, even under the assumption that articulatory gestures and their acoustic manifestations are relatively immutable and mainly slide in time with respect to one another (Bell-Berti & Krakow, 1991; Boyce et al., 1990), the acoustic measures may be sufficiently noisy to weaken the obtained functions. Great care was taken in the correction of the automatically-generated formant tracks and the measurement of ΔTAR, but even small measurement errors could have large effects on regression functions involving relatively small Ns. A second explanation, again under the assumption of a strict coproduction account, is that our estimate of the effect of gesture sliding—that is, of segmentalization—is incomplete. We originally measured both F_2 and F_1 onset frequencies in an attempt to provide a more complete index of segmentalization. When the F_1 onset frequency was included with F_2 as a predictor of ΔTAR, however, there was no improvement in prediction efficiency. Perhaps F_3 onset frequencies, when added to or combined with F_2 onsets, would constitute a more complete index of segmentalization and thus provide for more reliable prediction of ΔTAR.[4] A third explanation is that the acoustic estimate of termination of the gesture—the vowel target—is not entirely appropriate. We evaluated the prediction with two estimates of the vowel target, and the one reflecting the greatest deviation of vocal tract configuration from the neutral vowel position produced superior results when compared to one based on a strict temporal criterion (Table 2). This is a satisfying result because the kinds of CV transitions explored in gesture theory (Boyce et al., 1990; Browman & Goldstein, 1991) are more in line with a definition of gesture termination based on a spatial, rather than temporal goal. Finally, the significant functions may leave so much variance unexplained simply because the strict coproduction view of speech production phenomena in these speakers is incorrect. Speakers with motor speech disorders, including those with apraxia of speech, often produce abnormally shallow formant transition slopes (Kent & Rosenbek, 1983; Liss & Weismer, 1994; Weismer,

---

[4]Other acoustic measures also might reflect the extent of gesture sliding. For example, there could be a relationship between the duration of obstruent intervals and segmentalization, although our data to date fail to support this possibility.

Martin, Kent, & Kent, 1992); this also appeared to be the case for the present apraxic speakers (Table 1). These shallow slopes reflect relatively slow changes in vocal tract configuration, suggesting that articulatory gestures are not immutable in at least some speakers with motor speech disorders. Thus an account of variability in formant trajectories produced by these speakers would most likely have to include factors in addition to segmentalization. For example, prediction of the interval between vowel onset and vowel target might be improved by adding to F_2 onset an index of articulatory slowness, such as the transition slope.[5]

We favor this third explanation of the data, but this does not mean that we reject the gesture theory approach to understanding articulatory variability in motor speech disorders. Indeed, the finding of several significant functions does suggest some influence of gesture sliding on the variability of formant trajectories produced by some speakers with motor speech disorders. The relatively large amount of unexplained variance in these significant functions simply indicates that this influence is not exclusive. Moreover, the failure to find significant regressions for some conditions is also not a good reason to reject this theoretical perspective, because we would not expect segmentalization to be a prominent component of every patient's speech production, nor even of every articulatory sequence produced by a given patient. The speech production deficit in motor speech disorders is exceedingly complex, and any theoretical perspective that makes a sensible prediction that is then confirmed (in a statistical sense) should be pursued. Thus, the general framework of gesture theory, as opposed to translation theories with their timeless constructs, appears to be a productive way to organize and test predictions concerning articulatory variability in motor speech disorders.

Finally, the finding of a significant relationship between F_2 onset frequency and ΔTAR in the normal speaker was somewhat unexpected. Weismer et al. (in press) argued that the phasing of overlapping articulatory gestures in normal speakers was probably sufficiently precise to prevent statistical confirmation of the F_2 onset- ΔTAR relationship (i.e., because of the highly constrained variances of the measures). Moreover, Weismer et al. pointed to motor speech disorders as an excellent test case, simply because the assumed greater variability of gesture phasing among these speakers would create a more favorable statistical environment in which to test the predictions of gesture theory. Unexpectedly, the F_2 onset- ΔTAR function was significant for the

---

[5]This acoustic estimate of articulatory slowness should not come from the transition connecting the F2 onset with the F2 target frequency. In a regression model, a transition slope calculated from the same segment used to estimate ΔTAR, the dependent variable, would raise the possibility of obtained functions that are partly or largely due to a part-whole correlation (i.e., the numerator of the ratio used as an index of slope is the same as ΔTAR). The transition slope could be obtained from the same utterance yielding ΔTAR, but not from the same segment.

normal speaker in both the 'toy' (Weismer et al., in press) and 'see' analyses. Moreover, the expectation of greater variability in gesture sliding for apraxic, as compared to normal speakers received inconsistent support in the two data sets. Under the assumption that the standard deviation of the F_2 onset frequency is an estimate of the variability of gesture phasing, apraxic speakers were more variable than normal in Weismer et al. but not in the current set of data. What is needed are articulatory data on the variability of gesture phasing in a number of normal speakers,[6] and the relationship of that phasing to acoustic measurements. In addition, further work on the acoustic manifestations of the coproduction view of articulation in normal speakers (see, for example, Boyce & Espy-Wilson, in press) should reveal the extent to which acoustic studies can contribute to the development of this speech production theory in both normal and disordered speakers.

ACKNOWLEDGMENT

This work was supported by NIDCD Award DC00316.

REFERENCES

Bell-Berti, F., & Krakow, R. A. (1991). Anticipatory velar lowering: A coproduction account. *Journal of the Acoustical Society of America, 90,* 112-123.

Boyce, S., & Espy-Wilson, C. (in press). Coarticulatory stability in American English /r/. *Speech Group Working Papers* (Research Laboratory of Electronics, MIT).

Boyce, S. E., Krakow, R. A., Bell-Berti, F., & Gelfer, C. E. (1990). Converging sources of evidence for dissecting articulatory movements into core gestures. *Journal of Phonetics, 18,* 173-188.

Browman, C., & Goldstein, L. (1986). Towards an articulatory phonology. *Phonology Yearbook, 3,* 219-254.

Browman, C., & Goldstein, L. (1991). Gestural structures: Distinctiveness, phonological processes, and historical change. In I. G. Mattingly & M. Studdert-Kennedy (Eds.), *Modularity and the motor theory of speech perception* (pp. 313-338). Hillsdale, NJ: Lawrence Erlbaum Associates.

Daniloff, R. G., & Hammarberg, R. (1973). On defining coarticulation. *Journal of Phonetics, 1,* 239-248.

---

[6]We are aware that data on multiple speakers are required to validate our conclusion that the F2 onsets reviewed here reveal segmentalization. Our current data could reflect nothing more than differences in vocal tract size across these few speakers, but the combined CV and VC evidence, plus other published data (Weismer, Martin, Kent, & Kent, 1992) (Table 1) doesn't favor this interpretation of the data.

Fowler, C. A. (1980). Coarticulation and theories of extrinsic timing. *Journal of Phonetics, 8,* 113-133.

Fujimura, O. (1986). Relative invariance of articulatory movements. An iceberg model. In J. S. Perkell & D. H. Klatt (Eds.), *Invariance and variability in speech processes* (pp. 226-234). Hillsdale, NJ: Lawrence Erlbaum Associates.

Fujimura, O. (1987). A linear model of speech timing. In R. Channon & L. Shockey (Eds.), *In honor of Ilse Lehiste* (pp. 109-123). Dordrecht, Holland: Foris Publications.

Kelso, J. A., Saltzman, E., & Tuller, B. (1986). The dynamical perspective on speech production: Data and theory. *Journal of Phonetics, 14,* 29-59.

Kent, R. D. (1983). The segmental organization of speech. In P. F. MacNeilage (Ed.), *The production of speech* (pp. 57-89). New York: Academic Press.

Kent, R. D., & Adams, S. G. (1989). The concept and measurement of coordination in speech disorders. In S. A. Wallace (Ed.), *Perspectives on the coordination of movement* (pp. 415-450). Amsterdam: Elsevier Science Publishers.

Kent, R. D., & Minifie, F. D. (1977). Coarticulation in recent speech production models. *Journal of Phonetics, 5,* 115-133.

Kent, R. D., & Rosenbek, J. C. (1983). Acoustic patterns of apraxia of speech. *Journal of Speech and Hearing Research, 26,* 231-249.

Klatt, D. H. (1976). Linguistic uses of segmental duration in English: Acoustic and perceptual evidence. *Journal of the Acoustical Society of America, 59,* 1208-1221.

Lehiste, I., & Peterson, G. E. (1961). Transitions, glides, and diphthongs. *Journal of the Acoustical Society of America, 33,* 268-277.

Lindblom, B. (1963). Spectrographic study of vowel reduction. *Journal of the Acoustical Society of America, 35,* 1773-1781.

Liss, J. M., & Weismer, G. (1992). Qualitative acoustic analysis in motor speech disorders. *Journal of the Acoustical Society of America, 92,* 2984-2987.

Liss, J. M., & Weismer, G. (1994). Acoustic characteristics of contrastive stress production in control geriatric, apraxic, and ataxic dysarthric speakers. *Clinical Linguistics and Phonetics, 8,* 45-66.

Milenkovic, P. (1992). CSpeech version 4.0 [Computer Program]. University of Wisconsin-Madison, Madison, WI.

Munhall, K., & Löfqvist, A. (1992). Gestural aggregation in speech: laryngeal gestures. *Journal of Phonetics, 20,* 111-126.

Perkell, J. S. (1980). Phonetic features and the physiology of speech production. In B. Butterworth (Ed.), *Language production* (pp. 337-373). London: Academic Press.

Saltzman, E., & Kelso, J. A. S. (1987). Skilled actions: A task dynamic approach. *Psychological Review, 94,* 84-106.

Saltzman, E. L., & Munhall, K. G. (1989). A dynamical approach to gestural patterning in speech production. *Ecological Psychology, 1,* 333-382.

Weismer, G. (1991). Assessment of articulatory timing. In J. Cooper (Ed.), *Assessment of speech and voice production: Research and clinical applications*. NIDCD Monograph Volume 1 (pp. 84-95). Bethesda: NIH.

Weismer, G., & Liss, J. M. (1991). Acoustical-perceptual taxonomies in speech production deficits in motor speech disorders. In C.A. Moore, K. Yorkston, D. R. Beukelman (Eds.), *Clinical dysarthria: Perspectives on management* (pp. 245-270). Baltimore: Brookes.

Weismer, G., & Martin, R. (1992). Acoustic and perceptual approaches to the study of intelligibility. In R. D. Kent (Ed.), *Intelligibility in speech disorders* (pp. 67-118). Amsterdam: John Benjamin.

Weismer, G., Martin, R., Kent, R. D., & Kent, J. F. (1992). Formant trajectory characteristics of males with ALS. *Journal of the Acoustical Society of America, 91,* 1085-1098.

Weismer, G., Tjaden, K., & Kent, R. D. (in press). Can articulatory behavior in motor speech disorders be accounted for by theories of normal speech production? *Journal of Phonetics*.

4 Towards an Explanation of Speech Sound Disorders in Children

Carole E. Gelfer
William Paterson College

Sarita Eisenberg
Teachers' College, Columbia University

INTRODUCTION

There has been a considerable amount of research that has looked at speech patterns of children who produce multiple sound errors. These children demonstrate severely restricted sound inventories as well as apparent sound deletions and substitutions, although these are not necessarily mutually exclusive. As a consequence, their speech is characterized by the presence of homonymous surface forms (Ingram, 1976) and reduced intelligibility.

In the 1970s, the term *phonological disorder* was suggested as a label for children who demonstrated multiple errors in the production of speech sounds. The traditional label of *articulation disorder* was considered inadequate because it focused exclusively on the motoric aspects of speech production. The term phonological disorder was conceptualized as a generic term encompassing all aspects of the sound system, including the underlying linguistic organization and the articulation (Locke, 1983; Shriberg & Kwiatkowski, 1982). In actuality, however, the term phonological disorder has become narrowly construed as a view in which "organizational principles are the focus, while articulatory concerns are virtually excluded" (Elbert, 1992, p. 241). What those principles might be, however, seems to vary considerably (e.g., Dinnsen & Chin, 1994; Stampe, 1969). Thus, a dichotomy has been created in which multiple speech sound errors are viewed as systematic and labeled as phonological, whereas the presence of one or two errors is considered to be random and labeled articulatory (Edwards, 1986).

We disagree with this characterization, and we consider the *a priori* rejection of an articulatory explanation for multiple speech sound errors to be unwarranted. We suggest that speech sound deficits could originate at either a linguistic or an articulatory level and that the origin of the deficits must be determined for each individual child.

DESCRIPTIVE CHARACTERISTICS

The productions of children with multiple speech errors can be summarized using four general categories (cf. Ingram, 1976): those affecting manner of production, those affecting place of articulation, those affecting voicing, and those affecting syllable structure. The most frequently reported manner patterns involve the production of stops instead of fricatives and affricates and production of glides (usually [w], but sometimes [j]) instead of liquids. Problems with place typically involve velar and palatal sounds, with production of alveolars in place of both of these back sound classes. Voicing errors tend to take the form of the voicing of syllable-initial voiceless stops and the devoicing of word-final voiced obstruents. Syllable structure patterns typically include the absence of word-final consonants, the absence of unstressed syllables, and the production of singleton consonants instead of clusters. There are also some children whose speech patterns are considered to be idiosyncratic (e.g., Camarata & Gandour, 1984) and/or are characterized by systematic sound preferences (Weiner, 1981).

LINGUISTIC APPROACHES

The traditional method of analyzing speech sound errors was to look at production sound-by-sound (cf. Van Riper, 1963). The presence of patterns that characterized the speech errors of children was first observed more than 25 years ago, and was triggered by developments in linguistic theory (e.g., Chomsky & Halle, 1968). Weber (1970), for instance, found similarities in production for groups of sounds that were phonetically similar based on traditional place, manner, and voicing features. Other investigators have identified segmental patterns based on distinctive features (e.g., Oller, 1973; Pollack & Rees, 1972). Phonological process analysis (e.g., Grunwell, 1975; Ingram, 1976) added patterns involving syllable structure as well as looking at substitution patterns for sound segments.

Stampe's Theory of Natural Phonology (1969) posited "phonological processes" as phonetically motivated innate mental operations that are imposed upon and that simplify an adult underlying representation. Over the course of phonetic development, a child is presumed to learn to suppress or limit these processes until her inventory matches that of the ambient community (Stampe, 1969). In this view, an English-speaking child would learn to suppress the developmental process of "final consonant deletion;" this suppression would allow consonants that are presumed to exist within the child's underlying system to emerge in word-final position. In contrast, given the absence of final consonants in Japanese, it would not be necessary for a Japanese-speaking child to suppress this process. Thus, English-speaking children whose speech lacks word-final consonants would be considered to have failed to "suppress" this process in a timely fashion. In this scheme, while the child is presumed to take an active role in extinguishing these processes, that role would appear to be relatively passive in terms of deriving the linguistic units of the ambient language and then generating underlying lexical forms.

Unlike Natural Phonology (Stampe, 1969), other characterizations of phonological disorders based on generative phonology (e.g., Dinnsen & Chin, 1994) make no *a priori* assumptions regarding a child's underlying representation, but rather infer that representation empirically from a child's actual productions. The basis for the speech sound errors may reflect nonadult constraints on either the rules or representations within the child's linguistic system (Dinnsen & Chin, 1994). In this view, nonoccurrence of a sound in any phonetic environment is evidence for an inventory constraint. For example, a child who does not produce any fricatives would be presumed to be missing this sound class in underlying representation. In contrast, if a sound is produced regularly in some instances, but not others, that sound would be assumed to exist in the child's underlying representation, and the error attributed to a phonological rule that specifies a positional or sequence constraint (Dinnsen & Chin, 1994). A child who produces stops word-initially but not word-finally would be presumed to include stops in underlying representation and to impose a positional constraint on this class of sounds.

Generative phonology represents each phoneme unit as a bundle of unordered features that include predictable as well as redundant information (Chomsky & Halle, 1968). This contrasts with nonlinear, or autosegmental, accounts that posit a feature geometry that includes only unpredictable properties and arranges these features hierarchically (Clements, 1985). One advantage to this latter account is that predictions can be made about the impact that the absence of one feature might have on the phonological system based on where that feature occurs within the hierarchy. For example, Bernhardt and Stoel-Gammon (1994) suggest that, if oral place features fail to be activated, [h] or [ʔ] might be produced in place of fricatives or stops. That is, if the place node is unspecified within a child's underlying representation, a laryngeal feature could be inserted in its stead through the operation of feature spreading, since place and laryngeal nodes are on the same plane. However, while this account provides a principled explanation of the substitution pattern, it still assumes, *a priori,* that the origin of the disorder is at the level of linguistic organization.

ARTICULATORY CONSIDERATIONS

The central issue that we are raising in this chapter concerns assumptions regarding the origin of disordered speech sound patterns. The linguistic characterizations of speech errors make several related assumptions. The first is that systematic patterns observed in a child's speech are inherently linguistic (Edwards, 1992; Fey, 1985). A second assumption is that the description of a child's speech corresponds to the status of the child's knowledge about the adult sound system. We would suggest that, while the linguistic methodologies do serve to describe patterns of speech sound production that occur across sound classes, syllable shapes, and phonetic contexts, we cannot assume that these analyses explain the patterns. Thus, we question the assumption that the linguistic unit drives the articulation, as if the child's own ability to achieve articulatory gestures were of no consequence.

Our view is that the motivation for the speech errors can come from either direction. We suggest that articulatory aspects of sound production might also involve patterns. For example, a child could have difficulty producing a constricted, but not totally occluded, vocal tract and therefore be unable to produce fricatives. We see this as no less feasible than positing an inventory constraint on fricatives or a phonological process that substitutes for fricatives. The observation of patterns is, in and of itself, merely a description of a child's production. It is an empirical question for each individual child whether this pattern reflects the underlying linguistic organization or whether the patterns relate to articulatory factors.

Articulatory phonology (cf. Browman & Goldstein, 1993) differs from both the phonological process and generative phonology accounts in that, among other things, the cognitive and physical levels are not viewed as independent, thus requiring an intermediate level of rules to relate them to each other. Rather, phonological units are identified with gestures or tract variables; that is, "dynamically specified units of articulatory action" (Browman & Goldstein, 1993, p. 51). Thus, because the gesture itself is an intrinsic specification of the linguistic unit, the need to posit a mechanism by which the phonological representation is converted into the phonetic realization becomes more or less unnecessary. Articulatory and contrastive properties of a sound system reflect an arrangement of mutual constraint, with the "gestural structures...providing a principled link between phonological and physical description" (Browman & Goldstein, 1986, p. 219).

This account raises a number of interesting questions regarding the nature of these gestural structures for children with speech sound disorders. For example, why do these gestures differentiate in development for some children (Browman & Goldstein, 1989), but fail to do so for others? And, is it possible to separate the specification of the gesture from the means to achieve it?

Kent (1992) emphasizes the relevance of biological development and sensorimotor coordination to phonological development. This approach promotes the notion that developing phonetic inventories are constrained, as are other motor behaviors, by the general properties of emerging motor control. In terms of sound classes, for example, the early emergence and continued predominance of stops (and homorganic nasals) in children's inventories may reflect a developmental tendency toward ballistic movements as opposed to the motorically more advanced regulation of force required for the production of most fricatives (Kent, 1992), although the production of [h] is apparently less motorically complex (Vihman, 1992). The increased occurrence of glides in developing inventories may reveal a preference for longer movements of constant velocity (Kent, 1992). Similarly, the systematic preference for anterior places of articulation, evident both in children's first words (de Boysson-Bardies et al., 1992) as well as in the productions of articulation-disorderd children, suggests a limited ability to regulate lingual articulation (Kent, 1992). The presence of appropriate voicing distinctions is clearly based on the regulation of glottal adjustments relative to vocal tract events, which is often not achieved successfully by children with articulation disorders (Catts & Jensen, 1983; Forrest & Rockman, 1988). Finally, citing evidence from studies of limb movement (Hay, 1979), Kent (1992) has also suggested that the absence of final consonants may reflect the failure to achieve the final phase of a movement, particularly when the target is small. Thus, speech sound disorders might be characterized as disruptions of the process by which a child applies limited resources in an attempt to achieve a motorically difficult goal (Kent, 1992).

Of course, many of the inferences regarding a child's articulation or underlying representation are based on whether or not s/he is perceived to produce the target sound in at least some context. However, there is evidence to suggest that children who ostensibly fail to produce some sounds in any contexts may in fact be producing contrasts that are not perceptible to the listener. This has motivated a search for imperceptible phonetic features that may relate to differences in the child's underlying representation; that is, subphonemic cues to which transcription techniques are insensitive, but which shed light on the organization of a child's sound system (e.g., Gibbon, 1990; Oller & Eilers, 1975).

Acoustic measures have been used to infer the presence of underlying phonemic contrasts for children who fail to make perceptible voicing contrasts (Tyler & Saxman, 1991) and/or place contrasts (Forrest, Weismer, Hodge,

Dinnsen & Elbert, 1990). Palatographic data have also demonstrated appropriate articulator placement in the absence of perceptible place contrasts for alveolar and velar stops (Gibbon, 1990). In fact, even when phones appear to be deleted entirely, they may be marked phonetically. For example, some children may mark a missing final consonant by producing reliable differences in vowel duration as a function of the voicing characteristic of the 'deleted' stop consonant (Weismer, Dinnsen, & Elbert, 1981) or by making appropriate F_2 transitions (Weismer, 1984).

These studies provide phonetic evidence that, even when never actually producing a particular sound target, a child may be attempting to produce some contrast involving that target. This contrasts with the view of Dinnsen (1984) that a child should not be credited with "underlying representations the details of which are never evidenced phonetically" (pp. 10-11). Our position is that it is possible to credit a child with an underlying representation if contrastive use of a sound or sound class is evident, even if it is not actually produced. But to do that one must not only use phonetic inventories to look for the presence or absence of a sound, but also look at contrastive analyses as evidence of what a child does when attempting a particular target.

DETERMINING THE ORIGIN OF SPEECH SOUND DISORDERS: PRODUCTIONS VS. REALIZATIONS

Based on the nature of the errors that children evidence clinically and the fact that even children with very similar error patterns respond with varying degrees of success to the same therapy approaches, we find it necessary to make some distinction between the "units of contrast" and the "units of articulatory organization" (Browman & Goldstein, 1990, p. 422). We believe that the ability to separate the phonological unit from the gesture could be useful in determining the origin of speech sound errors. In what follows, it is more for the sake of convenience than a reflection of our theoretical bias that we will use the terms *linguistic* and *articulatory*.

While the distinction may seem to be a traditional one, our approach is based on several assumptions that set it apart from those it would seem to resemble at first glance. Most of all, we argue that a pattern of errors across sound targets within a class does not necessarily imply a linguistic level disorder. Thus, both linguistic and articulatory errors may involve an entire class of speech sounds. The occurrence or nonoccurrence of sounds in a child's production is, in and of itself, not sufficient for determining whether a child's speech pattern is consistent with either an underlying or articulatory problem.

We suggest as an alternative that the focus turn to realizations. In other words, two sound targets that do not represent an underlying contrast in a child's system will share the same articulatory realization by the child. This means that two sounds or sound classes are the same to the child in the underlying representation, and s/he will produce them the same way in overt speech. This production might look like one of the confounded sounds, or it might be some different sound used for both targets. In contrast, differential realizations by a child of the two sound targets constitute evidence for knowledge of a contrast in the child's underlying representation, regardless of whether or not either sound is actually produced correctly in any context.

Based on these assumptions, we define two possible bases for speech sound errors: (1) A linguistic level problem in which some sound target A is confounded with some sound target B, so that both targets form a single meaning category and are realized in the same way, and (2) an articulatory level problem in which some sound target A is realized differently from all other targets, even if that sound target is produced incorrectly or is never actually produced at all.

We will illustrate this approach with two case studies of children who produced virtually identical sound inventories. For the purposes of this discussion, we focus only on one sound class, fricatives, and not on individual speech sounds.

The first child was a 5.4 year old girl, MS, whose limited inventory of sounds included only labial and alveolar nasals, stops, and glides, and the glottal fricative. Her inventory thus included [m n p b t d w h]. For lingual stop, fricative, and affricate targets, she produced an alveolar stop in word-initial position and lacked an overt sound production in final position (Figure 1). She, therefore, showed no evidence of distinguishing among these sounds since they were realized in the same way. We suggest that the absence of fricatives from her sound inventory originates at the linguistic level. Although she produced the glottal fricative [h], we do not see this as evidence for a class of fricatives, since all other fricative targets patterned with the alveolar stops (as did velar stops and affricates). Therapeutic goals for MS focused on the establishment of feature classes for fricatives.

The second child, BW, a 3.9 year old boy, evidenced an almost identical sound inventory. He produced front nasals and stops, the palatal glide, and the glottal fricative, limiting his inventory to [m n p b t d j h]. For all lingual stop and affricate targets, he also produced alveolar stops. For fricative targets, however, he produced mostly [h] in initial word position with occasional production of stops, and he produced stops for fricatives in word-final position (Figure 2). Unlike MS, this child showed evidence for a class of fricatives that appeared to be represented by [h] and which was clearly differentiated from the class of totally obstructed sounds (i.e., stops). In contrast to the feature-spreading accounts described above (Bernhardt & Stoel-Gammon, 1994), we suggest that BW experiences articulatory difficulty in achieving a narrow constriction within the oral cavity, but not at the level of the glottis (Vihman, 1992). Instead of the linguistically-based goal proposed for MS, we propose an articulatory goal for BW that focuses not on establishing a fricative sound class, but on developing the articulatory gestures for achieving frication within the oral cavity.

Error Analysis: Initial Position

	Bilabial	Labio-Dental	Lingua-Dental	Alveolar	Palatal	Velar	Glottal
Nasal	*m* √			*n* √			
Stop	*p b* √ √			*t d* √ √		*k g* t d	
Fricative		*f v* d d	*θ ð* d d	*s z* d d	*ʃ* d		*h* √
Affricate					*tʃ dʒ* t d		
Liquid				*l* d	*r* d		
Glide	*w* √				*j* d		

Error Analysis: Final Position

	Bilabial	Labio-Dental	Lingua-Dental	Alveolar	Palatal	Velar	Glottal
Nasal	*m* √			*n* √		*ŋ* n	
Stop	*p b* √ √			*t d* ø √		*k g* ø ø	
Fricative		*f v* ø ø	*θ ð* ø ø	*s z* ø ø	*ʃ ʒ* ø ø		
Affricate					*tʃ dʒ* ø ø		
Liquid				*l* ø	*r* ø		
Glide							

FIGURE 1. Error analysis of the speech of Subject MS. Within each cell, the upper symbols in italics indicate adult forms; the lower symbols indicate the subject's inventory items (√ indicates that MS produced the adult form).

Error Analysis: Initial Position

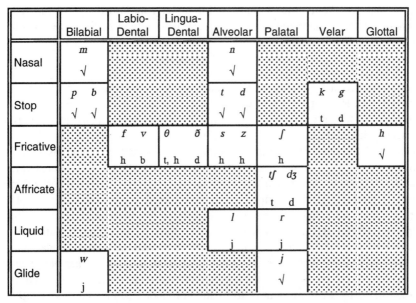

Error Analysis: Final Position

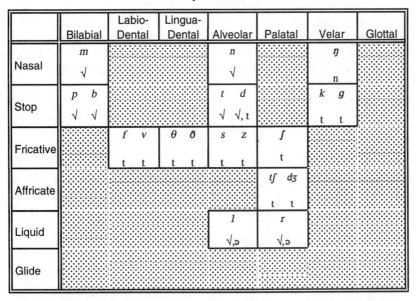

FIGURE 2. Error analysis of the speech of Subject BW. Within each cell, the upper symbols in italics indicate adult forms; the lower symbols indicate the subject's inventory items (√ indicates that BW produced the adult form).

We are, therefore, suggesting that, in spite of limited phonetic realizations, a child such as BW manifests considerable underlying knowledge of the sound system but, at the same time, has articulatory difficulty in realizing gestures that reflect this knowledge.

CONCLUSION

Certainly, linguistic theory has provided much of the impetus for looking at patterns in the errors of children with speech sound disorders. However, we cannot automatically assume that because we have adopted linguistic methods to describe disordered speech, these methods necessarily explain the observed patterns. As Locke (1983) pointed out more than a decade ago, there is a danger in equating description with explanation. Among other things, this can result in a failure to consider more than one explanation or therapy approach for children with limited sound inventories. We believe that it is imperative to establish criteria that will better differentiate speech sound disorders in children as either linguistically or articulatorily based. Furthermore, we hold that this distinction should motivate the choice of therapy approach chosen for each particular child.

Also of concern is that speech pathologists seem to hold a rather old view of articulation in general and speech sound disorders in particular. This is exemplified by what we believe to be the unfounded comments of Fey (1992) that articulation-based therapy approaches lead only to change in the targeted sound(s), while phonological intervention approaches can be expected to lead to system-wide change. We believe the practice of speech-language pathology needs to become more informed about current research in articulation and its relationship to both motor control and phonology. At the same time, we also believe that the study of developmental speech disorders has implications for models of articulation and linguistic organization. This chapter is an attempt towards these ends.

ACKNOWLEDGMENT

It is fitting that a chapter of this sort be included in a volume dedicated to Katherine Harris. Her strong conviction that clinical practice should be informed by theory, and vice versa, has influenced and enriched our lives in our roles as teachers, researchers, and clinicians.

REFERENCES

Bernhardt, B., & Stoel-Gammon, C. (1994). Nonlinear phonology: Introduction and clinical application: Tutorial. *Journal of Speech & Hearing Research, 37,* 123-143.

de Boysson-Bardies, B., Vihman, M. M., Roug-Hellichius, L., Durand, C., Landberg, I., & Arao, F. (1992). Material evidence of infant selection from target language: A cross-linguistic phonetic study. In C. A. Ferguson, L. Menn, & C. Stoel-Gammon (Eds.), *Phonological development: Models, research, implications* (pp. 369-392). Timonium, MD: York Press.

Browman, C. P., & Goldstein, L. (1986). Towards an articulatory phonology. *Phonology Yearbook, 3,* 219-252.

Browman, C. P., & Goldstein, L. (1989). Articulatory gestures as phonological units. *Phonology, 6,* 201-249.

Browman, C. P., & Goldstein, L. (1990). Representation and reality: Physical systems and phonological structure. *Journal of Phonetics, 18,* 411-424.

Browman, C. P., & Goldstein, L. (1993). Dynamics and articulatory phonology. *Haskins Laboratories Status Report on Speech Research, SR-113,* 51-62.

Camarata, S., & Gandour, J. (1984). On describing idiosyncratic phonologic systems. *Journal of Speech & Hearing Disorders, 49,* 262-266.

Catts, H. W., & Jensen, P. J. (1983). Speech timing and phonologically disordered children: Voicing contrasts of initial and final stop consonants. *Journal of Speech & Hearing Research, 26,* 500-509.

Chomsky, N., & Halle, M. (1968). *The sound pattern of English.* New York: Harper & Row.

Clements, G. (1985). The geometry of phonological features. *Phonology Yearbook, 2,* 305-328.

Dinnsen, D. A. (1984). Methods and empirical issues in analyzing functional misarticulation. In M. Elbert, D. Dinnsen, & G. Weismer (Eds.), *Phonological Theory and the Misarticulating Child. ASHA Mongraphs, 22,* 5-17.

Dinnsen, D. A., & Chin, S. B. (1994). Independent and relational accounts of phonological disorders. In M. Yavas (Ed.), *First and second language phonology* (pp. 135-148). San Diego: Singular Publishing Group, Inc.

Edwards, M. L. (1986). Phonological disorders: Assessment and remediation. Course presented at William Paterson College, Wayne, NJ.

Edwards, M. L. (1992). In support of phonological processes. *Language, Speech, & Hearing Services in Schools, 23,* 233-240.

Elbert, M. (1992). Consideration of error types: A response to Fey. *Language, Speech, & Hearing Services in Schools, 23,* 241-246.

Fey, M. E. (1985). Articulation and phonology: Inextricable constructs in speech pathology. *Human Communication Canada, 9,* 7-16. Reprinted in *Language, Speech, & Hearing Services in the Schools* (1992), *23,* 225-232.

Fey, M. E. (1992). Articulation and phonology: An addendum. *Language, Speech & Hearing Services in the Schools, 23,* 277-282.

Forrest, K., & Rockman, B. K. (1988). Acoustic and perceptual analysis of word-initial stop consonants in phonologically disordered children. *Journal of Speech and Hearing Research, 31,* 449-459.

Forrest, K., Weismer, G., Hodge, M., Dinnsen, D. A., & Elbert, M. (1990). Statistical analysis of word-initial /k/ and /t/ produced by normal and phonologically disordered children. *Clinical Linguistics & Phonetics, 4,* 327-340.

Gibbon, F. (1990). Lingual activity in two speech-disordered children's attempts to produce velar and alveolar stop consonants: Evidence from electropalatographic (EPG) data. *British Journal of Disorders of Communication, 25,* 329-340.

Grunwell, P. (1975). The phonological analysis of articulation disorders. *British Journal of Disorders of Communication, 10,* 31-42.

Ingram, D. (1976). *Phonological disability in children.* New York: Elsevier.

Kent, R. D. (1992). The biology of phonological development. In C.A. Ferguson, L. Menn, C. Stoel-Gammon (Eds.), *Phonological development: Models, research, implications* (pp. 65-90). Timonium, MD: York Press.

Locke, J. L. (1983). Clinical phonology: The explanations and treatment of speech sound disorders. *Journal of Speech and Hearing Disorders, 48,* 339-341.

Oller, D. K. (1973). Regularities in abnormal child phonology. *Journal of Speech and Hearing Disorders, 38,* 36-47.

Oller, D. K., & Eilers, R. E. (1975). Phonetic expectations and transcription validity. *Phonetica, 31,* 288-304.

Pollack, E., & Rees, N. (1972). Disorders of articulation: Some clinical applications of distinctive feature theory. *Journal of Speech & Hearing Disorders, 37,* 451-461.

Shriberg, L. D., & Kwiatkowski, J. (1982). Phonological disorders I: A diagnostic classification system. *Journal of Speech & Hearing Disorders, 47,* 226-241.

Stampe, D. (1969). The acquisition of phonetic representation. In R. I. Binnick, A. Davison, G. Green, & J. L. Morgan (Eds.), Papers from the Fifth Regional Meeting of the Chicago Linguistic Society (443-454). Chicago: University of Chicago Department of Linguistics.

Tyler, A. A., & Saxman, J. H. (1991). Initial voicing contrast acquisition in normal and phonologically disordered children. *Applied Psycholinguistics, 12,* 453-479.

Van Riper, C. (1963). *Speech correction: Principles and methods* (4th ed.). Englewood Cliffs, NJ: Prentice-Hall.

Vihman, M. M. (1992). Early syllables and the construction of phonology. In C. A. Ferguson, L. Menn, & C. Stoel-Gammon (Eds.), *Phonological development: Models, research, implications* (pp. 393-422). Timonium, MD: York Press.

Weber, J. L. (1970). Patterning of deviant articulation behavior. *Journal of Speech & Hearing Disorders, 35,* 135-140.

Weiner, F. F. (1981). Systematic sound preference as a characteristic of phonological disability. *Journal of Speech & Hearing Disorders, 46,* 281-286.

Weismer, G. (1984). Acoustic analysis strategies for the refinement of phonological analysis. In M. Elbert, D. Dinnsen, & G. Weismer (Eds.), *Phonological Theory and the Misarticulating Child, ASHA Monographs, 22,* 30-53.

Weismer, G., Dinnsen, D., & Elbert, M. (1981). A study of the voicing distinction associated with omitted, word-final stops. *Journal of Speech & Hearing Disorders, 46,* 320-327.

5 Timing of Lip, Jaw, and Laryngeal Movements Following Speech Disfluencies

Michael D. McClean, Mary T. Cord, and
Donna R. Levandowski
Audiology and Speech Center, Walter Reed Army Medical Center

INTRODUCTION

A view that has guided considerable research on the physiology of movement is that distinct neuroanatomic centers are associated with the control of different temporal stages of the motor control process. These stages are believed to involve selection, programming, and execution of intended movements (Gracco & Abbs, 1987; Wise et al., 1991). Although current understanding is still very limited regarding how different components of the motor system contribute to these processes (e.g., supplementary motor area and cerebellum), this view continues to provide an important framework for research on motor control.

Focusing attention on different stages of the speech movement process may be particularly useful in the case of stuttering. Speech disfluency is most likely to occur on the initial sound segment of speech utterances (e.g., Peters & Hulstijn, 1987). Thus, disordered neuromotor processes would be expected during the organizational or programming stages that occur prior to the onset of muscle activation for speech. This issue has been addressed by investigations that describe the relative timing of movement preparation within speech reaction-time intervals as a function of such variables as utterance complexity and length, level of prior response cueing, and duration of the period prior to the response stimulus (Peters & Hultsijn, 1987; Peters, Hultsijn, & Starkweather, 1989; Watson & Alfonso, 1987). In general, stutterers are reported to differ as a group from nonstutterers during fluent speech on several variables studied.

At present there is little information on the nature of movements immediately following speech disfluencies, but two recent acoustic studies provide some data in this area. Prosek, Montgomery, and Walden (1988) compared the relative timing of several acoustic measures in phonetically-matched utterances taken from passages with and without disfluent speech. They observed that the temporal characteristics of fluent speech in the disfluent passages did not differ from phonetically-matched segments in fluent passages. Viswanath (1991) measured the durations of words occurring prior to and following disfluent words as they were produced on repeated readings of the same passage. The durations of words immediately following disfluencies in the first reading were approximately equivalent to their durations in the stutterers' fluent utterances produced in subsequent readings.

If structural movements during the period immediately following speech disfluencies are equivalent to those in phonetically-matched fluent utterances, it would suggest that neural control processes underlying disfluency are relatively distinct from those operating during subsequent movement execution within the same utterance. This would further imply that the mechanisms of speech disfluency tend to be more episodic or intermittent in nature rather than continuously present during stutterers' speech. The purpose of the present research was to address these issues through description of the timing of speech structure movements following speech disfluencies. Timing of speech kinematic events is generally viewed as an important control parameter of target specification in the speech neuromotor planning process (Bell-Berti & Harris, 1981; Gracco & Abbs, 1988; Weismer & Fennell, 1985). This implies that activation across different muscle systems is planned and regulated by the CNS to achieve relatively invariant timing among different respiratory airway structures. The specific question addressed in the present study concerns whether the timing of lip, jaw, and laryngeal movements immediately following utterance-initial speech disfluencies differs from phonetically-matched fluent speech utterances.

METHODS

Subjects

The subjects included 10 adult male stutterers ranging in age from 24 to 46 years. Four of the subjects had previously been involved in intensive speech therapy programs that targeted prolongation of sound segment duration. Using video tape recordings, the percentage of disfluent words was determined for each subject for reading, telephone conversation, and self description. The mean disfluency levels across these three speaking situations for the 10 subjects ranged from 1% to 29% with a median of 9.5%. These subjects were part of a larger study dealing with the kinematics of the fluent speech of stutterers. They were selected for the present study because each showed a sufficient number of disfluencies during the physiologic recording procedures to permit comparison of fluent and disfluent utterances.

Procedures

Movements of the lips, jaw, and larynx were recorded while subjects produced a set of simple speech utterances in response to visual cueing. Superior-inferior displacements of the upper lip, lower lip, and jaw were transduced with a head-mounted strain-gauge system. Strain-gauge cantilever beams were coupled to the lips and jaw by means of small beads attached to the tissue with double-sided biomedical tape. The degree of vocal fold contact was transduced indirectly by means of an electroglottograph (Glottal Enterprises). During the experiment, each of the three strain-gauge signals and the electroglottograph signal were digitized at 500 Hz and stored in a microcomputer.

Following satisfactory placement of the transducers, subjects were positioned in front of an oscilloscope screen that served as the speech instruction and response cueing device. Five words, "pop, bop, top, fop, op," were printed in a column at the left of the screen. An LED was positioned at the top-center of the screen, and the horizontal line produced by the oscilloscope sweep was vertically aligned with one of the five words. Individual speech trials were initiated by the experimenter pressing a switch. This produced a low level click, caused the LED to come on for 200 ms and caused the instruction line to move to a different word. At the same time the computer began digitizing the physiologic signals for a period of 4 seconds. Subjects were instructed to produce the target word in the carrier phrase "_____is the word" when the light came on and they had determined the appropriate word from the line position. Subjects were told that this was not a reaction time task, and that they did not have to produce the speech utterances as quickly as possible. On each trial a different instructional tag voltage was digitized along with the physiologic signals, indicating the particular word produced. Each of the five test words was presented in an unpredictable order at least ten times to each subject.

During the recording procedure a speech-language pathologist sat next to the oscilloscope facing the subject and judged each trial as to whether the utterance was fluent or disfluent. After making her decision, she asked the subject to make a similar judgment and then recorded each of their responses. In addition to sound repetitions and prolongations, obvious struggle behavior immediately before the sentence-initial test word was treated as disfluent speech. After the disfluency judgments had been recorded, the experimenter waited three to five seconds before initiating the next trial.

Data Analysis

Utterances judged by either the speech-language pathologist or the subject as disfluent were treated as disfluent utterances for this study. Of the total number of disfluencies recorded, 53% were identified by both observers, 18% only by the speech language pathologist, and 29% only by the subject. Kinematic analyses were performed on a total of 292 fluent and 230 disfluent utterances across the ten subjects. Based on reports of the speech-language pathologist and subject, disfluencies were associated with the initial test word and/or obvious struggle behavior before its production. The disfluencies were distributed among the five test words as follows: op, 32; fop, 59; top, 29; bop, 44; and pop, 66.

Measurement of kinematic signals involved off-line display of the multichannel signals and cursor positioning on specific events. Figure 1 illustrates the types of signal display used for positioning the measurement cursor. For purposes of this study, the upper and lower lip signals (top signal in Figure 1) were combined to provide a single lip displacement signal. This was done to simplify the measurement procedures and reduce the number of dependent measures. Combining the upper and lower lip signals was judged appropriate, because examination of these signals indicated that the lips functioned in a highly synergistic manner to produce the movements studied. The combination of the two lip signals was derived by subtracting the jaw signal from the lower lip signal and adding an inverted version of the upper lip signal. Timing measures were obtained for each of the six events labeled in Figure 1. These were defined as follows:

1. The onset of the first glottal cycle associated with the test word vowel.
2. The maximal point of inferior jaw displacement associated with the test word vowel.
3. The point of maximal combined upper and lower lip displacement associated with lip closure on the final /p/ sound in the test word.
4. The onset of the first glottal cycle associated with the word "is" in the carrier phrase.
5. The point of maximal jaw elevation associated with production of lingual consonants in "is the."
6. The point of maximal lip displacement associated with lip rounding for the word "word" in the carrier phrase.

Each of these measures was stored in a text file along with the instructional tag voltage. Separate text files of each subject's fluent and disfluent utterances were read into Minitab statistical software. Six derived measures were calculated within the statistical spreadsheet, three reflecting intrastructural timing and three reflecting interstructural timing. These are defined as the time intervals, in milliseconds, between the following time points identified in Figure 1:

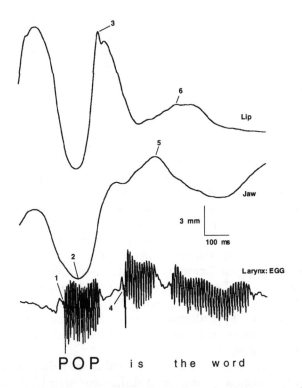

FIGURE 1. Example of the types of signals used to obtain timing measures of lip and jaw movements and vocal fold contact. In this case, a fluent production of "Pop is the word" is displayed. The numbers correspond to the six measures taken on each utterance. See the text for further description of these measures and how they were used to obtain derived duration measures.

L-L: lip closure to lip rounding (points 3 and 6)
J-J: jaw depression to jaw elevation (points 2 and 5)
LA-LA: voicing onset between two words (points 1 to 4)
L-J: lip closure to jaw elevation (points 3 to 5)
LA-J: voicing onset to jaw elevation (points 4 to 5)
L-LA: lip closure to voicing onset (points 3 to 4)

The selections of various time measures and derived durations were based on the ease and reliability with which the measures could be made, the association of the measures with significant phonetic events, and the desire to evaluate both inter- and intrastructural timing. It is assumed that durations associated with the six derived measures reflect important features of the temporal programming underlying the test utterances.

In making statistical comparisons of the fluent and disfluent utterances, the six timing measures were averaged across the different test words and repetitions for eight subjects. For the other two subjects, disfluencies occurred on only one of the test words, and comparisons between fluent and disfluent utterances were restricted to repetitions of the particular test word for each subject. The average number of disfluencies across subjects was 23 with a range of 6 to 46.

RESULTS

When the mean durations for the fluent and disfluent utterances were paired for each subject, the correlation coefficient across subjects exceeded 0.98 (df = 8) for each of the six timing measures. Thus, it was judged appropriate to compare the differences between the fluent and disfluent utterances for the stutterers as a group by performing matched-pairs t-tests on each of the six duration measures. The means of the six duration measures associated with the fluent and disfluent utterances were paired within each subject, and t-tests were performed evaluating the null hypothesis that the mean difference across subjects was zero. While there was a slight tendency for mean durations to be greater for the disfluent utterances, the difference was statistically significant ($p < 0.05$) in only one case, measure LA-LA. For the six measures the mean difference ranged from 3 ms to 23 ms. It should be noted that the LA-LA measure was most likely to have been affected by disfluencies associated with the initial speech sound, because this measure spanned the duration of the initial vowel and final stop consonant in the test word.

The above results indicate that the temporal characteristics of lip, jaw, and vocal fold movements following speech disfluencies were highly similar to the corresponding segments of fluent speech utterances. In order to assess the pattern and extent of this trend in terms of individual subjects, multiple t-tests were performed comparing the mean durations of the multiple repetitions of fluent and disfluent utterances for each subject. In 51 of 60 t-tests (ten subjects by six measures) there was no significant difference ($p > 0.05$) between the means of the timing measures of the fluent and disfluent utterances.

The relative durations or rank orderings of the six timing measures were equivalent for the fluent and disfluent utterances for all ten subjects. This result is illustrated in Figure 2 which shows the means and standard deviations of the six timing measures obtained for two subjects. Neither of these subjects had previously undergone intensive speech therapy. As indicated in Figure 2, no significant differences in the fluent and disfluent utterances were observed for Subject GT, while Subject TB showed significant differences in three timing measures.

There was a high degree of intersubject variability in the overall mean durations of the six timing measures. This was related in large part to the performance of four subjects who had previously participated in a speech therapy program that targeted prolongation of speech segment durations.

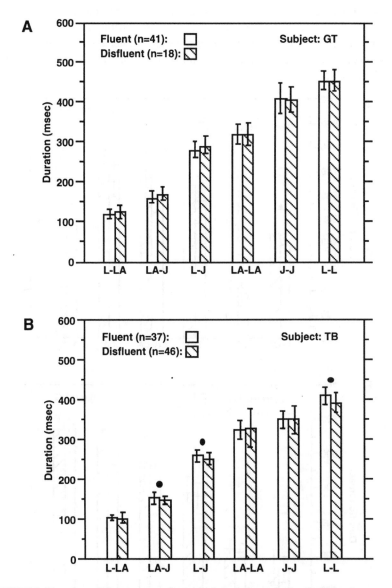

FIGURE 2. Bar plots of the means and standard deviations obtained on the six measures from Subject GT (A) and Subject TB (B). Filled circles indicate cases where the difference in the mean levels were statistically significant at the 0.05 level.

Comparing combined mean durations for the fluent and disfluent utterances across all six timing measures, these four subjects showed an average duration that was 125 ms longer than the other subjects. A t-test revealed this difference to be significant at the 0.05 level. Figure 3 illustrates the timing data obtained for two subjects who had previously undergone speech therapy. In general the absolute durations of the six timing measures are greater for these subjects than the subjects illustrated in Figure 2, who had not previously participated in speech therapy.

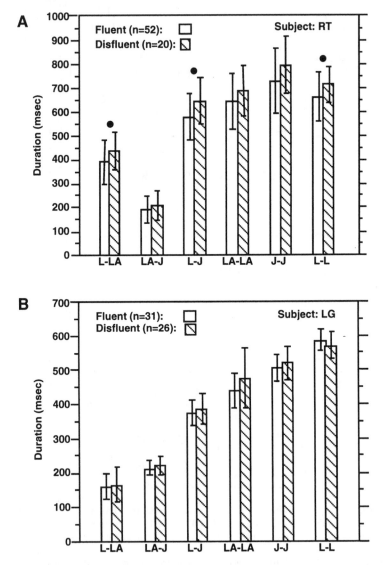

FIGURE 3. Bar plots of the means and standard deviations obtained on the six measures from Subject RT (A) and Subject LG (B). Filled circles indicate cases where the difference in the mean levels were statistically significant at the 0.05 level.

Subject RT's data, shown in Figure 3A, displayed significantly longer durations in the disfluent utterances on three of the six measures. The opposite trend was seen for Subject TB (Figure 2B), who showed significant reductions in duration on the disfluent utterances in the case of three measures. These two are the only subjects who showed consistent differences between the fluent and disfluent utterances. Of the nine instances across all subjects and measures where significant differences were noted in timing measures between fluent and disfluent utterances, six occurred in Subjects TB and RT. Three other subjects each showed a single case of a significant difference in mean duration.

The various timing measures might be expected to vary with the initial target sound (/p, b, t, f, ɑ/). Thus, statistical analyses were performed on individual subject data in order to determine whether there were any significant interactions between target sound and fluency. In order to evaluate this possibility, two-way analyses of variance were performed for each of the six timing measures using the factors of target sound and fluency. Six of the ten subjects had a sufficient number of fluent and disfluent utterances associated with the various target sounds to permit such comparison. Five of these subjects showed a significant main effect for target sound on both the LA-LA and J-J measures, the most consistent tendency being increased duration for utterances initiated with /b/. However, of primary importance here, a significant interaction between target sound and fluency was seen in only one of 36 cases (six subjects and six timing measures).

DISCUSSION

The primary observation in the present study is that the timing of speech structure movements immediately following stutterers' disfluent speech was for the most part equivalent to that of their fluent speech. For all subjects, the temporal sequencing of kinematic events following the initial test word was equivalent in the fluent and disfluent utterances, and eight of the ten subjects showed little or no significant durational differences between the fluent and disfluent utterances. These results are consistent with previous acoustic studies (Prosek et al., 1988; Viswanath, 1991), although the data are not directly comparable because of differences in study design.

Useful parallels may exist between the present findings and the results of recent load perturbation studies by Gracco and Abbs (1988, 1989). In their 1988 study, loads of 45 grams were unpredictably applied to the lower lip at the approximate onset and throughout the production of multisyllabic words produced on each trial. On load trials, lip and jaw displacements varied in a manner that allowed the perioral system to achieve the spatial positioning necessary for production of the target sounds. The temporal sequencing and relative timing of upper lip, lower lip, and jaw movements were largely unaffected by the mechanical loads. In interpreting these findings, Gracco and Abbs suggest that somatosensory input resulting from the load perturbations was integrated with centrally generated motor patterns to compensate for the mechanical disruptions produced by the load. Similar processes may operate in temporal proximity to speech disfluencies and partially explain the invariances in movement timing observed here. An alternative interpretation of the present results is that the temporal programming and execution processes associated with the carrier phrase were largely unaffected by disruptive neural input underlying the disfluencies. According to this view, disruption of movement initiation was followed by normal execution of the intended movements, with integration of somatosensory input playing a minimal role. This type of behavior is similar to that observed in simple neural oscillators that rapidly return to their normal state of rhythmic output following a perturbing input (e.g., Friesen & Stent, 1978).

The present results suggest some degree of independence of speech movement initiation and execution processes, and further point to movement initiation as being critical in speech disfluency. In this regard, it is notable that recent efforts at realistic computer modeling of neural systems underlying motor control assume that movement initiation involves unique processes that are distinct from movement preparation and execution (Kien & Altman, 1992). Kien and Altman view motor control as a distributed process in which diverse motor centers function in parallel as behavior progresses from preparation to execution. Movement initiation is seen as involving the temporal coincidence of inputs to brain stem or spinal cord motor centers from a number of converging parallel pathways, each of which may contribute to distinct features of movement output. It may be that disruption of the normal timing of these inputs is a central feature of the mechanisms of speech disfluency.

A variety of neurophysiologic studies indicate that marked changes occur in the distributed nature of neural activity at the approximate time of initiation of normal movements. Among these are human studies involving methods that could be extended to the analysis of speech-movement initiation processes in stutterers. For example, studies of readiness potentials suggest a gradual build up of neural activity in the premotor cortex that is rapidly followed by high levels of discharge in motoneuron pools at the onset of movement (Deecke, Grozinger, & Kornhuber, 1976; Wohlert & Larson, 1991). Also, reflex and evoked potential studies suggest that somatosensory input from muscle systems undergo systematic modulation prior to movement (Cohen & Starr, 1987; MacKay & Bonnet, 1990). Continued refinements and wider application of the methods employed in these various studies may enhance understanding of how speech movement initiation processes differ between fluent and disfluent utterances.

ACKNOWLEDGMENTS

This research was supported by the Department of Clinical Investigation, Walter Reed Army Medical Center, under Work Unit no 2510, and was approved by the

Center's Human Use Committee. All subjects enrolled into the study voluntarily agreed to participate and gave written informed consent. The opinions or assertions contained herein are the private views of the authors and are not to be construed as official or as reflecting the views of the Department of the Army or the Department of Defense.

REFERENCES

Bell-Berti, F., & Harris, K. S. (1981). A temporal model of speech production. *Phonetica, 38,* 9-20.

Cohen, L., & Starr, A. (1987). Localization, timing specificity of gating of somatosensory evoked potentials during active movement in man. *Brain, 110,* 451-467.

Deecke, L., Grozinger, B., & Kornhuber, H. H. (1976). Voluntary finger movement in man: Cerebral potentials and theory. *Biological Cybernetics, 23,* 99-119.

Friesen, W. O., & Stent, G. S. (1978). Neural circuits for generating rhythmic movements. *Annual Review of Biophysics and Bioengineering, 7,* 37-61.

Gracco, V. L., & Abbs, J. H. (1987). Programming and execution processes of speech movement control: Potential neural correlates. In E. Keller & M. Gopnik (Eds.), *Motor learning and sensory processes of language* (pp. 163-201), Hillsdale, NJ: Lawrence Erlbaum Associates Inc.

Gracco, V. L., & Abbs, J. H. (1988). Central patterning of speech movements. *Experimental Brain Research, 71,* 515-526.

Gracco, V. L., & Abbs, J. H. (1989). Sensorimotor characteristics of speech motor sequences. *Experimental Brain Research, 75,* 586-598.

Kien, J., & Altman, J. S. (1992). Preparation and execution of movement: Parallels between insect and mammalian motor systems. *Comparative Biochemistry and Physiology, 103A,* 15-24.

MacKay, W., & Bonnet, M. (1990). CNV, stretch reflex and reaction time correlates of preparation for movement direction and force. *Electroencephalography and Clinical Neurophysiology, 76,* 47-62.

Peters, H., & Hulstijn, W. (1987). Programming and initiation of speech utterances in stuttering. In H. Peters & W. Hulstijn (Eds.), *Speech motor dynamics in stuttering* (pp. 185-195). New York: Springer-Verlag.

Peters, H. F., Hulstijn, W., & Starkweather, C. W. (1989). Acoustic and physiological reaction times of stutterers and nonstutterers. *Journal of Speech and Hearing Research, 32,* 668-680.

Prosek, R. A., Montgomery, A. A., & Walden, B. E. (1988). Constancy of relative timing for stutterers and nonstutterers. *Journal of Speech and Hearing Research, 31,* 654-658.

Viswanath, N. (1991). Temporal structure is reorganized when an utterance contains a stuttering event. In H. Peters, W. Hulstijn, & C. W. Starkweather

(Eds.), *Speech motor control and stuttering* (pp. 341-353). Amsterdam: Elsevier Science Publications.

Watson, B. C., & Alfonso, P. J. (1987). Physiological bases of acoustic LRT in nonstutterers, mild stutterers, and severe stutterers. *Journal of Speech and Hearing Research, 30,* 434-447.

Weismer, G., & Fennell, A. (1985). Constancy of (acoustic) relative timing measures in phrase-level utterances. *The Journal of the Acoustical Society of America, 78,* 49-57.

Wise, S. P., Alexander, G. E., Altman, J. S., Brooks, V. B., Freund, H. J., Fromm, C. J., Humphrey, D. R., Sasaki, K., Strick, P. L., Tanji, J., Vogel, S., & Wiesendanger, M. (1991). Group report: What are the specific functions of the different motor areas? In D. R. Humphrey & H. J. Freund (Eds.), *Motor control: concepts and issues* (pp. 463-485). New York: John Wiley & Sons.

Wohlert, A., & Larson, C. (1991). Cerebral averaged potentials preceding oral movements. *Journal of Speech and Hearing Research, 34,* 1387-1396.

Anticipatory and Carryover Effects: Implications for Models of Speech Production

6

Fredericka Bell-Berti
St. John's University
Haskins Laboratories

Rena A. Krakow
Temple University
Haskins Laboratories

Carole E. Gelfer
William Paterson College

Suzanne E. Boyce
Boston University
M. I. T.

Katherine Harris has been our mentor since the days we each met her. We have decided that one way of thanking Kathy with this volume is by providing a chapter aimed at the new generation of Speech Science students that can serve as a tutorial on coproduction, a model that Kathy helped to formulate. To the extent that this and the rest of our work contributes to increased understanding of speech behavior, Kathy must be credited—for having questioned and deliberated and forced us to think and explain more clearly. We hope that this chapter reflects our mentor's fine hand.

INTRODUCTION

The term "coarticulation" is used to describe the fact that speech is produced as a sequence of sounds in a smooth flow of articulatory movements—that is, there is "blurring of the edges" of segmental articulations as the vocal tract moves from one articulatory configuration to the next. Since studies of speech production have shown the vocal tract to be nearly always on the move from one segment to another, with occasional periods in which some, but not all, articulators can be seen to maintain static positions, it seems obvious that articulatory and acoustical patterns for any one segment must reflect characteristics of (at least) its adjacent segments.

A major goal of speech research has been to discover the spatial-temporal domain of coarticulatory influences, in the belief that they reflect the domain over which the motor system organizes production of a sequence of segmental articulations.[1] By this logic, the observed extent of coarticulatory influence provides us with an estimate of the size of the organizational units of speech production, units whose nature has been much discussed for many years. One problem in establishing the temporal and spatial domains of coarticulatory influence has been that the literature has appeared to provide evidence simultaneously for at least two types of conflicting models of articulatory organization (i.e., "feature-spread" and "coproduction"), in part, we believe, because of a failure to control for intrinsic segmental movements (e.g., Gelfer, Bell-Berti, & Harris, 1989). A second problem is that the studies have assumed that anticipatory and carryover coarticulation are produced via fundamentally different mechanisms, with anticipatory coarticulation envisaged as cognitively controlled, intentional, and large-scale, while carryover coarticulation has been viewed as the small-scale effect of mechanical and inertial forces acting on the articulators (e.g., Lindblom, 1963; MacNeilage, 1970), even in the face of evidence that both carryover and anticipatory coarticulation can occur over similar time spans (Daniloff & Hammarberg, 1973; Fowler, 1984). A third problem has been the failure to acknowledge that coarticulation results from both intersegmental interaction and higher-level prosodic organization. In a 1977 review paper on coarticulation research and models, Kent and Minifie concluded that none of the models of the time were adequate, because none of the theories that were segmentally based could handle the range of reported segmental

---

[1]As developing technologies made it possible to study the production of natural speech, the pervasive nature of coarticulation was viewed as problematic because it contradicted researchers' intuitions that the speech stream is composed sequences of (relatively) invariant segments; it had been, after all, in the domain of speech production that researchers had finally expected to find the invariant units of speech that had eluded them in the acoustical domain. Continuing to maintain the notion that speech is segmentally structured, in the face of the apparently extended range of coarticulatory effects, required an account of speech production that takes a string of discrete segments as its input and outputs a stream of overlapping and asynchronous gestures.

effects, that few attempted to include prosodic effects, and those that did lacked the specificity necessary to allow assessment of their predictions. Although many years have elapsed since the review by Kent and Minifie, their paper remains important for the insights it offers and the gaps it highlights, as the theories continue to require modification and improvement.

As we consider coarticulation research and theory in the mid-1990s, it seems clear that important progress has been made in the intervening years, most notably perhaps in our understanding of the nature of intersegmental organization (e.g., Bell-Berti & Harris, 1981; Browman & Goldstein, 1991; Fowler, 1980; Fowler & Saltzman, 1993). Research has increasingly shown the importance of prosody and syllable organization in predicting patterns of articulatory overlap (e.g., Beckman, Edwards, & Fletcher, 1992; Browman & Goldstein, 1991; de Jong, 1991; Fujimura, 1990; Krakow, 1993; Nittrouer, Munhall, Kelso, Tuller, & Harris, 1988; Turk, 1994; Vaissière, 1988). However, we still lack an adequately detailed model of the combined effects of segmental, syllabic, and prosodic organization.

This chapter reviews the results of several studies that we have conducted on the nature of speech motor organization. Our primary focus is on intra- and intersegmental organization, but we will also demonstrate the importance of the prosodic component by showing how speaking rate variation shapes segmental organization. These studies investigate three different kinds of data (movement, acoustic, and electromyographic) related to the activities of three different articulators (velum, lips, and tongue). While velar movement data were used by Kent and Minifie (1977) to reveal the inadequacies of earlier theories, we have shown that the patterns they observed conform to the predictions of a model based on the temporal overlap among characteristic movements for adjacent and near-adjacent segments, that is, a coproduction model. The apparent conflicts in the data on both velar and labial coarticulation, previously taken as supporting either the "feature-spread" or the "coproduction" model, are resolved when intrinsic segmental characteristics are considered and the coproduction model is appropriately applied (e.g., Bell-Berti, 1993; Bell-Berti & Krakow, 1991; Boyce, Krakow, & Bell-Berti, 1991; Boyce, Krakow, Bell-Berti, & Gelfer, 1990; Gelfer et al., 1989). EMG data related to tongue activity provide additional support for the segmental interactions of speech that were predicted by the coproduction model. In addition, we have shown that velar function is affected by a variety of nonsegmental factors, including position of the segment within a syllable and within a sentence, as well as by stress and speaking rate (Krakow, 1989, 1993; Krakow, Bell-Berti, & Wang, this volume).[2] Similarly, Krakow (1989, 1993)

---

[2]Furthermore, in stark contrast to previous reports of the velum as a functionally simple articulator, we believe that the segmental and nonsegmental characteristics drawn together here make it clear that the velum is a complex articulator that must be treated as such in any viable model of speech production.

has shown differences between labial articulations for consonants occurring syllable-initially and -finally (see also Browman and Goldstein, this volume, who have shown such differences for lingual articulations), as well as for consonants in stressed and unstressed syllables. These patterns support the notion that the development of a model of speech organization that accounts for the co-occurrence of segmental, syllabic, and prosodic characteristics is imperative.

INTRA- AND INTERSEGMENTAL ORGANIZATION

Intrinsic Segmental Movements

Understanding the nature of intersegmental organization requires an understanding of intrinsic positions for segments, that is, intrasegmental organization. In our work, we have stressed that theories of coarticulation must attend to such information. Data supporting our arguments can be found for a number of different articulators, including the velum (e.g., Bell-Berti, 1980, 1993; Bell-Berti & Krakow, 1991), the lips (e.g., Bell-Berti & Harris, 1979; Boyce, 1988, 1990; Gelfer et al., 1989), and the tongue (e.g., Bell-Berti & Harris, 1974; Harris & Bell-Berti, 1984), and confirmed by other investigators (Marchal, 1988; Silverman & Jun, 1994). Additional demonstrations have been found for the glottis (Munhall & Löfqvist, 1992; Saltzman & Munhall, 1989) and for the pharynx (Parush & Ostry, 1993). These data cover a wide range of measurement techniques, including opto-electrical tracking of lip and velum movement, EMG, acoustic analysis, electropalatography, ultrasound, and transillumination.

Consider the schematic diagrams (Figure 1) that represent two typical velar lowering patterns one is likely to observe during CV_nN sequences (where C represents an obstruent; V_n, some number of vowels; and N, a nasal consonant).[3] Three accounts of these patterns—all widely discussed in the literature—can be shown to fail precisely because they do not consider what each segment, whether phonemically oral or nasal, brings to the sequence.

Moll and Daniloff (1971) applied the feature-spreading model of Henke (1966) to velar movement data. In this account, segments that do not exhibit a contrast for a specific feature simply take on the feature value of the next specified segment in the sequence. Typically, English vowels have been viewed as unspecified with respect to the feature NASAL, as there is no oral-nasal distinction for vowels in English. Under these conditions—that is, assuming vowels to be unspecified for nasality and that Henke's model applies—evidence of velar lowering during the transition between an oral consonant and a vowel, or during a vowel sequence preceding a nasal consonant, is taken as evidence of anticipatory feature spreading from the nasal consonant. Examining the patterns

---

[3]Data reflecting the patterns seen in Figure 1 have been reported elsewhere (e.g., Bell-Berti & Krakow, 1991; Bell-Berti, 1993; Bladon & Al-Bamerni, 1982; Boyce et al., 1990).

presented in Figure 1, the earliest onset of velar lowering in both sequences would be identified as the onset of anticipatory coarticulation for the nasal consonant. Velar lowering would be considered to have spread through the vocalic portions of both utterances. It is important to note that this account does not predict the different patterns shown in the schematic representations of Figures 1(a) and 1(b). Furthermore, this account is largely unidirectional—that is, only the anticipated segment is considered to be important.

In a more recent proposal, Keating (1988a) examined movement patterns like those shown in Figure 1(a). She argued that a pattern in which there is smooth movement between two segments with extreme and opposite positions, despite the occurrence of one or more intermediate segments, should be taken as indicating interpolation through those intermediate segments, and that such segments would then be classified as phonetically unspecified. Although they introduce a timing unit, unspecified segments in Keating's theory contribute nothing to the articulatory trajectory. The notion "unspecified" is shared with Moll and Daniloff (1971), but the nature of coarticulation is different; in Keating's view, the mechanism seems inherently bidirectional (both the start and end points, and the timing of all intervening segments, contribute to the shape of the interpolated trajectory). According to Keating's model, movement patterns in CV_nN sequences like that shown in Figure 1(a) would be taken as evidence of lack of specification of the feature NASAL for English vowels.

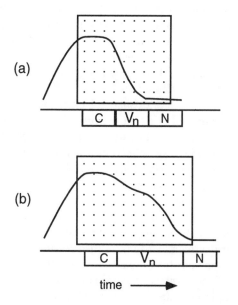

FIGURE 1. Schematic representations of predictions of coproduction model of velar movement patterns for CV_nN sequences.

Keating goes on to say that if a segment shows what looks like a target (i.e., it shows an articulatory position predicted by its featural composition), then it is safe to assume that it enters the motor program with categorical specification (+ or –). In this view, the movement pattern shown in Figure 1(b) would indicate specification of the intervening vowel or vowels with respect to velum position and/or nasalization. However, in a separate paper, Keating (1988b) specifically noted the occurrence of velar movement patterns like that shown in Figure 1(b), where she said that despite being "unspecified," some segments may have limits in the range of articulatory (in this case velar position) variation that they will tolerate; she referred to such limits as "windows." She claimed that this (albeit a wide) spatial window required the velum to remain in an intermediate position, for a period of time, between the extreme high and low positions occurring on either side, giving rise to the pattern presented in Figure 1(b). Keating (1988b) says nothing, however, about when or why one sometimes observes pattern 1(a) and sometimes pattern 1(b) for similar, if not the same, sequences.

Bladon and Al-Bamerni (1982) described the variation in velar movement patterns represented in Figures 1(a) and 1(b), and suggested that speakers simply have the choice between two alternative patterns for the same sequences. Describing pattern 1(a), they said that such movements conformed to the predictions of feature-spreading models, with the time between the onset of velar lowering and the beginning of the nasal murmur being correlated with the duration of the intervening vowel string. Describing pattern 1(b), they said that while the shallow movement onset conformed to the temporal predictions of the feature-spreading models (i.e., it migrates to the beginning of the vocalic portion), the onset of the steeper lowering movement seemed to be in close temporal proximity to the nasal consonant. They offered no way to predict the occurrence of 1(a) vs. 1(b) differentially.

Taken together, the three theories share a lack of insight into what, precisely, phonemically oral segments bring to the articulatory sequence. Fortunately, however, the speech production literature offers considerable insight into the relation between velar height/velopharyngeal port opening and such segmental characteristics as vowel height, and consonant place, manner, and voicing. Velar position is lowest for nasal consonants, being somewhat higher for nasal vowels than nasal consonants (see Henderson, 1984). Furthermore, velar position varies directly with vowel height (Brucke, 1856; Czermak, 1869; Passavant, 1863; Bell-Berti, Baer, Harris, & Niimi, 1979; Fritzell, 1969; Moll, 1962), and the velum is higher for obstruent consonants than for high vowels (for a detailed discussion of intrasegmental velar organization, see Bell-Berti, 1980, 1993).

An alternative to feature-spread theories is offered by coproduction approach that posits the simple overlap of segmental gestures as the source of most coarticulatory phenomena. We will discuss coproduction in some detail and then show how this approach accounts for the patterns in schematics 1(a) and 1(b). Bell-Berti and Harris (1981), for example, propose that (in the absence of articulatory conflict) the period of time taken up by a given articulator movements (i.e., its "articulatory period") begins at a relatively constant time before the period when the segment dominates the acoustic signal. (Note that this prediction holds even in conditions where articulatory overlap may be expected to increase, and thus affect the measured acoustic period, conditions like faster speaking rates.) The Bell-Berti and Harris theory predicts that an articulator's movements will overlap substantially with the articulatory periods of neighboring segments. In some cases, overlap will extend into the acoustic periods of neighboring segments.[4] This coproduction theory also assumes that intrasegmental organization takes into account the different periods for different articulators so that they are synchronized with respect to each other for the segment's acoustic period. Thus, the timing of articulatory periods bear a constant relation to each other.

A further aspect of the theory is that gestures will combine in an additive fashion for the period of time during which they overlap. Data to this effect were first reported by Bell-Berti (1980), who noticed that when oral consonants in a sequence were tightly overlapped, velar position was often higher than the segments' supposed intrinsic positions might suggest. More extensive work by Boyce (1988) and Munhall and Löfqvist (1992) supports this view of addition as the mechanism for combining simultaneous commands to the articulators. Munhall and Löfqvist (1992) also note that additive combination is frequently found in studies of nonspeech movement control.

In 1991, we tested this model with velar movement data and found that the model predicted the distribution of occurrence of patterns like those in schematics 1(a) and 1(b). We provided clear evidence of a distinct velar movement for the vowel segment (or segments) in CV_nN strings and support for the notion that increasing the duration of the vowel portion will function to separate the overlapping velar lowering gestures for the vowel(s) from that for the nasal consonant (Bell-Berti & Krakow, 1991). Figure 2 presents actual movement data[5] that match the schematic patterns of Figures 1(a) and 1(b). Figures 2(a) and 2(b) show that separate vowel- and nasal consonant-related velar lowering gestures are observed when there is sufficient time between the fricative and the

---

[4]In addition, Bell-Berti and Harris (1981; also Bell-Berti, 1993) suggest that the speech production mechanism adjusts overlap between gestures when the outcome would be unacceptable, and that this adjustment takes the form, when possible, of adjusting the timing of movements (i.e., timing of articulatory periods) to delay their onset. They noted that a delayed onset is more likely to occur when two articulatory gestures that are in conflict occur close together in time. On the other hand, the evidence of the complete movement trajectory is more likely to be seen when two competing gestures are separated by increased temporal intervals.

[5]In the speech samples collected for the Bell-Berti and Krakow (1991) study, /l/ was assumed to be vocalic in nature in terms of velar height because of evidence that it is produced with a velar position more like that of vowels than oral consonants in English (see Kuehn, 1976; Moll & Daniloff, 1971; Ohala, 1971; Schourup, 1973).

nasal, for example, when the vowel sequence is extended by the addition of vocalic segments. Figure 2(c) shows a minimally contrastive sequence without a nasal consonant, from the same data set. Such non-nasal control sequences must be considered in order to distinguish intrinsic velar positions from coarticulatory effects. The data in Figure 2(c) show that the velum lowers for a vowel following a fricative consonant even when there is no nasal consonant in the vicinity. Furthermore, Bell-Berti and Krakow (1991, Figure 8, p. 119) showed that the early part of the velar lowering gesture in CV_nN sequences matches the lowering gesture in minimally contrastive CV_nC sequences—that is, in sequences ending with an obstruent rather than a nasal consonant.

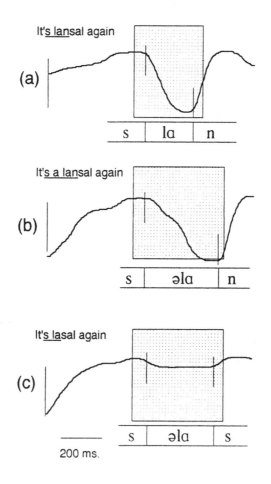

FIGURE 2. Observed anticipatory velar movement patterns for one subject producing "It's lansal" and "It's a lansal;" the relevant segments ([slɑn], [səlɑn], and [səlɑs]) are indicated (adapted from Bell-Berti & Krakow, 1991).

As discussed above, most studies of coarticulation have focused on anticipatory effects, and the examples that we have just provided are limited in the same way. To test the predictions of the coproduction model further, we have recently begun to investigate carryover coarticulation in sequences of the form NV_nC, comparing them to sequences of the form CV_nC. The patterns that we have observed parallel those reported (and described above) for anticipatory coarticulation, lending further support to the model. This work, however, is clearly in its infancy.

Figure 3 provides examples of velar raising patterns when there is sufficient time between the nasal and oral consonants to allow separate vowel- and obstruent-related velar raising gestures to be observed. In the first sequence, [ɪtsmastɪk], the vowel provides a sufficiently long interval (about 300 ms) for the vowel and obstruent gestures to be distinguishable. In the second sequence, [ɪtsmɑlistɪk], separate raising gestures are evident for [l], [i], and [s], leading to the sort of multi-stage pattern shown in Figure 3(b). (In this second sequence, there is a slightly higher position for [ɑ] than [m], a midposition for [l], a high position for [i], and an even higher position for [s]).

As noted above, other types of data confirm the predictions of our model. In what follows, we discuss other aspects of the model and offer evidence from acoustic and articulatory data related to lip movement and then EMG data related to tongue movement.

Articulatory Periods

The studies described above found evidence that individual vowels and consonants have intrinsic velar positioning patterns, that these patterns become visible as time is added to the sequence, and that intrinsic patterns appear to contribute additively to the overall sequence. In a series of studies spanning more than a decade (e.g., Bell-Berti & Harris, 1979, 1982; Boyce, 1988, 1990; Boyce et al., 1991; Boyce et al., 1990; Gelfer et al., 1989), we have examined orbicularis oris EMG activity and lip movement for lip rounding gestures for similar effects, and found confirming evidence for each of these points. In fact, Boyce (1988, 1990) provided evidence that naturally occurring patterns of lip movement for particular words could be synthesized by adding together intrinsic segmental movements extracted from a variety of words with different phonetic contexts. An example of the results of this procedure is shown in Figure 4. Here a synthetic movement trace for the word [kuktluk] has been constructed by adding (ensemble-averaged) movement traces for two other words, [kuktlik] and [kiktluk], that were obtained from a speaker of American English (who produced them in identical carrier phrases). As Figure 4 shows, the "constructed" trace bears a strong resemblance to the true ensemble-averaged [kuktluk] trace.

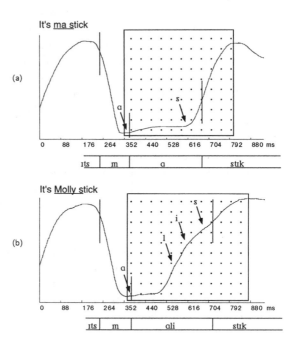

FIGURE 3. Observed carryover velar movement patterns for one subject: (a) producing [mɑ] and (b) [mɑli] utterances; discontinuities assumed to be the result of the onsets of successive segments are indicated by arrows.

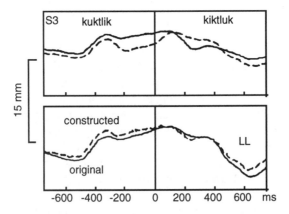

FIGURE 4. Overlaid averaged lip protrusion traces for [kuktlik] and [kiktluk] (top panel); naturally produced averaged lower lip protrusion trace for [kuktluk] with superimposed trace (lower panel) constructed by adding together traces in the upper panel and subtracting averaged trace for [kiktlik]. In all cases, anterior movement is upward (adapted from Boyce, 1990).

One aspect of the model not addressed in our velar and lip studies is whether the acoustic and articulatory periods for a segment have a stable relation in time. We describe here a previously unreported study whose results bear on this issue. Two subjects produced 20 repetitions of each of eight utterance types, comprising four minimal pairs, of the form iC_nV_2, where C_n was [t], [s], [sts], or [stst], and V_2 was [i] or [u]. The sequences were embedded in the carrier phrase, "It's a [__] again." The two subjects were selected from among subjects of our earlier studies (e.g., Bell-Berti & Harris, 1982; Gelfer et al., 1989) because they exhibited different lip activity patterns for [s] and [t]. One subject, S1, used substantial lip activity during [s] and [t] only when the consonants preceded a rounded vowel. This pattern fit the traditional definition of coarticulation from a contextual rounded vowel. If such coarticulation results from overlap of a stable articulatory gesture into the consonant, we might expect that rounding activity to begin at a consistent time. Further, the intrinsic rounding specification for the consonant is "unrounded" (i.e., no intrinsic rounding movement or intrinsic lip-spreading movement). The second subject, S2, produced alveolar consonants with substantial labial activity regardless of the phonetic properties of the following vowel. Such a pattern suggests that for this subject, the intrinsic rounding specification for alveolar consonants is "rounded." Again, we expect rounding activity to begin at a consistent time, but that time should be earlier in the case of S2.

We measured the frequency of resonances in the region of F2 from the beginning of V2 (the release burst of [t] or the end of frication of [s]), at 40 ms intervals back through the consonant sequence to the end of the first vowel, [i].[6] In all cases, the earliest measurement was made at the end of the first vowel. We took the ensemble average of the frequency measures at each time interval (the average of the 20 repetitions of each type, separately for each subject), and used t-tests to compare the significance of the differences between averages for the members of each minimal pair (i.e., [iC_ni] vs. [iC_nu]) at each interval.

Since lip protrusion should have the effect of lowering vocal tract resonances (Fant, 1960; Stevens & House, 1961), anticipation of lip rounding should result in lower frequency F2 resonances in consonants preceding a rounded vowel than in consonants preceding a nonrounded vowel. Since [s] and [t] share the same place of articulation, the acoustical resonances for these segments should be similar in the same phonetic contexts. Based on the observation that S2 does not differentiate lip activity for consonants preceding [i] and [u], we expected the two subjects to differ in the effect of the upcoming vowel on resonances in the frication. In particular, we expected S1 to show higher resonances before [i].

As expected, for both subjects there were significant differences at the onset of the second vowel for all utterance pairs, although this difference was larger for S1; these acoustic data are plotted in Figure 5. In addition, S1 showed a

---

[6]No measurements were made of a 40-ms point if it fell during a [t] closure.

significant effect in the predicted direction only for /isV/, and only for the frication within 80 ms of V2. On the other hand, S2 showed significant differences in the predicted direction for /isV/ and /istsV/ sequences, with the differences occurring as early as 160 ms before the second vowel. S2 also showed a significant difference in the predicted direction in the /iststV/ sequence, but only as early as 80 ms before V2. For both subjects, there were also scattered and inexplicable differences in the direction opposite to that predicted; that is, where resonances preceding [u] were higher than those preceding [i]. We would conclude that overall these acoustic data reflect the coproduction of the vowels [i] and [u] with the preceding consonant string. Specifically, the coarticulatory effects (i.e., resonances preceding [u] being lower than those preceding [i]) began no more than 160 ms before V2, even when the consonant sequences were more than twice that duration.

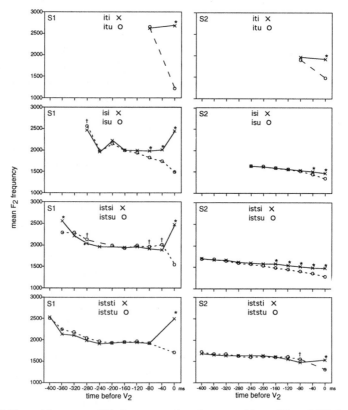

FIGURE 5. Ensemble-averaged F_2 frequency plots for two subjects (S1 and S2) for four minimally contrastive [iC$_n$i] and iC$_n$u] utterances; F_2 measurements occur at 40 ms intervals. Significantly different means are marked with '*' for comparisons in which the means of V2=[i] were higher, and '†' for comparisons in which the means of V2=[u] were higher.

In their 1982 paper, Bell-Berti and Harris reported not only that the onset of orbicularis oris EMG activity for [u] showed limited anticipatory effects, but also that the offset of orbicularis oris activity for the first vowel in [uC$_n$i] and [uC$_n$u] sequences did not extend any further into longer sequences than shorter ones (i.e., carryover coarticulation was temporally limited). Furthermore, they also reported that lip position for the following vowel did not influence the timing of the end of the first vowel gesture. Hence, these early EMG labial data complement the velar carryover data described above. They also reflect the pattern, described immediately below, in which articulatory activity for the vowels in a VCV sequence is suppressed during the consonantal articulation.

Articulatory Overlap

Among the earliest studies in this series were some that called into question the feature-spread models in general, and Henke's (1966) lookahead model in particular. In 1974, Bell-Berti and Harris reported EMG recordings from the genioglossus muscle, which is active for the forward, bunching and raising gesture for [i]. That study examined data for three speakers who produced a series of four-syllable nonsense utterances of the form [əpipiCə], where C was [p] or [b] and stress was systematically varied between the second and third syllables. The EMG signals were ensemble averaged, using the moment of closure for the second [p] as the reference point for aligning the signals of the 15-20 tokens of each of the eight utterance types. For the subject whose data are presented in Figure 6, as for the other two, the data are striking for the biphasic nature of the genioglossus activity for each utterance type, with one burst for the first [i] and a separate burst for the second [i]. Furthermore, closer inspection reveals that, for each minimal comparison, the duration of the interval between bursts was longer and the ensemble-average EMG minimum between the vowels was lower when stress fell on the second [i]. As one might expect, the acoustic duration of the medial [p] closure was longer when it immediately preceded the stressed vowel. This fits with what we know about the effects of stress on consonants in prestressed position: acoustic silent durations are longer, bursts are stronger. Moreover, movement displacements are greater; and stressed vowels are longer. The effect of these differences is to alter the temporal separation of the [i] gestures, and, as a consequence, their overlap is decreased. Or, in more current usage, we would say that there was less coproduction of the [i] gestures of adjacent syllables, and less overlap means more time for the tongue to relax from its [i]-related position. Thus, the troughs in the EMG traces are lower when the second vowel is stressed.

Additional confirmation of this pattern can be found in Boyce (1988), who found that double-peaked "trough" patterns occurred for lip movement and EMG orbicularis oris activity for sequences like [kuktluk], [kuktuk], and [kukuk], and that the troughs became shallower (i.e., the lip displacement or EMG activity minimum between peaks decreased) when fewer consonants intervened between rounded vowels. This effect is shown in Figure 7, which shows ensemble-averaged lip movement (S1, above) and EMG activity (S3, below) traces for the three words, spoken by one male (S1) and one female (S3) speaker of American English. Statistical significance of this effect was established by comparing coefficients of quadratic trend in peak-to-peak portions of the traces.

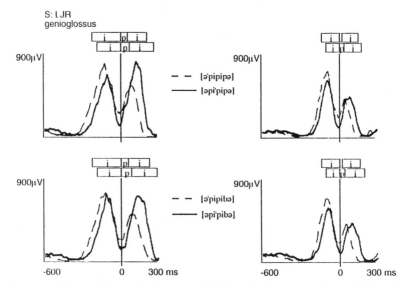

FIGURE 6. Ensemble-average genioglossus EMG activity for one speaker producing four minimally contrastive utterances at normal rate (left-hand panels) and four at fast rate (right-hand panels) (adapted from Bell-Berti & Harris, 1974).

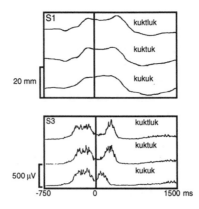

FIGURE 7. Ensemble-averaged lower-lip protrusion movement for [kuktluk], [kuktuk], and [kukuk], produced by S1 (above), anterior movement is upward; ensemble-averaged orbicularis oris EMG activity for [kuktluk], [kuktuk], and [kukuk], produced by S3 (below), (adapted from Boyce, 1988).

SUMMARY

Observations of multistage velar lowering before nasal consonants (anticipation) and multistage velar raising after nasal consonants (carryover) are, we believe, equivalent to the observation of lingual EMG and displacement minima, or "troughs," for [iCi] sequences (e.g., Bell-Berti & Harris, 1974; Harris and Bell-Berti, 1984) and of labial EMG and displacement minima for [uC$_n$u] sequences (e.g., Bell-Berti & Harris, 1982; Boyce, 1988; 1990). That is, these troughs, along with the stable time intervals found for the beginnings of segmental gestures (when appropriate control utterances are used), are reflections of the unitary nature of segments. Taken together, these data on velar, labial, and lingual articulation all share a common characteristic: they reflect segment-by-segment organization within overlapping temporal windows.

It is also important to examine the temporal cohesion of the gestures that comprise a segment, since an important assumption of the Bell-Berti and Harris (1981) model is the synchronization of a segment's component gestures. Happily, support for the idea of tightly constrained within-segment timing of component articulatory gestures can be found in research on the effects of perturbations of one articulator on the relative timing of gestures of another (see, for example, Saltzman, Löfqvist, Kinsella-Shaw, Kay, & Rubin, this volume).

PROSODIC ORGANIZATION: SPEAKING RATE

Because there are other sources of influence on the resulting movement ensemble, identification of a segmental level of gestural organization and an understanding of the manner in which gestures for successive segments combine provide only a partial understanding of the nature of speech motor organization. These influences include, but are not limited to, variations in speaking rate, syllable structure, stress, and the location of a syllable within a phrase or sentence. Of these, we have selected the effects of speaking rate on velar movements for discussion in this chapter.

The data on rate variation shed light on precisely the sorts of patterns we have just been addressing, and on the nature of variation between the patterns shown in the schematics in Figures 1(a) and 1(b). The coproduction model predicts that when speech rate is increased, there will be increased overlap of gestures for adjacent and near-adjacent segments. Hence, increasing the speaking rate in the production of a CV$_n$N sequence ought to have an effect that resembles that of decreasing the number of intervening vowels—that is, a smoother lowering movement ought to be evident. Figure 8(a) shows this effect on the movement patterns of a speaker producing the same utterance at two different rates (adapted from Bell-Berti & Krakow, 1991). We would also expect similar variation to occur when two different speakers produce the same sequence at different rates, and Figure 8(b) shows movement patterns of two speakers, one whose "comfortable" rate was much faster than the "comfortable" rate of the other (adapted from Bell-Berti and Krakow, 1991).

Other studies on the effects of speaking rate show a reduction in the magnitude of velar movements at faster rates (e.g., Kent, Carney, & Severeid, 1974; Kuehn, 1976). Both types of observation (i.e., a decrease in multi-stage gestures and a reduction in positional extremes) are predicted by the coproduction model, as both are outcomes of increased temporal overlap of gestures, although studies have generally examined only one or the other effect. Subsequent to the publication of Bell-Berti and Krakow (1991), we reexamined the data and measured the height of the velum at the release of the /s/ of "It's" or "It's say" for each sequence containing a nasal consonant. The results revealed that the velar peak was consistently higher in the normal than the fast rate productions, indicating a reduction in overlap between the gesture for the oral consonant and that for the nasal consonant at the normal rate. Still further support for this interpretation of these data can be found in the fact that the velar peak was higher when more vocalic segments intervened between the /s/ and the /n/ as shown in Figure 9.

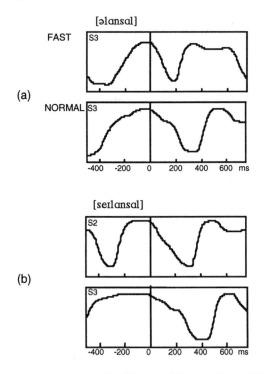

FIGURE 8. Velar movement patterns for (a) one subject producing "It's a lansal" at her "normal and "fast" speaking rates; (b) two subjects producing "It's say lansal" at their normal speaking rates (adapted from Bell-Berti & Krakow, 1991).

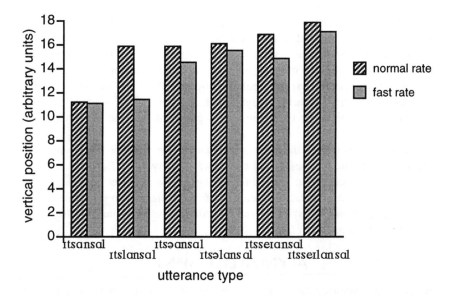

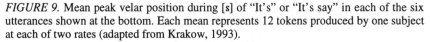

utterance type

FIGURE 9. Mean peak velar position during [s] of "It's" or "It's say" in each of the six utterances shown at the bottom. Each mean represents 12 tokens produced by one subject at each of two rates (adapted from Krakow, 1993).

These results show us how to predict when continuous single-stage velar lowering will occur and when complex, multi-stage velar lowering will occur. These results also make it possible to reconcile the data reported by Bladon and Al-Bamerni (1982) and by Keating (1988a, b) with those reported by Bell-Berti and Krakow (1991), and to refute the conflicting interpretations of Bladon and Al-Bamerni (1982) and of Keating (1988a, b) in favor of the coproduction model.

CONCLUSION

Drawing upon studies of three articulators, studies that employed very different types of measurements, we have found that there is actually rather little "anticipation" of articulatory activity. Whether in the "multi-stage" gestures observed in velar movement patterns, or the failure to find extended acoustical differences for consonant sequences preceding rounded and nonrounded vowels, or the adjustment of gesture onset timing to acoustic output constraints on lingual and labial EMG and displacement (corresponding to the duration of the intervening consonant occlusions), segmental articulatory behaviors do not extend very far from the segments for which they were intended. That is, all of the data discussed here, and much of the data presented in the literature, whether displacement, acoustical, or EMG data, support the predictions of the coproduction model.

That this is true for the extensive data on anticipatory coarticulation seems fairly obvious; we hold that the, thus far, rather limited data on carryover coarticulation are also consistent with a coproduction interpretation. The preliminary carryover data on velar function presented here appear to reflect the same time-stable patterns as those reported by Bell-Berti and Harris (1982) for lip-rounding gestures. Clearly, though, much additional work needs to be done to describe adequately the time course of segment offsets and their effects on nearby segments.

We have also examined the interaction between one prosodic effect, speaking rate, with intra- and intersegmental behaviors. Clearly, one effect of increased speaking rate is increased temporal overlap of the gestures for successive segments. But additional data are needed before we are able to provide a model that integrates segmental, syllabic, and prosodic factors to account for the range of observed coarticulatory effects.

ACKNOWLEDGMENT

The research reported here and the preparation of this manuscript were supported by NIDCD grant DC-00121 to the Haskins Laboratories.

REFERENCES

Beckman, M., Edwards, J., & Fletcher, J. (1992). Prosodic structure and tempo in a sonority model of articulatory dynamics. In G. Docherty & D. Ladd (Eds.), *Papers in laboratory phonology II: Gesture, segment, tone* (pp. 68-86). Cambridge: Cambridge University Press.

Bell-Berti, F. (1980). A spatial-temporal model of velopharyngeal function. In N. J. Lass (Ed.), *Speech and language: Advances in basic research and practice, Volume IV* (pp. 291-316). New York: Academic Press.

Bell-Berti, F. (1993). Understanding velic motor control: Studies of segmental content. In M. K., Huffman & R. A. Krakow (Eds.), *Nasals, nasalization, and the velum (Phonetics and Phonology,* Vol. 5, pp. 63-85). New York: Academic Press.

Bell-Berti, F., & Harris, K. S. (1974). More on the motor organization of speech gestures. *Haskins Laboratories Status Report on Speech Research, SR-37/38,* 73-77.

Bell-Berti, F., & Harris, K. S. (1979). Anticipatory coarticulation: Some implications from a study of lip rounding. *Journal of the Acoustical Society of America, 65,* 1268-1270.

Bell-Berti, F., & Harris, K. S. (1981). A temporal model of speech production. *Phonetica, 38,* 9-20.

Bell-Berti, F., & Harris, K. S. (1982). Temporal patterns of coarticulation: Lip rounding. *Journal of the Acoustical Society of America, 71,* 449-454.

Bell-Berti, F., & Krakow, R. A. (1991). Anticipatory velar lowering: A coproduction account. *Journal of the Acoustical Society of America, 90,* 112-123.

Bell-Berti, F., Baer, T., Harris, K. S., & Niimi, S. (1979). Coarticulatory effects of vowel quality on velar function. *Phonetica, 36,* 187-193.

Bladon, R. A. W., & Al-Bamerni, A. (1982). One-stage and two-stage temporal patterns of velar coarticulation. *Journal of the Acoustical Society of America,* 72, S104. (A)

Boyce, S. E. (1988). *The influence of phonological structure on articulatory organization in Turkish and in English.* Unpublished doctoral dissertation, Yale University.

Boyce, S. E. (1990). Coarticulatory organization for lip rounding in Turkish and in English. *Journal of the Acoustical Society of America, 88,* 2584-2595.

Boyce, S. E., Krakow, R. A., & Bell-Berti, F. (1991). Phonological underspecification and speech motor organization. *Phonology, 8,* 219-236.

Boyce, S. E., Krakow, R. A., & Bell-Berti, F., & Gelfer, C. E. (1990). Converging sources of evidence for dissecting articulatory movements into gestures. *Journal of Phonetics, 18,* 173-188.

Browman, C. P., & Goldstein, L. (1991). Tiers in articulatory phonology, with some implications for casual speech. In J. Kingston & M. E. Beckman (Eds.), *Papers in laboratory phonology I: Between the grammar and physics of speech* (pp. 341-376). Cambridge: Cambridge University Press.

Brucké, E. W. (1856). *Grundzuge der Physiologie und Systematik der Sprachlaute.* Vienna: Carl Gerold's Sohn (cited in Fritzell, 1969).

Czermak, J. N. (1869). Wesen und Bildung der Stimmund Sprachlaute. In *Czermak's gesammelte Schriften, Vol. 2.* Leipzig: Wilhelm Engelman (cited in Fritzell, 1969).

Daniloff, R. G., & Hammarberg, R. E. (1973). On defining coarticulation. *Journal of Phonetics, 1,* 239-248.

deJong, K. (1991). An articulatory study of consonant-induced vowel duration changes in English. *Phonetica, 48,* 1-17.

Fant, C. G. M. (1960). *Acoustic theory of speech production.* The Hague: Mouton.

Fowler, C. A. (1980). Coarticulation and theories of extrinsic timing control. *Journal of Phonetics, 8,* 113-133.

Fowler, C. A. (1984). Current perspectives on language and speech production: A critical review. In R. Daniloff (Ed.), *Recent advances in speech, hearing, and language, Vol. 3* (pp. 195-278). Boston: College-Hill Press.

Fowler, C. A., & Saltzman, E. (1993). Coordination and coarticulation in speech production. *Language and Speech, 36,* 171-195.

Fritzell, B. (1969). The velopharyngeal muscles in speech: An electromyographic-cinéradiographic study. *Acta Otolaryngologica, Supplement 250.*

Fujimura, O. (1990). Methods and goals of speech production research. *Language and Speech, 33,* 195-258.

Gelfer, C. E., Bell-Berti, F., & Harris, K. S. (1989). Determining the extent of anticipatory coarticulation: Effects of experimental design. *Journal of the Acoustical Society of America, 86,* 2443-2445.

Harris, K. S., & Bell-Berti, F. (1984).On consonants and syllable boundaries. In L. J. Raphael, C. B. Raphael, & M. R. Valdovinos (Eds.), *Language and cognition* (pp. 89-95). New York: Plenum Publishing.

Henderson, J. B. (1984). *Velopharyngeal function in oral and nasal vowels: A cross-language study.* Unpublished doctoral dissertation, University of Connecticut, Storrs.

Henke, W. (1966). *Dynamic articulatory model of speech production using computer simulation.* Unpublished doctoral dissertation, MIT.

Keating, P. A. (1988a). Underspecification in phonetics. *Phonology, 5,* 275-292.

Keating, P. A. (1988b). The window model of coarticulation: Articulatory evidence. *UCLA Working Papers in Phonetics, 69,* 3-29.

Kent, R. D., Carney, P. J., & Severeid, L. R. (1974). Velar movement and timing: Evaluation of a model of binary control. *Journal of Speech and Hearing Research, 17,* 470-488.

Kent, R. D., & Minifie, F. D. (1977). Coarticulation in recent speech production models. *Journal of Phonetics, 5,* 115-133.

Krakow, R. A. (1989). *The articulatory organization of syllables: A kinematic analysis of labial and velar gestures.* Unpublished doctoral dissertation, Yale University, New Haven, CT.

Krakow, R. A. (1993). Nonsegmental influences on velum movement patterns: Syllables, sentences, stress, and speaking rate. In M. K., Huffman & R. A. Krakow (Eds.), *Nasals, nasalization, and the velum* (*Phonetics and Phonology,* Vol. 5, pp. 87-114). New York: Academic Press.

Kuehn, D. P. (1976). A cineradiographic investigation of velar movement in two normals. *Cleft Palate Journal, 13,* 88-103.

Lindblom, B. E. F. (1963). Spectrographic study of vowel reduction. *Journal of the Acoustical Society of America, 35,* 1773-1781.

MacNeilage, P. F. (1970). Motor control of serial ordering of speech. *Psychological Review, 77,* 182-196.

Marchal, A. (1988). Coproduction: Evidence from EPG data. *Speech Communication, 7,* 287-295.

Moll, K. (1962). Velopharyngeal closure in vowels. *Journal of Speech and Hearing Research, 5,* 30-77.

Moll, K. L., & Daniloff, R. G. (1971). Investigation of the timing of velar movements during speech. *Journal of the Acoustical Society of America, 50,* 678-684.

Munhall, K. G., & Löfqvist, A. (1992). Gestural aggregation in speech: laryngeal gestures. *Journal of Phonetics, 20,* 111-126.

Nittrouer, S., Munhall, K., Kelso, J. A. S., Tuller, B., & Harris, K. S. (1988). Patterns of interarticulator phasing and their relation to linguistic structure. *Journal of the Acoustical Society of America, 84,* 1653-1661.

Parush, A., & Ostry, D. J. (1993). Lower pharyngeal wall coarticulation in VCV syllables. *Journal of the Acoustical Society of America, 92,* 715-722.

Passavant, G. (1863). Ueber die Verschliessung des Schlundes beim Sprechen. Frankfurt a.M.: J. D. Sauerlander (cited in Fritzell, 1969).

Silverman, D., & Jun, J. (1994). Aerodynamic evidence for articulatory overlap in Korean. *Phonetica, 51,* 210-220.

Saltzman, E., & Munhall, K. (1989). A dynamical approach to gestural patterning in speech production. *Ecological Psychology, 1,* 333-382.

Stevens, K. N., & House, A. S. (1961).An acoustical theory of vowel production and some of its implications. *Journal of Speech and Hearing Research, 4,* 303-320.

Turk, A. (1994). Articulatory-phonetic clues to syllable affiliation: Gestural characteristics of labial stops. In P. Keating (Ed.), *Papers in laboratory phonology III: Phonological structure and phonetic form* (pp. 107-135). Cambridge: Cambridge University Press.

Vaissière, J. (1988). Prediction of velum movement from phonological specifications. *Phonetica, 45,* 122-139.

7 Laryngeal Mechanisms and Interarticulator Timing in Voiceless Consonant Production

Anders Löfqvist
Haskins Laboratories

INTRODUCTION

During speech, parts of the vocal tract are briefly coupled in a functional manner to produce the acoustic characteristics of speech sounds. For example, the production of the bilabial voiceless stop /p/ requires the following set of actions. The lips are closed by joint activity of the jaw and the lips. The velum is elevated to seal off the entrance into the nasal cavity. The glottis is widened and the longitudinal tension of the vocal folds is often increased to prevent glottal vibrations. These articulatory actions all contribute to the period of silence in the acoustic signal and the increase in oral air pressure that are associated with a voiceless stop consonant. Speech production thus involves control and coordination of different parts of the vocal tract. Variations in their timing and coordination are commonly used to produce linguistic contrasts. For example, in stop consonants timing is used to control voicing and aspiration. In the following, we shall review aspects of voiceless consonant production with particular emphasis on laryngeal mechanisms and the coordination between laryngeal and supralaryngeal events. The material will be discussed in relation to control of coordinated movements in general.

99

LARYNGEAL MECHANISMS

The larynx serves as the source for voiced sounds, controlling the amplitude, the fundamental frequency, and the spectral properties of the source. For voiceless consonants, the source has to be momentarily turned off and then turned on again for the following sound. Turning off the source is commonly made by opening the glottis. The glottal abduction-adduction movement is controlled by the intrinsic laryngeal muscles. The posterior cricoarytenoid (PCA) muscle controls abduction, while the interarytenoid (INT), the vocalis (VOC), and the lateral cricoarytenoid (LCA) muscles adduct the vocal folds. The cricothyroid (CT) muscle regulates the length and tension of the folds. Figure 1 presents records of glottal opening, obtained by transillumination of the glottis, and the electromyographic activity of the PCA and the INT muscles in the production of a voiceless fricative and a voiceless stop consonant; the signals represent averages of 10-15 repetitions. The glottal opening is constantly changing, first increasing and then decreasing. Shortly before the start of the opening gesture, the activity of the PCA begins to increase, while the INT shows a reciprocal decrease in activity. The latency between the recorded electromyographic activity and the resulting mechanical action is in the order of 50-100 ms. For the glottal adduction, the pattern of activity in the two muscles is reversed with a decrease in PCA and an increase in INT.

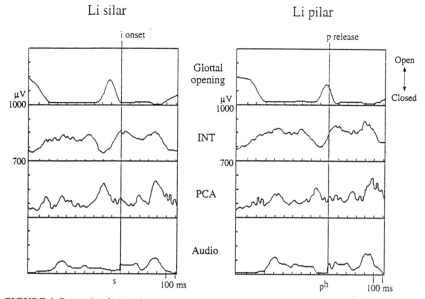

FIGURE 1. Records of glottal opening, the activity of the INT and the PCA muscles, and the audio envelope during the production of utterances with a voiceless fricative and a voiceless stop. The vertical line marks the line-up point used for averaging 10-15 repetitions. (Modified from Löfqvist & Yoshioka, 1980.)

In clusters of voiceless consonants, the glottis may show several opening movements (e.g., Löfqvist & Yoshioka, 1980, 1981; Pétursson, 1976a; Yoshioka, Löfqvist, & Collier, 1982; Yoshioka, Löfqvist, & Hirose, 1981). Such a pattern is illustrated in Figure 2, showing glottal opening and PCA and INT activity in the clusters /sts/ and /sts#p/. Note, in particular, that separate openings occur for the voiceless fricative /s/ and word initial voiceless aspirated stop /p/; between these opening movements, the glottis is narrowed without complete closure. In these clusters, the activity of both the INT and the PCA changes to produce the variations in glottal opening.

Figure 3 shows the activity of the other intrinsic laryngeal muscles in the production of a voiceless and a voiced bilabial stop. The LCA, the VOC, and the CT muscles all show a peak of activity around the line-up point used for averaging. This increase is related to the stress pattern of the utterance, where sentential stress occurs on the first syllable of the second word; this syllable has a high fundamental frequency. The difference in the activity of the muscles related to the voicing distinction can be seen in a decrease of activity of the LCA, the VOC, and the INT for the voiceless consonant. Note, however, that the CT activity does not appear to decrease for the voiceless stop.

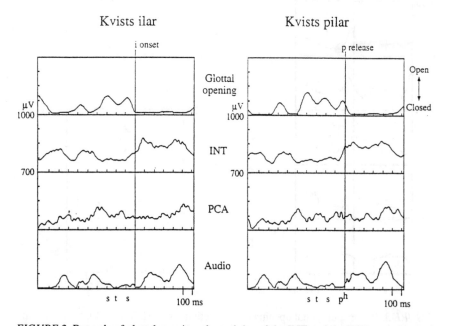

FIGURE 2. Records of glottal opening, the activity of the INT and the PCA muscles, and the audio envelope during the production of utterances with voiceless consonant clusters. The vertical line marks the line-up point used for averaging 10-15 repetitions. (Modified from Löfqvist & Yoshioka, 1980.)

In fact, a closer examination of the CT muscle suggests that it is activated more for voiceless than for voiced consonants, most likely to increase the longitudinal tension of the vocal folds and thus assist in the suppression of glottal vibrations (cf. Löfqvist, Baer, McGarr, & Seider Story, 1989). This mechanism can also account for the commonly observed high fundamental frequency at the beginning of a vowel following a voiceless stop or fricative.

The activity of several intrinsic laryngeal muscles in the production of voiceless consonant clusters is shown in Figure 4. From this figure, it is apparent that the activity of the VOC muscle successively increases and decreases in synchrony with the changes in the glottal opening. The continuous glottal adjustments in such consonant clusters are thus under active motor control as exemplified in Figures 2 and 4.

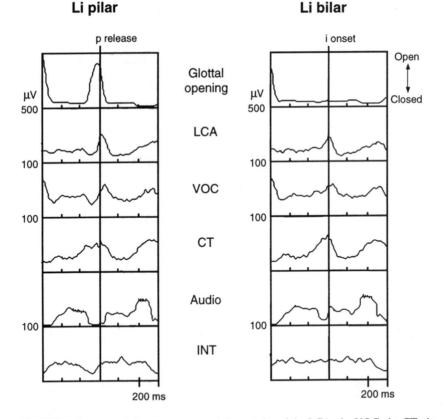

FIGURE 3. Records of glottal opening and the activity of the LCA, the VOC, the CT, the INT muscles, and the audio envelope during the production of utterances with a voiceless and voiced stop. The vertical line marks the line-up point used for averaging 10-15 repetitions. (Modified from Löfqvist, McGarr, & Honda, 1984.)

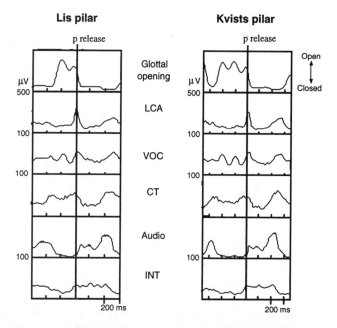

FIGURE 4. Records of glottal opening and the activity of the LCA, the VOC, the CT, the INT muscles, and the audio envelope during the production of utterances with voiceless consonant clusters. The vertical line marks the line-up point used for averaging 10-15 repetitions. (Modified from Löfqvist, McGarr, & Honda, 1984.)

There appear to be some differences in the glottal opening movement between voiceless stops and fricatives. Figure 5 shows glottal opening as a function of time during the production of a single voiceless fricative, a single voiceless aspirated stop, and a cluster of a voiceless fricative plus a voiceless unaspirated stop. The velocity of the opening change is also plotted. The size[1] of the opening is larger for the fricative than for the stop (cf. Munhall & Ostry, 1983; Gracco & Löfqvist, in press). Note that the movement is virtually identical for the single fricative and the cluster of fricative plus stop. In addition to the size difference, there is also a difference in the timing of the maximum opening relative to the offset of the preceding vowel: the maximum opening occurs closer to the vowel offset in the fricative than in the stop. The peak glottal opening corresponds to the onset of glottal adduction; it is under motor control and can be used as a reference point in studies of interarticulator timing. In the next section, we shall examine the timing between the laryngeal and oral articulatory movements in more detail.

---

[1]It should be kept in mind that the transillumination signal is uncalibrated. Care is thus necessary in drawing conclusions about the size of the glottal opening. Measurements obtained by transillumination, however, show good agreement with those obtained using ultrasound (Munhall & Ostry, 1983).

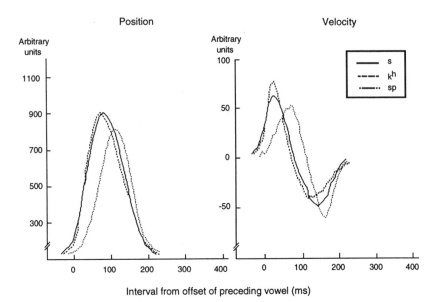

Interval from offset of preceding vowel (ms)

FIGURE 5. Records of the glottal opening movement during the production of different voiceless consonants. (Modified from Löfqvist & Yoshioka, 1981.)

INTERARTICULATOR TIMING

The production of voiceless consonants requires a tight coordination between the oral and laryngeal movements. First, at the transition from a preceding vowel to a voiceless consonant, the onset of glottal abduction has to be coordinated with the onset of the oral closure or constriction. Second, at the transition from the voiceless consonant to a following vowel, the onset of glottal adduction has to be coordinated with the release of the closure/constriction so that the glottal vibrations start at the appropriate time. It is well established that variations in the relative timing of the oral and laryngeal events are used to control voicing and aspiration in stop consonants. For example, if the onset of the glottal gesture precedes the formation of the oral closure, the last part of the preceding vowel is produced with a voice source that is breathy. This pattern of coordination is observed in languages that have preaspirated stops, such as Icelandic and Irish (e.g., Pétursson, 1976b; Löfqvist & Yoshioka, 1981). The opposite pattern of coordination occurs in voiced aspirated stops, as in Hindi (Dixit, 1989). Here, the glottis begins to open near the end of the oral closure; the closure is thus voiced. The abduction-adduction gesture is then completed after the release of the oral closure, making the onset of the following vowel characterized by a breathy mode of phonation. Yet another timing pattern is observed in voiceless postaspirated stops. For these sounds, the onset of glottal abduction is synchronized with the onset of the oral closure; the closure is thus voiceless.

Glottal adduction starts at about the oral release. After the release, the vocal folds are being adducted while there is a high rate of air flow out of the vocal tract. As a result, the onset of glottal vibrations relative to the oral release is delayed while noise is being generated in the glottis.

Variations in laryngeal-oral coordination thus produce different acoustic and aerodynamic patterns in voiceless consonants (cf. Löfqvist & McGowan, 1992). Such differences are illustrated in Figure 6. Comparing the single voiceless stop /p/ with the stop in the cluster /sp/, we see that the glottal condition at the oral release is different in the two cases. For the single stop, the glottis is open at the release, while for the stop in the cluster, the glottis is closed at the release. The air flow after the release is higher for the single stop and the onset of glottal vibrations relative to the release is delayed compared to that following the stop in the cluster.

There has been some controversy about the nature of control of aspiration and Voice Onset Time (VOT, cf. Lisker & Abramson, 1964) in stop consonants (for reviews, see Abramson, 1977; Löfqvist, 1980; Dixit, 1989). The arguments have centered on the roles of laryngeal-oral timing and the size of the glottal opening. The currently received view would seem to suggest that differences in interarticulator timing are always found between stop categories differing in VOT, but that differences in the size of the glottal opening can also occur. However, most of the evidence for the role of interarticulator timing in the control of aspiration and VOT has been collected in studies comparing different stop categories, e.g., unaspirated vs. aspirated stops, mostly using average values.

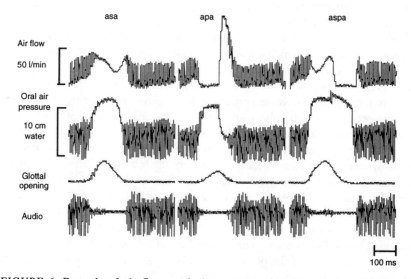

FIGURE 6. Records of air flow, oral air pressure, glottal opening, and the audio waveform during the production of different voiceless consonants. (Modified from Löfqvist, 1992.)

In fact, when the relation between oral-laryngeal coordination and VOT is examined across individual tokens, it turns out that the correlation between a measure of interarticulator timing and VOT is low, and that interarticulator timing explains less than 50% of the variance in VOT (Löfqvist, 1992). This finding suggests that, besides timing, aerodynamic and myodynamic factors also play a role in determining the onset of glottal vibrations following stop consonants. Among these factors are most likely the size of the glottal opening as well as the thickness and the viscosity of the vocal folds. Another implication is that fine-grained control of VOT may not be possible. This is probably not necessary, since languages do not appear to use fine-grained control of VOT for making linguistic distinctions. In the well-known case of Korean postaspirated stops, where there is a three-way distinction of VOT (short, medium, and long), an additional dimension commonly referred to as 'tensity' appears to be used (cf. Hirose, Lee, & Ushijima, 1974; Dart, 1987). The dimension of tensity is related to the force of glottal contact and is indexed by the activity of the VOC muscle.

Figure 7 plots the relation between the duration of the oral closure/constriction and the interval between onset of closure/constriction and peak glottal opening in voiceless stops and fricatives spoken at two different rates and under two different stress conditions (see Löfqvist & Yoshioka, 1984, for a more detailed description). For both the stops and the fricatives, there is a positive relation between the two temporal intervals, suggesting that they change together to maintain the relative phasing between the oral and laryngeal events. Also note that the relation between the two intervals differs between stops and fricatives. The slope of the regression between the variables is steeper for the stops than for the fricatives. Another difference in interarticulator timing is also evident from Figure 7. In the stops, which are all postaspirated, the onset of glottal adduction consistently occurs at about the release of the oral closure. For the fricatives, the same event occurs just before the middle of the oral constriction. These differences in interarticulator phasing are due to the different requirements in stop and fricative production, particularly the delay in voice onset after the stop release.

The temporal coordination between articulatory movements must be maintained within certain limits for speech to be intelligible, across changes in speaking rate. This requirement explains why the relations plotted in Figure 7 are positive. How this temporal cohesion is achieved, however, is not well understood. It has been suggested that variations in speaking rate result in a scaling between the different articulatory movements that are involved in the production process. This suggestion is based on the following theoretical view. If someone is writing a word on a paper with a pencil or on a blackboard with a piece of chalk, different parts of the body are used. When the word is written on paper, writing involves movements of the hand around the wrist; when it is written on the blackboard, the arm moves around the shoulder joint. Since the written pattern on the blackboard can be seen as a scaled version of the one on paper, it has generally been argued that there is a single underlying representation of the movement pattern that is instantiated by different parts of the body using a scaling relation (cf. Wright, 1990). The alternative view, that each pattern is stored as a separate entity, is at least intuitively implausible and inefficient. Thus, the claim is that the pattern is stored as a "generalized motor program" that can be reparameterized (see Schmidt, 1975). A generalized motor program predicts that when variations in speed and amplitude of a movement complex occur, the relations among the individual movements should remain virtually unchanged.

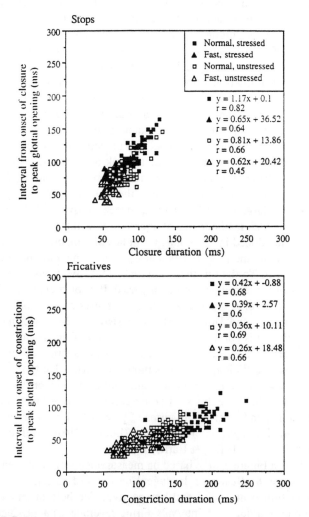

FIGURE 7. Plots of the duration of the oral closure/constriction and the interval from onset of closure/constriction to peak glottal opening in voiceless stops and fricatives (Modified from Löfqvist & Yoshioka, 1984.)

The reason is that a submovement interval should maintain a constant proportion of the whole movement interval. Hence, the model is usually referred to as a proportional duration model (see Heuer, 1991, for a general discussion of such models). Initially, several studies claimed that proportional timing was indeed found for motor activities like locomotion (Shapiro, Zernicke, Gregor, & Diestel, 1981), handwriting (Viviani & Terzuolo, 1980), typing (Terzuolo & Viviani, 1979), and speech (Tuller & Kelso, 1984).

Gentner (1987) proposed a stronger test of proportional duration by examining if the ratio between the duration of one movement interval and that of the whole movement sequence is unrelated to the duration of the whole movement sequence. The proportional duration model predicts that these should be related, since the duration of all the components of a movement sequence should maintain a constant proportion of the overall duration. Studies applying this statistical analysis suggest that proportional timing does not occur in speech or any other motor activity that has been examined (cf. Löfqvist, 1991; Sock, Ollila, Delattre, Zilliox, & Zohair, 1988; Wann & Nimmo-Smith, 1990). The slope of the regression usually deviates from zero. An examination of the data plotted in Figure 7 for constant proportionality shows that the slope of the regression between closure duration and the ratio of time to peak opening and closure duration is significantly different from zero for the stressed stops and the fricatives but not for the unstressed stops; for the latter, the lack of a statistically significant effect is due to a large variability in the data (see Löfqvist, 1991, for a more detailed discussion). One methodological uncertainty facing students of speech timing should be mentioned in this context. Studies of temporal phenomena by necessity must break up the flow of articulatory movements into discrete intervals for measurement. To delimit these intervals, movement onset and offset, and peak velocity of movement are commonly used. It is, of course, possible that these events are not the ones that the nervous system uses for controlling movements. Kelso, Saltzman, and Tuller (1986) suggested that the proper metric for constant relative timing is phase as measured on a phase plane, rather than ratio of articulatory intervals, and presented some evidence in support of this notion. In a phase-plane representation, position is plotted against velocity. In a vowel-labial consonant-vowel sequence, a phase-plane plot of the jaw or the lower lip shows an elliptical orbit. Using this kind of representation, movement onsets for different articulators can be defined in terms of phase relations. Further studies have, however, failed to replicate their findings (Lubker, 1986; Nittrouer, 1991; Nittrouer, Munhall, Kelso, Tuller, & Harris, 1988).

The data plotted in Figure 7 were based on measurements of movement onsets and offsets. Figure 8 illustrates laryngeal-oral coordination in voiceless consonant production from a different perspective. The bottom panel of this figure shows a plot of articulatory intervals during several normal productions of the utterance "It's a sifting again." The top panel shows how these intervals have been defined. The interval plotted along the x axis is measured from the peak glottal opening for the voiceless consonant cluster in "It's" (identified by the vertical dotted line in the top panel) to the peak velocity of the lower lip raising movement for the labiodental fricative /f/ in "sifting." On the y axis is plotted the interval from the same instance of peak glottal opening to the peak velocity of the glottal abduction movement for the fricative+stop sequence in "sifting." The movements of the lower lip and the glottis are both integral parts of the fricative. The intervals plotted in Figure 8 are thus temporally related to each other in the production of the specific utterance, and one would expect that they should covary. As variations in the overall duration of the utterance occur between productions, the intervals measured for the lower lip and the larynx should change together; remember that they have been measured from the same temporal reference point, peak glottal opening for the voiceless consonant cluster in "It's." As is evident from Figure 8, this is indeed the case. Their covariation can be indexed by the high correlation between them. At the same time, it is also apparent from Figure 8 that they do not scale proportionally, since the intercept of the regression is not at, or close to, zero.

A comparison of the correlation coefficients in Figures 7 and 8 shows that they are higher in Figure 8. The measurements shown in Figure 8 were made using peak velocity of movements as reference points, while those in Figure 7 were based on movement onsets and offsets. As noted above, it is not clear what peripheral events are the most revealing from the point of view of motor control. The higher correlations between temporal intervals defined by peak velocities may suggest that they are more useful for studying speech motor control. It is possible, however, that this is simply due to a methodological problem. That is, peak velocities may be more reliably identified in movement records than onsets and offsets, which are usually defined by zero crossings in the velocity signals.

Studying speech movements across changes in stress and speaking rate can provide insights into the mechanisms controlling the movements. Another valuable experimental paradigm for understanding movement coordination and control is to introduce unexpected perturbations to motor acts in a systematic manner. In a standard experiment, a subject is attached to a small motor that can be activated during some trials to generate a brief load. The rationale for this research is that the nature and time course of the response to the load may reveal the motor organization and reflex structure of the motor act. This paradigm has been applied to different types of motor behavior in humans such as posture control (e.g., Nashner & McCollum, 1985), hand and finger movements (e.g., Abbs, Gracco, & Cole, 1984; Rothwell, Traub, & Marsden, 1982; Traub, Rothwell, & Marsden, 1980), and respiratory control (Newsom Davis & Sears, 1970). A number of studies have also used this method to study speech motor control (Abbs & Gracco, 1984; Folkins & Abbs, 1975; Folkins & Zimmermann, 1982; Gracco & Abbs, 1985, 1988, 1989; Kelso, Tuller, Vatikiotis-Bateson, & Fowler, 1984; Kollia, Gracco, & Harris, 1992; Munhall, Löfqvist, & Kelso, 1994; Shaiman, 1989; Shaiman & Abbs, 1987).

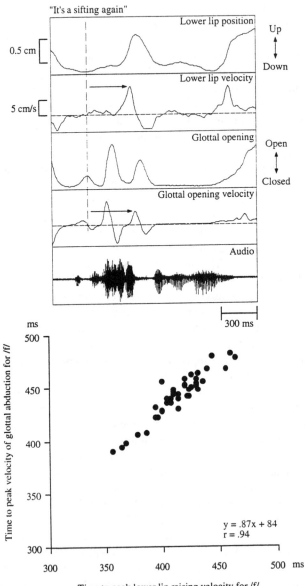

FIGURE 8. The top panel shows a single production of the utterance "It's a sifting again." The signals represent lower lip, glottal opening, and audio. Two articulatory intervals are defined that are related to the production of the labial fricative /f/ in the word "sifting." The bottom panel plots these two articulatory intervals. See text for further details.

From these speech perturbation studies, some general conclusions can be drawn. First, compensations are rapid. Electromyographic responses can occur 20-30 ms after load onset. The latency is not fixed, however, but depends on when the load was applied with respect to onset of activity in the muscles responsible for the movement in question (Abbs et al. 1984). Such short latencies suggest that the responses are not due to reaction time processes. Second, compensations are mostly task-specific. That is, they are neither stereotypic nor evident throughout the system, but rather tailored to the needs of the ongoing motor act. For example, when the jaw is loaded during the transition from a vowel to a bilabial stop, compensatory responses are made in the upper and lower lips to achieve the labial closure. On the other hand, when the jaw is loaded during the transition from a vowel to a dental fricative or a dental stop, a response is seen in the tongue (Kelso et al., 1984; Shaiman, 1989). We should add a word of caution here, however, since task specificity is not always consistent across speakers. In particular, one of the subjects in the study by Shaiman (1989) showed increased lower lip movement in addition to jaw and tongue compensatory movements when the jaw was perturbed during the utterance /ædæ/, which does not require lip activity. Similarly, Kelso et al. (1984) found increased upper lip EMG activity, for the alveolar fricative, in perturbed productions of /bæz/. Third, compensations are flexible and distributed among articulators involved in a specific task. Thus, when the jaw is loaded in the production of a bilabial stop, responses can occur in the jaw itself and/or in the upper and lower lips (Shaiman, 1989). Fourth, compensations are mostly functional and effective in the sense that the intended goal is normally achieved. For example, Munhall et al. (1994) perturbed the lower lip at the transition from the first vowel to the medial bilabial voiceless stop in the utterance /i'pip/. The system was able to overcome the load, making the intended closure of the vocal tract and increasing the air pressure in the oral cavity: Recordings of oral pressure revealed no differences in pressure between load and control productions.

While the results of these studies clearly indicate that the articulatory system is capable of rapid and functional responses to external loads, what happens to the larynx if the lower lip is mechanically perturbed while it is making the oral closure for a bilabial voiceless consonant? Is the phasing between the lips and the larynx maintained in spite of the load? A few studies have examined this question (e.g., Löfqvist & Gracco, 1991; Saltzman, Löfqvist, Kinsella-Shaw, Rubin, & Kay, 1992). Munhall et al. (1994) and Saltzman, Löfqvist, Kinsella-Shaw, Kay, and Rubin (this volume) examined laryngeal responses to lower lip perturbations during the production of a voiceless bilabial stop. In addition to lip and jaw actions to achieve the spatial target of a labial closure, a laryngeal response was evident in the perturbed trials by a delay of the onset of glottal abduction, measured relative to the acoustic onset of the preceding vowel. This delay was presumably made to maintain lip-larynx coordination at the transition from the vowel to the voiceless stop, and resulted in an increased acoustic duration of the preceding vowel. However, the period of bilabial closure for the stop was shortened by the perturbation, while the laryngeal abduction-adduction movement increased in duration. Consequently, the normal phasing between the oral and laryngeal movements was disrupted at the release of the oral closure. As a result, Voice Onset Time increased in the perturbed trials since it depends in part, as we have seen, on the timing between the oral and laryngeal events in stop production. These results suggest that interarticulator timing may be more affected by perturbations than spatial targets such as a closure or constriction in the vocal tract.

SUMMARY

The experimental material reviewed in this paper provides evidence that all the parts of the vocal tract involved in the production of a specific sound are controlled as a unit. In the case of the voiceless consonants discussed here, the laryngeal and oral movements necessary for their production are tightly linked. When one part of the articulatory system is perturbed, the movements of the other parts of the system are flexibly adjusted to attain the spatial goal of closing or constricting the vocal tract, while interarticulator timing may be affected by the perturbation. Under variations in speaking rate, the oral and laryngeal events change together, although they do not scale proportionally. An important task for speech motor control is to define the metric that governs temporal relations among speech movements.

ACKNOWLEDGMENTS

I am grateful to Elliot Saltzman for comments on an earlier version of this paper. Preparation of this manuscript was supported by Grant DC-00865 from the National Institute on Deafness and Other Communication Disorders and in part by Esprit-BR Project 6975-Speech Maps through Grant P55-1 from the Swedish National Board for Industrial and Technical Development.

REFERENCES

Abbs, J., & Gracco, V. L. (1984). Control of complex motor gestures: Orofacial muscle responses to load perturbations during speech. *Journal of Neurophysiology, 51*, 705-723.

Abbs, J., Gracco, V. L., & Cole, K. (1984). Control of multimovement coordination: Sensorimotor mechanisms in speech motor programming. *Journal of Motor Behavior, 16*, 195-232.

Abramson, A. S. (1977). Laryngeal timing in consonant distinctions. *Phonetica, 34*, 295-303.

Dart, S. (1987). An aerodynamic study of Korean stop consonants: Measurements and modeling. *Journal of the Acoustical Society of America, 81,* 138-147.

Dixit, P. (1989). Glottal gestures in Hindi plosives. *Journal of Phonetics, 17,* 213-237.

Folkins, J., & Abbs, J. (1975). Lip and jaw motor control during speech: Responses to resistive loading of the jaw. *Journal of Speech and Hearing Research, 18,* 207-220.

Folkins, J., & Zimmermann, G. (1982). Lip and jaw interaction during speech: Responses to perturbation of lower-lip movement prior to bilabial closure. *Journal of the Acoustical Society of America, 71,* 1225-1233.

Gentner, D. (1987). Timing of skilled movements: Test of the proportional duration model. *Psychological Review, 94,* 255-276.

Gracco, V. L., & Abbs, J. (1985). Dynamic control of the perioral system during speech: Kinematic analyses of autogenic and nonautogenic sensorimotor processes. *Journal of Neurophysiology, 54,* 418-432.

Gracco, V. L., & Abbs, J. (1988). Central patterning of speech movements. *Experimental Brain Research, 71,* 515-526.

Gracco, V. L., & Abbs, J. (1989). Sensorimotor characteristics of speech motor sequences. *Experimental Brain Research, 75,* 586-598.

Gracco, V. L., & Löfqvist, A. (1994). Speech motor coordination and control: Evidence from lip, jaw, and laryngeal movements. *Journal of Neuroscience, 14,* 6585-6597.

Heuer, H. (1991). Invariant timing in motor-program theory. In J. Fagard & P. Wolfe (Eds.), *The development of timing control and temporal organization in coordinated action* (pp. 37-68). Amsterdam: Elsevier.

Hirose, H., Lee, C. Y., & Ushijima, T. (1974). Laryngeal control in Korean stop production. *Journal of Phonetics, 2,* 145-152.

Kelso, J. A. S., Saltzman, E., & Tuller, B. (1986). The dynamical perspective on speech production: Data and theory. *Journal of Phonetics, 14,* 29-59.

Kelso, J. A. S., Tuller, B., Vatikiotis-Bateson, E., & Fowler, C. (1984). Functionally specific articulatory cooperation following jaw perturbations during speech: Evidence for coordinative structures. *Journal of Experimental Psychology: Human Perception and Performance, 10,* 812-832.

Kollia, B., Gracco, V. L., & Harris, K. S. (1992). Functional organization of velar movements following jaw perturbation. *Journal of the Acoustical Society of America, 91,* 2474. (A)

Lisker, L., & Abramson, A. S. (1964). A cross-language study of voicing in initial stops: Acoustic measurements. *Word, 20,* 384-422.

Lubker, J. (1986). Articulatory timing and the concept of phase. *Journal of Phonetics, 14,* 133-137.

Löfqvist, A. (1980). Interarticulator programming in stop production. *Journal of Phonetics, 8,* 475-490.

Löfqvist, A. (1991). Proportional timing in speech motor control. *Journal of Phonetics, 19,* 343-350.

Löfqvist, A. (1992). Acoustic and aerodynamic effects of interarticulator timing in voiceless consonants. *Language and Speech, 35,* 15-28.

Löfqvist, A., Baer, T., McGarr, N. S., & Seider Story, R. (1989). The cricothyroid muscle in voicing control. *Journal of the Acoustical Society of America, 85,* 234-239.

Löfqvist, A., & Gracco, V. L. (1991). Discrete and continuous modes in speech motor control. Papers from the symposium Current phonetics research paradigms: Implications for speech motor control. *PERILUS, XIV,* 27-34. Stockholm: Institute of Linguistics.

Löfqvist, A., McGarr, N. S., & Honda, J. (1984). Laryngeal muscles and articulatory control. *Journal of the Acoustical Society of America, 76,* 951-954.

Löfqvist, A., & McGowan, R. S. (1992). Influence of consonantal environment on voice source aerodynamics. *Journal of Phonetics, 20,* 93-110.

Löfqvist, A., & Yoshioka, H. (1980). Laryngeal activity in Swedish obstruent clusters. *Journal of the Acoustical Society of America, 68,* 792-801.

Löfqvist, A., & Yoshioka, H. (1981). Laryngeal activity in Icelandic obstruent production. *Nordic Journal of Linguistics, 4,* 1-18.

Löfqvist, A., & Yoshioka, H. (1984). Intrasegmental timing: Laryngeal-oral coordination in voiceless consonant production. *Speech Communication, 3,* 279-289.

Munhall, K., Löfqvist, A., & Kelso, J. A. S. (1994). Lip-larynx coordination in speech: Effects of mechanical perturbations to the lower lip. *Journal of the Acoustical Society of America, 95,* 3605-3616.

Munhall, K., & Ostry, D. (1983). Ultrasonic measurements of laryngeal kinematics. In I. Titze & R. Scherer (Eds.), *Vocal fold physiology: Biomechanics, acoustics, and phonatory control* (pp. 145-161). Denver: The Denver Center for the Performing Arts.

Nashner, L., & McCollum, G. (1985). The organization of human postural movements: A formal basis and experimental synthesis. *Behavioral and Brain Sciences, 8,* 135-172.

Newsom Davis, J., & Sears, T. A. (1970). The proprioceptive reflex control of the intercostal muscles during their voluntary activation. *Journal of Physiology, 209,* 711-738.

Nittrouer, S. (1991). Phase relations of jaw and tongue tip movements in the production of VCV utterances. *Journal of the Acoustical Society of America, 90,* 1806-1815.

Nittrouer, S., Munhall, K., Kelso, J. A. S., Tuller, B., & Harris, K. S. (1988). Patterns of interarticulator phasing and their relation to linguistic structure. *Journal of the Acoustical Society of America, 84,* 1653-1661.

Pétursson, M. (1976a). Jointure au niveau glottal. *Phonetica, 35,* 65-85.

Pétursson, M. (1976b). Aspiration et activité glottal. Examen experimental à partir des consonnes islandaises. *Phonetica, 33,* 169-198.

Rothwell, J., Traub, M., & Marsden, C. (1982). Automatic and 'voluntary' responses compensating for disturbances of human thumb movements. *Brain Research, 248,* 33-41.

Saltzman, E., Löfqvist, A., Kinsella-Shaw, J., Rubin, P., & Kay, B. (1992). A perturbation study of lip-larynx coordination. In J. Ohala, T. Neary, B. Derwing, M. Hodge, & G. Wiebe (Eds.), *ICSLIP 92 Proceedings, Addendum* (pp. 19-22). Edmonton: The University of Alberta.

Schmidt, R. (1975). A schema theory for discrete motor skill learning. *Psychological Review, 82,* 225-260.

Shaiman, S. (1989). Kinematic and electromyographic responses to perturbation of the jaw. *Journal of the Acoustical Society of America, 86,* 78-88.

Shaiman, S., & Abbs, J. (1987). Sensorimotor contributions to the temporal coordination of oral and laryngeal movements. *SMLC Preprints* (Speech Motor Control Laboratories, University of Madison) *Spring-Summer 1987,* 185-202.

Shapiro, D., Zernicke, R., Gregor, R., & Diestel, J. (1981). Evidence for generalized motor programs using gait pattern analysis. *Journal of Motor Behavior, 13,* 33-47.

Sock, R., Ollila, L., Delattre, C., Zilliox, C., & Zohair, L. (1988). Patrons de phases dans le cycle acoustique de détente en français. *Journal Acoustique, 1,* 339-345.

Terzuolo, C., & Viviani, P. (1979). The central representation of learned motor patterns. In R. Talbot & D. Humphrey (Eds.), *Posture and movement* (pp. 113-121). New York: Raven Press.

Traub, M., Rothwell, J., & Marsden, C. (1980). A grab reflex in the human hand. *Brain, 103,* 869-884.

Tuller, B., & Kelso, J. A. S. (1984). The timing of articulatory gestures: Evidence for relational invariants. *Journal of the Acoustical Society of America, 76,* 1030-1036.

Viviani, P., & Terzuolo, C. (1980). Space-time invariance in learned motor skills. In G. Stelmach & J. Requin (Eds.), *Tutorials in motor behavior* (pp. 525-533). Amsterdam: North-Holland.

Wann, J., & Nimmo-Smith, I. (1990). Evidence against the relative invariance of timing in handwriting. *Quarterly Journal of Experimental Psychology, 42A,* 105-119.

Wright, C. (1990). Generalized motor programs: Reexamination of claims of effector independence in writing. In M. Jeannerod (Ed.), *Attention and performance XIII* (pp. 294-320). Hillsdale: Lawrence Erlbaum.

Yoshioka, H., Löfqvist, A., & Collier, R. (1982). Laryngeal adjustments in Dutch voiceless obstruent production. *Annual Bulletin* (Research Institute of Logopedics and Phoniatrics, University of Tokyo), *16,* 27-35.

Yoshioka, H., Löfqvist, A., & Hirose, H. (1981). Laryngeal adjustments in the production of consonant clusters and geminates in American English. *Journal of the Acoustical Society of America, 70,* 1615-1623.

8 Intermediate Values of Voice Onset Time

Lawrence J. Raphael
Herbert H. Lehman College, CUNY, The Graduate School, CUNY, and Haskins Laboratories

Yishai Tobin
Ben-Gurion University of the Negev and University of Tel Aviv

Alice Faber
Haskins Laboratories and Wesleyan University

Tova Most
University of Tel-Aviv

H. Betty Kollia
Brooklyn College, CUNY, and Haskins Laboratories

Doron Milstein
The Graduate School, CUNY

This potpourri of authors includes not only Katherine Harris' colleagues and her students, but also her students' students. We are all grateful to her for teaching us to think clearly, to explain data rather than explaining them away, and to adopt a broad perspective even when conducting a narrow experiment.

Beginning with the publication of Lisker and Abramson's (1964) landmark paper, it has become conventional to assign the Voice Onset Time (VOT) of a prevocalic stop consonant to one of three categories commonly found in the natural languages of the world: (1) voicing lead (approximately −30 ms. or more VOT); (2) zero onset/short-lag (approximately 0 to +30 ms VOT); and (3) long-lag (approximately +50 ms or more VOT). In general, these categories of VOT correspond, respectively, to the phonetic descriptions of stop consonants as (1) voiced, unaspirated, (2) voiceless, unaspirated, and (3) voiceless, aspirated.[1]

Lisker and Abramson's data also revealed that in languages employing a single phonological voicing opposition, speakers realize stops phonetically using VOTs from adjacent categories (i.e., from the voicing lead and zero onset/short-lag categories, or from the zero onset/short-lag and long-lag categories) and not from the extreme categories (i.e., voicing lead and long-lag). The generality of this observation, however, may require some modification, as suggested by some recent studies (Caramazza, Yeni-Komshian, Zurif, & Carbone, 1973; Keating, Mikos, & Ganong, 1981; Recasens, 1985; Yeni-Komshian, Caramazza, & Preston, 1977), including the one we present here.

Our initial studies of VOT in Hebrew, more than a decade ago (Raphael & Tobin, 1983; Raphael, Tobin, Faber, & Most, 1989; Raphael, Tobin, & Most, 1983), were, in fact, prompted by the conflicting opinions and data provided by phoneticians and phonologists who have studied Modern Hebrew. Rosén's (1966) phonological study, for example, maintains that Hebrew /ptk/ are never aspirated, even in syllable-initial position, which is the expectation for stops in languages like Spanish and Italian, which draw these stops from the zero onset/short-lag category of VOT. Chayen (1973) is somewhat more equivocal, appealing to the so-called fortis/lenis distinction as the basis for distinguishing Hebrew /ptk/ from /bdg/. He does, however, indicate that both voicing and aspiration play a minor role in the distinction, and mentions that heavy aspiration can sometimes occur in association with syllables bearing emphatic stress. He also observes that aspiration is stereotypically associated with British- or American-accented Hebrew.

In contrast, other commentators have suggested that it is not uncommon for some degree of aspiration to be associated with Hebrew /ptk/. Blanc (1956, 1964) presents relatively narrow phonetic transcriptions of Hebrew and comments that aspiration (especially of /t/) is evidenced by a number of speakers.[2] Laufer (1972), in his synthesis of Hebrew stops, used VOT values of +20 ms for /p/ and /t/, and +30 ms for /k/. These values may be taken to represent a slight degree of aspiration. Indeed, Laufer, a native speaker of

---

[1]The phonetic category of voiced, aspirated stops is not distinguished by a unique range of VOT values (Lisker & Abramson, 1964).

[2]Blanc's studies established the distinction between the dialect of Modern Hebrew spoken by those with a central or eastern European linguistic heritage (Ashkenazic) and the dialect spoken by those with a Mediterranean/Middle-Eastern linguistic heritage (Sephardic). He reports that the two dialects differ in their use of aspiration, but does not amplify further.

Hebrew, reports that aspiration "may or may not follow the release of /ptk/" in his own speech.[3]

More recent, instrumental, studies of VOT in Hebrew reinforce the observations of Blanc and Laufer. Devens' (1978) study reported values of about +30 ms VOT for Hebrew /pt/ and a value of about +65 ms for /k/. Obler's (1982) study reported mean VOT values of +25.6 ms, +33.9 ms, and +63.7 ms for /ptk/, respectively, for five monolingual speakers of Hebrew. All other things being equal, these data might lead us to infer a moderate degree of aspiration accompanying the release of /p/ and /t/, and a degree of aspiration approaching that found in long-lag languages (e.g., English) for /k/. Obler's data for 10 balanced bilingual speakers of Hebrew and English present a more extreme picture, with mean VOT values of +58.0 ms, +73.8 ms, and +93.2 ms for Hebrew /ptk/, respectively. These values fall well into the ranges evidenced by native English speakers, and suggest that the stops are aspirated (at least by some of the speakers) in much the same way they might be in English.[4] Obler notes that the relatively long VOTs for Hebrew /ptk/ are accompanied by typical voicing lead values for Hebrew /bdg/, a finding reported by other investigators for bilingual speakers of voicing lead/short-lag languages who have acquired and short-lag/long-lag language as their L2 (Flege, 1987, 1991). It would appear, then, that some bilinguals, when using L2, are not constrained (or able) to draw their stops from adjacent VOT categories, using voicing-lead values from their L1 for /bdg/, and voicing-lag values either intermediate to their L1 and L2 for /ptk/ or equivalent to the values of native speakers of their L2 (in the case of balanced bilinguals).[5]

It was our hope that using a larger pool of subjects than those of the previous studies might help to resolve some of the uncertainty regarding VOT and aspiration in Hebrew /ptk/.

STUDY I

Procedure

Our subjects were 23 students at the University of Tel Aviv and Ben-Gurion University of the Negev, and seven Israelis who have lived in the New York

---

[3]The impression of several of the authors of this paper who are either native speakers of Hebrew or who have heard Hebrew spoken in Israel (and in the United States) is that /ptk/ are produced with a moderate degree of aspiration by many speakers, and, by some speakers, with a degree of aspiration typical of long-lag languages like English.

[4]Indeed, the mean value for the velar stop (+93.2 ms) exceeds that found by Obler for five monolingual speakers of English (89.4 ms). Moreover, the bilinguals' VOT values for /ptk/ equal or exceed those values reported by Lisker and Abramson (1964) for English.

[5]We might assume that the bilinguals maintain voicing lead in /bdg/ because native English-speaking listeners do not distinguish voicing lead from zero-onset/short-lag.

City Metropolitan area for between two and five years. All were native speakers of Hebrew. Of the 30 subjects, 10 were male and 20 were female. All spoke English fluently, but with an easily discernible accent.[6] They recorded 30 minimal pairs of Hebrew words, ten pairs for each of the three places of articulation. The minimal pair list used in Israel was arranged to provide maximal contrast, with each token of a /ptk/ word immediately preceding a token of its respective voiced cognate; the list used for recording in the United States was randomized.

The speakers' productions were analyzed on a Kay Elemetrics Sonograph, model 6061B, using a wideband filter. In addition to the standard wideband display, we generated a waveform at the top of each spectrogram. VOT's were first measured from the wideband display, using the stop burst and the first discernible vocal pulse at the level of the baseline as points of reference for /ptk/. In instances of voicing lead, VOT was measured from the burst only as far back as there was evidence of continuous vocal pulsing. Measurements were estimated to the nearest 5 ms. A second set of measurements was made from the accompanying waveform, using the burst and the onset of periodicity as points of reference. Both sets of measurements were made by each of two of the authors. Agreement between measurements was better than 90 percent. Differences were resolved by a third set of measurements performed jointly by the two authors who made the original measurements.

Results

The distribution of VOT values for the Hebrew stops recorded in Israel is described below. The mean values for these data are shown in Table 1. The VOT ranges and means for Hebrew are typical of those found in languages that use voicing lead for one phonological category of stops. These data are in general agreement with Obler's for the same sounds. The mean VOT for /b/, shown in Table 1, is −91.9 ms, with a clear mode in the data at −90 ms. Values ranged from −145 ms to −30 ms. The −90 ms mean for /d/ was accompanied by a nominal mode at −85 ms and a range of values from −135 ms to −40 ms.[7] The mean VOT for /g/ was −81.3 ms, with a nominal mode at −80 ms and a range of values from −145 ms to −25 ms.

---

[6]Languages other than English that were spoken by our subjects included German, Hungarian, Yiddish, Russian, and Polish.

[7]We have excluded from these data three intended productions of /d/, which we perceived as /t/: one at +5 ms and two at +10 ms. These values overlapped those for /t/—the only instance of overlap in our data.

TABLE 1. *Mean VOT values (Hebrew-speaking adults) from data recorded in Israel.*

	HEBREW (ISRAEL) MEAN VOICE ONSET TIME (ms)		
/b/	−91.9	+28.5	/p/
/d/	−90.9	+35.2	/t/
/g/	−81.3	+54.2	/k/

The data for /ptk/ are generally intermediate to those found in languages that contrast short-lag and long-lag stops. The mean VOT for /p/ is 28.5 ms, with a mode at +15 ms and a range of −5 ms to +85 ms. For /t/, the mean and mode were +35.2 ms and +30 ms, respectively, with a range of 0 to 65 ms The velar, /k/, displayed a mean VOT of +54.2 ms with a mode at +50 ms, and a range of −10 ms to +105 ms. A comparison of these data with those of Obler's study reveals general agreement between our bilingual subjects and Obler's monolinguals. In contrast, Obler's (balanced) bilingual subjects, as we have mentioned, displayed VOT values much closer to those of native English speakers and, thus, considerably greater than those of our (not-balanced) bilinguals. Devens' data, also for bilinguals,[8] is in close agreement with ours (and with Obler's for monolingual speakers).

The distributions of VOT values for the 7 US-based subjects were similar in every regard to those of the Israeli-based subjects. The mean VOT values for these subjects are shown in Table 2. We have pooled the values for the two sets of subjects in Table 3.

TABLE 2. *Mean VOT values (Hebrew-speaking adults) from data recorded in the United States.*

	HEBREW (USA) MEAN VOICE ONSET TIME (ms)		
/b/	−82.2	+25.4	/p/
/d/	−91.5	+34.2	/t/
/g/	−77.4	+66.1	/k/

---

[8]Devens (1980) does not specify whether or not her 10 subjects were balanced bilinguals. The first language of some of her subjects might have been Arabic, and they may also have been speakers of English or French, as well as Hebrew.

TABLE 3. *Mean VOT values (Hebrew-speaking adults) pooled across data recorded in Israel and in the United States.*

	HEBREW (ALL SS) MEAN VOICE ONSET TIME (ms)		
/b/	−89.8	+28.3	/p/
/d/	−90.9	+35.6	/t/
/g/	−79.2	+55.5	/k/

The values in Tables 1–3 indicate that bilingual speakers of Hebrew produce /bdg/ with VOT values that are quite typical of languages that employ voicing lead in a two-way stop opposition. The values for /ptk/, however, do not provide a good fit with the voiceless inaspirates of such languages, nor with the voiceless aspirates of languages like English, which distinguish short-lag from long-lag stops. That is, they are somewhat closer to the mean values of the former group of languages, but appear to constitute an independent set of means and ranges.

One possible explanation for the distribution of VOT values that we found is that modern Hebrew is in the midst of a shift from short-lag to long-lag values for /ptk/. Another possibility is that VOT values in modern Hebrew were much like they are now when the language was revived about a century ago. One possible way of gaining insights about these alternatives might be to investigate the VOT values of children born in Israel who speak Hebrew as a first language. It is possible that, if a shift in values were in progress, the children might evidence greater values of voicing lag than adults for the same stop consonants. Accordingly, we turned our attention to a younger generation of speakers.

STUDY II

Procedure

Our subjects were 21 10- to 11-year old children (range 10.0-11.1 years) who were students at Be'eri elementary school in Be'er Sheva, Israel. Eight of the students were male and 13 were female. All were native speakers of modern Hebrew. Four of the students had some knowledge of another language (Romanian—two children; French and English—one each), although none could be classified as skilled bilinguals. The children recorded a list of 30 monosyllabic minimal pairs containing some of the words read by the adults in Study I. Their first reading was from a randomized list, the second from a list in which each minimal pair constituted a sequence, the words beginning with /ptk/ preceding those beginning with /bdg/. The methods of analysis and measurement were the same as those used in Study I, with the exception that measurements were made on a Kay Elemetrics Digital Sonagraph Model 7800.

Results

Because of the well-known variability in speech production evidenced by children, relative to adults, we expected that the data for the children would be somewhat less systematic than those from Study I. Indeed, the ranges of the children's VOT values for each of the stop sounds were greater than those of the adults, and, as a result, there was more overlap between the VOT values for each pair of cognates.[9] There were no clear modes in the distribution of VOT values for /bdg/ and the modes for /ptk/ were consistently smaller than the means for those sounds. Table 4 shows the mean VOT values for the stops read in words from the randomized list, the minimal pairs read as pairs, and the two conditions pooled.

As shown in Table 5, the children produced /bdg/ with values of voicing lead that were somewhat lower than those of the adults in all three places of articulation. Nonetheless, the mean voicing lead values are quite typical for languages which employ this VOT category. The means of the children and the adults for /ptk/ are extremely similar for the labial and velar stops, showing differences of 0.3 ms and 1.4 ms, respectively. The means for /t/ differ by about 7 ms. The children's VOT means for /ptk/, like those of the adults, fall at values intermediate to the short- and long-lag values typical of many languages.

TABLE 4. *Mean VOT values (Hebrew-speaking children) from randomized word lists, minimal pair lists, and data pooled across both conditions.*

		MEAN VOICE ONSET TIME (ms)			
	RANDOM	−78.4	+25.1	RANDOM	
/b/	PAIRS	−75.2	+28.2	PAIRS	/p/
	POOLED	−76.9	+26.9	POOLED	
	RANDOM	−67.7	+22.5	RANDOM	
/d/	PAIRS	−76.9	+28.3	PAIRS	/t/
	POOLED	−72.5	+25.1	POOLED	
	RANDOM	−80.7	+65.4	RANDOM	
/g/	PAIRS	−70.9	+55.6	PAIRS	/k/
	POOLED	−75.4	+60.6	POOLED	

---

[9]The number of cases of overlap was still quite small in relation to the total number of measurements: 75 out of more than 2400 measurements.

TABLE 5. *Comparison of mean VOT values for minimal pair word lists produced by Hebrew-speaking adults and children.*

			MEAN VOT (PAIRED WORDS)			
	/b/	/p/	/d/	/t/	/g/	/k/
ADULTS	−91.9	+28.5	−90.9	+35.2	−81.3	+54.2
CHILDREN	−75.2	+28.2	−76.9	+28.3	−70.9	+55.6

DISCUSSION

The existence of VOT values intermediate to the short- and long-lag categories has, by this time, a considerable amount of precedent. Such an intermediate category has long been known to exist in Korean (Lisker & Abramson, 1964), although it exists between the short-lag and long-lag stops of that language. Caramazza et al. (1973) studied VOT in French bilinguals in Canada. The bilingual, native French speakers in their study produced /ptk/ in French with intermediate VOT values.[10] Our own study of native speakers of Puerto Rican Spanish (Raphael et al., 1983) and Greek (Kollia, 1993) revealed mean VOTs for /ptk/ that were slightly smaller than those reported here for speakers of Hebrew, but which fall into the intermediate category. Similar results have also been found for Polish (Keating et al., 1981), Catalan (Recasens, 1985), and Lebanese Arabic (Yeni-Komshian et al., 1977). The mean VOTs from these studies for /ptk/, and from the present study, are presented in Table 6.

TABLE 6. *Mean VOT values for /p t k/ from studies of seven languages.*

STUDY	LANGUAGE	/p/	/t/	/k/
Present Study (adults)	Hebrew	28.3	35.6	55.5
Obler, 1982 (Monolinguals)	Hebrew	25.6	33.9	63.7
Raphael et al., 1983	Spanish	20.2	27.6	39.3
Caramazza et al., 1973	Québec French	20	28	35
Keating et al., 1981	Polish	21	28	52
Recasens, 1985	Catalan	23	27	47
Yeni-Komshian et al., 1977	Lebanese Arabic	——	25	30
Kollia, 1993	Greek	18.5	27.1	49.2

---

[10]They also found a considerable amount of overlap between the two classes of French stops for both bilingual and monolingual speakers of French. This overlap, coupled with their perceptual data for French-speaking monolinguals led them to conclude that VOT is not an "important variable for voicing distinctions...in Canadian French." We make no such claim for Modern Hebrew.

It is tempting to attribute the existence of an intermediate category of VOT to effects of bilingualism. (See Flege, 1987, 1991 for some compelling reasons why this can be so.) In most of the studies we have cited, including ours, many or all of the speakers were bilingual. For instance, Kollia's (1993) subjects were Greek students tested in New York; all spoke English, but Greek was clearly dominant for them. Moreover, studies of English spoken by non-native speakers have presented data with VOT values that can be termed intermediate (Flege & Eefting, 1987; Gass, 1984). However, it seems clear that bilinguals are not the only speakers who produce intermediate VOT values. Monolingual speakers of Hebrew (including the children in the present study) and Obler's (1982) adult subjects, Polish (Keating et al., 1981) and Catalan (Recasens, 1985) have produced such values. The subjects in the study of Lebanese Arabic (Yeni-Komshian et al., 1977) were either students who were studying English or employees of the American University of Beirut, whose "knowledge of English was either very limited or nonexistent." Moreover, even though many of our Hebrew-speaking subjects were bilingual, the other language(s) they spoke, whether natively or as an L2, were not in all cases languages which employ long-lag stops. We do not mean to imply that the linguistic pressures exerted on speakers who are bilingual cannot or do not directly influence VOT or any other aspect of language. But such pressures cannot serve as a general explanation for all of the VOT data we have presented here.

Finally, considering the ranges and means for VOT that we and others have reported, as well as the fact that both child and adult monolinguals evidence intermediate values of VOT, it seems to us that any attempts to link VOT categories causally to auditory sensitivities (e.g., Diehl & Kluender, 1989; Pisoni, 1977) are largely unwarranted. Rather, it appears more likely that the perceptual mechanisms, especially of children, are tuned to the productions of skilled speakers, no matter what category of VOT those skilled speakers may employ.

ACKNOWLEDGMENTS

The research reported herein and the preparation of the manuscript were supported by grants HD-01994, DC-00403, and DC-00121 from the National Institutes of Health to Haskins Laboratories. We are grateful to Loraine Obler for discussion of the issues treated in this paper and for her helpful comments on an earlier version thereof.

REFERENCES

Blanc, H. (1956). A selection of Israeli Hebrew Speech. *Leshonenu, 21*, 33–39.
Blanc, H. (1964). Israeli Hebrew texts. *Studies in Egyptology and linguistics in honor of H. J. Polotsky* (pp. 132–152). Jerusalem: Israel Exploratory Society.

Caramazza, A., Yeni-Komshian, G., Zurif, E., & Carbone, E. (1973). The acquisition of a new phonological contrast: The case of stop consonants in French-English bilinguals. *Journal of the Acoustical Society of America, 54,* 421–428.

Chayen, M. (1973). *The phonetics of Modern Hebrew.* The Hague: Mouton.

Devens, M. (1978). *The phonetics of Israeli Hebrew: Oriental vs. General Israeli Hebrew.* Unpublished doctoral dissertation, UCLA.

Devens, M. (1980). Oriental Israeli Hebrew: A study in phonetics. *Afroasiatic Linguistics, 7,* 127-142.

Diehl, R. L., & Kluender, K. R. (1989). On the objects of speech perception. *Ecological Psychology, 1,* 121–144.

Flege, J. E. (1987). The production of "new" and "similar" phones in a foreign language: Evidence for the effect of equivalence classification. *Journal of Phonetics, 15,* 47–65.

Flege, J. E. (1991). Age of learning affects the authenticity of voice onset time (VOT) in stop consonants produced in a second language. *Journal of the Acoustical Society of America, 89,* 395–411.

Flege, J. E., & Eefting, W. (1987). Production and perception of English stops by native Spanish speakers. *Journal of Phonetics, 15,* 67–83.

Gass, S. (1984). Development of speech perception and speech production abilities in adult second language learners. *Applied Psycholinguistics, 5,* 51–74.

Keating, P. A., Mikos, M. J., & Ganong, W. F. III. (1981). A cross-language study of range of voice onset time in the perception of initial stop voicing. *Journal of the Acoustical Society of America, 70,* 1261–1271.

Kollia, H. B. (1993). Segmental duration changes due to variations in stress, vowel, place of articulation, and voicing of stop consonants in Greek. *Journal of the Acoustical Society of America, 93,* 2298(A).

Laufer, A. (1972). *Synthesis by rule of a Hebrew dialect.* Unpublished doctoral dissertation, University of London.

Lisker, L., & Abramson, A. S. (1964). A cross-language study of voicing in initial stops: acoustical measurements. *Word, 20,* 384–422.

Obler, L. (1982). The parsimonious bilingual. In L. Obler & L. Menn (Eds.), *Exceptional language and linguistics* (pp. 339–346). New York: Academic Press.

Pisoni, D. B. (1977). Identification and discrimination of the relative onset of two component tones. *Journal of the Acoustical Society of America, 61,* 1352–1361.

Raphael, L. J., & Tobin, Y. (1983). Perceptual and acoustic studies of voice onset time in Hebrew. In A. Cohen & M. P. R. van den Broeke (Eds.), *Abstracts of the Tenth International Congress of Phonetic Sciences* (p. 516). Dordrecht, The Netherlands: Foris Publications.

Raphael, L. J., Tobin, Y., Faber, A., & Most, T. (1989). Voice onset time in Hebrew: acoustic and perceptual studies. Paper presented at the International Conference on Linguistic Approaches to Phonetics, Ann Arbor Michigan.

Raphael, L. J., Tobin, Y., & Most, T. (1983). Atypical VOT categories in Hebrew and Spanish. *Journal of the Acoustical Society of America*, 74, S89A.

Recasens, D. (1985). *Estudis de fonètica experimental del Català Oriental Central.* Barcelona: Publicacions de l'Abadia de Monserrat.

Rosén, H. B. (1966). *A textbook of Israeli Hebrew.* Second, corrected edition. Chicago: University of Chicago Press.

Yeni-Komshian, G., Caramazza, A., & Preston, M. S. (1977). A study of voicing in Lebanese Arabic. *Journal of Phonetics, 5*, 35–48.

9 English /w, j/: Frictionless Approximants or Vowels Out of Place?

Leigh Lisker
Haskins Laboratories
University of Pennsylvania

The existence of strictly alphabetic representations of spoken language, of the kinds practiced by linguists, has sometimes been taken to mean that any speech signal may be described as a sequence of discrete sounds, and that a difference in the spelling of two utterances implies a dissimilarity, audible and "phonetic," that is potentially of semantic significance. Sounds spelled alike are heard to be linguistically, and perhaps also phonetically, the same. With the advent of methods of rapid spectrographic analysis some researchers have come to believe that speech cannot be analyzed into acoustically specifiable segments isomorphic with their alphabetic spellings. The linguist's spelling of a speech signal, therefore, is not equivalent to an acoustic description of its linguistically relevant properties.[1] The basis for this conclusion, not universally accepted, is

---

[1]Nor is it very possibly isomorphic with any physiological account, despite hopes expressed by some speech researchers. We may remember that at least one linguist, not primarily a phonetician, long ago remarked that "Since the representation of an utterance or its parts is based on a comparison of utterances, it is really a representation of distinctions. It is this representation of differences which gives us discrete combinatorial elements (each representing a minimal difference). A noncomparative study of speech

that sounds spelled alike, and auditorily judged to be the same phonetically,[2] are always found to differ acoustically even when produced by the same vocal tract.[3] Some of these differences are no doubt auditorily irrelevant, but others can be shown to be systematically related to differences in context and rate of articulation (whatever their perceptual significance might be). Those who are convinced that a purely acoustic account of speech perception is unachievable have raised the question as to whether there might not be a closer fit between the discrete model of speech as sounds in sequence and the movements in the vocal tract by which speech is generated. These movements, the "tract variables" they affect, and perhaps the motor commands that activate them, have been described as discrete and sometimes temporally overlapping (Browman & Goldstein, 1992; Harris, 1984; Liberman & Mattingly, 1985) and it is yet too early to decide whether they can serve as the elements of a description that either 1) more directly matches alphabetic perception than does one based on selected acoustic properties, or 2) replaces the alphabetic model with one that better accounts for the motor as well as some perceptual aspects of speech communication. It must be noted, however, that a model which features overlapping elements can hardly be said to serve as an explanation of the central fact of speech perception that is the basis of an alphabetic representation.[4] Of course, the processes of speech production, whatever their relation to perception and alphabetic writing might be, are inherently interesting to those of an inquiring turn of mind. If one's purpose is to relate production to perception, it is not clear what is the "best" way to describe production,– whether in terms of electromyographically determined "motor commands," related movements of the mobile parts of the respiratory tract, or the size and shape of the cavities of that tract. With respect to articulatory activity, moreover, one may feel compelled to decide whether a dynamic description of movement trajectories or one couched more abstractly in terms of static target shapes is more revealing.[5] Indeed, it is not clear that we have here a properly posed question, since a physiological description in terms of variously phased vocal tract movements is not strictly incompatible with one that posits vocal tract shapes as stationary

behavior would probably deal with complex continuous changes, rather than with discrete elements" (Z. S. Harris, 1951, p. 367).

[2]They would therefore be written alike no matter how "narrow" the representation. In broader spellings, where phonetically differentiable sounds are sometimes written alike, it is hardly surprising to find acoustically diverse sounds represented by the same letter.

[3]The acoustic identity of two intervals of speech would strongly suggest that they were recordings of a single phonetic event.

[4]Though, as Harris years ago pointed out in a statement that still holds true, "we appear to have failed to find a simple, absolutely invariant correlate of the phoneme, at the peripheral levels thus far investigated," (Harris, 1974, p. 2298).

[5]Browman and Goldstein (in press), for example, report that the management of lip closure for [p, b, m] varies with vocalic contexts, but that the lip aperture achieved is "relatively" context-independent.

targets toward and away from which those movements are directed. Moreover, the phonetic literature suggests the possibility, inelegant though it might be, that the same mode of description may not be equally expedient for all the phonetic elements (whether thought of as sounds or articulatory events) attributed to speech. Thus some elements may conceivably be characterized as movements without targets (e.g., the discussion of English schwa by Browman & Goldstein, 1992), while others involve targets without invariant characteristic movements.[6]

The inventory of sounds reported for English includes two vowels often now represented phonologically as /u/ and /i/. /u/ is conventionally said to be a high back rounded vowel, while /i/ is high front unrounded. Such vocal tract shapes are by definition characteristic of the IPA cardinal vowels [u] and [i], respectively, and we may suppose that for some varieties of English these IPA norms are not very far from the targets of tongue and lip movements used to produce the vowels in, for example, the words *ooze* and *ease*.[7] It might be supposed that all the other sounds of English are somehow different from them in their articulatory properties. However, included in the English sound inventory are two other items, /w/ and /j/ (not presumably very different from IPA [wj]),[8] that are often described (though not invariably, cf. Lehiste, 1964) as sounds involving vocal tract shapes identical, or very nearly identical, with those of /ui/ (= [ui]$_{Eng}$) : /w/ is high back rounded like /u/, while /j/ is like /i/ in being high front unrounded. Indeed, in some treatments of English phonology, it has been questioned whether /w/ and /j/ should be recognized as phonemes distinct from /u/ and /i/ (Swadesh, 1947). If, despite the fact (?) that /w/ = /u/ and /j/ = /i/ with respect to the target vocal tract shapes ascribed to them, /wj/ are nevertheless perceived and treated as sounds differing from /ui/, then it would seem that while target identity is sometimes proposed to explain the perceptual identity of acoustically different sounds, target identity does not always guarantee perceptual invariance.

While the pairs /uw/ and /ij/ are sometimes said to be characterized by only one target shape each, a common way to differentiate /wj/ from /ui/ is to call them "glides," suggesting that they are "dynamic" in a way or to a degree that all other sounds are not, despite the often enough expressed observation that speech is in general produced by a vocal tract of continuously changing shape. The boundary between glides and vowels is somewhat fuzzy, since /wj/ may involve, at least in initial position, brief steadystate intervals (Lehiste, 1964;

---

[6]Thus monophthongal vowels and fricatives might be characterized by target vocal tract postures, but not with invariant movements, no matter how much information the listener may extract from the context-specific transitional intervals of connected speech.

[7]In this paper "/ui/" will refer to "[ui]$_{Eng}$," i.e., those pronunciations of the English phonemes /ui/ that are phonetically closest to IPA [ui]. Other realizations of /u/ and /i/, e.g., [ɨw] and [ɨj] (cited by Trager & Smith, 1951), will not be considered.

[8]According to Ladefoged, though, "English /w/ is in between the French sounds /w/ and /ɥ/" (Ladefoged, 1982, p. 209).

O'Connor, Gerstman, Liberman, Delattre, & Cooper, 1957), while in the vowels of fluent speech steadystate intervals are not invariably present (Lindblom & Studdert-Kennedy, 1967). Since in fluent speech transitions as surely mark vowels as they do glides, and their contribution to perception is not negligible (Nearey, 1989; Strange, 1989) then the difference seems to be that only the vowels can be produced in isolation, while glides occur only in juxtaposition to vowels.[9] The production of /wj/, then, necessarily involves articulatory movements with their acoustic consequences, that is, transitions, while /ui/ can be produced with the vocal tract held fixed over an appreciable time interval, even if in speech transitions too will necessarily be present. Catford goes so far as to makes the point that /wj/ "cannot be prolonged" (Catford, 1977, p. 128), though this seems to be only doubtfully true.[10]

Another basis for the distinction drawn between glides and homorganic vowels refers to a difference in degree of prominence within syllables: "...the entire difference between the English phonemes /wj/ and /ui/ is definable only in these terms: where vocoids of the high back rounded or high front unrounded types are the most prominent elements in syllables, they are instances of the phonemes /ui/, but where they occur as marginal elements in a syllable with something else as the most prominent part, they are instances of the phonemes /wj/" (Hockett, 1958, p. 82).[11] Such a difference would of course be prosodic in nature rather than a matter of vocal tract shaping. It should be noted that, oddly enough, this description of the phonetic basis of the contrast between /wj/ and /ui/ is virtually the same as the one that Swadesh (1947) used to argue that the glides do *not* constitute a phonologically distinct class of phonetic events.

Other differences between /wj/ and /ui/ have been adduced. Thus not everyone is agreed that these glides and vowels share precisely the same targets, though /wj/ are both labial-velar and /ui/ are both nonlabial-palatal. Acoustically the two sets have been said to differ in their formant structures, particularly the second formant (F_2), while in production the glides are characterized by significantly greater constrictions (Chomsky & Halle, 1968; Lehiste, 1964), even to the point of generating audible frication, at least for /j/ (Heffner, 1950). Furthermore, the targets postulated for the glides have sometimes been said to be context-determined, so that /w/ always involves movement from a position that is more high-back-rounded than that of a

---

[9]"Steadystate /w/ and /j/ are quite out of the question since the absence of transitions can only result in /u/ and /i/" (O'Connor et al., 1957, p. 31)

[10]Languages with long or geminated [w,j] are not unknown, e.g., in Cairene Arabic (I.P.A., 1949, p. 34).

[11]For Hockett, in line with general American practice, English /ui/ are closer to IPA [ʊɪ] than to cardinal [ui], while the vowels more like [ui] are represented as the sequences /uw iy/. Assertions that /wy/ have the same target positions as /ui/ may therefore sometimes be understood to equate the English glides with the "lax" vowels [ʊɪ].

following vowel, while /j/ is a movement from a more high-front-unrounded position (Trager & Bloch, 1941, p. 234).

If the basis for calling /wj/ "glides" is that they cannot be produced and recognized as isolated utterances, then of course stop consonants must also be recognized as glides. In fact, stops are even more deservedly called glides (though they are not) since not only their intelligibility, but their status as speech sounds depends on the presence of transitions. (Given no transitions, any isolated signal emitted during a stop closure will be only doubtfully identified as speech.) On the other hand, any /wj/ steadystate intervals, if extracted and presented in isolation, are likely to be identified as vowels, most likely /ui/ or possibly /ʊɪ/. One difference between glides and stops is in the nature of the F_2 targets specified: the target frequencies may be achieved in the case of the glides, while for the stops the so-called "loci" are frequencies that the F_2 transitions presumably aim toward but regularly fall short of (Delattre, Liberman, & Cooper, 1955). Measurements of natural speech as well as experiments in synthesis seem to bear out this suggested difference. Thus F_2 transitions with the same terminal frequencies adjacent to different vowels do not as a rule evoke the same stop percept (Sussman, McCaffrey, & Matthews, 1991), but intervocalic /wj/ can be successfully synthesized with F_2 steadystate frequencies ("targets") that *are* independent of their vocalic contexts (Lisker, 1957). These happen to be frequencies whose closest matches in natural English speech are the values commonly reported for /ui/ (Lehiste, 1964).

There is a further similarity between English /wj/ and /ui/: just as there is no overlap in the F_2 formant frequency ranges reported for /u/ and /i/ (Peterson & Barney, 1952), so too is there a wide separation between the F_2 target frequencies for the two glides. This is of course not at all unexpected, since it would seem from the literature that the articulatory differences between the glides are as great, or even greater, than those between the vowels. This /w/-/j/ difference should hold even if we compare the two glides across different vocalic contexts. *If,* for example, the /w/ in /iwi/ requires tongue backing and lip rounding movements, while the /j/ in /uju/ calls for tongue fronting and lip unrounding, then these /w/ and /j/ allophones should show quite distinct F_2 frequency ranges. F_2 frequency determinations of /wj/ over a wide range of vowel contexts indicate that this is so (Lehiste, 1964), as do the results of several experiments in synthesis (Lisker, 1957; O'Connor et al., 1957).

Because spoken /w/ and /j/ are so different acoustically, reflecting their very different tongue and lip positions, it might be supposed that glides in selected contexts, as in the sequences /iwi/ and /uju/, are also necessarily disparate, both acoustically and articulatorily. However, the perceived difference between /w/ in /i/–/i/ and /j/ in /u/–/u/ need not crucially depend on such large acoustic and articulatory differences. Thus we can imagine results of a very simple operation using an old fashioned but readily available piece of equipment—the experimenter's own vocal tract—which would make a straightforward account of /wj/ perception problematical. The sequence /iwi/ is usually described as one in which the tongue moves from front to back and front again, while the lips shift from unrounded position to rounded and back to unrounded. In the sequence /uju/, on the other hand, the tongue first moves from back to front and then back again, while the initially rounded lips unround and then round again. Thus we have:

	i	w	i	u	j	u
tongue	front	back	front	back	front	back
lips	–rnd	+rnd	–rnd	+rnd	–rnd	+rnd

Let us now instead imagine an /i/–glide–/i/ sequence in which only the lips change position, while the tongue remains fixed. The glide target is then the vocal tract shaped as for the high front rounded vowel [y]. Let us also imagine another sequence, /u/–glide–/u/, this time one in which the tongue moves from back to front to back, while the lips maintain a rounded position. The glide target is again that of a front rounded vowel.

	i	ɥ	i		u	ɥ	u
tongue	f	r o n	t		back	front	back
lips	– rnd	+rnd	–rnd		+ r	n	d

If these sequences are performed, at least one speaker-listener (the present author) is convinced that the resulting acoustic signals will be heard by English-speaking listeners as /iwi/ and /uju/.[12] If this expectation is in fact fulfilled, then we may conclude that a satisfactory /w/, at least in the /i/–/i/ environment, requires no more than a labial movement, while a phonologically adequate /uju/ can be achieved with no more than a tongue fronting maneuver.

At this point, of course, all we have is little more than a rather crude Gedanken experiment, except for the fact that two utterance types were in fact articulated and identified a number of times by at least one subject (the author). The intent was to produce [iɥi] and [uɥu], but the phonetic interpretations of the acoustic results of the articulatory movements sharing the single goal were not the same: the disyllables heard were /iwi/ and /uju/. Of course, we have no guarantee that in producing [iɥi] only the lips moved between the vowels, or that in [uɥu] only the tongue shifted position. In fact, acoustic measurements of a number of tokens of the two sequence types suggest that the effort to achieve the same target [y] (or [ɥ]) in the two contexts was a failure: the apparent F_2 target of the glide was about 500 Hz higher in one context than in the other. This failure, moreover, can hardly be explained by the phenomenon of undershoot (Lindblom and Studdert-Kennedy, 1967), since F_2 frequency at the glide midpoint was higher in the /u/–/u/ context than the /i/–/i/. If the acoustic

---

[12]In the tests in which listeners were asked whether or not they heard the palatal glide [j], the label suggested was "y."

mismatch is acceptable evidence of a failure by the speaker to perform as intended, then it appears that the glides in one or both sequences were produced with considerable *overshoot* of the single intended high-front-rounded target position. At this point, moreover, we cannot know whether the context-dependent difference in interpretation of the intended single glide is a consequence of the acoustic difference.

Since the acoustic evidence suggests that the speaker, despite his intention, did not approximate the same glide target in the two contexts, the technique of terminal analog synthesis was resorted to, in order to determine whether acoustic identity of the glide targets would elicit the same phonetic percept in the two contexts. A terminal analog software synthesizer set for series operation was used to generate two sets of patterns having the frequency and time characteristics shown in Figure 1(A and B). In each set the variables were the F_2 "target" frequency and flanking transitions: 18 acoustically different patterns in all.

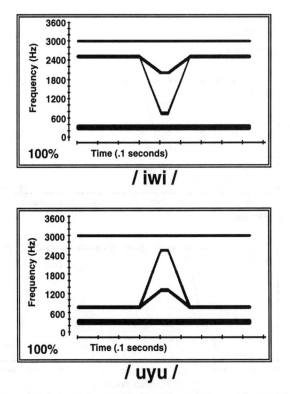

FIGURE 1. Two sets of spectral patterns tested. In each F_1 and F_3 were fixed, while F_2 was varied as shown, so that it moved linearly to and from each of 18 "target" frequencies. In the /iwi/ set F_2 targets ranged from 750 to 2400 Hz, while for /uyu/ the F_2 targets ranged from 800 to 2500 Hz.

On the basis of preliminary trials it was thought appropriate to test the two stimulus sets separately. In one, the members of the A set were each presented five times in random order, and listeners were asked to decide whether or not they heard /iwi/, responding with either "w" or "x." In the other test the judgments elicited were either "y" (for /uju/) or "x." Thus the labeling tests did not force a choice between two competing phonemes, allowing instead for the easy "out" of an "x" response." It could therefore be expected that stimuli reported as "w" or "y" would be phonologically very convincing imitations of /iwi/ and /uju/. The responses from two linguist/phoneticians, both native American English speakers, are shown in Figure 2. Both show a broad range of F_2 frequencies judged as "w" in /i/–/i/ and as "y" in /u/–/u/. It may be noted that neither listener, in post-test discussion, reported hearing a [ɥ] glide in any of the stimuli.

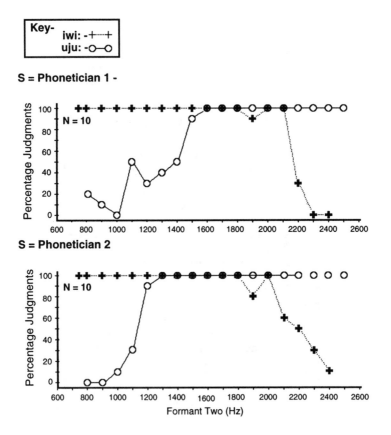

FIGURE 2. "w"/ "not-w" and "y" / "not-y" judgments by two phonetically trained listeners. Each point represents 10 independent judgments per stimulus by one listener.

These tests were also given to 20 other English-speaking subjects, none of them phoneticians, but all with some experience as subjects of speech perception experiments. Their responses (Figure 3) show a much reduced range of overlap between "w" and "y" responses, reflecting more the "noisiness" of the data than any systematic difference between their percepts and those of the two phonetically trained subjects. It may be noted that the crossover between the *w* and *y* curves falls near 80%, another indication that the two glides, phonetically and phonologically distinct as they are perceptually, do not occupy distinctly different F_2 frequency ranges.

From this experiment in synthesis it must be concluded that the failure to hear the same glide in the naturally produced (intended) sequences [iɥi] and [uɰu] is not to be explained by the speaker's failure to perform glides over the different contexts that approximated a single acoustic (primarily F_2) target. There was, to be sure, a failure by the speaker to produce glides allowing us to infer a single target, but speaker success would not have led listeners to divine the speaker's intention.

The failure to produce glides having the same acoustic target, given the speaker's intention to do just that, suggests a failure of proprioceptive judgment. Conceivably, in trying to produce [iɥi] the lips achieved a more closely rounded position than that maintained over the [uɰu] sequence. Or perhaps the tongue reached a more forward position during the glide in [uɰu] than in [iɥi]. Because of uncertainty as to the basis for the speaker's apparent failure to realize his intention, a somewhat more elaborate experiment in articulation was performed.

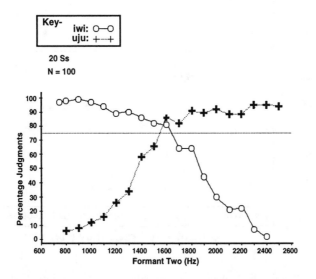

FIGURE 3. "w"/ "not-w" and "y" / "not-y" judgments by 20 phonetically untrained English-speaking listeners, each of whom made 5 independent judgments per stimulus.

Ten tokens of each of six phonetic sequences were produced by the author. Four glide targets were his goals: 1) a high back rounded shape, that of standard English /w/; 2) the high front unrounded shape of English /j/; 3) a high front rounded shape, that of French [ɥ]; and 4) a high central rounded shape, that of IPA [ɥ]. In aiming to produce the glide [ɥ] in the context of /u/–/u/ only the tongue was consciously activated, but in /i/–/i/ the lips were very active, while the tongue moved over a restricted range.

The F_2 frequencies achieved in these articulatory exercises are given in Figure 4. In the first pair /w/ and /j/ have the widely separated F_2 frequencies we expect of those glides, where both tongue and lips are active. In the second pair, for which a single target, that of the vowel [y], was intended, F_2 frequencies for the glides were closer than for intended [wj], but differ significantly in mean values. They represent a replication of the initial attempts to achieve the sequences [iɥi uɥu], with no greater success than before. With the third pair of sequences, however, the same intended target in the different contexts, that of a high central rounded vowel [ɥ], appears to have been hit, in that there is no significant difference between the mean F_2 frequencies achieved in the two vocalic environments. However, although here we may thus have some warrant for speaking of a single glide, informally tested listeners showed no hesitation in identifying the glides in /i/–/i/ and /u/–/u/ as "w" and "y," respectively.

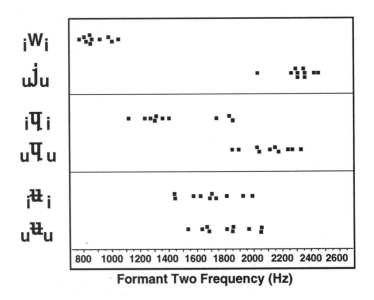

FIGURE 4. "Target" F_2 frequencies measured at the points of inflection (minimum in /i/—/i/; maximum in /u/—/u/); means of 10 productions of the (intended) phonetic strings indicated.

The extensive overlap in F_2 values found for synthesized intervocalic /w/ in /i/–/i/ and /j/ in /u/–/u/ is not consistent with the results of earlier work on glides, either in medial or in initial position, nor is it compatible with the view that /w/ and /j/ must have as targets the tongue and lip positions of the vowels /u/ and /i/. If we suppose that in the disyllables /iwi/ and /uju/ there is a syllable division between the initial vowel and the glide, then we might expect about the same perceptions to hold if we delete the initial steadystate and following transition intervals of the disyllabic patterns of Figure 1. However, when the same kind of labeling tests were carried out on the truncated patterns, rather different results emerged. From Figure 5 it appears that the "w"–"y" ranges along F_2 did not overlap at all. Instead, there is a considerable gap, about 500 Hz wide, between the highest F_2 frequency heard as "w" in /wi/ and the lowest F_2 frequency that is heard as "y" in /ju/. We may note also that the F_2 ranges heard as "w" and "y" in initial position are very much narrower than those perceived as the two glides in intervocalic position.

From the results of these experiments in synthesis and deliberate articulation it appears that even if English /w/ is normally produced as a labial-velar and /j/ is an unrounded palatal approximant, at least in some intervocalic positions the labial and lingual components of the two glide articulations are not of equal perceptual weight. Thus English /iwi/ can be satisfactorily produced by a lip gesture with the tongue maintaining a fixed palatal constriction, while for /uju/ the tongue fronting gesture need not be accompanied by lip movement.[13]

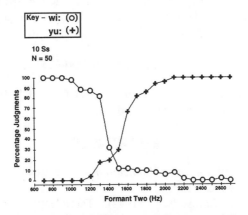

FIGURE 5. Judgments of synthesized glide-vowel patterns as "w"/ "not-w" before /i/ and "y" / "not-y" before /u/. Each point represents means of 50 judgments per stimulus, 5 by each of 10 phonetically untrained English-speaking listeners.

---

[13] Its description as an "unrounded palatal semi-vowel" (Jones 1960, p. 209) surely exaggerates the role of lip position in its formation. According to Bronstein (1960) the lip position for /j/ is determined by the following vowel, being rounded before back vowels (p. 123).

This suggests that English /w/ is primarily a labial, while /j/ is a palatal approximant. Of course, if /w/ is not distinctively back and /j/ is not distinctively unrounded, then they are not to be equated phonologically with the vowels /u/ and /i/, for each of which both lip and tongue positions must presumably be specified. If the /iwi/ and /uju/ sequences are managed with the degree of gestural economy achieved in realizing them as [iɥi] and [uɥu], then the included glides should have the same target vocal tract shape, that of the front rounded vowel [y]. Why, then, do listeners presented with [iɥi]–[uɥu] seem not to know that the lip gesture in [iɥi] and the tongue gesture in [uɥu] are aimed at the same goal, inasmuch as they do not perceive the glides to be phonetically the same? After all, if the same glide *were* perceived, an explanation appealing to the single articulatory target might be readily applied. In the case of the [iɥ̨i]–[uɥ̨u] pair, where the gestural difference is not as stark, an explanation based on acoustic invariance would also be viable, but again only if only a single glide, not two, were reported. Since in neither of the pairs [iɥi]–[uɥu] and [iɥ̨i]–[uɥ̨u] was the single articulatory goal of the glides perceived, not even by the listener privy to the speaker's intention (the speaker himself), then clearly neither an invariant spectrum nor a single articulatory target serves any explanatory function in accounting for perception. Instead, in both pairs of phonetic sequences listeners perceived the different maneuvers as two distinct phonetic messages, despite the fact that they were aimed at a single articulatory target. The speaker's distinct lip and tongue movements may be construed as different ways of achieving the same goal, but the speaker's successful execution of the articulatory task, even to the point of achieving acoustic invariance, does not mean that listeners will appreciate it.

ACKNOWLEDGMENT

The support for this work provided by NIH Grant HD–01994 to Haskins Laboratories is gratefully acknowledged.

REFERENCES

Bronstein, A. J. (1960). *The pronunciation of American English.* New York: Appleton-Century-Crofts.

Browman C. P., & Goldstein, L. (1992). "Targetless" schwa: An articulatory analysis. In D. Docherty & D. R. Ladd (Eds.), *Papers in laboratory phonology II: Gesture, segment, prosody* (pp. 26-56). London: Cambridge University Press.

Browman, C. P., & Goldstein, L. (in press). Dynamics and articulatory phonology. In T. van Gelder & B. Port (Eds.), *Mind as motion.* Cambridge MA: MIT Press.

Catford, J. C. (1977). *Fundamental problems in phonetics.* Edinburgh: Edinburgh University Press.

Chomsky, N., & Halle, M. (1968). *The sound pattern of English.* New York: Harper and Row.

Delattre, P. D., Liberman, A. M., & Cooper, F. S. (1955). Acoustic loci and transitional cues for consonants. *Journal of the Acoustical Society of America, 27,* 769-773.

Harris, K. S. (1974). Physiological aspects of articulatory behavior. In T. A. Sebeok (Ed.), *Current trends in linguistics* (Vol. 12, pp. 2281-2302). The Hague: Mouton.

Harris, K. S. (1984). Coarticulation as a component in articulatory description. In R. G. Daniloff (Ed.), *Articulatory assessment and treatment issues* (pp. 147-167). San Diego, CA: College-Hill Press.

Harris, Z. S. (1951). *Methods in structural linguistics.* Chicago: University of Chicago Press.

Heffner, R-M. S. (1950) *General phonetics.* Madison WI: University of Wisconsin Press.

Hockett, C. F. (1958). *A course in modern linguistics.* New York: Macmillan.

I. P. A. (International Phonetic Association). (1949). *The principles of the International Phonetic Association.* London.

Jones, D. (1960). *An outline of English phonetics* (9th ed.). Cambridge: Cambridge University Press.

Ladefoged, P. (1982). *A course in phonetics* (2nd ed.). New York: Harcourt Brace Jovanovich.

Lehiste, I. (1964). Acoustic characteristics of selected English consonants. *Publication 34 of the Indiana University Research Center in Anthropology, Folklore, and Linguistics.* Bloomington: Indiana University Press.

Liberman, A. M., & Mattingly, I. G. (1985). The motor theory of speech perception revised. *Cognition, 21,* 1-36.

Lindblom, B. E. F., & Studdert-Kennedy, M. (1967). On the role of formant transitions in vowel recognition. *Journal of the Acoustical Society of America, 42,* 830-843.

Lisker, L. (1957). Minimal cues for separating /w,r,l,y/ in intervocalic position. *Word, 13,* 256-267.

Nearey, T. M. (1989). Static, dynamic, and relational properties in vowel perception. *Journal of the Acoustical Society of America, 85,* 2088-2113.

O'Connor, J. D., Gerstman, L. J., Liberman, A. M., Delattre, P. D., & Cooper, F. S. (1957). Acoustic cues for the perception of initial /wjrl/ in English. *Word, 13,* 24-43.

Peterson, G. E., & Barney, H. L. (1952). Control methods used in a study of the vowels. *Journal of the Acoustical Society of America, 24,* 175-184.

Sussman, H. M., McCaffrey, H. A., & Matthews, S. A. (1991). An investigation of locus equations as a source of relational invariance for stop place categorization. *Journal of the Acoustical Society of America, 90,* 1309-1325.

Swadesh, M. (1947). On the analysis of English syllabics. *Language, 23,* 137-150.

Strange, W. (1989). Evolving theories of vowel perception. *Journal of the Acoustical Society of America, 85,* 2081-2087.

Trager, G. L., & Bloch, B. (1941). The syllabic phonemes of English. *Language, 17,* 223-246.

Trager, G. L., & Smith, H. L., Jr. (1951). *An outline of English structure* (*Studies in Linguistics: Occasional Papers,* 3). Norman: OK: Battenburg Press.

10 How the Tongue Takes Advantage of the Palate During Speech

Maureen Stone
The Johns Hopkins University

The vocal tract is often modeled as a sound source and a filter. The articulators play a role in both aspects of the model. They function as a source when two articulators oppose each'other to create frication or a burst release, and they also function as a filter by moving synchronously to shape the resonating chamber. This chapter is concerned with how the tongue and hard palate contribute to these two speech functions of the vocal tract.

The tongue is the major articulator of the vocal tract. In English the tongue is crucial to the formation of all sounds except [p,b,m,f,v,ʔ,h]. Even phones that do not use the tongue as a primary articulator are affected by it, since the tongue directs the airstream toward the primary constriction or shapes the airstream beyond the constriction. Since the hard palate is immobile, the tongue—or rather the tongue-jaw unit—does all the active shaping or constricting in the central part of the tract. The immobility of the hard palate can be a real advantage to the tongue because the tongue does not have to deal with palatal variability in the motor planning for their contact.[1] The tongue has many contact options, because

---

[1]While the lower lip has a similar relationship with the upper teeth for [f] and [v], that constriction is limited solely to close approximation with or without rounding.

small regions of the tongue can contact portions of the long, wide palate. Moreover, the part of the tongue not touching the palate appears to be affected by the nature of the tongue-palate contact. The complexity of contact patterns and the interaction between contact and tongue configuration lead to the questions of why and how the tongue uses the palate to transform its bulk into the various unique shapes required for speech.

The "why" is easy. The more shapes the tongue can create, the richer the phonology of the language, but answering the "how" requires information from the direct examination of the vocal tract to help us better understand the phonetic nature of speech. There appear to be three obvious ways for the tongue to use the palate during speech: (1) to create specific tongue shapes, (2) to accommodate aerodynamic changes, and (3) to coordinate movements of the tongue with those of the jaw.

The first use the tongue makes of the palate, creating specific tongue shapes, refers to speech gestures in which one part of the tongue pushes against the palate to help position or shape another part. In other words, local tongue-palate "bracing" enables the entire tongue to make shapes it otherwise could not make.

Vowels, particularly nonhigh vowels, use little or no tongue-palate contact. Their tongue shapes are, therefore, almost exclusively the result of muscular contraction. In such cases the tongue is very like the muscular hydrostat proposed by Smith and Kier (1989). In that model the muscles of the tongue have two roles: first to position the tongue; second to act as a skeleton by stiffening parts of the tongue. Hydrostatic animals, such as squid or octopus, contain fluid filled sacks that act as skeletons. Hydrostats are independent, individual entities that do not need the support of rigid structures to facilitate their activity. Instead they use the internal fluid sack for support.

During the production of nonhigh vowels, the tongue behaves as a muscular hydrostat. It has neither a bony skeleton nor a fluid sack to give it shape. Therefore, the tongue uses some of its muscles to stiffen and shape itself. The stiffening has the secondary effect of reducing its degrees of freedom of movement. When the tongue behaves like a hydrostat, compression in one location leads to expansion in at least one other location.

As a muscular hydrostat, expansion and compression of the tongue are relatively systematic. Indeed, for vowels, the shape and location components of tongue movement are extremely interdependent, because deformation is the only way a noncompressible, volume preserving, fluid structure like the tongue can be self-propelled. Therefore, tongue shape and position during vowels (Figure 1) are correlated and predictable: a higher tongue has a steeper posterior slope ([i] vs. [ɑ]), a more anterior tongue body uses a more anterior tongue root ([i] vs. [u]), and a more anterior tongue root is made with a deeper posterior groove.

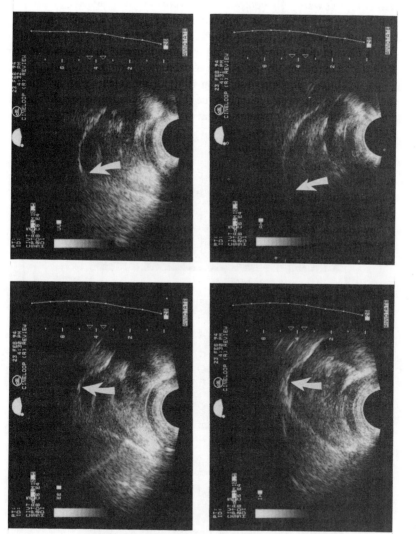

FIGURE 1. Sagittal tongue shapes for the vowels [i], [a], [u].

The result is that models of tongue behavior during vowels, especially the "point" vowels, abound (Hashimoto & Sasaki, 1982; Jackson, 1988; Ladefoged & Lindau, 1989; Maeda, 1990; Mermelstein, 1973; Perkell, 1974), because knowing how the point vowels behave allows us to predict quite accurately how the intermediate vowels will behave. Moreover, the more similar in tongue location two vowels are, the more similar their shape will be, and the more similar their acoustic spectra will be.

Consonants, on the other hand, do not show this systematic location-to-shape relationship. Thus, many consonants use a lingua-alveolar constriction (e.g., /t,l,s,n/). The commonality of constriction location, however, is of no use in predicting the shape of the rest of the tongue. The posterior tongue shape for [l], for example, is quite different from that of [s] (Figure 2) (Stone, 1990).

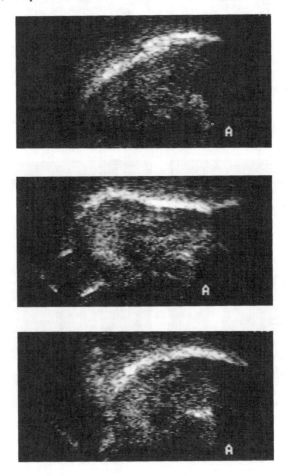

FIGURE 2. Sagittal tongue shapes for the consonants [t], [l]. [s].

Tongue shapes for consonants are less predictable from constriction location than those for vowels primarily because of the way in which the tongue uses the palate. Thus, lingual consonants tend to have greater palatal contact than vowels, and the tongue-palate contacts for lingual consonants require a much finer motor coordination than do those for vowels (in the same way that an elephant picking up a peanut with its trunk uses finer motor coordination than it uses when simply lifting its trunk in the air (Wilson, Mahajan, Wainwright, & Croner, 1991)). During consonants, the tongue touches the palate locally in ways that relate directly to the creation of complex tongue shapes (cf. Stone, Faber, & Cordero, 1991; Stone, Faber, Raphael, & Shawker, 1992). Moreover, the tongue appears to contain semi-independent segments that can oppose each other when one or more are braced or stabilized by local palatal contact (Stone, 1990).

In the case of consonants, the tongue does not behave as a muscular hydrostat; i.e., it does not control all its behaviors solely using its muscles. Instead, it uses the resistance afforded by the palate to fine-tune its shape (cf. Stone & Lundberg, 1994). A tongue model for consonants needs to consider both the tongue's ability to move in segments and how it uses local palatal contact to shape itself. Such a model must also consider the reciprocal effects between the tongue using palatal contact to shape itself, and to shape the vocal tract. The use of local tongue-palate contact to shape the tongue during consonantal, but not vocalic, sounds is consistent with the view that consonants and vowels are different subsystems (cf. Öhman, 1966).

What of tongue-palate contact during high vowels? For high vowels like /i/ and /u/, tongue-palate contact, although as great as during many consonants, does not appear to increase the complexity of tongue shape. Tongue-palate contact patterns among vowels are different from each other in magnitude, more than in shape (Recasens, 1991). Articulatory models have been able to generate accurate tongue shape and position for /i/ using tongue muscular contraction alone (Fujimura & Kakita, 1979; Perkell, 1974). In the case of /i/, the palate appears to be simply a limiting factor, providing an upper boundary for the tongue just as the teeth and floor of the mouth provide lateral and lower boundaries for other vowels. These boundaries provide both outer limits and tactile feedback, apparently for fine motor tongue positioning rather than tongue shaping. Gross positioning of the tongue, of course, is often accomplished by the jaw. Thus, the first use of tongue-palate contact is to shape the tongue locally, especially during consonants.

The second use of tongue-palate contact is to help the freestanding part of the tongue adapt to changing aerodynamic conditions during speech movements. In continuous speech, the tongue has to deal with aerodynamic changes that may be both gradual or abrupt in nature. In a stop-to-vowel movement, for example, the tongue must respond to large and rapid changes in intraoral air pressure, while maintaining smooth movement from the strong tongue-palate pressure of the stop to the more free-standing position of the vowel. Electropalatography (EPG) reveals (Figure 3) that when the tongue releases its contact for [t] into a vowel, the tip is released before the sides, and a midsagittal channel is formed (Farnetani, Vagges, & Magno-Caldognetto, 1985; Hardcastle, Jones, Knight, Trudgeon, & Calder, 1989; Manuel & Vatikiotis-Bateson, 1988; Mizutani, Hashimoto, Hamada, & Miyura, 1988).

For vowels with a high back and low tip this release pattern is consistent with the effects of jaw lowering. Less explainable is that the same pattern occurs in a vowel that uses a low tongue back and a relatively high tip like [ɛ] (Manuel & Vatikiotis-Bateson, 1988; Mizutani et al., 1988). It would seem that the easiest way to reach the [ɛ] position would be to lower the posterior tongue simultaneously with the tongue tip and jaw. Yet the posterior tongue lags behind. Therefore, a better explanation than simple jaw lowering is that the lateral margins of the tongue maintain palatal contact to provide stability, just as a person descending a steep stairway might touch the wall for balance.

For aerodynamic purposes, the release of [t] must be controlled in a number of ways: The magnitude and duration of the puff of air released as the tongue lowers must be appropriate to the degree of aspiration required (from fully - to weakly- to unaspirated); the path of the airflow must be appropriate to the phonetic context (anterior if the following sound is a vowel, lateral if the following sound is [l], nasal if the following sound is [n]). Lowering of the tip of the tongue before the body directs the airflow appropriately for a vowel and provides prolonged tactile feedback via lingua-palatal contact, concerning the duration and direction of release. In addition, the maintenance of upward force in the lateral margins of the tongue body braces the tongue against the air rushing through the opening.

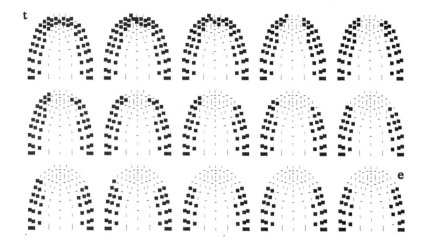

FIGURE 3. EPG contact patterns for [t].

It is apparent that the same role, the manipulation of airflow, may be served by tongue-palate contact during the production of fricatives. Here it is critical that the direction, quantity, and duration of airflow be accurate (Shadle, 1990). In most /VsV/ utterances the tongue rises for the constriction from posterior to anterior and lowers for release from anterior to posterior (Farnetani et al., 1985; Hardcastle et al., 1989; Stone et al., 1991). While this simple case can be explained in part by jaw raising and lowering, more complex coarticulatory patterns suggest an additional functional component, i.e., the effects of acoustic requirements. A good example of this occurs in an [sl] cluster. Jaw position is higher for an [s] than an [l]. Therefore, one might expect the tongue to release anteriorly-to-posteriorly, following jaw lowering. However, movement from [s]-to-[l] requires the lateral margins of the tongue to lower and the tip to elevate and contact the palate. If the tongue tip touches the alveolar ridge before the tongue margins are lowered, an intrusive [t] might be produced. Therefore, for acoustic reasons, lateral lowering should occur first. To test this theory, two subjects repeated "a slot machine" 15 times each. The two predominant transition patterns supported the prediction. In 9 repetitions for Subject 1 (S1) and 8 for Subject 2 (S2), the lateral margins of the tongue-palate contact lowered before complete anterior closure was made (see Figure 4a). The second major pattern was simultaneous back lowering and tip contact. This occurred in five repetitions for S1 and four for S2 (Figure 4b). The least common pattern, occurring only once for S1 and three times for S2, showed tongue tip contact before back lowering (Figure 4c).

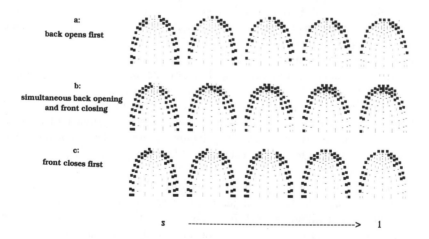

FIGURE 4. EPG contact patterns for three [sl] patterns.

In other words, in the majority of cases (26/30) the lateral margins lowered before or simultaneously with tip elevation, indicating that the tongue transition pattern was determined by acoustic requirements rather than jaw opening. Thus it would appear that local tongue-palate bracing can be used in three ways: to overcome articulatory effects (jaw lowering), to facilitate various (acoustic) phonetic and aerodynamic demands, (as in [sl]; see Figure 4), and to control subtle tongue maneuvers preventing inappropriate audible effects (e.g., an excrescent [t] in the [sl] transition).

There is yet another facet to the use of the palate by the tongue during speech. The tongue uses the stability and tactile feedback provided by palatal contact to coordinate its movements with the jaw. Many studies have discussed the two-way trading relation between the jaw and the tongue during speech (cf. Kent, 1972; Kent & Netsell, 1971; Maeda, 1991). In actuality, however, there is a three-way relationship, which includes the jaw, the tongue and the palate. As the tongue is the most flexible of the three, it is the best able to execute subtle compensatory maneuvers. By using the tactile feedback and support provided by tongue-palate contact, the tongue can create appropriate vocal tract shapes while remaining in synchrony with jaw movement. This can been seen when the tongue reaches a consonantal target position despite variations in jaw height that result from vowel coarticulation.

Coarticulatory constraints can be viewed as small articulatory perturbations (Fowler & Saltzman, 1993). Large perturbations created during laboratory experiments result in immediate compensation for the perturbation of one articulator (the jaw) by a related one (the tongue) (Fowler & Turvey, 1980). Stone and Vatikiotis-Bateson (1993) offered evidence to substantiate the theory that the tongue may sometimes compensate for jaw perturbations caused by coarticulation. Their study considered the effects of jaw height on the narrow palatal air channel needed for [i]. They examined tongue height, tongue-palate contact, and jaw height during [i] in [s] and [l] context. During [l] the jaw is held in a lower position than during [s]. Therefore, an [i] adjacent to [l] has a lower jaw position than an [i] adjacent to [s]. This jaw "perturbation" during [i] might either be ignored by the speaker or compensated for by the tongue in order to achieve the required narrow palatal air channel. The study found that two of three subjects maintained a narrow air channel and a constant first formant during [i] despite the lowered jaw. They did this by arching the tongue blade and advancing the tongue root to a greater extent during [i] in [l] context than in [s] context.

In other words, even though the jaw was lowered, providing less support for the elevated tongue body, the tongue balanced the conflicting (coarticulatory) demands of the small air channel (for [i]) and the low jaw (for [l]). This degree of accuracy in tongue placement would have had to take advantage of the feedback provided by tongue-palate contact.

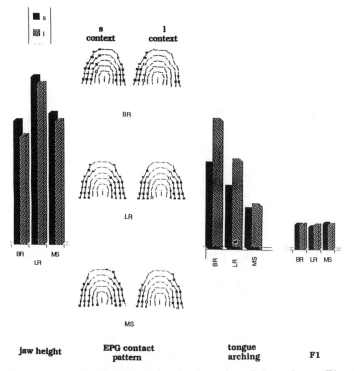

jaw height   EPG contact   tongue   F1
                    pattern   arching

FIGURE 5. Jaw height, EPG contact patterns/total number of electrodes on, F1 patterns, and tongue curvature for the vowel [i] in [s] and [l] context. (From Stone & Vatikiotis-Bateson, 1993, with permission.)

In conclusion, this chapter has reflected on how the tongue might take advantage of its contact with the hard palate in shaping the vocal tract during speech. First, local tongue-palate contact facilitates the production of tongue shapes that would otherwise be difficult or impossible to create. This is particularly true for consonants. Second, EPG records of consonant occlusion and release patterns indicate that the tongue may use tongue-palate contact to control not only the varying airflows and pressures required by speech, but also its own behavior in the face of those aerodynamic changes. Finally, the tongue can use the tactile information available through tongue-palate contact to adjust its position and shape finely so that it can create appropriate constrictions despite the coarticulatory effects of jaw position.

REFERENCES

Farnetani, E., Vagges, K., & Magno-Caldognetto, E. (1985). Coarticulation in Italian VCV sequences. *Phonetica, 42,* 78-99.

Fowler, C., & Saltzman, E. (1993). Coordination and coarticulation in speech production. *Language and Speech, 36,* 171-195.

Fowler, C., & Turvey, M. (1980). Immediate compensation in bite-block speech. *Phonetica, 37,* 306-326.

Fujimura, O., & Kakita, Y. (1979). Remarks on quantitative description of the lingual articulation. In B. Lindblom & S. Öhman (Eds.), *Frontiers of speech communication research* (pp. 17-24). London: Academic Press.

Hardcastle, W., Jones, W., Knight, C., Trudgeon, A., & Calder, G. (1989). New developments in electropalatography: A state-of-the-art report. *Clinical Linguistics and Phonetics, 3,* 1-38.

Hashimoto, K., & Sasaki, K. (1982). On the relationship between the shape and position of the tongue for vowels. *Journal of Phonetics, 10,* 291-299.

Jackson, M. (1988). Phonetic theory and cross-linguistic variation in vowel articulation. *UCLA Working Papers in Phonetics, 71.* Los Angeles: UCLA.

Kent, R. (1972). Some considerations in the cinefluorographic analysis of tongue movements during speech. *Phonetica, 26,* 16-32.

Kent, R., & Netsell, R. (1971). Effects of stress contrasts on certain articulatory parameters. *Phonetica, 24,* 23-44.

Ladefoged, P., & Lindau, M. (1989). Modeling articulatory-acoustic relations: A comment on Stevens' "On the Quantal Nature of Speech." *Journal of Phonetics, 17,* 99-106.

Maeda, S. (1990). Compensatory articulation during speech; evidence from the analysis and synthesis of vocal-tract shapes using an articulatory model. In W. Hardcastle & A. Marchal (Eds.), *Speech production and speech modelling* (pp. 131-150). Dordrecht: Kluwer Academic Publishers.

Maeda, S. (1991). On articulatory and acoustic variabilities. *Journal of Phonetics, 19,* 321-333.

Manuel, S., & Vatikiotis-Bateson, E. (1988). Oral and glottal gestures and acoustics of underlying /t/ in English. *Journal of the Acoustical Society of America, 84,* S84. (A)

Mermelstein, P. (1973). Articulatory model for the study of speech production. *Journal of the Acoustical Society of America, 53,* 1070-1082.

Mizutani, T., Hashimoto, K., Hamada, H., & Miyura, T. (1988). Analysis of tongue motion for the dental consonants based on high-speed palatographic data. *Journal of the Acoustical Society of America, 84,* S127-8. (A)

Öhman, S. (1966). Coarticulation in VCV utterances: Spectrographic measurements. *Journal of the Acoustical Society of America, 39,* 151-168.

Palmer, J. (1973). Dynamic palatography. *Phonetica, 28,* 76-85.

Perkell, J. (1974). *A physiologically-oriented model of tongue activity in speech production.* Unpublished doctoral dissertation, MIT, Cambridge.

Recasens, D. (1991). An electropalatographic and acoustic study of consonant-to-vowel coarticulation. *Journal of Phonetics, 19,* 177-192.

Shadle, C. (1990). Articulatory-acoustic relationships in fricative consonants. In W. Hardcastle & A. Marchal (Eds.), *Speech production and speech modelling* (pp. 187-209). Dordrecht: Kluwer Academic Publishers.

Smith, K. K., & Kier, W. M. (1989). Tongues, trunks and tentacles: Moving the skeletons of muscle. *American Scientist, 77,* 28-35.

Stone, M. (1990). A three-dimensional model of tongue movement based on ultrasound and X-ray microbeam data. *Journal of the Acoustical Society of America, 87,* 2207-2217.

Stone, M., & Lundberg, A. (1994). Tongue-palate interactions in consonants vs. vowels. *Proceedings of the International Conference on Spoken Language Processing, 1,* 49-52, Yokohama, Japan.

Stone, M, Faber, A., & Cordero, M. (1991). Cross-sectional tongue movement and tongue-palate movement patterns in [s] and [ʃ] syllables. *Proceedings of the XII International Conference on Phonetic Sciences, Aix-en-Provence, 2,* 354-357.

Stone, M., Faber, A., Raphael, L., & Shawker, T. (1992). Cross-sectional tongue shape and lingua-palatal contact patterns in [s], [S], and [l]. *Journal of Phonetics, 20,* 253-270.

Stone, M., & Vatikiotis-Bateson, E. (in press). Trade offs in tongue, jaw, and palate contributions to speech production. *Journal of Phonetics.*

Wilson, J., Mahajan, U., Wainwright, S., & Croner, L. (1991). A continuum model of elephant trunks. *Journal of Biomechanical Engineering, 113,* 79-84.

11 Laryngeal Timing in Karen Obstruents

Arthur S. Abramson
The University of Connecticut
Haskins Laboratories

BACKGROUND

While doing fieldwork at The University of Connecticut with students of mine on two dialects of Karen, one in 1987 and the other in 1992, I found that among the obstruents the systems of plosive consonants of both of them are largely characterized by a three-way contrast at each place of articulation, while the affricates of one and the sibilants of the other have a two-way contrast. A search of the scant literature revealed one comprehensive source of information on the major dialects and subdialects of this Southeast Asian language (Jones, 1961). Jones' description of the obstruents in terms of voicing and aspiration, together with our own auditory impressions, suggested that this three-way distinction for the stops and a two-way distinction for affricates in one dialect and for sibilants in the other, could well be explained by a mechanism of laryngeal timing (Abramson, 1977).

Although such a model has been widely accepted and used, a brief review may be in order. In a series of publications, starting 30 years ago, with supporting data from acoustic (Lisker & Abramson, 1964) and physiological (e.g., Sawashima, Abramson, Cooper, & Lisker, 1970) studies of production and the

155

results of experiments in perception (e.g., Lisker & Abramson, 1970), Leigh Lisker and I presented data in support of the proposal that the timing of glottal pulsing relative to supraglottal articulation will account for the great majority of homorganic consonantal distinctions traditionally said to depend on voicing, aspiration, "tensity," and the like. Because in the early days we concentrated on stop consonants, for which such distinctions are more plentifully found in word-initial position in a great many languages, we measured the onset of glottal pulsing relative to stop-release and called that measurement voice onset time (VOT). In general, this model of laryngeal timing stands up well, although its efficacy is low or nil in some contrasts, as in plain vs. glottalized stops and voiced aspirated vs. voiceless aspirated stops.

Karen is a tone language. It belongs to the Sino-Tibetan family, within which it is commonly assigned to the Tibeto-Burman branch, although some argue for a separate Karennic branch (Jones, 1961). It is spoken by a sizable minority of people on both sides of much of the border between Myanmar (Burma) and Thailand and in parts of northern Thailand. Two of its major dialects, Pho and Sgaw, each of which has a number of subdialects, are of concern to us here.

The Pho variety treated here is spoken in the village of Paa Sangngaam in the province of Chiangrai, northern Thailand. The single native speaker available to me was a woman in her twenties with a secondary-school education and some training in a vocational school. Although her first language was Karen, she also spoke Northern Thai and Standard Thai and was studying English at The University of Connecticut, while her American husband worked on a Master's degree. Her dialect has the following obstruents that seemed amenable to the proposed approach:

Voiced unaspirated:	b	d		
Voiceless unaspirated:	p	t	c	k
Voiceless aspirated:	ph	th	ch	kh

My hypothesis for these stops and affricates was simply that the distinctions of voicing and aspiration would lie along the VOT dimension, although an articulatory complication with regard to the affricates /c ch/ will have to be discussed.

Our Sgaw variety is spoken in Taunggyi, Southern Shan States, Myanmar (Burma). My one native informant was a woman in her thirties doing graduate work toward a Master's degree at The University of Connecticut. She also spoke Burmese and English. Her dialect is the one mentioned earlier as having a two-way distinction for sibilants along with a three-way distinction for stops:

Voiced unaspirated:	b	d		
Voiceless unaspirated:	p	t	s	k
Voiceless aspirated:	ph	th	sh	kh

The stop systems, including the lack of a dorsovelar voiced stop, are the same in both dialects. Indeed, they are the same as in neighboring Thai and Shan. A striking trait of the Sgaw variety is the pair of sibilants /s sh/ distinguished by aspiration. This is rare in the languages of the world. Again, my hypothesis was that in speech production the distinctions of voicing and aspiration would lie along the VOT dimension.

PROCEDURE

The eliciting of utterances was done in the same way for both dialects, one in one semester of the graduate course Field Methods in Linguistics and the other in another semester five years later. We worked with each informant for three or four hours a week over a period of four months or so. All sessions were recorded on tape. A few hundred isolated words were elicited from the informant with a view toward obtaining a basic vocabulary and determining the system of tonal, consonantal, and vocalic contrasts. Usually three utterances of each word were recorded. Many of these lexical items were then used to form phrases and sentences for syntactic analysis. In addition, narrative passages were recorded and later analyzed linguistically with the help of the informant.

For both Karen dialects all occurrences of words containing the consonants of interest were digitized at a sampling rate of 22, 254 Hz and analyzed acoustically by means of the Signalyze™ program on Macintosh computers. Data for citation forms and for words embedded in running speech were recorded separately.

Both wave forms and wideband spectrograms, time-aligned in the displays, were used for making measurements of the timing of glottal pulsing relative to acoustic indices of articulatory closure and, in the case of sibilants, constriction. For the analysis of stops and affricates, the conventions of voice onset time (VOT) were used (Lisker & Abramson, 1964). That is, with the value of zero assigned to the moment of release, onsets of voicing before the release ("voicing lead") were recorded as negative values in milliseconds, while onsets after the release ("voicing lag") were recorded as positive values. For intervocalic /bdg/ voicing leads were measured only if there was a break in the glottal pulsing between the offset of the vowel formants and voicing onset in the stop closure; if there was no break, the item was described as "unbroken."

For the special case of sibilants in the Sgaw dialect, in order to facilitate comparison between /s/ and /sh/, two measurements were made. One was the duration of the frication noise and the other, VOT from the offset of the frication. Of course, given the spectral nature of turbulence, especially in the transition from frication to aspiration, the best possible estimates were made through inspection of the spectrograms.

RESULTS

Stop Consonants in Citation Forms

The data for the stop consonants of the Pho dialect of Karen in citation forms of words from the speech of one informant are displayed in Table 1. Means, standard deviations, and numbers of tokens are shown on the left for word-initial position and on the right for medial intervocalic position. Unfortunately, statistical tests for levels of significance seemed futile because of the very small numbers of tokens, as few as 1–3 in many of the cells of the five tables. In addition, neither the original informants nor others have been available to me for some time. Speakers of these two languages are hard to find away from their homelands. For the instances of unbroken voicing, of course, the cells for standard deviations are blank.

It is quite clear, even without significance levels, that the separation between categories along the VOT dimension is good. Note that in spite of the absence of a voiced dorso-velar stop in the system,[1] the voiceless unaspirated /k/ stays in the short-lag region, even in intervocalic position.

TABLE 1. *Initial and medial intervocalic Pho Karen stop consonants in isolated words: "Unbr" = unbroken glottal pulsing.*

				VOT in ms			
		Initial			Intervocalic		
	/b/	/p/	/ph/		/b/	/p/	/ph/
M	−64	5	59		Unbr	11	48
SD	−31	4	13		—	5	14
N	12	6	6		2	3	9
	/d/	/t/	/th/		/d/	/t/	/th/
M	−65	9	56		Unbr	13	65
SD	−17	10	18		—	2	6
N	9	3	15		3	3	3
		/k/	/kh/			/k/	/kh/
M		18	64			21	128
SD		6	19			10	45
N		6	8			3	3

---

[1]The phone [g] may occasionally appear as a variant of the dorso-velar voiced fricative /ɣ/.

The data for the stop consonants of the Sgaw dialect of Karen in citation forms of words from the speech of one informant are displayed in Table 2. Here, too, the categories are well separated along the VOT dimension. Aside from making the same generalizations here as for the Pho stops, I am prevented by the paucity of data from making quantitative comparisons between the two dialects.

Pho Affricates in Citation Forms

If the Pho alveolo-palatal affricates, voiceless unaspirated /c/ and voiceless aspirated /ch/, are to be included in this laryngeal-timing framework, certain conceptual problems have to be faced. Let us consider them with the help of Figure 1.

In Figure 1 note that I have shown the time from the release of the full closure to the onset of glottal pulsing to be, in this particular pair, 41 ms for the syllable /ci/ and 152 ms for the syllable /chi/. In the general sense of this paper, these may be taken to be two spans of voicing lag that, if typical, surely differentiate the two categories much as VOT does for the stops. Looking closely at the spectrograms, however, we are hard put to claim that we have anything substantial other than the local turbulence of the fricative portions of the affricates, with the "aspirated" one showing a considerably longer duration of frication.

TABLE 2. *Initial and medial intervocalic Sgaw Karen stop consonants in isolated words: "Unbr" = unbroken glottal pulsing.*

				VOT in ms		
		Initial			Intervocalic	
	/b/	/p/	/ph/	/b/	/p/	/ph/
M	−102	9	92	Unbr	16	80
SD	−19	6	13	—	6	13
N	6	17	9	2	3	9
	/d/	/t/	/th/	/d/	/t/	/th/
M	−121	13	94	Unbr	13	98
SD	−11	11	12	—	6	—
N	3	16	16	7	5	1
		/k/	/kh/		/k/	/kh/
M		10	96		14	109
SD		5	11		8	18
N		8	9		5	16

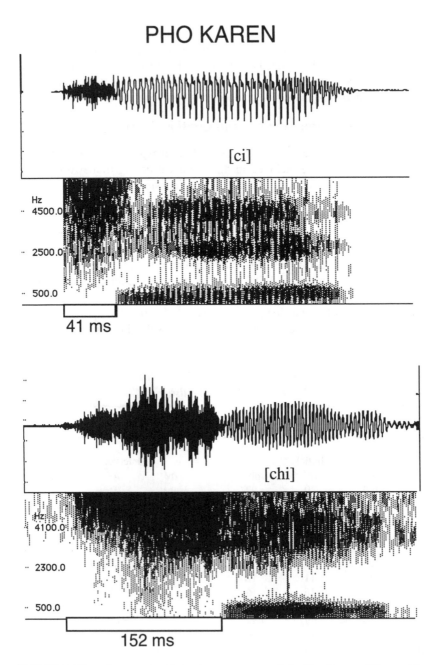

FIGURE 1. Wave forms and wideband spectrograms of the Pho Karen affricates. Top: voiceless unaspirated /c/. Bottom: voiceless aspirated /ch/.

That is, even if, toward the end of the frication, as the articulators open into the vocalic span, there might be a bit of aspiration, (i.e., noise-excitation of the overall supraglottal vocal tract by glottal turbulence), it alone surely could not serve to differentiate the two affricates. Admittedly, then, there is a gestural difference between "aspirates" of this type and aspirated stops. That is, the fricative portion of the "aspirated" affricate is longer than that of the "unaspirated" affricate, so one could argue for a length distinction that is unique in the phonology to this pair of consonants. On the other hand, given the full consonant system, it is tempting to align them with the two voiceless-stop categories. If so, we find that the two sets of values of voicing lag measured from the occlusive release, as seen in Table 3, do indeed separate the two categories, but we must understand that in this case VOT is the output of a gestural complex. The situation is the same in neighboring Thai (Abramson, 1989).

SGAW SIBILANTS IN CITATION FORMS

Although the sibilants /s sh/ of Sgaw Karen are indeed obstruents, they do not figure among the plosives discussed so far, yet they appear to belong to the same subsystem in that they exploit the property of aspiration. Wave forms and wideband spectrograms of such a minimal pair of syllables are illustrated in Figure 2. In the lower spectrogram an estimate is marked of the transition point between the frication ([s]) and the aspiration ([h]) of the token of /sh/. In contrast to the case of the Pho aspirated affricate, here one can not only also see the overall cavity turbulence with vowel formants going through it but can also easily hear the latter as aspiration distinct from the preceding frication.

The data of Table 4 are arranged to show the duration of the [s]-frication for both consonants. It is longer for /s/ than for /sh/. For the latter we see also the duration of the aspiration (labeled [h]) which constitutes true voicing lag, and so, in the conventional sense, distinguishes the two consonants; it is virtually impossible to find measurable aspiration in plain /s/. For /sh/ in the two contexts, the final column gives the total duration, labeled [s^h], of voiceless turbulence.

TABLE 3. *Initial and medial intervocalic Pho Karen affricates in isolated words.*

	VOT in ms			
	Initial		Intervocalic	
	/c/	/ch/	/c/	/ch/
M	35	108	36	85
SD	12	41	4	33
N	8	6	3	5

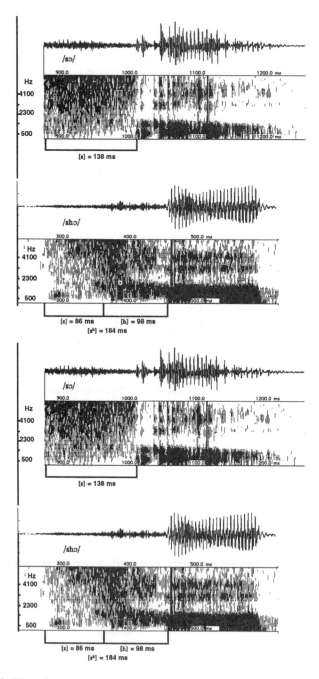

FIGURE 2. Wave forms and wideband spectrograms of the Sgaw Karen sibilants. Top: unaspirated /s/. Bottom: aspirated /sh/.

TABLE 4. *Initial and medial intervocalic Sgaw Karen sibilants in isolated words. [h] = aspiration.*

	Duration of Turbulence and Aspiration in ms							
	Initial				Intervocalic			
	/s/	/sh/			/s/	/sh/		
	[s]	[s]	[h]	[sʰ]	[s]	[s]	[h]	[sʰ]
M	156	126	69	195	189	148	74	222
SD	8	14	12	23	9	13	14	23
N	3	14	14	14	3	7	7	7

Brief Treatment of Running Speech

Regrettably, failing to look beyond the scope of the Field Methods course to the possibility of an acoustic study, I did not have my two informants record a great deal of running speech. Some of the tokens of relevant consonants in the narrative material could not be used because of low signal level or intrusive noises. I have limited myself in Table 5 to data from the medial intervocalic aspirated consonants of the two dialects. My thought was that they might be the ones most vulnerable to variation in running speech. In addition, there were fewer tokens for most of the rest of the consonants and none for some.

The entries of Table 5 are arranged to show a comparison of the aspirated consonants in running speech with their counterparts in citation forms of words. The latter values are taken from the preceding four tables. Out of eight comparisons, five voicing lags are longer for citation forms than for running speech. (This includes Sgaw /sh/ whether one takes the [h]-portion alone or the total duration of turbulence.) With so few data, however, I cannot as yet attach significance to this finding. In general, the data do allow us to say that the characteristic long voicing lags of the voiceless aspirates are well maintained.

CONCLUSION

The sufficiency of laryngeal timing, with voice onset time as its index, appears to be clear as a differentiating acoustic property for the sets of homorganic consonants of Karen treated here. Extending this criterion to the two voiceless alveolo-palatal affricates of Pho Karen is, in my view, justified by their place in the system of obstruents and, of course, by the data, although one does have to provide for gestural complexity in the account. In the case of the Sgaw Karen sibilants the relevance of this temporal property seems quite firm. This finding is especially interesting, given the rarity of such a distinction in the languages of the world.

It would certainly be desirable to validate the perceptual efficacy of voice timing through experiments with suitable stimuli and a number of native speakers as subjects. Perhaps on a forthcoming visit to Southeast Asia I can do that for at least one of the dialects.

TABLE 5. *Comparison of medial intervocalic voiceless aspirated consonants in words and running speech. [h] = aspiration.*

Intervocalic /ph/

	Pho		Sgaw	
	Words	Running Sp.	Words	Running Sp.
M	48	56	80	94
SD	14	16	13	9
N	9	9	9	3

Intervocalic /th/

	Pho		Sgaw	
	Words	Running Sp.	Words	Running Sp.
M	65	52	98	102
SD	6	19	—	6
N	3	19	1	3

Intervocalic /kh/

	Pho		Sgaw	
	Words	Running Sp.	Words	Running Sp.
M	128	58	109	101
SD	45	23	18	9
N	3	8	16	2

Intervocalic /ch/ and /sh/

	Pho /ch/		Sgaw /sh/					
	Words	Run. Speech	Words			Running Sp		
			[s]	[h]	[sʰ]	[s]	[h]	[sʰ]
M	108	85	148	74	222	136	62	198
SD	41	33	13	14	23	14	15	6
N	6	5	7	7	7	3	3	3

ACKNOWLEDGMENTS

The fieldwork on Sgaw Karen was done in the spring of 1987 with my students Aliaa A. Abdel Moneim and the late Nianqi Ren. Our gracious informant was Ms. Htoo Htoo Shwe. The fieldwork on Pho Karen was done in the spring of 1992 with my students Lisa M. Ferro and Yi Xu. Our very responsive informant was Ms. Jankaew Kangyang.

This is a somewhat elaborated revision of an oral paper given at the 25th International Conference on Sino-Tibetan Languages and Linguistics, University of California, Berkeley, 14-18 October 1992.

Support for the acoustic analysis and reduction of the data was provided by NIH Grant HD-01994 to Haskins Laboratories.

Having known Kathy Harris for 39 years, I am very happy to offer this modest contribution in her honor. I first met her in 1955 when she overwhelmed me with a lucid lecture on psychophysics in a Columbia University graduate course taught by Franklin S. Cooper. Not too long thereafter I became her colleague and then, to my delight, her friend.

REFERENCES

Abramson, A. S. (1977). Laryngeal timing in consonant distinctions. *Phonetica, 34*, 295–303.

Abramson, A. S. (1989). Laryngeal control in the plosives of Standard Thai. *Pasaa, 19*, 85–93.

Jones, R. B. (1961). *Karen linguistic studies: Description, comparison, and texts.* University of California Publications in Linguistics 25. Berkeley: University of California Press.

Lisker, L., & Abramson, A. S. (1964). A cross-language study of voicing in initial stops: Acoustical measurements. *Word, 20*, 384–422.

Lisker, L., & Abramson, A. S. (1970). The voicing dimension: Some experiments in comparative phonetics. In *Proceedings of the 6th International Congress of Phonetic Sciences*, Prague, 1967 (pp. 563–567). Prague: Academia.

Sawashima, M., Abramson, A. S., Cooper, F. S., & Lisker, L. (1970). Observing laryngeal adjustments during running speech by use of a fiberoptics system. *Phonetica, 22*, 193–201.

SECTION 2

Producing Speech: The Source

12 Dynamic Analysis of Speech and Nonspeech Respiration

Robert J. Porter
University of New Orleans
Kresge Hearing Research Laboratory
Louisiana State University Medical Center

David M. Hogue
University of New Orleans

Emily A. Tobey
Louisiana State University Medical Center

INTRODUCTION

Nonspeech (quiet) breathing consists of roughly equal inspiration and expiration intervals, uses a small range of lung volumes, and employs a relatively small set of muscles (e.g., Grassino & Goldman, 1986; Mead, 1960; Newsom Davis & Stagg, 1975). Quiet breathing can be modeled using muscle-force-optimized equations expressed in terms of airway resistance, alveolar ventilation, lung volume, and system compliance (e.g., McGregor & Becklake, 1961; Rodarte & Rehder, 1986; Roussos & Campbell, 1986; Widdicombe & Nadel, 1963). Respiration during speech, on the other hand, exhibits unequal inspiratory and expiratory durations and a wide range of volumes, with breathing patterns clearly influenced by linguistic structure and other communication-relevant factors (e.g., Hixon, 1973, 1987; Weismer, 1985). In fact, speech is often considered to be a task in which communicative constraints are *imposed* upon the respiratory system, forcing it to operate outside of its normal, force-optimized mode. From this point of view, speech breathing is a specially constrained behavior optimized in terms of communicative demands rather than (or in addition to) energy optimization requirements. We challenge this assumption and argue, instead, the temporal and volume characteristics of speech breathing reflect optimization of the same system variables as nonspeech breathing. This argument implies "communication" variables are not imposed on the respiratory system but are, together with physical and physiological variables, represented in common metrics of the system's state-space.

CONTROL OF SPEECH AND NONSPEECH BREATHING

As noted above, nonspeech breathing appears force-optimized across physical variables. However, the rate and pattern of nonspeech breathing also is related to physiological variables and may be predicted from metabolic measures, such as blood gas levels (e.g., Whipp & Ward, 1980). Thus, the nonspeech respiratory system may be described in terms of muscle force optimized over both physical and physiological variables.

Speech breathing also appears to reflect optimization of muscle force in spite of differences in volumes and temporal patterns. For example, during speaking, speakers typically inspire to approximately 60% vital capacity (VC) and expire, across the utterance, to approximately 35-40% VC, which is near the resting expiratory level (Weismer, 1985). The subglottal pressure resulting from muscle relaxation at the 60% VC point approaches the pressure necessary for normal phonation (3-5 cm H_2O), thereby minimizing the need for muscle action. As lung volume approaches the 35-40% VC point, the need for active muscle force increases to maintain the subglottal pressures necessary for speech. If inspirations were higher than 60% VC level, subglottal pressures would be greater than those needed for speech and inspiratory musculature would have to expend energy counteracting the additional passive forces (Weismer, 1985). Conversely, if lung volume falls below resting level, the expiratory musculature must expend energy to counteract an elastic recoil effect that would produce an inspiratory movement (Weismer, 1985). Since physiological need also is met during speech breathing, it appears speech breathing, like nonspeech breathing, also may be described in terms of muscle force optimized over physical and physiological variables.

Speech breathing also must accommodate factors arising from communication-relevant variables, notably the presence of varying airway resistances generated by articulation and source requirements, as well as the temporal variations characterizing speech prosody. One possible mechanism for dealing with communication variables was proposed by Hixon (1973; 1987) who suggested that brief, speech-specific, expiratory muscular efforts ("pulses") were superimposed on a more general, constant muscular effort in order to maintain necessary pressures and flows. Hixon's proposed mechanism would operate by monitoring pressure indirectly via lung volume; volume could be sensed by muscle stretch receptors (which are sensitive to length and rate of change in length) located in the intercostal muscles.

An alternative proposed mechanism for respiratory control during speech suggests a possible role for direct pressure and flow receptors, in addition to muscle stretch receptors (Davis, Bartlett, & Luschei, 1992). Such receptors are speculated to be part of a reflex system maintaining stable pressures by counteracting speech-produced perturbations. However, the amount of evidence in support of such reflex control is meager. In fact, it has been suggested that the proposed reflexes and fast muscles acting as a continuously active, high-gain servomechanism with significant nerve conduction delays would result in an uncontrollable oscillation of the system rather than in the desired stability (Davis et al., 1992).

An underlying theme of the proposed models of speech respiration control is that communicatively relevant aspects of speech require special, or additional, mechanisms compared to those required to meet physical and physiological demands during nonspeech breathing. We suggest, from another perspective, speech breathing may be viewed as one of many adaptive behaviors of a dynamic biological control system. (This perspective is further elaborated in the following section.)

A DYNAMIC SYSTEMS VIEW OF RESPIRATORY CONTROL

The dramatic differences across respiration patterns across speech and non-speech tasks may reflect the adaptive behavior of a complex, nonlinear, energy-optimized control system. Such systems are common in biology and several recent reports suggest important aspects of their nature may be revealed by experimental approaches more commonly applied to physical systems described by complex, nonlinear, state equations (e.g., Lauterborn & Parlitz, 1988). One approach, *allometry,* involves generating an energy-optimized state equation (see below) relating several physical and physiological parameters participating in system behavior. Experimental data then may be analyzed either to discover the values of parameters' exponents or to examine the correspondence between theoretical and actual exponents.

Another approach, *graphic,* provides a visualization of biological systems' behaviors in their state-spaces using, for example, phase portraits and return maps (e.g., Crutchfield, Farmer, Packard, & Shaw, 1986). These methods demonstrate how nonlinear deterministic systems display several different, but stable behavioral patterns in space or time, depending upon the relationships among system parameters. These patterns may reflect underlying system "attractors" that can have characteristic topological shapes (morphologies) depending on the nature of the underlying dynamics. In such systems, attractor morphology may change dramatically at particular "boundary-values" of the parameters and, in some cases, the attract may become "strange" or "chaotic." In the latter case, *analytic* techniques reveal a fractional dimensionality (fractal), as well a positive rate of information loss (entropy increase) associated with system operation in that region of the state-space (e.g., Fraser & Swinney, 1986). In physical systems, these methods allow researchers to examine, for example, changes in the spatial or temporal patterns of fluid flow (e.g., from laminar to turbulent) as fluid velocity or other parameters are varied.

The *allometric, graphic,* and *analytic* methods have been successfully applied to a variety of biological system behaviors including normal and arrhythmic cardiac function (Skinner, Goldberger, Mayer-Kress, & Ideker, 1990), changes in the pattern of EEG activity with stimulation (Molnar & Skinner, 1992), olfactory neural organization changes during learning (Skinner et al., 1989), and the control of respiration in the rat (Sammon & Bruce, 1991). Our own recent work suggests these techniques also may be applied in the case of human speech and nonspeech breathing.

A dynamic systems analysis first requires the generation of a state equation relating the variables involved in respiration.

The State Equation for Breathing

Quiet breathing (as well as breathing in other nonspeech conditions, e.g., during exercise) can be mathematically represented by a state equation expressing muscle force as applied pressure, with amplitude of pressure for a given tidal volume being a product of the amplitude of flow and magnitude of the mechanical impedance. The equation, adapted from Roussos and Campbell (1986), is (Equation (1)):

$$\hat{P} = \hat{V} |Zm| \tag{1}$$

(where $\hat{P}$ = Applied Pressure, $\hat{V}$ = Flow, $|Zm|$ = Mechanical Impedance)

and

$$(\hat{V}) = 2\pi f V_T \text{ and } |Zm| = \sqrt{R^2 + \frac{1}{(2\pi f C)^2}}$$

(where V_T is Tidal Volume, R is Airway Resistance, and C is System Compliance).

If approximately sinusoidal breathing is assumed, and the equation is re-expressed in terms of alveolar ventilation and dead-space volume, rather than tidal volume, the equation can be differentiated with respect to frequency, and an equation for breathing frequency, when muscular force is minimized, can be derived (Hogue, 1992; Hogue & Porter, 1992; Roussos & Campbell, 1986).

Under normal pressure and temperature conditions, and with tasks within the normal range of breathing, we may assume system compliance remains relatively constant for a speaker. In addition, because dead-space volume and alveolar ventilation are linear, positively increasing functions of tidal volume, we may substitute the more easily measured tidal volume for a required ratio of alveolar flow over dead-space volume in the optimized equation and express the result in terms of the inverse of breathing frequency, i.e., respiratory period, T. The resulting equation is (Equation (2)), in which

$$\hat{V} \text{ and } C \text{ and } 2\pi \text{ are combined into the constant, } K, \text{ and}$$

$$V_T \text{ is substituted for } V_{DS} \text{ and } T = \frac{1}{f}$$

$$\text{yielding } T = K' V_T^{\frac{1}{3}} R^{\frac{2}{3}}. \tag{2}$$

The appropriateness of Equation (2) may be investigated by measuring each of the parameters (i.e., T, V_T, and R) under a variety of respiratory tasks and determining if the derived relation of the measured parameters does, in fact, hold. Several *nonspeech* breathing studies suggest the derived relation holds, and alternative formulations representing different optimized quantities, such as minimal work, do not model non-speech breathing behavior as closely as a force-optimized equation (Roussos & Campbell, 1986).

Equation (2) represents a general relation among respiratory system variables that is maintained, apparently, by a dynamic control system optimized in terms of minimal force. The question arises whether the same equation also might apply in the case of speech breathing. As previously discussed, there is some reason to suspect that it might. In any case, the applicability of the state equation may be directly investigated by determining the relationship among the relevant parameters during speech breathing.

Allometric Analysis of Speech Breathing Data

In order to investigate the applicability of Equation (2) to speech breathing, we asked speakers to produce speech in a variety of tasks selected to provide a wide range of resistances and lung volumes. In one set of tasks, nonsense phrases of different lengths (e.g., "Abba, do." vs. "Abba abba, do.") were used to elicit different respiratory periods. The voiced, stop consonant, /b/, was contrasted with the unvoiced, lower resistance, articulatory homolog, /p/, in two variants of the nonsense sentences. A second stimulus condition, the first paragraph of the California Passage (Hoit & Hixon, 1987), provided a more "natural" speaking condition. In addition, whispered (low laryngeal resistance), naturally voiced (high laryngeal resistance), and monotone (less variable laryngeal resistance) productions of all materials were obtained. Four males and two females, mean age 26.3 years, of average height and weight, served as subjects. All subjects, by self-report, were nonsmokers, native American English speakers, and had negative histories for cardiovascular, neurological, respiratory, or laryngeal disease.

In an initial session prior to collecting the speech respiration data, estimates of airway resistance were obtained for the 24 consonants and 15 vowels in American English which occur in the stimulus materials. Airway resistances for each phonetic event were estimated from airflow measures of VCV (e.g., /AbA/) or CVC (e.g., /hAd/) utterances (spoken in the frame: "That's ___ again."). Flow was measured using a calibrated Rothenberg mask. Relative resistance was estimated using a variant of the method of Smitheran and Hixon (1981) and was expressed as a proportion referenced to the maximal flow (1.0) measured with a whispered /ha/. Measures of flow were made at the temporal midpoint of the target phonetic event. Three productions of each nonsense word were performed by each subject in each of the speaking conditions. The mean of the measures in a given condition were used in subsequent calculations for testing the model. Lung-volume measures (expressed in percent vital capacity, %VC) were obtained with a Respitrace. Calibration and baseline measures of volume were obtained at the beginning of a session using standardized procedures. Respitrace output and audio recordings were digitized (10 kHz). The voice signal was obtained via electret microphone placed 2 cm from the speakers lips and was low-pass filtered at 5 kHz.

Periods between the peaks of inspiration and the greatest points of expiration prior to the next inspiration (henceforth, *breath groups*) were the units of analysis. Measures of inspiratory duration, breath-group duration, inspiratory and expiratory volume, initial and final fundamental frequency, and number of syllables produced during the breath group were obtained. Broad phonetic transcriptions were made of all breath groups, and the total resistance for each period was calculated as the sum of the resistance values for each phonetic event, as estimated from the earlier flow measures.

Overall analysis of our data indicated the respiration and speech measures were comparable to those of prior studies, suggesting that our speaking conditions were comparable to those of earlier studies (e.g., Hixon, 1987; Horii & Cooke, 1978; Stathopolous, Hoit, Hixon, Watson, & Solomon, 1991; Weismer, 1985).

The fit of Equation (2) to the data was determined using the logarithmic form of Equation (2) (Equation (3)):

$$\log(T) = K' + 0.333\log(V_T) + 0.667\log(R). \tag{3}$$

Logarithmic analysis converts the exponential form of the allometric state equation to a linear one in which the exponents become slopes and the multiplicative components appear in the intercept. In this case, the intercept contains a numerical constant (2π), together with the influence of compliance, a multiplicative constant introduced by the use of tidal volume instead of alveolar ventilation/dead-space volume ratio, and any multiplicative effects of unknown factors. Because of the unknown values of these multiplicative factors, neither prediction nor interpretation of the intercept value is possible. However, a three-dimensional regression of log(breath group duration) vs. log(%VC) and LOG(total resistance during the breath group) should reveal a plane with slopes of 0.333 and 0.667 in the volume-by-duration and in the resistance-by-duration projections, respectively.

Comparison of regression results for the nonsense sentences and the California passage did not reveal systematic differences, nor were there any differences among sentences of different lengths. Data for all speaking conditions for one representative subject are presented in Figure 1; one point is plotted for each breath group. Regression equations for all speakers are shown in Table 1; all are significant ($p < 0.05$). The mean coefficients for the resistance component are all very close to the equation predictions, and are not significantly different across voicing conditions. This finding suggests that the resistance effects of articulator position on airflow are, as expected, relatively consistent regardless of voicing (Melcon, Hoit, & Hixon, 1989). The intercepts also are not significantly different across conditions. Mean volume coefficients, however, are somewhat smaller than the 0.333 expected.

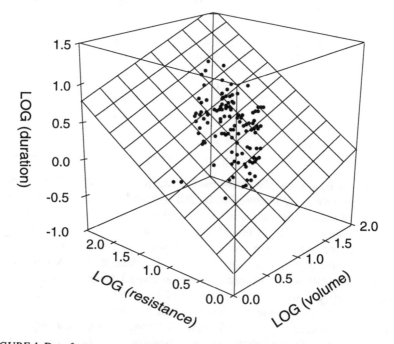

FIGURE 1. Data from one representative subject; all experimental conditions. Each point represents measures taken from one respiration period.

TABLE 1. *Regression equations and mean coefficients for all subjects in all conditions.*

Normal Subject	Intercept	Volume Coefficient	Resistance Coefficient	R^2
#1	−0.571	0.147	0.669	93.0
#2	−0.744	0.099	0.856	94.5
#3	−0.695	0.219	0.713	95.7
#4	−0.551	0.277	0.550	86.7
#5	−0.533	0.450	0.448	94.6
#6	−0.669	0.286	0.607	92.0
MEAN	−0.627	0.246	0.641	---

Monotone Subject	Intercept	Volume Coefficient	Resistance Coefficient	R^2
#1	−0.773	0.432	0.508	94.2
#2	−0.714	0.199	0.779	95.1
#3	−0.686	0.232	0.746	96.5
#4	−0.759	0.163	0.810	94.6
#5	−0.544	0.243	0.645	94.4
#6	−0.629	0.199	0.722	94.0
MEAN	−0.684	0.245	0.702	---

Whisper Subject	Intercept	Volume Coefficient	Resistance Coefficient	R^2
#1	−0.749	0.379	0.542	93.2
#2	−0.745	0.071	0.926	97.6
#3	−0.612	0.196	0.721	96.6
#4	−0.763	0.296	0.684	96.6
#5	−0.640	0.684	0.286	90.3
#6	−0.762	0.226	0.823	96.7
MEAN	−0.712	0.309	0.664	---

Overall, the results reveal a close match between the slopes obtained and those predicted on the basis of the force-optimized equation. The close match supports the choice of assumptions made in the derivation of the formula and in the data collection. The match also is remarkable given the differences among speakers in speaking rate and the range of phonetic types, resistances, and breath group durations provided by the stimulus materials. The control of speech breathing, thus, appears to reflect the operation of the same mechanisms involved in non-speech breathing, in spite of the "demands" of communication variables such as phonetic type, speaking style, prosodic structure, and phrase length.

The match between expected and observed exponents of the state equation is quite good. The volume coefficients are, however, somewhat smaller than expected. This may be the result of substituting tidal volume for the dead-space volume/alveolar flow term in Equation (2); the substitution was justified by the positive, linear regression of tidal volume on both dead-space volume and alveolar ventilation (Gray, Grodins, & Carter, 1956). The slopes of the two regression lines are not parallel, however. Alveolar ventilation increases at a faster rate than dead-space volume. Therefore, tidal volume tends to underestimate the ratio. The underestimation consequently generates observed volume/ventilation coefficients proportionally smaller than the 0.333 expected. The systematically smaller-than-expected volume coefficients therefore may be due to this underestimation.

The dead-space volume/alveolar flow coefficient could be more accurately determined if measures of dead-space volume and alveolar ventilation could be obtained. Unfortunately, both are difficult to measure directly, especially during speech. The increase in ventilation during speech has indirect effects, however, in that it tends to produce hyperventilation (Proctor, 1986). Interestingly, the respiratory response to hyperventilation is the appearance of very brief exhalations within an utterance and periods of apnea lasting as long as 18 seconds after the communication task has ended (Proctor, 1986). Examination of the Respitrace data revealed that four of our six subjects displayed such behavior. This suggests the possibility of replacing alveolar ventilation, in the state equation, with a relative measure of blood oxygen/carbon dioxide levels. Such a possibility could be investigated by measuring (or manipulating) relative blood gas levels and using the Gray et al. (1956) data to estimate relative dead-space volume from tidal volume. The possibility of this type of analysis reminds us the components of allometric state equations for biological systems need not be expressed only in terms of physical variables but also may include physiological variables, and, as our data suggest, communication (or psychological) variables, both of which have meaning in the context of the dynamic control system underlying a behavior (Kugler & Turvey, 1987).

Our results demonstrate how an allometric, state-equation approach may clarify understanding of the interrelation of physical, biological, and communication variables in complex, dynamic systems such as speech breathing. Our results also suggest *graphical* methods and *analytic* techniques might be useful in examining the nature of the dynamic *attractor(s)* characterizing respiratory speech and nonspeech behaviors.

Graphic and Analytic Analysis of the Respiratory Attractor

The flexible way in which the respiratory system accommodates the different constraints of speech and nonspeech breathing suggests that respiration may be viewed as a self-organizing, nonlinear, dynamic control-system (e.g., Kugler & Turvey, 1987). From this perspective, it is possible the dynamic behavior of the respiratory system under a variety of conditions might be represented by an "attractor."

For the present discussion, an attractor is a collection of states (observed or theoretical) toward which a dynamic system tends to move. An attractor is, thus, a topological entity aiding in the conceptualization and visualization of system behavior. For example, the surface of a toroidal (doughnut) attractor is a two-dimensional, continuous, closed surface bounding the inside and outside spaces. The points in the three-dimensional space embedding the toroidal attractor represent different possible states of the system produced by different combinations of variables. As an attractor, the torus surface represents the domain of states to which the dynamic process tends to move as it operates within its constraints. The attractor attracts, or traps, nearby points, or states, guiding their paths into the attractor's domain. Variations in conditions may metrically (i.e., quantitatively) "distort" the attractor (e.g., stretch the torus, put a "bump" in it, or move its center relative to the frame of reference of the observer) but they do not (by definition) change its topology, or *qualitative* form.

The positive results of our allometric analysis suggest the topology of the attractor we presume to underlie speech respiration may be the same as that involved in nonspeech breathing. Consequently, we sought to visualize graphically the presumed attractor, and observe what "metric distortions" the attractor might display across conditions, and to estimate analytically the dimension of the attractor.

Our interest in the dimension of the attractor was stimulated by the research of Sammon and Bruce (1991) who reported that the attractor in quiet breathing may have a fractal dimension. These investigators examined the respiration of anesthetized rats, breathing pure O_2, before and after vagotomy. Using the correlation dimension, the authors found that the normal rats' respiration had a fractal dimension ranging from 2.25 to 3.71, whereas the vagotomized rats' respiration dimension was not fractal and was equal to 1.0. The authors concluded that the fractal dimension in intact rats reflected the operation of a chaotic, optimizing, dynamic system with a central nervous system center as a critical component. Chaotic, or "strange," attractors often are associated with systems where many variables must be flexibly accommodated: i.e., chaotic systems tend to be especially adaptable systems (e.g., Kelso & Tuller, 1984). Consequently, severing the vagus presumably removes the central, optimizing component, leaving a nonchaotic, one-dimensional, inflexible, peripheral-system attractor. We, therefore, examined the possibility the attractor presumed to be represented in speech and nonspeech breathing also might have a fractal or chaotic aspect.

Visualizing the Attractor. Visualization of the attractor was achieved by plotting respiratory traces in a phase space of instantaneous volume (expressed as %VC) by instantaneous flow (expressed as the change in %VC between subsequent measures of lung volume) for three subjects. Results for one representative subject are shown in Figure 2.

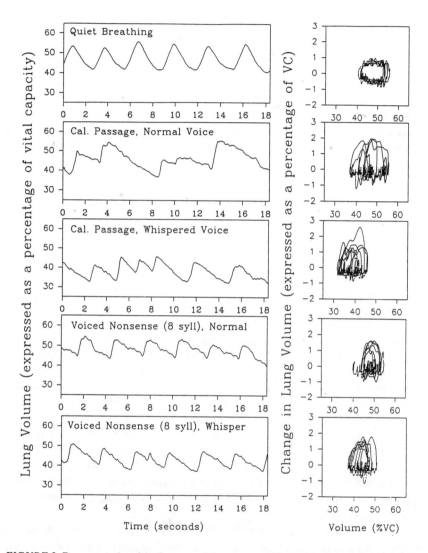

FIGURE 2. Representative data for one subject (same subject as in Figure 1). To the left are samples of the temporal course of respiration under each of the indicated conditions. To the right are "attractor" plots of percent change in lung volume vs. volume (dV/dt vs. V) for successive time points over an entire condition.

The different patterns of speech breathing compared to quiet breathing illustrate metrical distortion. That is, examination of the quiet breathing (QB) figures revealed a pattern *qualitatively* similar but *quantitatively* different from the other respiration conditions. The differences are as expected: The flattened and longer expiratory component in the speaking tasks represents expiration over the communication-constrained breath group. The large, arching parts of the trajectories reflect higher-velocity inspirations typical of speech. Thus, the "attractor" in speaking appears qualitatively (i.e., topologically) similar to quiet breathing but metrically "distorted" in the volume and/or flow dimensions.

Estimating the Attractor Dimension. Our first choice of an analytic method involves a measure of the changes in "relative dispersion" (%RD) of points in a time series (Glenny, Robertson, Yamashiro, & Bassingthwaighte, 1991). In a system with a fractal character, the %RD decreases with increases in the number of analyzed points (partition, p), as a function of exponential form (Equation (4)):

$$D_p(N_{p0}) = \%RD_p(p_0)N^{1-D}, \tag{4}$$

where D is the fractal dimension. The exponent 1-D will be the slope of a straight line in logarithmic coordinates. The steeper the slope, (1-D), the greater the scale-independent irregularity, or fractal dimension. Slopes will range from zero (yielding a D of 1.0, which indicates no fractal dimension) to –0.5 (yielding a D of 1.5, which suggests a random process).

The dimension, D, was calculated for the respiration time-series of the different speaking conditions, 18.35 second portions of which are shown in Figure 2 (367 points sampled at 20 Hz). Calculation of the dimension was automated using a standard statistical package (Minitab, version 7.1). All points recorded for each subject in each condition, except whisper, were analyzed. Because of an initially large inspiration sometimes occurring for the whisper condition, points for that condition were obtained from a portion of the record beginning at the second inspiration. Table 2 provides a summary of the results for the several conditions. Inspection reveals D to be close to 1.0 in all cases, suggesting no fractal structure. (It is important to note that the finding of D = 1.0 occurs for any system existing in an integer-dimension space. That is, the data suggest that the respiratory attractor has an integer dimensionality *of at least* 1.0.) The %RD analysis, thus, suggests that the speech-related variation in phase space may be orthogonal to the underlying behavior of a (nonchaotic) attractor.

A second analytic method, the "FFT" method, was also investigated (Glenny et al., 1991). In this analysis, the fast fourier transform (FFT) for the time series is calculated and the log of the power at each FFT derived frequency is plotted vs. the log of the frequency. Fractal dimensionality of the time series is revealed in a constant ratio of power to frequency, regardless of resolution (i.e., frequency). A negative slope suggests fractal self-similarity. Table 3 provides a summary of the results for the FFT analysis. The obtained Ds are comparable to those obtained using the %RD method. (The tendency for the values to be less than one may reflect limitations of our analysis technique rather than the method itself.)

TABLE 2. *Fractal dimension using the Relative Dispersion (RD) method.*

	Subject #1	Subject #3	Subject #5
Quiet breathing	1.0401	1.0438	1.0567
California passage, normal voice	1.0509	1.0732	1.1100
California passage, whispered	1.0718	1.0831	1.0708
8 syllable nonsense, normal voice	1.0583	1.0358	1.0695
8 syllable nonsense, whispered	1.0592	1.0524	1.0920

TABLE 3. *Fractal dimension using the FFT method.*

	Subject #1	Subject #3	Subject #5
Quiet breathing	0.7148	1.0388	1.0161
California passage, normal voice	0.8373	0.8419	0.9113
California passage, whispered	0.6747	0.7619	0.8443
8 syllable nonsense, normal voice	0.9431	0.8618	0.8868
8 syllable nonsense, whispered	0.6015	0.8888	0.9123

The data suggest speech and quiet breathing attractors to have a 1-D (limit cycle) or, possibly, a 2-D (torus) topology. In addition, little evidence for a "strange" or chaotic attractor was found, nor does the attractor appear to change qualitatively across conditions. These findings raise the question of why our results are not the same as those of Sammon and Bruce (1991). We believe this may be because the human breathing rate is much slower than that of rats and our analysis, therefore, did not include a sufficient number of respiratory cycles to reveal a fractal dimensionality. Work in progress seems to support this conclusion.

CONCLUDING COMMENTS

The allometric, graphic, and analytic explorations of speech and nonspeech respiration discussed in this article illustrate important theoretical and experimental differences between the *dynamic* and the *classical* perspectives on biological systems. From the classical perspective, biological systems have a finite but large number of controlling variables. Because of this, understanding the behavior of the system only is practical when the number of variables is constrained to a manageable number (experimentally or statistically) and/or the variables are subdivided into smaller, "functional" sets within which investigations are carried out. This practical consequence of a classical perspective may be referred to as a "linear" approach since it seeks to reduce the context of investigation to those situations in which the behavior of the system is a linear function (in an engineering sense) of a few, manipulable variables.

Thus, a classical perspective on respiration during speech subdivides research into experimental areas investigating physical, physiological, muscle/kinetic, or communicative variables. Each experimental area has its "own" set of metrics associated with its variables (e.g., resistance and flow, gas balance and pH level, inspiratory and expiratory muscles, phonetic and syllabic, respectively, for the four areas in respiration). If studies combine variables from different areas, one set is usually considered to comprise independent variables and the other set the dependent variables (e.g., the effects of varying sentence stress on subglottal pressure).

Implicit in the classical approach is the view that a hierarchy exists among the different experimental areas, with each area lying in a higher or lower (e.g., the communicative) functional relation to other areas. Constraints defined in terms of a higher level are presumed to be ultimately manifest in terms of lower levels. This view presumes that communication variables impose on physiological and physical variables that, in turn, define the "real" operation of the respiratory system. Thus, communicative variables have implications for physical and physiological variables but not vice versa.

In contrast to the classical view, the dynamic system perspective does not focus on constraining variables either experimentally or statistically, nor does it partition variables into areas. Instead, the system's behavior is observed in terms of one or more variables and the interrelation of the variables is examined in terms of a state equation and/or in terms of graphical or mathematical aspects of state-space trajectories (Kugler & Turvey, 1987). Importantly, the dynamic approach assumes that behavioral structure emerges from the operation of an adaptive dynamic system defined over a single, finite set of control variables, or "control parameters" that *may, or may not,* map directly onto variables with the classical metrical identities (Kelso, Ding, & Schoner, 1992). Experimental measurement or manipulation of a classical variable does not, from this perspective, mean the selected variable is in any sense higher or lower in a functional hierarchy than another variable, nor does selection designate a variable to be independent or dependent. In a dynamic system, the system structure interrelates *all* variables in such a way that *any* variable reflects the control structure (attractor topology) of the system. In the absence of a hierarchy, all variables have implications for all others regardless of their classical metrical identities (Crutchfield et al., 1986).

In this article, we illustrated the application of the dynamic systems approach in respiration during speech and nonspeech. Our illustration began by developing a state equation based upon consideration of some variables (physical and physiological) suspected to reflect system optimizing behavior in terms of energy expenditure. Our allometric, graphical, and analytic investigations suggest that this biophysical formulation captures system behavior in speech as well as nonspeech tasks: "manipulation" of communicative variables results in system behavior changes of the same topological type, under the same optimizing conditions, as those observed for physical and physiological variables. Recasting these observations from a dynamical perspective, we conclude that physical, physiological, muscular/kinetic, and communicative variables all reflect participation of the same control parameters, with common metrics, defining a common state-space. An exciting challenge of this conclusion is that the heretofore rather mysterious interrelation of "phonetic," "physiological," and "physical" aspects of communication may reveal itself, as future investigations, guided by the dynamic perspective, define the nature of the metrics of the state-space coordinates.

ACKNOWLEDGMENTS

Some of the data reported in this chapter were part of a thesis submitted by D. M. Hogue in partial fulfillment of requirements for a Master's degree in the Department of Psychology, University of New Orleans. The authors acknowledge the support provided by a Chancellor's Fellowship to the second author, and support from the Departments of Psychology and Communication Disorders. Additional support was provided by an NIH Academic Research Enhancement Award to E. Tobey. Some of the data reported were presented at meetings of the Acoustical Society of America (Hogue & Porter, 1992; Porter & Hogue, 1993).

REFERENCES

Crutchfield, J. P., Farmer, J. D., Packard, N. H., & Shaw, R. S. (1986). Chaos. *Scientific American, 255,* 46-57.

Davis, P. J., Bartlett, D., & Luschei, E. S. (1992). Coordination of the respiratory and laryngeal systems in breathing and vocalization. *NCVS Status and Progress Report, 2,* 59-81.

Fraser, A. M., & Swinney, H. L. (1986). Independent coordinates for strange attractors from mutual information. *Physical Review A, 33,* 1134-1140.

Glenny, R. W., Robertson, H. T., Yamashiro, S., & Bassingthwaighte, J. B. (1991). Applications of fractal analysis to physiology. *Journal of Applied Physiology, 70,* 2351-2367.

Grassino, A. E., & Goldman, M. D. (1986). Respiratory muscle coordination. In A. P. Fishman (Ed.), *Handbook of physiology, section 3: The respiratory system, volume III, Mechanics of breathing, part 1* (pp. 463-480). Bethesda, MD: American Physiology Society.

Gray, J. S., Grodins, F. S., & Carter, E. T. (1956). Alveolar and total ventilation and the dead-space problem. *Journal of Applied Physiology, 9,* 307-320.

Hixon, T. J. (1973). Respiratory function in speech. In F. D. Minifie, T. J. Hixon, & F. Williams (Eds.), *Normal aspects of speech, hearing, and language* (pp. 73-125). Englewood Cliffs, NJ: Prentice-Hall.

Hixon, T. J. (1987). Respiratory function in speech. In T. J. Hixon (Ed.), *Respiratory function in speech and song.* Boston, MA: College-Hill Press.

Hogue, D. M. (1992). *Examining a respiratory control model using an allometric analysis of speech breathing.* Unpublished Master's thesis, University of New Orleans.

Hogue, D. M., & Porter, R. J. (1992). Examining a respiratory control model using an allometric analysis of speech breathing. *Journal of the Acoustical Society of America, 92,* 2389. (A)

Hoit, J. D., & Hixon, T. J. (1987). Age and speech breathing. *Journal of Speech and Hearing Research, 30,* 351-366.

Horii, Y., & Cooke, P. A. (1978). Some airflow, volume, and duration characteristics of oral reading. *Journal of Speech and Hearing Research, 21,* 470-481.

Kelso, J. A. S., Ding, M., & Schoner, G. (1992). Dynamic pattern formation: A primer. In J. Mittenthal & A. Baskin (Eds.), *Principles of organization in organisms, SFI studies in the sciences of complexity, Proceedings. Vol. XIII* (pp. 397-439). Reading, MA: Addison-Wesley.

Kelso, J. A. S., Saltzman, E., & Tuller, B. (1986) The dynamical perspective on speech production: Theory and data. *Journal of Phonetics, 14,* 29-60.

Kelso, J. A. S., & Tuller, B. (1984). Converging evidence in support of common dynamical principles for speech and movement coordination. *American Journal of Physiology, 246,* R928-R935.

Kugler, P.N., & Turvey, M. T. (1987). *Information, natural law, and the self-assembly of rhythmic movement.* Hillsdale, NJ: Lawrence Erlbaum Associates.

Lauterborn, W., & Parlitz, U. (1988). Methods of chaos physics and their application to acoustics. *Journal of the Acoustical Society of America, 84,* 1975-1993.

McGregor, M., & Becklake, M. (1961). The relationship of oxygen cost of breathing to respiratory mechanical work and respiratory force. *Journal of Clinical Investigations, 40,* 971-980.

Mead, J. (1960). Control of respiratory frequency. *Journal of Applied Physiology, 15,* 325-336.

Melcon, M. C., Hoit, J. D., & Hixon, T. J. (1989). Age and laryngeal airway resistance during vowel productions. *Journal of Speech and Hearing Research, 54,* 282-286.

Molnar, M., & Skinner, J. E. (1992). Low-dimensional chaos in event-related brain potentials. *International Journal of Neuroscience, 66,* 263 -276.

Newsom Davis, J., & Stagg, D. (1975). Interrelationships of the volume and time components of individual breaths in resting man. *Journal of Physiology, 245,* 481-498.

Proctor, D. F. (1986). Modifications of breathing for phonation. In A. P. Fishman (Ed.), *Handbook of physiology, section 3: The respiratory system, volume III, Mechanics of breathing, part 1* (pp. 597-604). Bethesda, MD: American Physiology Society.

Porter, R. J., & Hogue, D. M. (1993). Searching for attractors in speech and nonspeech respiration. *Journal of the Acoustical Society of America, 94,* 1881. (A)

Rodarte, J. R., & Rehder, K. (1986). Dynamics of respiration. In A. P. Fishman, (Ed.), *Handbook of physiology, section 3: The respiratory system, volume III. Mechanics of breathing, part 1* (pp. 131-144). Bethesda, MD: American Physiology Society.

Roussos, C., & Campbell, E. J. M. (1986). Respiratory muscle energetics. In A. P. Fishman (Ed.), *Handbook of physiology, section 3: The respiratory system, volume III, Mechanics of breathing, part 1* (pp. 481-509). Bethesda, MD: American Physiology Society.

Sammon, M. P., & Bruce, E. N. (1991). Vagal afferent activity increases dynamical dimension of respiration in rats. *Journal of Applied Physiology, 70,* 1748-1762.

Skinner, J. E., Goldberger, A. L., Mayer-Kress, G., & Ideker, R. E. (1990). Chaos in the heart: Implications for clinical cardiology. *Biotechnology, 8,* 1018-1024.

Skinner, J. E., Martin, J. L., Landisman, C. E, Mommer, M. M., Fulton, K., Mitra, M., Burton, W. D., & Saltzberg, B. (1989). Chaotic attractors in a model of neocortex: Dimensionalities of olfactory bulb surface potentials are spatially uniform and event related. In E. Basar & T. H. Bullock (Eds.), *Springer series in brain dynamics 2* (pp. 158-173). Berlin: Springer-Verlag.

Smitheran, J. R., & Hixon, T. J. (1981). A clinical method for estimating laryngeal resistance during vowel production. *Journal of Speech and Hearing Research, 46,* 138-146.

Stathopolous, E. T., Hoit, J. D., Hixon, T. J., Watson, P. J., & Solomon, N. P. (1991). Respiratory and laryngeal function during whispering. *Journal of Speech and Hearing Research, 34,* 761-767.

Weismer, G. (1985). Speech breathing: Contemporary views and findings. In R. G. Daniloff (Ed.), *Speech science: Recent advances* (pp. 47-72). San Diego, CA: College-Hill Press.

Whipp, B. J., & Ward, S. A. (1980). Ventilatory control dynamics during muscular exercise in man. *International Journal of Sports Medicine, 1,* 146-159.

Widdicombe, J. G., & Nadel, J. A. (1963). Airway volume, airway resistance, and work and force of breathing: theory. *Journal of Applied Physiology, 18,* 863-868.

13 Speech Breathing Processes of Deaf and Hearing-Impaired Persons

Dale Evan Metz and Nicholas Schiavetti
State University of New York

INTRODUCTION

Respiratory system behavior that supports speech production (speech breathing processes) is a complex physiological activity that, under normal circumstances, provides a relatively constant alveolar pressure for speech production (Hixon, 1973). In addition to providing the primary driving force for speech production, the respiratory system (in concert with laryngeal mechanisms) makes significant contributions to the instantiation of suprasegmental aspects of speech including contrastive syllabic stress, variations in vocal intensity and fundamental frequency, and to the overall prosody of connected utterances (Isshiki, 1964; Ladefoged, 1974; Stathopoulos & Sapienza, 1993a,b).

It is well established that deaf speakers may exhibit a profound lack of control over suprasegmental aspects of speech production (Hood & Dixon, 1969; Murphy, McGarr, & Bell-Berti, 1991; Rubin-Spitz & McGarr, 1990). Frequently reported speech production control problems demonstrated by deaf persons involving vocal intensity regulation, fundamental frequency regulation, and control of the overall prosody of speech suggest the possibility of respiratory system involvement. The extent to which abnormal respiratory function has a negative impact on the speech intelligibility of persons who are deaf has long been a topic of interest, with early work in this area consisting of descriptive research studies conducted in the 1930s and 1940s. Collectively, the results of these studies suggested that abnormal respiratory support of speech may contribute to reduced speech intelligibility of deaf persons (Hudgins, 1934; Hudgins & Numbers, 1942; Rawlings, 1935).

In an extensive examination of the factors contributing to reduced speech intelligibility of deaf persons, Hudgins and Numbers (1942) demonstrated that the variance associated with reduced speech intelligibility of deaf children was nearly equally distributed between two general error categories. The first error category was characterized by an "inaccuracy or failure of articulatory processes" (i.e., consonant and vowel production errors) and the second error category was characterized by a lack of coordination between "articulation processes and the breathing muscles" (p. 378).

The conclusions drawn by Hudgins and Numbers constituted a rather obvious invitation to explore in more detail the respiratory system's role in degrading the intelligibility of the speech of deaf persons. With rare exception, however (e.g., Woldring, 1968), investigations designed to delineate the specific nature of respiratory system activity and its specific role in the reduced intelligibility of the speech of deaf persons were conspicuously absent in the research literature for more than 30 years following Hudgins and Numbers' (1942) seminal investigation.

Although recent research efforts have partially rectified the dearth of information in this area (e.g., Cavallo, Baken, Metz, & Whitehead, 1991; Feudo, Zubick & Strome, 1982; Forner & Hixon, 1977; Itoh, Horii, Daniloff, & Binnie 1982; Whitehead, 1983), it can still be argued convincingly that a systematic delineation of respiratory function and its specific contribution to the degraded speech intelligibility of deaf persons is still far from complete. This assertion takes on additional significance when one considers the restricted pool from which many of the recent studies mentioned above have drawn subjects (typically deaf male college students) and recent findings regarding systematic variations in speech breathing processes associated with gender (Hoit & Hixon, 1989; Stathopoulos & Sapnienza, 1993a), age (Hoit & Hixon, 1987; Stathopoulos & Sapnienza, 1993b), and body type (Hoit & Hixon, 1986). Additionally, the relative lung volumes at which speech is produced appears to influence at least one segmental feature, voice onset time (Hoit, Solomon, & Hixon, 1993), that is known to have a significant relationship to the speech intelligibility of deaf persons (Monsen, 1978; Metz, Samar, Schiavetti, Sitler, & Whitehead, 1985; Metz, Schiavetti, Samar, & Sitler, 1990).

Given the early recognition of the potential importance of abnormal respiratory support and of the reduced speech intelligibility of the deaf, the lack of empirical attention is somewhat surprising. A partial answer to the question of why more attention has not been devoted to the specific relation between respiratory function and speech intelligibility of deaf persons lies, perhaps, in Weismer's (1985) observation that respiratory function underlying normal speech production has received relatively less research attention than laryngeal and supralaryngeal functions of speech. Weismer suggested that the paucity of research in this area may be attributed to the confluence of a belief (a) that speech breathing processes are uninteresting and straightforward, and (b) that research conducted in the 1950s and 1960s by Draper and his colleagues (e.g., Draper, Ladefoged, & Whitteridge, 1959) essentially exhausted the topic. In light of research conducted primarily by Hixon and his colleagues (e.g., Hixon, Goldman, & Mead, 1973; Hixon, Mead, & Goldman, 1976), Weismer (1985) asserts that neither of these two positions is convincing or tenable, and that expanded and extended efforts in this area are clearly needed.

Arguably, the work of Hixon and his colleagues, referred to by Weismer (1985), has served as the impetus underlying a recently renewed interest in respiratory behaviors associated with both normal and disordered speech. We believe this assertion to be particularly true with respect to research regarding speech breathing processes of deaf persons (cf. Forner & Hixon, 1977), although such research continues to be underrepresented. In this chapter, we will examine speech breathing processes of deaf persons with respect to prephonatory posturing of the respiratory apparatus, lung-volume regulation, and selected chest-wall adjustments and configurations that could potentially degrade intelligibility during speech production. Primary consideration will be given to inappropriate coordination between the respiratory apparatus and upper airway structures. In addition, we will argue for and suggest specific research avenues that are in need of extended consideration.

PREPHONATORY CHEST-WALL ADJUSTMENTS FOR SPEECH PRODUCTION

A sometimes overlooked, yet important, aspect of speech production is the biomechanical adjustments that occur prior to normal speech onset. Baken and his colleagues (Baken & Cavallo, 1981; Baken, Cavallo, & Weissman, 1979; Cavallo & Baken, 1985) have demonstrated that the prephonatory interval (approximately 100 ms before the acoustic realization of speech) is the time when coordinated respiratory, laryngeal, and supralaryngeal adjustments are made in preparation for a target utterance. The extent to which such coordinations are successful may be reflected ultimately in the intelligibility of the utterance. Of particular interest, here, is the biomechanical adjustment made by the chest wall during the prephonatory interval. Baken et al. (1979) have argued that chest-wall adjustment prior to speech onset is a veridical manifestation of the neuromotor organization of the respiratory apparatus. In normal-hearing persons, this adjustment involves an integrated simultaneous rib-cage expansion and abdominal contraction approximately 100 ms before the onset of phonation. Cavallo and Baken (1985) argue that this abdomen-to-rib cage volume shift is a speech-specific behavior that alters the biomechanical properties of the chest wall in such a manner as to facilitate the rapid aerodynamic pressure changes required for producing certain suprasegmental features.

Cavallo, Baken, Metz, and Whitehead (1991) investigated the patterns and timing of chest-wall postural adjustments and lung-volume management of seven congenitally severely-to-profoundly hearing-impaired males. Rib-cage and abdominal movements during the prephonatory interval were tracked with a mercury strain-gauge system (Baken & Matz, 1973) which was also used to estimate lung-volume change according to the procedures described by Baken (1977). The data from the hearing-impaired subjects were compared to data collected previously from eight normal-hearing males of comparable age (Baken et al., 1979). In both investigations, subjects were instructed to produce the vowel /ɑ/ as rapidly as possible following a stimulus cue.

The typical (90.9% of all responses) chest-wall posturing response of the normal-hearing individuals during the epoch following presentation of the stimulus cue and vocal response was composed of two distinct intervals. Immediately following stimulus presentation, ongoing ventilatory movements (either inspiratory or expiratory in direction) continued unaltered for about 250 ms. This epoch is labeled the "latency time" or "LT." Immediately following the LT, relative motion of the chest-wall components changes. The onset of these direction changes of the chest-wall components marks the beginning of the epoch labeled the "adjustment time" or "AT." The AT epoch is, on average, approximately 90 ms in duration and is terminated by phonation onset. During the 90 ms AT interval, a new chest-wall configuration is assumed and it is characterized by oppositional displacement of the two chest-wall components: rib-cage expansion is accompanied by a simultaneous contraction of the abdomen. The normal-hearing subjects did exhibit, infrequently, either purely inspiratory or purely expiratory adjustments (4.0% and 5.1% of the trials, respectively) during the AT, but the oppositional adjustment was clearly the dominant biomechanical pattern.

The Cavallo et al. (1991) study revealed that a surprising 86% of the chest-wall adjustments observed in the hearing-impaired individuals were qualitatively identical to the dominant chest-wall adjustment pattern observed in the Baken et al. (1979) study; that is, there was an oppositional shift of the chest-wall components. The only other chest-wall adjustment pattern observed by Cavallo et al. (1991) was a purely expiratory adjustment that occurred during 14% of the responses and was distributed among the subjects. Additionally, the time required to complete the AT phase was comparable in the hearing-impaired and normal-hearing subjects (104.1 ms and 93.3 ms, respectively).

Lung-volume adjustments (net increases or losses of volume) occur during chest-wall posturing prior to phonation. The direction of this volume change in normal-hearing persons is determined by the ventilatory phase in progress at the time of stimulus onset. Small lung-volume increases occur when phonation is elicited during the inspiratory phase of the tidal cycle and small lung-volume losses occur when phonation is elicited during the expiratory phase of the tidal cycle (Baken et al., 1979).

In sharp contrast to the above patterns, Cavallo et al. (1991) observed that the lung-volume adjustments of the hearing-impaired subjects during the AT was exclusively expiratory in direction, independent of ventilatory status at the time of stimulus presentation, and the net magnitude of lung-volume loss was considerable with respect to that of the normal-hearing subjects. On average, the hearing-impaired subjects lost 129.4 ml of lung volume during the AT compared with a 51.8 ml lung-volume change (net lung-volume loss or net gain) for the normal-hearing subjects (the net loss of lung volume associated with the expiratory ventilatory phase was 41.05 ml for the normal-hearing subjects).

These results suggest strongly that the neuromuscular organization of chest-wall preparation for phonation is intact in congenitally severely-to-profoundly hearing-impaired individuals. The qualitative and quantitative similarities of prephonatory chest-wall posturing are quite striking between the two groups of subjects. However, the observations of consistent and substantial lung-volume loss prior to the initiation of speech for the hearing-impaired subjects contrasts with the data for normal-hearing subjects. We will interpret the above findings in more detail later in this chapter in conjunction with respiratory kinematic data that also indicate abnormal lung-volume loss at speech initiation and during the connected utterances of deaf persons.

RESPIRATORY KINEMATICS

Considerable additional information regarding the speech breathing processes of deaf persons can be obtained by employing kinematic procedures designed to study respiratory apparatus behaviors (Hixon et al., 1973; 1976; Konno & Mead, 1967). Two studies conducted at the National Technical Institute for the Deaf, Rochester, NY (Forner & Hixon, 1977; Whitehead, 1983) directly address the kinematic details of chest-wall movement and relative lung-volume changes of deaf persons during speech production. Hixon et al. (1973; 1976) and Forner and Hixon (1977) discuss the theoretical and methodological aspects of such analyses, and as such, they will not be discussed in detail here, although a brief mention of the essential procedures underlying the technique is in order.

The centerpiece of respiratory kinematic analysis during speech is a magnetometer system that is capable of sensing independently anteroposterior changes of the rib cage and abdomen during speech (Hixon et al., 1973; 1976). Coils placed on the anterior midline surfaces of the rib cage and abdomen generate an electromagnetic field that is sensed by coils placed at the same axial levels on the posterior surfaces of the body. The strength of the signal received by the sensing coils varies inversely with the distance between the coil mates. Each part of the chest wall (i.e., rib cage and abdomen) exhibits a fixed shape at a given lung volume that is reflected by the relative signal strength between the coil mates. Assuming proper calibration of the magnetometer instrumentation, relative volume displacements can be determined by measuring the motion of the constituent parts of the chest wall. Graphic results of such analyses are displayed via relative motion diagrams in which the x-axis refers to abdominal shape change and the y-axis refers to rib-cage shape change. Examination of the slope of individual expiratory limbs (summation signals from the constituent chest-wall parts collected during speech) permits determination of the magnitude of independent rib-cage and abdominal contributions to lung-volume change, since the slope is proportional to the percentage contribution of both the rib cage and the abdomen. It should also be pointed out that net muscular activity (inspiratory or expiratory) can be inferred by contrasting chest-wall configuration deviations during speech to the chest-wall configuration assumed during relaxation of the respiratory muscles throughout the lung-volume range (Hixon et al., 1976; Weismer, 1985).

The first investigation that used magnetometry to study speech breathing processes of deaf persons was conducted by Forner and Hixon (1977). Ten severely-to-profoundly hearing-impaired college students enrolled at NTID served as subjects, all of whom had rated speech intelligibility levels that classified them as "unintelligible" speakers. Whitehead's (1983) investigation included 20 male college students: five normal-hearing *intelligible* speakers, five hearing-impaired *intelligible* speakers, and 10 hearing-impaired *unintelligible* speakers. In contrast to the Cavallo et al. (1991) study, the Forner and Hixon (1977) and Whitehead (1983) studies were concerned with respiratory function during a variety of connected speech tasks. We limit discussion of these two studies to findings regarding respiratory behaviors underlying continuous discourse, and we summarize major findings with only limited discussion of reported variability within and between the subjects and conditions. The specific respiratory behaviors of concern here are those discussed by Forner and Hixon (1977): a) the lung volumes used during connected speech, b) the relative volume of displacements of the rib cage and abdomen, and c) the separate volumes of the rib cage and abdomen.

Lung Volume

Forner and Hixon (1977) reported that their hearing-impaired subjects initiated and maintained speech within the midvolume range of the vital capacity. The majority of expiratory limbs were initiated above the functional reserve capacity (FRC) level with one-half being initiated from within the tidal volume range. Most limbs were terminated below FRC with four of the 10 subjects terminating their utterances well into the expiratory reserve volume.

Similar patterns of lung-volume usage were reported by Whitehead (1983). Speech initiation levels for Whitehead's normal-hearing subjects were, on average, 900 ml above FRC, in contrast to 750 ml and 125 ml above FRC for his intelligible and unintelligible hearing-impaired subjects, respectively. Expiratory limbs were typically terminated above FRC for the normal-hearing and intelligible hearing-impaired subjects in Whitehead's study, but the unintelligible speakers continued their utterances, on average, to 500 ml below FRC.

Forner and Hixon (1977) observed extreme variability in lung-volume excursions among subjects and across and within speaking tasks. Some lung-volume excursions were contained within the tidal breathing range whereas others consumed as much as two liters. Additionally, air wastage of magnitudes approaching 250 ml was frequently observed in the lung-volume excursion tracings. For example, one subject would initiate an utterance, then stop the utterance for a few seconds while continuing to expire air, then resume the utterance. Such behaviors could clearly contribute to the average syllabic air expenditures of 100 ml/syllable, several-fold greater than syllabic air expenditures of normal-hearing persons. Excessive air expenditure per syllable significantly limited the number of syllables produced during an uttered breath. Whitehead (1983), for example, reported an average of 14 syllables per expiratory limb for his normal-hearing subjects compared to three syllables per expiratory limb for his unintelligible hearing-impaired subjects.

Relative Volume Displacements of the Rib Cage and Abdomen

The majority of subjects in the Forner and Hixon (1977) study showed varying degrees of rib-cage predominance during connected discourse. Generally, there were no major dysynchronies observed between actions of the rib cage and abdomen. Occasionally, however, four of the ten subjects demonstrated paradoxical displacements of the rib cage and abdomen. Paradoxical behaviors are evidenced when the sign of the volume displacement for either the rib cage or abdomen is opposite that of the lung-volume change. The observed paradoxical behaviors were typically characterized by decreasing lung and abdomen volumes while rib-cage volumes increased.

Separate Rib-Cage and Abdominal Volumes

Forner and Hixon (1977) observed that the majority of the expiratory limbs were initiated at rib-cage volumes greater than those observed at the tidal end-expiratory level and abdominal volumes smaller than those at end-expiratory levels. More than half of the expiratory limbs were initiated at rib-cage levels greater than those at tidal end-inspiratory levels, consistent with the observations made by Cavallo et al. (1991). The majority of the expiratory limbs were terminated at rib-cage volumes greater than those at tidal end-expiratory level and abdominal volumes smaller than those at tidal end-expiratory level. Generally, utterances took place to the left of the tidal volume chest-wall configuration, indicating that the rib cage was relatively more expanded during speech than during tidal breathing and that the abdomen was relatively less expanded.

Collectively, the Forner and Hixon (1977) and Whitehead (1983) findings, as well as those of Cavallo et al. (1991) discussed earlier, suggest that some hearing-impaired speakers' most significant departures from normal speech breathing processes are related to the management of relative lung volumes, whereas other kinematic aspects of chest-wall activity are generally comparable with those of normal-hearing individuals. Lung-volume mismanagement was characterized predominantly by the tendency to initiate utterances at relatively low lung volumes; the tendency to terminate speech at levels below FRC; air wastage [observed also by Cavallo et al. (1991)]; and air expenditures per syllable that were several fold greater than those observed in normal-hearing individuals.

One final observation made by both Forner and Hixon (1977) and Whitehead (1983) was that many of the unintelligible deaf subjects paused at unexpected intervals during connected utterances for inspiratory refills. Normal-hearing individuals will, in general, pause for inspiratory refills at major (and sometimes minor) constituent boundaries in the utterance (i.e., clause and sentence endings), which is linguistically appropriate (Hixon, 1973; Grosjean & Collins, 1979). Many of the hearing-impaired subjects, according to Forner and Hixon (1977, p. 395) paused for abnormally long periods of time at junctures "other than those known to be linguistically appropriate (for example, in nonpunctuated word sequences)."

CONCLUSION

Inappropriate management of lung volumes during the speech of hearing-impaired persons seems likely to be a failure of coordination among the respiratory, laryngeal, and supralaryngeal subsystems supporting speech (Cavallo et al., 1991; Forner and Hixon, 1977; Harris & McGarr, 1980; Hudgins & Numbers, 1942; Itoh et al., 1982; Mahshie & Conture, 1983; Metz, Whitehead, & Whitehead, 1984; Whitehead, 1983; Whitehead & Metz, 1982). However, in spite of the wealth of evidence suggesting speech subsystem discoordination, there has been no systematic attempt to determine the specific influence of these aberrations on speech intelligibility.

Application of techniques to estimate laryngeal airway resistance (Hoit & Hixon, 1992; Smitheran & Hixon, 1981) in conjunction with respiratory kinematic and aerodynamic measurement techniques (Stathopoulos & Sapienza, 1993a, b) could be employed fruitfully in an attempt to delineate speech subsystem discoordinations among hearing-impaired individuals and relate them to degradations of speech intelligibility. Such research must, however, consider subject populations that differ in age, gender, early exposure to English and/or American Sign Language, degree of hearing loss, and age of hearing loss in order to improve the external validity of extant research findings in this area.

The techniques outlined above could also be employed to assess and/or track speech subsystem behaviors following cochlear implantation, not unlike those recently undertaken by Lane and his associates (Lane, Perkell, Svirsky, & Webster, 1991). Properly controlled physiological studies of young recent cochlear implant recipients could become the foundation for increasing our understanding of hearing loss in relation to speech-language production.

Finally, recent findings that show an effect of lung volume on VOT in normal-hearing speakers (Hoit, Solomon, & Hixon, 1993), provide unambiguous motivation to explore the well-documented segmental production errors of deaf persons at various temporal and lung-volume points in the expiratory cycle during speech. The application of advanced noninvasive measurement techniques could better elucidate the claim made over 50 years ago that reduced speech intelligibility among deaf persons is a function of "a lack of integration and coordination of the several component muscle groups that make up the complex speech mechanism" (Hudgins & Numbers, 1942, p. 379).

ACKNOWLEDGMENT

The late Dr. Joanne Davis Subtelny, a colleague and friend of Dr. Katherine Safford Harris, was instrumental in the initiation of research regarding the speech breathing processes of deaf persons at the National Technical Institute for the Deaf. We are indeed indebted to them for their pioneering efforts.

REFERENCES

Baken, R. J. (1977). Estimation of lung volume change from torso hemicumferences. *Journal of Speech and Hearing Research, 20,* 808-812.

Baken, R. J., & Cavallo, S. A. (1981). Prephonatory chest wall posturing. *Folia Phoniatrica, 33,* 193-203.

Baken R. J., & Matz, B. J. (1973). A portable impedance pneumograph. *Human Communication, 2,* 28-35.

Baken, R. J., Cavallo, S. A., & Weissman, K. L. (1979). Chest wall movements prior to phonation. *Journal of Speech and Hearing Research, 22,* 862-872.

Cavallo, S. A., & Baken, R. J. (1985). Prephonatory laryngeal and chest wall dynamics. *Journal of Speech and Hearing Research, 28,* 79-87.

Cavallo, S. A., Baken, R. J., Metz, D. E., & Whitehead, R. L. (1991). Chest wall preparation for phonation in congenitally profoundly hearing-impaired persons. *Volta Review, 93,* 287-300.

Draper, M. H., Ladefoged, P., & Whitteridge, D. (1959). Respiratory muscles in speech. *Journal of Speech and Hearing Research, 2,* 16-27.

Feudo, P., Jr., Zubick, H. H., & Strome, M. (1982). Air volumes during connected speech of normal-hearing and hearing-impaired adults. *Journal of Communication Disorders, 15,* 309-318.

Forner, L., L., & Hixon, T., J. (1977). Respiratory kinematics in profoundly hearing-impaired speakers. *Journal of Speech and Hearing Research, 20,* 373-408.

Grosjean, F., & Collins, M. (1979). Breathing, pausing, and reading. *Phonetica, 36,* 98-114.

Harris, K. S., & McGarr, N. S. (1980). Relationship between speech perception and speech production in normal-hearing and hearing-impaired subjects. In J. D. Subtelny (Ed.), *Speech assessment and speech improvement for the hearing impaired* (pp. 316-337). Washington, DC: A. G. Bell Association for the Deaf.

Hixon, T. J. (1973). Respiratory function in speech. In F. D. Minifie, T. J. Hixon, & F. Williams (Eds.), *Normal aspects of speech, hearing, and language* (pp. 73-125). Englewood Cliffs, NJ: Prentice-Hall.

Hixon, T. J., Goldman, M., & Mead, J. (1973). Kinematics of the chest wall during speech: Volume displacements of the rib cage, abdomen, and lung. *Journal of Speech and Hearing Research, 16,* 78-115.

Hixon, T. J., Mead, J., & Goldman, M. D. (1976). Dynamics of the chest wall during speech production: Function of the thorax, rib cage, diaphragm, and abdomen. *Journal of Speech and Hearing Research, 19,* 297-356.

Hoit, J. D., & Hixon, T. J. (1986). Body type and speech breathing. *Journal of Speech and Hearing Research, 29,* 313-324.

Hoit, J. D., & Hixon T. J. (1987). Age and speech breathing. *Journal of Speech and Hearing Research, 30,* 351-366.

Hoit, J. D., & Hixon, T. J. (1989). Speech breathing in women. *Journal of Speech and Hearing Research, 32,* 353-365.

Hoit, J. D., & Hixon, T. J. (1992). Age and laryngeal airway resistance during vowel production in women. *Journal of Speech and Hearing Research, 35,* 309-313.

Hoit, J. D., Solomon, N. P., & Hixon, T. J. (1993). Effect of lung volume on voice onset time (VOT). *Journal of Speech and Hearing Research, 36,* 516-520.

Hood, R., & Dixon, R. (1969). Physical characteristics of speech rhythm of deaf and normal hearing speakers. *Journal of Communicative Disorders, 2,* 20-28.

Hudgins, C. V. (1934). A comparative study of the speech coordinations of deaf and normal subjects. *Journal of Genetic Psychology, 44,* 1-48.

Hudgins, C. V., & Numbers, F. C. (1942). An investigation of the intelligibility of the speech of the deaf. *Genetic Psychology Monographs, 25,* 289-392.

Isshiki, N. (1964). Regulatory mechanism of vocal intensity variation. *Journal of Speech and Hearing Research, 7,* 17-29.

Itoh, M., Horii, Y., Daniloff, R. G., & Binnie, C. A. (1982). Selected aerodynamic characteristics of deaf individuals during various speech and nonspeech tasks. *Folia Phoniatrica, 34,* 119-209.

Konno, K., & Mead, J. (1967). Measurement of separate volume changes of rib cage and abdomen during breathing. *Journal of Applied Physiology, 22,* 407-422.

Ladefoged, P. (1974). Respiration, laryngeal activity and linguistics. In B. Wyke (Ed.), *Ventilatory and phonatory control systems* (pp. 299-306). London: Oxford University Press.

Lane, H., Perkell, J., Svirsky, M., & Webster, J. (1991). Change in speech breathing following cochlear implant in postlingually deafened adults. *Journal of Speech and Hearing Research, 34,* 526-533.

Mahshie, J. J., & Conture, E. G. (1983). Deaf speakers' laryngeal behavior. *Journal of Speech and Hearing Research, 26,* 550-559.

Metz, D. E., Samar, V. J., Schiavetti, N., Sitler, R. W., & Whitehead, R. L. (1985). Acoustic dimensions of hearing-impaired speakers' intelligibility. *Journal of Speech and Hearing Research, 28,* 345-355.

Metz, D. E., Schiavetti, N., Samar, V. J., & Sitler, R. W. (1990). Acoustic dimensions of hearing-impaired speakers intelligibility: Segmental and suprasegmental characteristics. *Journal of Speech and Hearing Research, 33,* 476-487.

Metz, D. E., Whitehead, R. L., & Whitehead, B. H. (1984). Mechanics of vocal fold vibration and laryngeal articulatory gestures produced by hearing-impaired speakers. *Journal of Speech and Hearing Research, 27,* 62-69.

Monsen, R. B. (1978). Toward measuring how well hearing-impaired children speak. *Journal of Speech and Hearing Research, 21,* 197-219.

Murphy, A., McGarr, N. S., & Bell-Berti, F. (1991). Acoustic analysis of stress contrasts produced by hearing-impaired children. *Volta Review, 93,* 80-91.

Rawlings, C. G. (1935). A comparative study of the movements of the breathing muscles in speech and quiet breathing of deaf and normal subjects. *American Annals of the Deaf, 80,* 147-156.

Rubin-Spitz, J., & McGarr, N. S. (1990). Perception of terminal fall contours in speech produced by deaf persons. *Journal of Speech and Hearing Research, 33,* 174-180.

Smitheran, J., & Hixon, T. J. (1981). A clinical method for estimating laryngeal airway resistance during vowel production. *Journal of Speech and Hearing Disorders, 46,* 138-146.

Stathopoulos, E. T., & Sapienza, C. (1993a). Respiratory and laryngeal function of women and men during vocal intensity variation. *Journal of Speech and Hearing Research, 36,* 64-75.

Stathopoulos, E. T., & Sapienza, C. (1993b). Respiratory and laryngeal measures of children during vocal intensity change. *Journal of the Acoustical Society of America, 94,* 2531-2543.

Weismer, G. (1985). Speech breathing: Contemporary views and findings. In R. G. Daniloff (Ed.), *Speech science* (pp. 47-69). San Diego: College-Hill Press.

Whitehead, R. L. (1983). Some respiratory and aerodynamic patterns in the speech of the hearing impaired. In I. Hochberg, H. Levitt, & M. Osberger (Eds.), *Speech of the hearing impaired: Research, training and personnel preparation* (pp. 97-116). Baltimore: University Park Press.

Whitehead, R. L., & Metz, D. E. (1982). The mechanics of abnormal laryngeal devoicing gestures exhibited by deaf persons. *Journal of the Acoustical Society of America, 71,* (S1), 555. (A)

Woldring, S. (1968). Breathing patterns during speech in deaf children. *American Annals of the Deaf, 155,* 206-207.

14 Respiratory Function in Stutterers

Peter J. Alfonso
The University of Illinois at Urbana-Champaign

INTRODUCTION

The primary respiratory demands for speech are essentially: 1) the suppression of the automatic metabolic motor plan by a voluntary speech motor plan (cf. Porter, Hogue, & Tobey, this volume), and 2) the achievement of a relatively narrow range of subglottal pressure targets in order to support a host of time-varying linguistic cues. The neuromotor solutions to either of these demands are not trivial. Shifts from metabolic to speech breathing involve general changes in the behaviors of not only the respiratory musculature but also certain laryngeal and supralaryngeal musculature from that of automatic and phasic activation to that of voluntary activation in specification of the variable movements of the structures of speech (e.g., Estenne, Zocchi, Ward, & Macklem, 1990; Onal, Lopata, & O'Connor, 1981a, b; Sawashima & Hirose, 1983). The voluntary activation of the respiratory musculature for speech is amended throughout the course of a speech utterance as a function of the differential between the subglottal pressure target and the available passive recoil pressure. There is no reason to believe that the speech motor control of the respiratory system is less complex than the speech motor control of the laryngeal and supralaryngeal

199

systems, although there are more published studies and presumably more interest in the latter. (For a detailed review of respiratory function in speech, see Hixon, 1987; Weismer, 1985.) Indeed, the literature on speech motor disorders in general, and in stuttering in particular, makes significant reference to aberrant respiratory behavior (e.g., Bloodstein, 1987; Van Riper, 1982), which is indicative of the inherent complexity of speech respiratory function.

EARLY STUDIES OF RESPIRATORY FUNCTION IN STUTTERERS

In the earliest days of our profession, clinicians and researchers observed that stuttering was accompanied by aberrant respiratory behavior. In fact, Bloodstein (1987) notes that "Disordered breathing is associated with stuttering so often and so conspicuously that it was one of the earliest factors to be investigated as a possible cause of stuttering." Beginning at about 1925, qualitative judgments were supported by quantitative physical measures through the use of the pneumograph. During the following next 30 to 35 years, a relatively large number, by contemporary standards, of respiratory experiments on stutterers was carried out, most of which used the pneumographic to transduce the movements of the chest wall. The majority of the early experiments reported that aberrant respiratory behaviors occurred more often in stutterers than in matched control subjects. Because the experimenters were not able to record the acoustic signal in parallel with the pneumographic signal permanently, it was difficult to determine whether aberrant respiratory behavior occurred only during overtly dysfluent speech or also during perceptually fluent speech. This and other technical difficulties aside, investigators reported that most aberrant respiratory behaviors produced by stutterers occurred during overtly dysfluent episodes, although a few of the early experiments concluded that aberrant respiratory behavior also occurred before or after dysfluent episodes (Van Riper, 1936), and during extended periods of perceptually fluent speech (Strother, 1937). Finally, it is worth noting that some of the aberrant respiratory behaviors observed in stutterers were also observed in the fluent speech of control subjects. Fossler (1930) concluded that the difference between the stuttering and control groups was in the relative frequency of the most commonly observed aberrant behaviors and in the extent of the departure from these behaviors by the stuttering group.

Table 1 summarizes the wide range of aberrant respiratory behaviors that was reported in these early experiments. It is important to note, first, that the early experiments documented that stutterers demonstrate different respiratory behaviors than controls during periods of metabolic breathing and speech breathing. Most notably, during metabolic breathing in stutterers, the early experiments reported that brief abnormal movements of the diaphragm occur throughout the respiratory cycle, and that stutterers are more variable than controls in inspiratory and expiratory phase duration and in the magnitude of lung-volume exchange. Thus, during speech breathing in stutterers, the early experiments reported that tremor occurs in the abdomen and thorax; the respiratory cycle is interrupted by tonic and clonic spasms; the temporal asynchrony between abdominal and thoracic movements is greater than that in controls; and, as was observed in metabolic breathing, stutterers are more variable than controls in inspiratory and expiratory phase duration and in the magnitude of lung-volume exchange. Stutterers as a group produced a wide range of aberrant respiratory behaviors, although idiosyncratic patterns were stable across time. Finally, there appeared to be no significant correlation between stuttering severity and specific patterns of aberrant respiratory behaviors.

TABLE 1. *Summary of early respiratory experiments on stutterers (1900-1960).*

I. Interruptions in respiratory cycle during fluent and dysfluent speech
    - Visible tremor in the abdomen and thorax during speech breathing
    - "Tonic and clonic spasms" as indicated by "flattening and fluctuating" pneumographic signals, respectively, during speech breathing
    - "Specific spasms" of the abdominal muscles and abdomen
      (Fletcher, 1914; Fossler, 1930; Gutzmann, 1908; Halle, 1900; Henrickson, 1936; Morley, 1937; Murray, 1932; Travis, 1927, 1936)

II. Inspiratory cycle
    - No group difference in duration of inspiratory phase or magnitude of inspiratory volume during metabolic breathing
    - Duration of inspiratory phase during speech breathing more variable and longer for stutterers than controls
    - Duration of inspiratory phase longer before dysfluent utterances compared to perceptually fluent utterances for stutterers
    - Magnitude of inspiration volume relative to expiration volume during speech breathing more variable for stutterers than controls
    - Stutterers attempted speech during inspiration
    - Stutterers produced "sharp inspiratory gasps" during speech breathing
      (Fletcher, 1914; Fossler, 1930; Halle, 1900; Henrickson, 1936; Hill, 1944; Morley, 1937; Murray, 1932; Ten Cate, 1902; Travis, 1927)

III. Expiratory cycle
    - No group difference in duration of inspiratory phase or magnitude of expiratory volume during metabolic breathing
    - Duration of expiratory phase during speech breathing more variable and longer for stutterers than control subjects
    - Expiratory phase duration shorter than inspiratory phase duration for stutterers, unlike controls
    - Stutterers "hold breath" after speaking
      (Fossler, 1930; Halle, 1900; Henrickson, 1936; Morley, 1937; Murray, 1932; Ten Cate, 1902; Travis, 1927)

IV. Temporal asynchrony
  - Inspiratory and expiratory phase duration and volume magnitude during metabolic breathing more variable for stutterers than control subjects
  - Asynchrony between thoracic and abdominal movements during speech breathing observed more often in stutterers than control subjects
  - Oppositional thoracic and abdominal movements observed more often in stutterers than in control subjects. (However, more recent studies have documented oppositional movements of the chest wall in control subjects. See, for example, Baken, Cavallo &, Weissman, 1979.)
  - "Disorganization of normal respiratory rhythm" in stutterers' speech breathing (e.g., the occurrence of respiratory gestures at inappropriate times during a speech utterance, such as inspiratory charge within a phrase)
    (Fletcher, 1914; Fossler, 1930; Gutzmann, 1908; Hill, 1944; Morley, 1937; Murray, 1932; Seth, 1934; Travis, 1927)
V. General observations
  - Brief abnormal diaphragmatic movements during metabolic breathing observed throughout the respiratory cycle in stutterers
  - Considerable variation in types of aberrant respiratory behavior during speech breathing among stutterers as a group
  - An individual stutterer's idiosyncratic aberrant respiratory behaviors during speech breathing are displayed consistently
  - No significant correlation between classes of stutterers (e.g., severity) and classes of respiratory aberrant behavior
  - Aberrant respiratory behavior that occurs during dysfluency can also prior to or following dysfluent episodes and during extended periods of perceptually fluent speech
    (Blackburn, 1937; Fletcher, 1914; Fossler, 1930; Murray, 1932; Schilling, 1960; Seth, 1934; Strother, 1937; Travis, 1927; Van Riper, 1936)

CONTEMPORARY STUDIES OF RESPIRATORY FUNCTION IN STUTTERERS

Ideally, a complete understanding of respiratory function in stutterers would be gleaned from the results of experiments that focus on: 1) metabolic breathing, speech breathing, and the transition from one mode to the other, and 2) electromyographic (EMG) activity of the respiratory musculature, chest wall kinematics and inferred lung-volume change, and the resultant subglottal pressure. It would also be useful to monitor the movements of the vocal folds and, in certain cases, the movements of supralaryngeal speech structures simultaneously with the movements of the chest wall, to determine the influence of laryngeal and vocal tract impedances on subglottal pressure and the resultant air flow. A second generation of studies of respiratory function, which began about 1970 as a result of technological advances, focused on some of these areas, and although the results of these studies have increased our understanding of the role that respiration plays in stuttering, much remains unresolved. For example, until recently very little work on ventilatory function and the shift from metabolic to speech breathing in stutterers has been carried out, even though it has been observed for many years that the occurrence of stuttering is greatest at speech onset. With the exception of one ongoing research project using surface EMG, which is discussed below, no EMG studies have been conducted of the respiratory muscles of stutterers. It would be particularly useful to have EMG studies of the intercostals and diaphragm, similar to that carried out, on singers (Leanderson, Sundberg, & von Euler, 1987); such studies should lead to a better understanding of the relation between the coordination of the respiratory musculature and stutterers' production of overtly dysfluent and perceptually fluent speech. There have been only three recent studies, each reviewed below, on the management of subglottal pressure in stutterers. Only a few experiments collected laryngeal and or supralaryngeal physiological data in parallel with respiratory data. On the other hand, much has been learned in recent years about chest wall kinematics and the inferred lung-volume change in stutterers, principally because of the availability of magnetometers, mercury strain gauges, and inductive plethysmography. (For a detailed discussion of these instruments, see Baken, 1987.)

The Management of Subglottal Pressure in the Speech of Stutterers

Subglottal pressure was measured directly by tracheal puncture in four stutterers and seven control subjects (Lewis, 1975), and recorded simultaneously with measures of oral pressure and flow. Although the planned-for within-stutterer comparison between dysfluent and perceptually fluent speech was not possible because of an insufficient number of dysfluent samples, Lewis was able to compare subglottal pressure in stutterers' perceptually fluent speech with that of the control subjects. In general, stutterers were found to generate lower of subglottal pressures and air flow rates than controls while reading aloud. By comparing subglottal pressure and flow during production of vowels and voiceless consonants in 28 CVC monosyllabic words embedded in reading passages, Lewis was able to infer the vocal tract DC impedance at the glottis during vowel production and at the oral constriction during syllable-initial voiceless consonant production. The results suggest that glottal and supraglottal resistances are greater for stutterers than for controls. Thus, the results of this study suggest that stutterers control the respiratory system in a clearly inefficient manner, that is, with lower driving force and with greater impedances to flow than do control subjects. The data also implicate aberrant functioning at laryngeal and supralaryngeal levels in parallel with respiratory dysfunction during stutterers' perceptually fluent speech production. Finally, these findings provide physiological support for a number of other observations of stutterers' speech production, including the ubiquitous observations made from acoustic reaction-time and VOT studies, and from studies employing more direct measures of vocal fold function, that the onset of vocal fold vibration in stutterers is delayed compared with data from control subjects.

Only one other study measured subglottal pressure directly in stutterers, in this case with a mini-pressure transducer positioned in the posterior commissure (Peters & Boves, 1988). Electroglottographic and speech acoustic signals were recorded simultaneously, with subglottal pressure. Patterns of subglottal pressure growth that were generated immediately before perceptually fluent production of isolated words were compared among ten stutterers and seven controls. Seven different categories of subglottal pressure growth, three of which were considered normal and four aberrant, differentiated the two groups. As a group, control subjects generally used the three normal patterns, in which the growth of subglottal pressure was smooth and monotonic, and the onset of phonation occurred simultaneously, or nearly simultaneously, with the peak in the pressure function or with a slight pressure drop, depending on the phonetic specification of the target word. As a group, stutterers generally used the four aberrant patterns. That is, stutterers produced smooth monotonic pressure-growth functions, but phonation onset occurred 100 ms or more after the pressure peak; or the pressure increase was to a level greater or less than the appropriate target pressure and phonation was delayed until appropriate subglottal pressure adjustments were made; or the subglottal pressure-growth functions were not monotonic.

More recently, a third study measured subglottal pressure indirectly in eight severe stutterers and five controls, with a balloon catheter positioned in the midesophagus and stomach (Zocchi, Estenne, Johnston, Del Ferro, Ward, & Macklem, 1990). Thus, in this study esophageal pressure was used to infer subglottal pressure, and the esophageal-gastric differential (transdiaphragmatic pressure) was used to infer diaphragmatic contraction. In addition, thoracic and abdominal wall movements were captured by means of Respitrace inductive plethysmography. Pressure was recorded during spontaneous conversation and during oral reading. All of the eight stutterers were dysfluent during production of the speech samples. An attempt was made to differentiate pressure recordings during dysfluent and perceptually fluent speech for only two of the eight stutterers. Thus, the results best reflect subglottal pressure management during dysfluent speech for the stutterers as a group. Whereas the control subjects maintained relatively constant subglottal pressure levels and a relaxed diaphragm during speech, subglottal pressures among the eight stutterers were both higher and lower than controls and fluctuated between the two extremes in an unpredictable fashion. Further, diaphragmatic contraction was observed in cases when dysfluent speech was identified.

Although the speech conditions in which subglottal pressure was measured directly or was inferred were quite different among the three experiments just discussed—that is, during the prephonatory period before single word utterances and during the production of spontaneous speech and oral reading—all of the studies observed that stutterers generate subglottal pressures in their dysfluent and perceptually fluent speech that differ significantly from controls. Taken together, the three studies indicate that in the production of utterances that are judged to be fluent by experienced listeners, stutterers: 1) generate subglottal pressure levels in preparation for speech in ways that differ significantly from, and presumably are less efficient than, those observed in controls; and 2) during the production of sustained speech, stutterers fail to maintain levels of subglottal pressure equal to that of control subjects. The problems associated with speech produced at inappropriately low levels of subglottal pressure are compounded further by: 1) having to meet the demands associated with high glottal and supraglottal resistances, and 2) inefficient expiratory air flow management associated with respiratory and laryngeal temporal asynchronies. During periods of dysfluent speech, stutterers failed to maintain a constant driving source, presumably because of discoordination of the respiratory musculature.

Chest Wall Kinematics and Lung-volume Change in Stutterers

The next set of experiments monitored the pre-phonatory posturing of the chest wall, that is, the chest wall kinematics during the period immediately before speech onset. The results of these studies shed some light on the ways in which stutterers' aberrant subglottal pressures are generated, although, as the following discussion indicates, certain questions remain unanswered about the relation between chest wall kinematics and aberrant subglottal pressures.

Baken, McManus, and Cavallo (1983) monitored thoracic and abdominal wall movements using Whitney gauges in five stutterers (see, Baken, 1987; Baken & Metz, 1973). The cue to initiate phonation for an isolated vowel was presented at random times during the tidal breathing cycle. Thus, the experimental protocol was designed to elicit phonation onset without the customary preparatory inhalation gesture to appropriate levels of the vital capacity. Rather, subjects were required to initiate phonation at vital capacity levels commensurate with their tidal breathing requirements and relative to the variable respiratory and laryngeal postures associated with the inspiratory or expiratory phases of their tidal breathing cycle. Although subjects were instructed to phonate as soon as possible upon presentation of an auditory tone, it was made clear that the intent was not to elicit minimal reaction time. Only data associated with perceptually fluent productions of the isolated vowels were analyzed and compared to data collected previously from eight control subjects using a similar protocol (Baken & Cavallo, 1980, 1981; Baken, Cavallo, & Weissman, 1979).

Analysis of the kinematic signals suggested to the experimenters that stutterers prepare the chest wall for phonation in a fashion that is "qualitatively identical" to that observed in the control subjects. Specifically, oppositional movements of the thoracic and abdominal walls, as opposed to both walls moving in the same direction, were observed consistently in both stutterers and controls. Oppositional chest wall movements are thought to increase rib cage stiffness and "tune" the diaphragm for speech (e.g., Weismer, 1985). However, other studies have shown that: 1) oppositional preposturing of the chest wall in control subjects does not occur as frequently as Baken suggests, and 2) it is difficult to determine the net effect of oppositional chest wall movements on lung-volume change (Hixon, Watson, Harris, & Pearl, 1988; McFarland & Smith, 1979).

Baken et al.'s (1983) analysis of the lung-volume signals revealed two significant group differences. First, stutterers usually lost lung volume before phonation, whereas control subjects lost lung volume only when the phonation stimulus was presented during the expiratory phase and gained lung volume when the phonation stimulus was presented during the inspiratory phase of the tidal breathing cycle. Second, the magnitude of the prephonatory volume depletion was greater for the stutterers than the controls. Because the qualitative scheme for classification of prephonatory chest wall movements (inspiratory, expiratory, or oppositional) did not differentiate the stutterers from controls, it was assumed that the lung-volume loss observed in stutterers was not related to respiratory dysfunction but rather to laryngeal dysfunction, that is, to the stutterers' delay in adducting the vocal folds for phonation. However, appropriate laryngeal data were not available to test this assumption.

A second and somewhat comparable experiment compared mild and severe stutterers' prephonatory chest-wall adjustments with those of control subjects (Watson, 1983; Watson & Alfonso, 1987). Here, the movements of the chest wall were transduced using Respitrace inductive plethysmography, and, in an important distinction between these and Baken et al.'s (1983) experiments, the simultaneous movements of the vocal folds were monitored using photoglottography. As in Baken et al. (1983) subjects were instructed to phonate an isolated vowel, although in Watson (1983) and Watson and Alfonso (1987) a simple reaction-time variable foreperiod protocol was used. Specifically, subjects were instructed to prepare for a known response at the presentation of a warning cue and to initiate phonation at the presentation of a phonation cue. The variable foreperiod is the temporal interval between the warning and phonation cues. Sufficiently long foreperiods provide subjects with adequate time to prepare for the known response. Thus, unlike the Baken et al. (1983) protocol, continuous prephonatory respiratory and laryngeal adjustments were experimentally segmented into two discontinuous prephonatory adjustment phases, preparation and initiation.

The results indicate that mild and severe stutterers differ in their pre-phonatory chest-wall and laryngeal adjustments, a result that supports previous acoustic RT experiments (Watson & Alfonso, 1982; 1983). During the preparation phase, mild stutterers and controls charge the respiratory system during short (100-700 ms) and long (900-2000 ms) foreperiods by thoracic and abdominal expansion, and increase the magnitude of the inspiratory charge as foreperiod increases. Severe stutterers, on the other hand, did not take advantage of the foreperiod to charge the respiratory system as often as did mild stutterers and controls, even at long foreperiods. During the initiation phase, both mild stutterers and controls initiate thoracic and abdominal compression before phonation onset. Severe stutterers, on the other hand, compress the abdominal cavity before phonation onset and compress the thoracic cavity after phonation onset. Mild and severe stutterers differed as well in regard to the latencies of the prephonatory adjustments. Severe stutterers' latencies for vocal fold adduction were significantly longer than those of controls at both short and long foreperiods. The delayed laryngeal adjustment, coupled with the poorly organized respiratory and laryngeal movements, resulted in greater loss of lung volume just before phonation for severe stutterers compared to either mild stutterers or controls. Mild stutterers, on the other hand, demonstrated delayed respiratory latencies, although the temporal organization of thoracic, abdominal, and laryngeal movements was similar to that of control subjects. Thus, at long foreperiods, when mild stutterers had sufficient time to prepare for phonation, their acoustic RT latencies approached normal values.

While the Baken et al. (1983) and the Watson and Alfonso (1987) experiments observed that stutterers lose lung volume immediately before phonation, they differ in their views of the role that prephonatory chest-wall adjustments play in this regard. Baken et al. (1983) concluded that the respiratory system, but not the laryngeal system, functions normally in stutterers, whereas Watson and Alfonso (1987) argue that neither the respiratory nor the laryngeal system functions normally in prephonatory adjustments, at least in the case of severe stutterers. Baken et al. (1983) relied on qualitative judgments of chest-wall posturing for speech and concluded that both stutterers and controls use similar oppositional chest-wall movements in preparation for speech in a consistent fashion, a conclusion that has not been supported by later research on control subjects. While the Watson and Alfonso (1987) experiment gathered data on nine stutterers, data were reported on only the four stutterers who fell unambiguously in the mild and severe categories. While it is difficult to reconcile the reported aberrant subglottal pressure data for stutterers without postulating a commensurate aberrant chest wall mechanism, the question of whether stutterers and controls differ in the way that they prepare the chest-wall for speech has not been resolved conclusively.

A final experiment is worth noting here because it suggests that aberrant habitual respiratory function in stutterers in amenable to speech therapy. A comparison of chest-wall kinematics during perceptually fluent phrase-length utterances immediately before and after completion of an intensive fluency enhancement program suggests that stutterers are able to modify the ways in which they habitually control the chest wall (Alfonso, Kalinowski, & Story, 1991; Story, 1990). Simultaneously monitoring chest wall, laryngeal, and lip and jaw movements, Story observed significant post-therapy alterations in speech movement characteristics across the three subsystems, alterations that coincided with post-therapy improvement in fluency. In regard to respiratory control, after successful completion of the therapy program, stutterers significantly increased inspiratory and expiratory volume, duration, and flow relative to the pretherapy values. In addition, fewer subperceptual respiratory abnormalities where observed post-therapy.

A NOVEL APPROACH IN THE STUDY OF RESPIRATORY FUNCTION IN STUTTERERS

Clearly then, aberrant respiratory behavior occurs in metabolic and speech breathing in stutterers; however, a critical question about respiratory function in stutterers remains unanswered. That is, how does the aberrant respiratory behavior relate to the aberrant behaviors at laryngeal and supralaryngeal levels during stutterers' dysfluent and perceptually fluent speech? And further, how can a mechanism that accounts for the motor disruptions associated with stuttering also take into account the majority, if not all, of the presumed factors (e.g., psycholinguistic and psychosocial) that affect the occurrence, frequency, and severity of stuttering (see, e.g., Alfonso, 1990; Smith, 1990; Smith & Weber, 1988; Zimmermann, 1980; Zimmermann, Smith, & Hanley, 1981)?

Recently, Smith and Denny have undertaken a novel approach to the study of respiration and the role it might play in dysfluent motor output across the speech production system by examining the patterns of bilaterally coherent, high-frequency oscillations that occur in the respiratory musculature (Denny, 1993; Denny & Smith, in preparation; Smith & Denny, 1990). Basically, the approach focuses on the neural brainstem circuitry that regulates metabolic breathing. Denny and Smith are exploring the possibility that deficits at this level of automatic motor control may contribute to the reported aberrant respiratory behaviors during metabolic breathing, as well as the reported aberrant respiratory, laryngeal, and supralaryngeal behaviors during speech breathing in stutterers.

Metabolic breathing requires precise coordination between the muscles of the abdomen and the thorax, and among the intrinsic and extrinsic muscles of the larynx and certain muscles of the supralaryngeal articulators so that the respiratory, laryngeal, and supralaryngeal structures behave as a functional unit. The requisite coordination is thought to be achieved by a common neural source, which most likely is the output of a brainstem central pattern generator (CPG) to the appropriate groups of motoneuron pools that underlie metabolic respiratory behavior. During speech production, the neural entrainment mediated by the CPG is suppressed or altered by voluntary neural inputs from the suprabulbar speech motor centers. Thus, the production of speech involves the overriding of automatic and entrained motoneural signals with voluntary motoneural signals to the same respiratory, laryngeal, and supralaryngeal effectors. Denny and Smith ask whether the discoordination observed at the respiratory, laryngeal, and supralaryngeal levels in stutterers' speech movements is associated with their inability to suppress the automatic metabolic signals during the period when the voluntary speech motor signals are executed.

Metabolic respiratory activity is usually associated with high-frequency oscillations in the 60-110 Hz band, which can be measured noninvasively with surface EMG of the diaphragm and the intercostal muscles. Evidence that the respiratory musculature is driven by a common CPG during metabolic breathing is supported by the observation of correlations in the power spectra between 60-110 Hz of the EMG signals from different respiratory muscles. The coherence statistic quantifies this relationship; that is, a high coherence value is indicative of correlated activity in the EMG spectra in the 60-110 Hz band among different respiratory muscles, which in turn suggests that the different muscles are driven by a common controller, in this example, the metabolic CPG. On the other hand, voluntary respiratory activity is not associated with high coherence in the 60-110 Hz band. For example, Smith and Denny (1990) have shown that normal subjects demonstrate significantly lower levels of coherence in speech breathing tasks compared to metabolic breathing. Thus, high coherence of the respiratory musculature in the 60-110 Hz band indicates that the metabolic CPG is dominant, whereas lower levels of coherence indicate that suprabulbar voluntary speech motor controllers are dominant.

The results of their work thus far on stutterers are complicated in that across subject variability is high, as is the case with the results of many other physiologically based experiments (e.g., Alfonso, 1991). While six of ten stutterers behaved like the majority of the controls, that is, they showed reduced levels of coherence during speech activity as compared to metabolic activity, four of the ten stutterers showed higher levels of coherence in the speech condition, a result not observed in any of the control subjects. Furthermore, the four stutterers who showed high levels of coherence in the speech condition demonstrated a wide range of fluency during the utterance, from perceptually fluent to severely dysfluent speech. Thus, the initial result seems to indicate that the occurrence of high levels of coherence, and by inference the high participation of the metabolic CPG during voluntary respiration, does not, in itself, predict the frequency or severity of dysfluent speech. Nevertheless, a significant group effect for coherence in the metabolic condition provides strong support for the notion that stutterers differ from controls in the way they manage the automatic and voluntary control systems for metabolic and speech respiration and suggests that some stutterers' voluntary speech motor plan is somehow disrupted or perhaps in competition with the metabolic respiratory control.

CONCLUSIONS

There is no doubt that the field has benefited from the physiologically based research conducted thus far on respiratory function in stutterers. We have learned that stutterers, as a group, do not have normal motor control of the respiratory system during metabolic breathing and during speech breathing in their dysfluent and perceptually fluent speech. It also appears quite likely that laryngeal, and perhaps supralaryngeal, deficits occur in parallel with aberrant respiratory function. The aberrant speech motor patterns that we observe throughout the speech production system are no doubt influenced by a multitude of other factors (e.g. psycholinguistic and psychosocial) that affect the occurrence, frequency, and severity of stuttering. In order to increase further our understanding of respiratory function in stutterers, we need to develop a testable framework that accounts for what we have observed, thus far, at the multiple physiological, linguistic, and psychological levels of speech. For example, we need to focus on those parts of the nervous system that have widely distributed outputs to the complete speech production system. Our experimental protocols should involve monitoring the state of the respiratory system in parallel with laryngeal and supralaryngeal states. We need to increase our focus on metabolic function and the shift from this state towards preparation and initiation of speech movements. We need to examine further the speech motor output that is associated with certain high risk linguistic markers, for example, phrase onset, where the likelihood of stuttering to occur is great. Clearly, study of the respiratory system in isolation, that is, in absence of a unified framework, no longer makes sense.

REFERENCES

Alfonso, P. J. (1990). Subject definition and selection criteria for stuttering research in adult subjects. In J. A. Cooper (Ed.), *Research needs in stuttering: Roadblocks and future directions. ASHA Reports 18* (pp. 15-24). Rockville, MD: ASHA Publications.

Alfonso, P. J. (1991). Implications of the concepts underlying task-dynamic modeling on kinematic studies of stuttering. In H. F. M. Peters, W. Hulstijn, & C. W. Starkweather (Eds.), *Speech motor control and stuttering* (pp. 79-100). Amsterdam: Elsevier Science Publishers.

Alfonso, P. J., Kalinowski, J. S., & Story, R. S. (1991). The effect of speech motor training on stutterers' speech physiology. In H. F. M. Peters, W. Hulstijn, & C. W. Starkweather (Eds.), *Speech motor control and stuttering* (pp. 513-526). Amsterdam: Elsevier Science Publishers. Also in pp. 111-122.

Baken, R. J. (1987). *Clinical measurement of speech and voice.* Boston, MA: College Hill Press.

Baken, R. J., & Cavallo, S. A. (1980). Chest wall preparation for phonation in untrained speakers. In V. Lawrence (Ed.), *Transcripts of the eighth symposium: Care of the professional voice* (Part 2). New York: Voice Foundation.

Baken, R. J., & Cavallo, S. A. (1981). Prephonatory chest wall posturing. *Folia Phoniatrica, 33,* 192-203.

Baken, R. J., Cavallo, S. A., & Weissman, K. L. (1979). Chest wall movements prior to phonation. *Journal of Speech and Hearing Research, 22,* 862-872.

Baken, R. J., McManus, D. A., & Cavallo, S. A. (1983). Prephonatory chest wall posturing in stutterers. *Journal of Speech and Hearing Research, 26,* 444-450.

Baken, R. J., & Metz, B. J. (1973). A portable impedance pneumograph. *Human Communication, 1973, 2,* 28-35.

Blackburn, B. (1931). Voluntary movements of the organs of speech in stutterers and nonstutterers. *Psychological Monographs, 41,* 1-13.

Bloodstein, O. (1987) *A handbook on stuttering.* Chicago, IL: The National Easter Seal Society.

Denny, M. E. (1993). *Respiratory high-frequency oscillations in normal and stuttering speakers.* Unpublished doctoral dissertation, Purdue University.

Denny, M. E., & Smith, A. (in preparation). Evidence for participation of metabolic respiratory control centers in stuttering.

Estenne, M., Zocchi, L., Ward, M., & Macklem, P. T. (1990). Chest wall motion and expiratory muscle use during phonation in normal humans. *Journal of Applied Physiology, 68,* 2075-2082.

Fletcher, J. M. (1914). An experimental study of stuttering. *American Journal of Psychology, 25,* 201-249.

Fossler, H. R. (1930). Disturbances in breathing during stuttering. *Psychological Monographs, 43,* 218-275.

Gutzmann, H. (1908). Die Athmungbewegung in ihrer Beziehung zu dem Sprachstorungen. *Monatschrift fur Sprachheilkunde, 18,* 179-201.

Halle, M. (1900). Ueber Storungen der Athmung bei Stottern. *Monatschrift fur Sprachheilkunde, 10,* 225.

Henrickson, E. H. (1936). Simultaneous recorded breathing and vocal disturbances of stutterers. *Archives of Speech, 36,* 133-149.

Hill, H. E. (1944). Stuttering II: A review and integration of physiological data. *Journal of Speech Disorders, 9,* 289-324.

Hixon, T. J. (1987). *Respiratory function in speech and song.* San Diego, CA: College-Hill Press.

Hixon, T. J., Watson, P. J., Harris, F. P., & Pearl, N. B. (1988). Relative volume changes of the ribcage and abdomen during pre-phonatory chest wall posturing. *Journal of Voice, 2,* 13-19.

Leanderson, R., Sundberg, J., & von Euler, C. (1987). Role of diaphragmatic activity during singing: a study of transdiaphragmatic pressures. *Journal of Applied Physiology, 68,* 259-270.

Lewis, J. I. (1975). *An aerodynamic study of "artificial" fluency in stutterers.* Unpublished doctoral dissertation, Purdue University.

McFarland, D. H., & Smith, A. (1992). The effects of vocal task and respiratory phase on pre-phonatory chest wall movements. *Journal of Speech & Hearing Research, 35,* 971-982.

Morley, A. (1937). An analysis of associated and predisposing factors in the symptomatology of stuttering. *Psychological Monographs, 49,* 50-107.

Murray, E. (1932). Dysintegration of breathing and eye-movements in stutterers during silent reading and reasoning. *Psychological Monographs, 43,* 218-275.

Onal, E., Lopata, M., & O'Connor, T. D. (1981a) Diaphragmatic and genioglossal electromyogram responses to CO_2 rebreathing in humans. *Journal of Applied Physiology, 50,* 1052-1055.

Onal, E., Lopata, M. & O'Connor, T. D. (1981b) Diaphragmatic and genioglossal electromyogram responses to isocapnic hypoxia in humans. *American Review of Respiratory Disease, 124,* 215-217.

Peters, H. F. M. & Boves, L. (1988). Coordination of aerodynamic and phonatory processes in fluent speech utterances of stutterers. *Journal of Speech and Hearing Research, 31,* 352-361.

Sawashima, M. & Hirose, H. (1983). Laryngeal gestures in speech production. In P. F. MacNeilage (Ed.), *The production of speech.* New York: Springer-Verlag.

Schilling, A. (1960). Rontgen-Zwerchfell-Kymogramme bei Stotterern. *Folia Phoniatrica, 12,* 145-153.

Seth, G. (1934). An experimental study of the control of the mechanism of speech and in particular that of respiration in stuttering subjects. *British Journal of Psychology, 24,* 375-388.

Smith, A. (1990). Factors in the etiology of stuttering. In J. A. Cooper (Ed.), *Research needs in stuttering: Roadblocks and future directions. ASHA Reports, 18* (pp. 39-47). Rockville, Maryland: ASHA Publications.

Smith, A., & Denny, M. (1990). High-frequency oscillations as indicators of neural control mechanisms in human respiration, mastication, and speech. *Journal of Neurophysiology, 63,* 745-758.

Smith, A., & Weber, C. M. (1988). The need for an integrated perspective on stuttering. *ASHA, 30,* 30-32.

Strother, C. R. (1937). A study of the extent of dyssynergia occurring during stuttering spasm. *Psychological Monographs, 49,* 108-127.

Story, R. S. (1990) *A pre-and post-therapy comparison of respiratory, laryngeal, and supralaryngeal kinematics of stutterers' fluent speech.* Unpublished doctoral dissertation, University of Connecticut, Storrs.

Ten Cate, M. J. (1902). Ueber die Untersuchung der Athmungsbewegung bei Sprachfehlern. *Monatschrift fur Sprachheilkunde, 12,* 247-259.

Travis, L. E. (1927). Studies in stuttering. I. Dysintegration of the breathing movements during stuttering. *Archives of Neurology and Psychiatry, 18*, 673-690.

Travis, V. (1936). A study of horizontal dysintegration of breathing during stuttering. *Archives of Speech, 1*, 157-169.

Van Riper, C. (1936). Study of the thoracic breathing of stutterers during expectancy and occurrence of stuttering spasm. *Journal of Speech Disorders, 1*. 61-72.

Van Riper, C. (1982). *The nature of stuttering.* Englewood Cliffs, NJ: Prentice-Hall, Inc.

Watson, B. C. (1983). *Simultaneous fiberoptic, transillumination, respitrace, and acoustic analysis of laryngeal reaction time in stutterers and nonstutterers.* Unpublished doctoral dissertation, The University of Connecticut, Storrs.

Watson, B. C., & Alfonso, P. J. (1982). A comparison of LRT and VOT values between stutterers and nonstutterers. *Journal of Fluency Disorders, 7*, 219-241.

Watson, B. C., & Alfonso, P. J. (1983). Foreperiod and stuttering severity effects on acoustic laryngeal reaction time. *Journal of Fluency Disorders, 8*, 183-206.

Watson, B. C. & Alfonso, P. J. (1987). Physiological bases of acoustic LRT in nonstutterers, mild stutterers, and severe stutterers. *Journal of Speech and Hearing Research, 30*, 434-447.

Weismer, G. (1985). Speech breathing: Contemporary views and findings. In R. G. Daniloff (Ed.), *Speech science: Recent advances* (pp. 47-72). San Diego, CA: College-Hill Press.

Zimmermann, G. (1980). Stuttering: A disorder of movement. *Journal of Speech and Hearing Research, 23*, 122-136.

Zimmermann, G., Smith, A., & Hanley, J. M. (1981). Stuttering: In need of a unifying conceptual framework. *Journal of Speech and Hearing Research, 24*, 25-31.

Zocchi, L., Estenne, M., Johnston, S., Del Ferro, L., Ward, M. E., & Macklem, P. T. (1990). Respiratory muscle incoordination in stuttering speech. *American Review of Respiratory Disease, 141*, 1510-1515.

15 Laryngeal and Extra-Laryngeal Mechanisms of F_0 Control

Kiyoshi Honda
ATR Human Information Processing Research Laboratories

INTRODUCTION

The characteristic patterns of vocal fundamental frequency (F_0) in speech are derived from the morphology of the human speech organs. Detailed illustrations of the larynx and supralaryngeal organs appear in the literature of anatomy and physiology, yet the mechanisms underlying phonetic diversity are not yet comprehensible. Phonological descriptions or current models of prosody suggest that the rules controlling F_0 are rather simple, while the relevant anatomy and physiology are overwhelmingly complex. For instance, laryngeal mechanisms that are used to tune vocal fold vibrations to conform to the rules of intonation are also involved in the adduction and abduction of the glottis for control of phonation. Further, their components are coordinated in anticipation of supra-laryngeal influences of articulatory movements. This chapter summarizes the anatomical and physiological bases of the multidimensionality of laryngeal and extra-laryngeal mechanisms involved in F_0 control. The approach used to describe the mechanisms involves morphological considerations of human laryngeal structures on the one hand, and interpretation of electromyographic data from the relevant muscles on the other.

Evolution of Laryngeal Function

Animal vocalization is a by-product of laryngeal function for protecting the airway from a foreign body or a bolus of food. In arboreal monkeys, hermetic closure of the glottis is used for air-trapping to support the thoracic cage for brachiation (Negus, 1949). The human larynx has evolved from a regulator for these animal functions to an acoustic apparatus used to generate a quasiperiodic sound source for human speech. The ballistic glottal action for brachiation is not a vital mechanism in terrestrial primates, and has been altered functionally into regulated actions for continuous vocal fold vibration in humans. Morphological changes in the vocal folds have also enhanced the capabilities of human vocalization. According to Fink (1962), the vocal folds are mainly ligamentous in most mammals, but human vocal folds consist almost entirely of muscle. Although this is not entirely accurate because some terrestrial primates do have human-like vocal folds, his suggestion that the evolutionary change from ligamentous to muscular vocal folds represents a significant advance in the efficiency of vocalization is still valid. These functional and morphological alterations of the sound generator have been further enhanced by the metamorphosis of the other vocal tract organs. The lower placement of the human larynx contributes to forming the uniquely human vocal tract structure with its wide pharynx. Obviously, the sound source acquired by humans is related to the entire set of morphological changes in the human body characterized by encephalization and bipedalism.

Figure 1 shows cross-sectional sketches of the orofacial structures of a human and a macaque. Most mammals have a larynx with a unified hyoid-thyroid complex, which inhibits independent control of tongue and larynx movements.

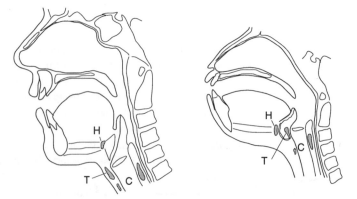

FIGURE 1. The morphologies of human and macaque speech organs. The separation of the thyroid cartilage (T) from the hyoid bone (H) provides the morphological basis for human speech, which is composed of segmental articulation superimposed on the melody of vocal fold vibration. The cricoid cartilage in humans (C) shows a caudal shift along the anterior curvature (lordosis) of the cervical spine.

In humans, the descent of the larynx and its separation from the tongue provide a morphological advantage for continuous vocal fold vibration during large displacements of the tongue and jaw. These structures are the morphological bases of human speech production, which includes segmental articulatory sequences coproduced with the melody of vocal fold vibration. The original biomechanical and neural connections among the structures still remain, however, and they provide some interaction between phonation and articulation.

Gross Anatomy of the Larynx with Respect to F_0 Control

The morphological characteristics of the larynx and its relation to the surrounding structures are critical to the control of F_0 by the musculoskeletal laryngeal system. Figure 2 depicts the anatomy relevant to the mechanisms of F_0 control. The laryngeal framework consists of the hyoid bone, the thyroid cartilage, and the cricoid cartilage. The thyroid cartilage, which encloses the vocal folds, is a new structure appearing in mammals; it articulates with the dorsocaudal corner of the cricoid cartilage through slender bilateral inferior horns. In humans, the thyroid cartilage is detached from the hyoid bone, but they are linked externally by the thyrohyoid muscle, membrane, and ligaments (the median and lateral thyrohyoid ligaments), and internally by the soft tissue supporting the epiglottis.

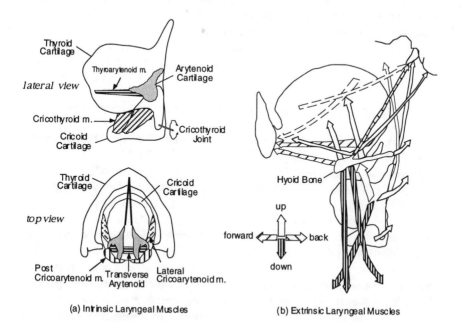

FIGURE 2. Gross anatomy of the larynx with respect to F_0 control, showing the intrinsic muscles (a) and the extrinsic muscles (b).

The superior horns of the thyroid cartilage stabilize the thyroid cartilage through the attachments of the para-pharyngeal muscles to this structure. The cricoid cartilage has a triangular shape in the lateral view with a thin arch anteriorly and a wide plate facing the cervical spine. In addition to direct and indirect muscular force, the cricoid cartilage receives force from the caudal pull of the trachea and the ventral push of the cervical spine via the post-cricoid soft tissue.

The laryngeal muscles are divided into two groups: the intrinsic laryngeal muscles that interconnect the laryngeal cartilages and the extrinsic laryngeal muscles that support the hyoid-larynx complex externally. The primary functions of the intrinsic muscles, shown in Figure 2(a), are the adduction and abduction of the glottis by antagonistic actions of the adductor muscles: the thyroarytenoid (TA), the lateral cricoarytenoid (LCA), and the transverse arytenoid (TrA); and the abductor muscle, the posterior cricoarytenoid (PCA). The second function of the intrinsic muscles is the rotation of the cricothyroid joint, for respiration, deglutition, and F_0 control. This rotation is accomplished by the antagonistic actions of the CT and the TA.

The extrinsic laryngeal muscles are numerous and their functions are varied. These muscles include not only the ones attached directly to the laryngeal structure, but also the extrinsic tongue muscles, when grouped by laryngeal-positioning functional category, as illustrated in Figure 2(b). The primary function of the extrinsic muscles is to control the vertical motion of the larynx via the vertical components of muscular forces. The secondary function is to maintain the horizontal position of the larynx, by acting on the hyoid bone and the thyroid cartilage. The actions of these muscles for speech articulation, as well as the effects of the ligamentous network connections, are believed to cause various interactions between the larynx and the supralaryngeal articulators.

INTRALARYNGEAL MECHANISMS OF F_0 CONTROL

In general, the frequency of vibration of a string or a bar is determined by the mass, the tension, and the length of the vibrating materials. In this account, these parameters are independent factors of the frequency of vibration. In the case of vocal fold vibration, these parameters are not independent but in fact interact with each other. It is well known that the principal biomechanical factor in the regulation of F_0 is the tension of the vocal folds. It is also believed that vocal fold tension is regulated by vocal fold length, which is determined by the angle of the cricothyroid joint. In the case of strings, longer vibrating materials have a lower frequency of vibration when the tension is constant. However, longitudinal stretching of the vocal folds causes a reduction in the mass per unit of length in addition to an increase in vocal fold tension. This effect overcomes the lengthening of vibrating material, resulting in a higher frequency of vibration.

The other intralaryngeal mechanisms of F_0 control, whose descriptions are rather obscure in the literature, are the regulation of effective length of vibration by medial compression forces, and vocal fold stretching by posterior sliding of the arytenoid cartilage. All of these tension-control mechanisms interact directly with subglottal air pressure to determine the dimensions of sound, i.e., frequency, intensity, and quality. Figure 3 illustrates possible functions of the intrinsic laryngeal muscles.

Cricothyroid Rotation

The rotation of the cricothyroid joint has been attributed to the action of the CT. This muscle runs from the arch of the cricoid cartilage to the inferior ridge of the thyroid cartilage. The main action of this muscle is the rotation of the cartilages around the cricothyroid joint. The rotation caused by the CT increases the distance between the thyroid and arytenoid cartilages, lengthening the vocal folds. Among the intrinsic laryngeal muscles, the CT is the largest muscle in length as well as in cross-sectional area. The contraction of the CT generates the largest torque on the joint, which can easily overcome the antagonistic action of adductors like the vocalis (Sonesson, 1983). The CT has two anatomical subdivisions: the *pars recta* (anterior portion) and the *pars obliqua* (posterior portion). In many textbooks, these subdivisions are described as causing different motions of the joint: rotation by the *pars recta* and translation by the *pars obliqua* (e.g., Fink & Demarest, 1978). Questions remain, however, with respect to the likelihood of joint translation because there are two opposing views: a negative view from a morphological investigation of the ligaments that stabilize the joint (Maue & Dickson, 1971), and a positive view from X-ray observations of singers' larynges (Sonninen, 1968).

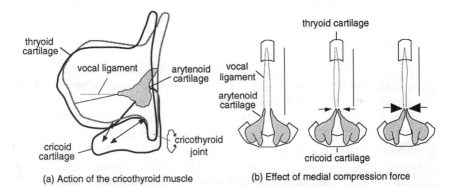

(a) Action of the cricothyroid muscle   (b) Effect of medial compression force

FIGURE 3. Functions of the intrinsic laryngeal muscles for F_0 control. The cricothyroid (CT) stretches the vocal folds by a rotation of the cricothyroid joint (a); while the adductor muscles, like the lateral cricoarytenoid, regulate effective length of vibration (b), which is represented by the lines to the right of the vocal ligaments.

Laryngologists have observed relative axial mobility of the two cartilages by manual operation of fresh cadaver larynges (Kirshner, *personal communication*). This mobility suggests a possibility of translation without joint dislocation because of the pliability of the inferior horn of the thyroid cartilage.

The two parts of the CT run in more-or-less oblique fashion despite the anatomical nomenclature for the *pars recta*. This orientation of the muscle is biomechanically inefficient for generating joint torque because the direction is not tangential to a momentum circle around the joint. This orientation does, however, provide a larger range of action, since, if the orientation were tangential, the muscle would be shorter. This would restrict rotation and counter-rotation, by contraction and relaxation, respectively. The original function of the CT was probably to stretch the vocal folds in maintaining the glottal airway during respiration. The oblique orientation is useful in achieving a large counter-rotation of the joint used during sphincteric glottal closure in deglutition. From a teleological perspective, it appears that the CT runs obliquely to maximize the range of vital action rather than the efficiency of F_0 control.

The EMG studies reporting CT activity in F_0 raising are numerous (Arnold, 1961; Atkinson, 1978; Gay, Strome, Hirose, & Sawashima, 1972; Hirano, Ohala, & Vennard, 1969; Shipp, Doherty, & Morrissey, 1979). These studies consistently support the proposition that the CT is the main F_0 control muscle. This belief is partly derived from laryngeal anatomy, which implies the self-evident mechanism of the joint rotation by CT contraction. The correlation between CT activity and F_0 is generally high in short utterances; it becomes less obvious in longer utterances. Considering the physiological data, along with the anatomical orientation of this muscle, the contribution of CT activity to F_0 control may be less definite than warranted by the previous descriptions. In addition to its F_0 control function, it is known that the CT also functions in voicing control (Löfqvist, Baer, McGarr, & Story, 1989). It is likely that a sudden stretch of the vocal folds by CT contraction enhances devoicing.

Other Intrinsic Mechanisms

Except for the CT, the intrinsic muscles function primarily for adduction (closing) and abduction (opening) of the glottis. There are three adductors (TA, LCA, and TrA) and one abductor (PCA). It can be assumed that these muscles contribute to F_0 control by their various effects on the parameters regulating F_0.

The TA muscle is conventionally described as comprising the medial thyroarytenoid (the vocalis muscle) and the lateral thyroarytenoid. The vocalis runs parallel to the vocal ligaments from the thyroid cartilage to the vocal process. The lateral thyroarytenoid inserts into the arytenoid cartilage near the muscular process. The contraction of the TA as a whole can shorten the vocal folds and serves as an antagonist to the CT with respect to the rotation of the cricothyroid joint. Conversely to the expected F_0 lowering effect, the TA can contribute to F_0 increase by stiffening the body of the vocal folds. These two effects are obviously opposite and may become effective in different situations: F_0 lowering in isotonic contraction and F_0 raising in isometric contraction. The electromyographic observations of the TA consistently indicate an increase in TA activity with an increased F_0 (Atkinson, 1978; Gay, Strome, Hirose, & Sawashima, 1972; Hirano, Ohala, & Vennard, 1969; Shipp, Doherty, & Morrissey, 1979). Fink (1962) believes that the TA may develop considerable tension during active isometric contraction, depending on the initial length. Titze (1989) shows evidence that an increase in F_0 is achieved by TA contraction in a low range of F_0. In several animal experiments that measured muscle response time, the TA was observed to be the fastest muscle (Sawashima, 1974, for review). This suggests that the TA may produce a rapid change in F_0 in certain situations.

The LCA, along with the lateral thyroarytenoid muscle, demonstrates increased activity for high F_0. The LCA, the smallest muscle, is believed to have some F_0 raising effect through the medial compression force on the vocal processes (Broad, 1973; Honda, 1985). The length of the vibrating portion of the vocal folds can be changed by increasing the force of medial compression, even when the anatomical length of the vocal folds remains unchanged. A tight compression of the left and right vocal processes causes a forward shift of the posterior end of vibration of the membranous part of the vocal folds, which results in shorter effective length of vibration (Arnold, 1961). This is a control for the length of vibration, independent of the other tension control mechanisms.

The PCA, the only abductor muscle of the larynx, not only functions for the glottal motion, but is also speculated to stretch the vocal folds by pulling the arytenoid cartilage backwards (Hollien, 1983). This effect needs to be assisted by other muscular components because excessive force of PCA contraction can cause glottal opening. The adductor muscles, such as the TrA muscle, appear to contribute to the action of the PCA by moving the arytenoid cartilages dorsocranially along the cricoarytenoid joints.

Extralaryngeal Mechanisms of F_0 Control

The muscles and the rigid structures supporting the laryngeal framework also function in F_0 control by supplying various external forces that modify the laryngeal configuration. Besides the action of the CT, the rotation of the cricothyroid joint is also produced by the extrinsic laryngeal muscles and the motion of the supporting structures. Since the mutual rotation of the cricoid and the thyroid cartilages occurs around the transverse axis of the joint, it is affected by any external force applied to these laryngeal cartilages in the sagittal orientation. This function of the extrinsic laryngeal muscles is described as "the external frame function" by Sonninen (1968). Several hypotheses have been proposed to account for F_0 control mechanisms by the external forces, although the details of the cause-effect relationship remain obscure.

The extrinsic laryngeal muscles are divided into two anatomical groups depending on their locations relative to the hyoid bone: the suprahyoid muscles and the infrahyoid muscles. The suprahyoid muscles are the jaw muscles, the tongue muscles, and the group of para-pharyngeal muscles. The infrahyoid muscles are the so-called "strap muscles." Most of these muscles have attachments to the hyoid bone, with a few exceptions (the sternothyroid muscle and two of the pharyngeal constrictors). The human hyoid bone is a unique structure that is connected to the rest of the skeleton by muscular and ligamentous network--that is, without articulations. The position of the hyoid bone is determined by the conditions of the other rigid structures and the relative balance of the forces in the entire system. The primary function of the hyoid bone in mammals is to elevate the larynx in deglutition. It is assumed that the role of the hyoid bone in speech production is to support the laryngeal cartilages. Hyoid bone movements, observed both in rotation and translation during dynamic speech gestures, are relatively independent of jaw motions (Westbury, 1988).

Extrinsic Laryngeal Muscles

The role of the extrinsic laryngeal muscles in F_0 control has been examined in many EMG studies (Erickson, 1983; Maeda, 1976; Sawashima, Kakita, & Hiki, 1973; Simada & Hirose, 1970). Most of the studies are concerned with the relationship between strap muscle activity and F_0 changes for accent and during intonation. The strap muscles, strictly defined, are the sternohyoid (SH), the sternothyroid (ST), and the thyrohyoid (TH) muscles. Anatomically, the SH and the ST are considered to be laryngeal depressors, and the TH to be a laryngeal elevator. Many researchers have been interested in the effect of these muscles' activity on F_0 lowering, since F_0 lowering is usually accompanied by larynx lowering. Questions remain, however, regarding the chain of mechanisms that account for vocal fold stretching or relaxation by vertical laryngeal movement.

Figure 4 depicts some potential mechanisms of F_0 control by the extrinsic laryngeal muscles. Among the strap muscles, the SH has been examined by many researchers in relation to F_0 lowering. The SH consistently shows various degrees of negative correlations (weak to moderate) with F_0 during speech utterances in various languages (Atkinson, 1978; Erickson, 1993; Simada & Hirose, 1970). In word accent, the activity of the CT is often followed by the activity of SH (Erickson, Baer, & Harris, 1983; Simada, Niimi, & Hirose, 1991). In sentences, the SH shows increased activity in the lower region of the F_0 range, mostly near the end of declarative sentences. Despite these local inverse correlations between SH activity and F_0, the integrated EMG curves of the SH do not resemble the F_0 curves (intonation patterns) in speech utterances. As a result, a high negative correlation is not necessarily observed between them (Atkinson, 1978). This inconsistency is found primarily because the SH also exhibits articulatory activities for jaw opening and tongue retraction gestures.

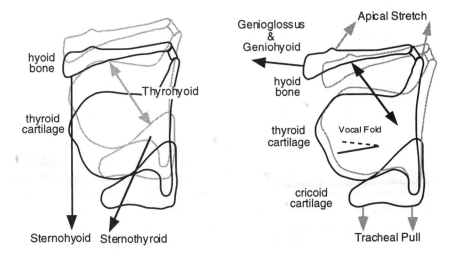

FIGURE 4. Actions of the extrinsic laryngeal muscles used for F_0 control. The strap muscles cause various effects on laryngeal configuration in the equilibrium of muscle contraction forces (a). The suprahyoid muscles and the tongue muscles exert a force to move the hyoid bone forward (b).

Under experimental conditions, the SH shows activity in test utterances containing open vowels, while it is nearly inactive for minimally contrastive sentences with close vowels.

The sternothyroid (ST) and the thyrohyoid (TH) muscles have also been examined by EMG studies. The ST is another laryngeal depressor, but EMG observations of this muscle do not show a consistent relation with F_0. The ST exhibits activity correlated with F_0 changes in speech (Atkinson, 1978; Maeda, 1976; Sawashima, Kakita, & Hiki, 1973) and in singing (Faaborg-Anderson & Sonninen, 1960). The muscle also shows a synchronous increase of activity with the CT in vibrato (Niimi, Horiguchi, & Kobayashi, 1991). The ST has an attachment to the thyroid cartilage and the contraction of this muscle should exert a direct influence on larynx lowering, thus contributing to F_0 lowering (Maeda, 1976). On the other hand, it is supposed that this muscle can cause F_0 raising by direct rotation of the thyroid cartilage (Sonninen, 1968). The attachment of this muscle to the thyroid cartilage is located on the oblique line anterior to the cricothyroid joint. Therefore, when the position of the cricoid is fixed, the ST can rotate the thyroid cartilage and stretch the vocal folds (Niimi, Horiguchi, & Kobayashi, 1991). This action probably requires co-contraction of the TH to pull the thyroid cartilage forward.

The thyrohyoid (TH) muscle connects the hyoid bone and the thyroid cartilage, and the contraction of this muscle causes approximation of these two rigid structures. Since both of them are floating, their anatomically expected muscle contractions are relative to each other. When the hyoid bone is fixed, TH contraction would presumably raise the thyroid cartilage. Likewise, when the thyroid cartilage is fixed, TH contraction should lower the hyoid bone. TH activity shows a pattern that follows the F_0 contour with segmental ripples for jaw opening (Sawashima, Kakita, & Hiki, 1973). This is consistent with the anatomical considerations.

The suprahyoid muscles have also been examined for their roles in F_0 control. The "functional chain" proposed by Zenker (1964) appears to be a possible function of the extrinsic laryngeal muscles for F_0 control. One of the component muscles in the chain is the geniohyoid (GH) muscle, which is known to show F_0 related activity (Erickson, Liberman, & Niimi, 1977). The function of the GH is to pull the hyoid bone anteriorly, and a forward shift of the hyoid bone applies a force to the thyroid cartilage to rotate along the cricothyroid joint in the direction that stretches the vocal folds (Honda, 1983). Thus, the GH can contribute to F_0 raising through a horizontal component of the muscular force on the hyoid-larynx complex. The counterbalancing force to the anterior pull seems to be the recoiling force of the hyoid-larynx complex. This structure receives vertical stretch from the tracheal and strap muscles, as well as from the vertically oriented suprahyoid muscles. The contribution of the pharyngeal constrictors to the recoil may be small because these muscles' contractions are mainly used for reducing the pharyngeal width in the coronal plane.

Tongue muscle activities are also involved in hyoid bone movements. The posterior genioglossus (GGP) muscle, which runs parallel to the GH, also shows increased activity for F_0 raising. The hyoglossus and the styloglossus show increased activity for F_0 lowering. The mechanism of these extrinsic tongue muscles for F_0 control can be accounted for by the horizontal force on hyoid bone displacement. The other supra-hyoid muscles, the mylohyoid and the anterior digastric, are not well understood with respect to F_0 control.

Another hypothesis by Zenker (1964), regarding a mechanism of F_0 control by external muscle forces, is that the action of the cricopharyngeus (CP) muscle pulls the cricoid cartilage posteriorly. The CP is the esophageal sphincter, and has an attachment to the cricoid cartilage near the cricothyroid joint. This muscle maintains a tonic continuous discharge to function as a sphincteric muscle and shows some increased activity as F_0 decreases. Zenker's explanation of the role of the CP in F_0 lowering is the rotation of the cricoid cartilage through a posterior pull from its point of attachment. However, the attachment of the CP on the cricoid cartilage is too close to the cricothyroid joint to generate an efficient torque on the joint. The alternative account of CP function by Honda and Fujimura (1991) concerns the thickening of the muscle belly in the post-cricoid space. Figure 5 shows the speculated effect of this thickening, which applies a force to push the apex of the cricoid cartilage forward and thus shorten the vocal folds. Both of the hypotheses need further experimental verification.

Vertical Laryngeal Movement

Vertical laryngeal movement associated with F_0 changes may be the most well-known observation of changes in laryngeal configuration. The larynx tends to be higher for higher F_0, and vice versa. EMG studies of strap muscle function have been conducted based on the presumed relationship between vertical larynx movement and F_0 change. Surprisingly, this well-known observation lacks a convincing causal explanation. Kakita and Hiki (1974), measuring the vertical movements of the larynx and the EMG of the strap muscles, found high correlations among F_0, vertical laryngeal movement, and activity of strap muscles. They made an anatomical model based on these data, in which the TH and the ST control vertical position of the larynx. The vertical translation of the cricothyroid joint is hypothesized in the model to account for the relationship between vertical laryngeal movement and vocal fold length.

Recent studies propose a plausible account for the causal relationship between vertical laryngeal movements and F_0 (Hirai, Honda, Fujimoto, & Shimada, 1994; Honda, Hirai, & Kusakawa, 1993). They used a magnetic resonance imaging (MRI) technique to measure the movements of the hyoid-larynx complex during sustained phonation at various F_0 levels. In their observation, the cricoid cartilage rotates with the vertical laryngeal movements while it slides along the anterior curvature (lordosis) of the cervical spine. The angle of the posterior plate of the cricoid cartilage remains parallel to the tangential line along the curvature of the cervical spine as the position of the cricoid cartilage changes. Figure 6 illustrates the mechanism by which vertical laryngeal movements are translated into changes in vocal fold length.

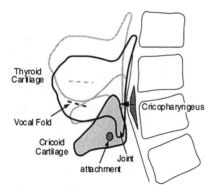

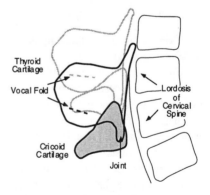

FIGURE 5. Speculation of the effect of the cricopharyngeus (CP) muscle. The contraction of the semicircular muscle produces a bulging of muscle belly, which pushes the cricoid cartilage forward.

FIGURE 6. Effect of vertical laryngeal movements on vocal fold length. Vertical movements of the cricoid cartilage along the cervical lordosis provide a control of vocal fold length.

This mechanism explains bi-directional effects for F_0 control, but it seems to be more effective for F_0 lowering than for F_0 raising. The human spine shows a sigmoid curvature with lordosis and kyphosis in cycle, and this is described as one of the results from human bipedalism and encephalization. The metamorphosis of the human body has resulted in laryngeal descent towards the point of the cervical spine's maximum lordosis, contributing to the efficiency of F_0 control in a wide frequency range.

Implications of Linguistic Problems

In spoken language, vocal fold vibration provides a meaningful distinction in speech utterances by local rise-fall patterns of F_0. It also provides information about linguistic structure and continuity to speech sounds beyond segments, connecting or disjoining words or phrases in a sentence. Sentential F_0 curves display many ripples on long-term downdrift with frequent breaks. In contrast, the auditory impression of such a pattern resembles a spotlight shed on particular sounds in a continuous flow of tonal melody. This may be crucial to the retrieval of the speaker's intention from the spoken message.

Vocal fold vibration also provides another dimension of F_0 used for speech, which is segmental F_0 information by articulatory perturbation. The universal phenomenon of "intrinsic vowel F_0" appears to contribute to vowel identification by a consistent tendency for high vowels to have a higher F_0. The "inverse effect" of the intrinsic vowel F_0, that is, the variation of vowel formant frequencies in different F_0, may also contribute to the robustness of vowel identification even in conditions involving deviation from normal vowel articulation.

Accent/Emphasis and Intonation

Both lexical (local) and sentential F_0 patterns in speech are determined by the laryngeal and extra-laryngeal mechanisms for F_0 control. A typical muscular strategy consists of the activation of the CT and other intrinsic muscles, commonly followed by activation of the strap muscles, to produce rise-fall patterns for accent and emphasis. Since the CT is a short muscle directly attached to the laryngeal cartilages, its action is relatively fast and effective for producing a local F_0 rise. In contrast, F_0 lowering muscles are mostly large, and the action of these muscles is slow and prolonged, involving the movements of the entire larynx. The size difference between the F_0 raising and lowering muscles may be part of the biological basis of such phonetic tendencies as "deferred fall" (Sugito, 1978), and "downstep" (Pierrehumbert & Beckman, 1988).

Compared with the clear physiological evidence for the causes of local F_0 changes, muscle activities for F_0 declination are surprisingly obscure. The correlation between F_0 and the CT, which is higher in short utterances, becomes less obvious in longer utterances (Atkinson, 1978). Even when sentential F_0 patterns have a typical pattern of declination, the CT does not show declining activities in parallel to the F_0 pattern (Collier, 1975; Gelfer, Harris, Collier, & Baer, 1983). Also, the strap muscles mainly show segmental activity and do not always demonstrate a gradual increase of activity towards the ends of utterances. These mysterious observations of muscle activities for F_0 declination have been discussed, leading to a common simplified explanation: local F_0 events are produced by laryngeal muscles and F_0 declination is caused by decreasing subglottal air pressure (Ps). This hypothesis seems to have two critical problems. The first problem is the range of Ps used for speech production. The rate of F_0 change with respect to Ps (r.f.p.) is about in the range of 2-6Hz/cm H_2O (e.g., Titze, 1989), and it is too small to explain the range of F_0 in a sentence, as Maeda (1976) pointed out. In his study, the magnitude of the baseline fall is between 20 Hz and 40 Hz, depending on the speaker. Assuming that the r.f.p value is about 5 Hz/cm H_2O and the Ps fall is 3 cm H_2O, the F_0 drop due to the Ps fall is only 15 Hz. Therefore, the Ps fall alone cannot account for the fall of the baseline (Maeda, 1976). The second problem is the change of vocal fold length in a lower F_0 range. The rate of vocal fold length change with respect to F_0 is sometimes larger in a lower F_0 range (Damste, Hollien, Moore, & Murry, 1968; Hollien & Moore, 1960; Nishizawa, Sawashima, & Yonemoto, 1988; Sonninen, 1954). This observation supports the hypothesis that vocal fold shortening is an effective parameter for lowering F_0, even in the range where the CT is relaxed and its action is not available.

These problems in the interpretation of the mechanisms for global F_0 changes need to be solved by future studies that cover a wide scope of F_0 control mechanisms. In proceeding with such studies, it should be recognized that the muscular controls over F_0 not only activate F_0 raising mechanisms, but also F_0 lowering mechanisms. It seems that the intra- and extra-laryngeal mechanisms for control over tension and length comprise the primary integrated factor. The precursor of F_0 declination pattern is found in the infant's cry of humans and other mammals, suggesting its origin in respiratory function. However, realization of linguistic rules does not strictly reflect physiological constraints of speech organs, permitting a supplementary role of Ps for maintaining vibration of tense and slack vocal folds.

Intrinsic Vowel F_0 and Its Inverse Effect

Intrinsic vowel F_0, a phonetic phenomenon in which F_0 correlates with tongue height, provides evidence of tongue-larynx interaction. This universal phenomenon has led researchers to search for its possible underlying physiological mechanisms. A contemporary view of the underlying mechanism is the biomechanical coupling between the larynx and the supralaryngeal articulators. The classical "tongue pull" theory (Lehiste, 1970), which describes a gross sketch of the tongue-larynx interaction, has been refined by a few physiological studies (Honda, 1983; Rossi & Autesserre, 1981). The tongue deformation for high vowel articulation causes anterior movement of the hyoid bone so as to rotate the thyroid cartilage around the cricothyroid joint for stretching the vocal folds. The "vertical tension" theory (Ohala & Eukel, 1978) explains another dimension of the effect of supralaryngeal articulation. The width of the laryngeal ventricle observed by x-ray pictures varies with vowel quality, which implies the existence of variation in the vertical tension on the vocal folds.

While these explanations are based on the influence of the supralaryngeal articulators on the configuration of the larynx, EMG studies have revealed the contribution of CT activity to intrinsic vowel F_0 (Honda & Fujimura, 1991; Vilkman, Aaltonen, Raimo, & Okasanen, 1989). The evidence of this active component obviously contradicts the explanation of intrinsic vowel F_0 by passive tongue-larynx interaction. Although this may be a problem beyond experimental verification, various explanations can be speculated. For example, a simple explanation could be that CT activity for high vowels is only a by-product of laryngeal muscle activity which compensates for the tendency of high vowels to have lower sound intensity. Conversely, the CT activity may be an active effort to realize vowel features. Honda and Fujimura (1991) explain this problem by the "phonologization" hypothesis, that is, an incorporation of phonetic tendencies into the phonology of languages. Diehl (1991) introduces the "speech enhancement" hypothesis to account for the perceptual effect that the production of the intrinsic vowel F_0 by the talker contributes to robust auditory separation of vowel quality by the listener. Another possibility is the "neural coupling" hypothesis. Sapir (1989) suggests a possible effect of sensory information from articulatory motion on laryngeal activity. Also, intrinsic vowel F_0 found in infants and deaf children implies some innate mechanism of neural connections which causes simultaneous muscle activities in the tongue and larynx. Thus, the phenomenon of intrinsic vowel F_0 appears simple and universal, although its implementation mechanisms may not be.

There is another example of the tongue-larynx interaction that may be described as the "inverse effect" of intrinsic vowel F_0, which is believed to result from tongue deformation or jaw movement used for F_0 control. While intrinsic vowel F_0 means a global correlation between vowel formant frequencies and F_0, the inverse effect describes a lesser degree of correlation, caused by articulatory variability for F_0 control. The mechanisms underlying the inverse effect are the same as those explained for the tongue pull theory, although the effects vary across vowels. For the vowel /i/, the activity of the genioglossus posterior (GGP) muscle, used for the articulation of the vowel, enhances F_0 raising. This activity of the GGP for F0 enhancement results in a further extreme-high front position of the tongue when this vowel is produced with high F_0. For the vowel /ɑ/, the hyoglossus (HG) and the styloglossus (SG) muscles that are active in producing a low-back position of the tongue can also contribute to F_0 lowering. When the vowel /ɑ/ is produced with high F_0, the HG and the SG become less active, resulting in a centralized tongue position. Contrarily, when the vowel is produced with low F_0, increased activity of these muscles creates an extreme low-back position of the tongue. The influence of these articulatory variations is small and do not interact with categorical boundaries of vowel quality, implying that these variations are of little value phonetically. However, the inverse effect may provide a certain naturalness to human speech. It may also be of importance to speech perception in certain situations, since the perceptual effects of the correlation between vowel formants and F_0 appear to be similar to those observed for speaker normalization.

COMMENTS

This chapter was intended to describe a contemporary summary of physiological mechanisms of F_0 control. The description is limited mainly to the aspects of the muscular components of the laryngeal and the extra-laryngeal mechanisms. The musculature of the larynx and its connections to the supralaryngeal articulators form a complex network that is quite different from the musculature of the other parts of the body. The anatomy of the actuators is also varied in shape and composition. Understanding of such a unique system requires close observation to find the relationship between its morphology and its function. Comparative morphology may help answer the questions, in a different domain, of why and how humans use the vocalization mechanisms for speech production. Ongoing research on the physiological mechanisms of F_0 lowering and the recent finding of the relationship between the larynx and the cervical spine appear to shed light on the organization of F_0 control mechanisms by the muscular network. Nonetheless, the complexity of the system defies definitive interpretation. EMG studies are very useful, but the number of muscles that can be measured at one time is always limited and does not always provide accurate information about the total system. From this view, a large-scale computer simulation using a realistic biomechanical model is a promising future approach to solving the problems. Many unexpected discoveries would result from simulations on the model, although it should be noted that the model's performance would reflect the implementer's understanding of the system.

REFERENCES

Arnold, G. E. (1961). Physiology and pathology of the cricothyroid muscle. *Laryngoscope, 71*, 687-753.

Atkinson, J. E. (1978). Correlation analysis of the physiological features controlling fundamental frequency. *Journal of the Acoustic Society of America, 63*, 211-222.

Broad, D. J. (1973). Phonation. In F. D. Minifie, T. J. Nixon, & F. Williams (Eds.), *Normal aspects of speech, hearing, and language* (pp. 127-167). Englewood Cliffs NJ: Prentice-Hall.

Collier, R. (1975). Physiological correlates of intonation patterns. *Journal of the Acoustic Society of America, 58,* 249-255.

Damste, P. H., Hollien, H., Moore, P., & Murry, T. (1968). An X-ray study of vocal fold length. *Folia Phoniatrica, 20,* 349-359.

Diehl, R. L. (1991). The role of phonetics within the study of language. *Phonetica, 48,* 120-134.

Erickson, D., Liberman, M., & Niimi, S. (1977). The geniohyoid and the role of the strap muscles. *Haskins Status Report on Speech Research, SR-49,* 103-110.

Erickson, D., Baer, T., & Harris, K. S. (1983). The role of the strap muscles in pitch lowering. In D. M. Bless, & J. H. Abbs (Eds.), *Vocal fold physiology* (pp. 281-285). San Diego: College-Hill Press.

Erickson, D. (1993). Laryngeal muscle activity in connection with Thai Tones. *Annual Bulletin of Research Institute of Logopedics and Phoniatrics, 27,* 135-149.

Faaborg-Anderson, K. (1957). Electromyographic investigation of intrinsic laryngeal muscles in humans. *Acta Physiologica Scandinavica, 41,* Suppl. 140.

Faaborg-Anderson, K., & Sonninen, A. (1960). The function of the extrinsic laryngeal muscles at different pitch. *Acta Otolaryngologica, 51,* 89-93.

Fink, B. R. (1962). Tensor mechanisms in the human larynx. *Acta Otol. Rhinol. Laryngol., 71,* 591-600.

Fink, B. R., & Demarest, R. J. (1978). *Laryngeal biomechanics.* Cambridge, MA: Harvard University Press.

Fink, B. R. (1975). *The human larynx: A functional study.* New York: Raven Press.

Gay, T., Strome, M., Hirose, H., & Sawashima, M. (1972). Electromyography of the intrinsic laryngeal muscles. *Annals of Otology, Rhinology & Laryngology, 81,* 401-409.

Gelfer, C. E., Harris, K. S., Collier, R., & Baer, T. (1983). Is declination actively controlled?. In I. Titze, & R. Scherer (Eds.), *Vocal fold physiology: Biomechanics, acoustics, and phonatory control* (pp. 113-126). Denver: Denver Center for Performing Arts.

Hirai, H., Honda, K., Fujimoto, I., & Shimada, Y. (1994). Analysis of magnetic resonance images on the physiological mechanisms of fundamental frequency control. *Journal of Acoustic Society of Japan, 50,* 296-304. [in Japanese]

Hirano, M., Ohala, J., & Vennard, W. (1969). The function of laryngeal muscles in regulating fundamental frequency and intensity of phonation. *Journal of Speech and Hearing Research, 12,* 616-628.

Hollien, H., & Moore, P. (1960). Measurements of the vocal folds during changes in pitch. *Journal of Speech and Hearing Research, 3,* 157-165.

Hollien, H. (1983). In search of vocal frequency control mechanisms. In D. M. Bless, & J. H. Abbs (Eds.), *Vocal fold physiology* (pp. 361-367). San Diego: College-Hill Press.

Honda, K. (1983). Relationship between pitch control and vowel articulation. In D. M. Bless, & J. H. Abbs (Eds.), *Vocal fold physiology* (pp. 286-297). San Diego: College-Hill Press.

Honda, K. (1985). Variability analysis of laryngeal muscle activities. In I. Titze & R. Scherer (Eds.), *Vocal fold physiology: Biomechanics, acoustics, and phonatory control* (pp. 127-137). Denver: The Denver Center for the Performing Arts.

Honda, K., Hirai, H., & Kusakawa, N. (1993). Modeling vocal tract organs based on MRI and EMG observations and its implication on brain function. *Annual Bulletin of Research Institute of Logopedics and Phoniatrics, 27,* 37-49.

Honda, K., & Fujimura, O. (1991). Intrinsic vowel F_0 and phrase-final F_0 lowering: phonological vs. biological explanations. In J. Gauffin & B. Hammerberg (Eds.), *Vocal fold physiology* (pp. 149-158). San Diego: Singular Publishing Group.

Kakita, Y., & Hiki, S. (1974). A study of laryngeal control for voice pitch based on anatomical model. *Proceeding of Speech Communication Seminar* (pp. 45-54). Stockholm, SCS-74.

Lehiste, I. (1970). *Suprasegmentals.* Cambridge, MA: MIT Press.

Löfqvist, A., Baer, T., McGarr, N. S., & Story, R. S. (1989). The cricothyroid muscle in voicing control. *Journal of Acoustic Society of America, 85,* 1314-1321.

Maeda, S. (1976). *A characterization of American English intonation.* Unpublished doctoral dissertation, MIT, Cambridge, MA.

Maue, W., & Dickson, D. R. (1971). Cartilages and ligaments of the adult human larynx. *Arch. Otolaryngologica, 94,* 432-439.

Negus, V. E. (1949). *The comparative anatomy and physiology of the larynx.* New York: Grune and Stratton.

Niimi, S., Horiguchi, S., & Kobayashi, N. (1991). F_0 raising role of the sternothyroid muscle: An electromyographic study of two tenors. In J. Gauffin & B. Hammerberg (Eds.), *Vocal fold physiology* (pp. 183-188). San Diego: Singular Publishing.

Nishizawa, N., Sawashima, M., & Yonemoto, K. (1988). Vocal fold length in vocal pitch change. In O. Fujimura (Ed.), *Vocal physiology: Voice production, mechanisms, and function* (pp. 75-82). New York: Raven Press.

Ohala, J. J., & Eukel, B. W. (1978). Explaining the intrinsic pitch of vowels. *Report of the Phonology Laboratory. Berkeley: University of California, 2,* 118-125.

Pierrehumbert, J. B., & Beckman, M. E. (1988). *Japanese tone structure.* Cambridge, MA: MIT Press.

Rossi, M., & Autesserre, D. (1981). Movements of the hyoid and the larynx and the intrinsic frequency of vowels. *Journal of Phonetics, 9,* 233-249.

Sapir, S. (1989). The intrinsic pitch of vowels: theoretical, physiological, and clinical considerations. *Journal of Voice, 3,* 44-51.

Sawashima, M., Kakita, Y., & Hiki, S. (1973). Activity of the extrinsic laryngeal muscles in relation to Japanese word accent. *Annual Bulletin of Research Institute of Logopedics and Phoniatrics, 7,* 19-25.

Sawashima, M. (1974). Laryngeal research in experimental phonetics. In A. T. Sebeok (Ed.), *Current trends in linguistics* (pp. 2303-2348). The Hague: Mouton.

Shipp, T., Doherty, E. T., & Morrissey, P. (1979). Predicting vocal frequency from selected physiological measures. *Journal of Acoustic Society of America, 66,* 678-684.

Simada, Z., & Hirose, H. (1970). The function of the laryngeal muscles in respect to the word accent distinction. *Annual Bulletin of Research Institute of Logopedics and Phoniatrics, 4,* 27-40.

Simada, Z., Niimi, S., & Hirose, H. (1991). On the timing of the sternohyoid muscle activity associated with accent in the Kinki dialect. *Annual Bulletin of Research Institute of Logopedics and Phoniatrics, 25,* 39-45.

Sonesson, B. (1983). Vocal fold kinesiology. In S. Grillner, B. Lindblom, J. Lubker, & A. Persson (Eds.), *Speech motor control* (pp. 113-117). Oxford: Pergamon Press.

Sonninen, A. (1954). Is the length of the vocal cords the same at all different levels of singing? *Acta Otolaryngologica,* Suppl. 118, 219-231.

Sonninen, A. (1968). The external frame function in the control of pitch in the human voice. *Annals of the New York Academy of Science,* 68-90.

Sugito, M. (1978). *Studies of Japanese accent.* Tokyo: Sanseido. [in Japanese]

Titze, I. R. (1989). On the relation between subglottal pressure and fundamental frequency in phonation. *Journal of the Acoustic Society of America, 85,* 901-906.

Titze, I. R., Luschei, E. S., & Hirano, M. (1989). Role of the thyroarytenoid muscle in regulation of fundamental frequency. *Journal of Voice, 3,* 213-224.

Vilkman, E., Aaltonen, O., Raimo, I., & Okasanen, H. (1989). Articulatory hyoid-laryngeal changes vs. cricothyroid muscle activity in the control of intrinsic F_0 of vowels. *Journal of Phonetics, 17,* 193-203.

Westbury, J. R. (1988). Mandible and hyoid bone movements during speech. *Journal of Speech and Hearing Research, 31,* 405-416.

Zenker, W. (1964). Questions regarding the function of external laryngeal muscles. In *Research potentials in voice physiology* (pp. 20-40). New York: State University of New York.

16 Between Organization and Chaos: A Different View of the Voice

R. J. Baken
New York Eye and Ear Infirmary

Stability is surely an important requirement of any physiologic system. Simple logic supports its necessity: any process must be resistant to disruption if its purposes are to be reliably accomplished. At least in oscillatory mechanisms, deeper-level stability is often gauged by the regularity of the output. Given a constant physiologic background, one expects heart rate, for example, to be perfectly steady. Similarly, absent variable communicative demands, one anticipates an unchanging vocal F_0.

A Newtonian view of the mechanical world, in fact, encourages the dismissal of irregularity as the result of the accidental or contaminating influence of irrelevant and uncontrolled variables that are identifiable in principle. Vocal function provides an excellent case in point. Vocal F_0 shows a distinct and seemingly-irreducible cycle-to-cycle irregularity, formally referred to as fundamental frequency perturbation and more colloquially called jitter. A powerful bias impels us to believe that this vocal "imperfection" must have identifiable proximate causes: unavoidable turbulence in the transglottal airstream, minute and random alterations of the subglottal driving pressure, or the residual muscle-tension "noise" that is the evidence of non-ideal integration of laryngeal motor-unit contractions. There is, we believe, a minimal stochastic-noise floor—in the generating mechanism or the measuring system or both—that clouds our view of the ideal and perfectly regular underlying function that, however unobservable in practice, we are convinced must really exist deep in the heart of the mechanism being observed.

It is undoubtedly true that physiologic imperfections such as nonideal airflow and the quantum-like nature of muscle contractions contribute significantly to (and perhaps even account for most of) the vocal irregularity that we measure and hence of the laryngeal instability that we infer. Yet a more recent understanding of the nature of complex systems strongly suggests that irregularity would persist even if all extraneous influences, imperfections, and observational limitations were to be accounted for. That understanding is embodied in the theory of nonlinear dynamics, commonly called the theory of chaos. It maintains that irregularity is inherent in the most basic aspects of many--perhaps most-- complex systems. It asserts that apparently-random variability is a feature of such systems even at the foundational level of their platonic ideal. Chaos theory encourages the hypothesis that fundamental frequency perturbation and even more dramatic changes of F_0 are an intrinsic feature of the voice and that they are evidence of a different form of functional integrity that might, for the moment, be described as unsteady stability. It is the purpose of this chapter to advance just such a hypothesis and to propose that such a situation may confer significant advantages even while it engenders distinct dangers that may be manifest in vocal disorder.

Chaos defined

Chaos theory has recently achieved wide exposure (not to say notoriety) and as some of its terminology has entered popular discourse it has suffered serious loss of precision or outright distortion. This is particularly true of the word "chaos" itself. It is therefore useful to pause to consider what "chaotic" implies.[1]

- A chaotic system's output is not predictable to any arbitrarily specifiable degree of resolution. That is, no matter how precisely it is measured, the system's behavior always has a residual amount of apparently-random error. Jitter, for example, is always present.
- A chaotic system is deterministic. In other words, its behavior is completely governed by some rule that, at least in principle, is specifiable. If this is
- so, then the apparently-random error in the output cannot be random at all, but must itself be deterministic. Jitter, to continue the example, is inherent and lawful.

---

[1]Good introductions to various aspects of chaos theory can be found in Crutchfield, Farmer, Packard, & Shaw, 1986; Glass & Mackey, 1988; Gleick, 1987; Holden, 1986; Moon, 1987; Stewart, 1989.

- A chaotic system is exquisitely sensitive to governing conditions. By this we mean that an infinitesimal change in some system parameter can cause a significant qualitative change in the system's behavior. Note that "infinitesimal" is to be understood in its mathematical sense: too small to be measured no matter how precise the measuring tool. Vocal production, for instance, might shift from modal register to glottal fry because of some change in laryngeal status that we could never measure. From the observer's point of view, the register shift would have occurred for no reason at all. Sudden and often dramatic alterations of behavior are referred to as "bifurcations." They are characteristic, and in fact diagnostic, of chaotic systems.

There are now convincing demonstrations that many important physiologic systems are, in fact, chaotic. Among them are the heart, respiratory, cerebrocortical, neuroendocrine, neuromotor, and cochlear systems (Beuter, Labrie, & Vasilakos, 1991; Coumel & Maison-Blanche, 1991; Goldberger, Bhargava, West, & Mandell, 1985; Kaplan & Talajic, 1991; Kronenberg, 1991; Kryger & Millar, 1991; Rapp et al., 1989; Sheldon & Riff, 1991; Skinner, Goldberger, MayerKress, & Ideker, 1990; Teich, Keilson, et al., 1991; Teich, Lowen, & Turcott, 1991). It is therefore abundantly clear that chaos is a characteristic of many important life processes, and it might well be essentially ubiquitous (Glass & Mackey, 1988; Goldberger, Rigney, & West, 1990; West, 1990). It would certainly not be surprising to find that phonatory function is chaotic.

Several investigators have found evidence of sudden alterations of the vocal signal that may reasonably be assumed to be bifurcations of glottal behavior. It has long been known, for example, that the cry of the normal infant commonly includes very abrupt frequency doublings or even more complex patterns of F_0 change (Herzel, Steinecke, Mende, & Wermke, 1991; Mende, Herzel, & Wermke, 1990; Robb & Saxman, 1988; Truby & Lind, 1965; Wasz-Höckert, Lind, Vuorenkoski, Partanen, & Valanné, 1968). Similarly, the sudden appearance of strange or complex patterns of F_0 or sound pressure is associated with normal glottal fry phonation (Hollien & Michel, 1968), vocal fold tumors (Askenfelt & Hammarberg, 1986; Koike, 1969), vocal nodules (Remacle & Trigoux, 1991), vocal fold paralysis (Hammarberg, Fritzell, Gauffin, & Sundberg, 1986; Ishizaka & Isshiki, 1976) and other neurologic disorders (Ludlow, Coulter, & Gentges, 1983; Ramig, Scherer, Titze, & Ringel, 1988; Titze, Baken, & Herzel, 1992; Wieser, 1984).

As examples of typical vocal-function bifurcations, Figure 1 presents the F_0 records of sustained vowels by two patients (A and B) with spasmodic dysphonia. In both cases the period of each vocal cycle is plotted (Y axis) against the period number in the sequence (X axis). The enormous F_0 instability gives these records a deceptive appearance of sound pressure waves, but they are not. Expansions of the records show normal and apparently-random variability at the start of each phonation followed by intervals of 2-valued and 4-valued alternations of F_0 (Sample A) or a complex 10-to-14 valued pattern (Sample B). The complexity of these patterns and the abruptness of their appearance sorely tax the explanatory power of any classical model of laryngeal function. It is hard to see how they could be anything but the bifurcations of a nonlinear dynamical system. They therefore provide circumstantial but compelling evidence that the larynx is, in fact, chaotic.

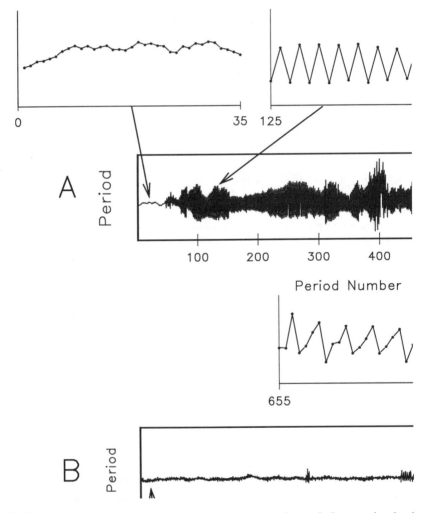

FIGURE 1. Sequential period durations during prolonged vowels by two dysphonic patients.

The difficulty, to the point of practical impossibility, of exploring the dynamics of laryngeal behavior in vivo or even in vitro has prompted the use of mathematical models as investigative tools. One of the earliest--the Ishizaka-Flanagan two-mass model (Ishizaka & Flanagan, 1972) has proven particularly useful and very considerable experience with it has established its validity. It is a quite simple model, reducing each vocal fold to just two shelf-like masses (upper and lower) that are linked to each other and constrained to move only mediolaterally. Various mechanical factors, such as driving pressure, tensions, and collisional recoils, are included in the model, and several of them are characterized by a certain amount of nonlinearity. The model makes no attempt to mimic laryngeal morphology nor to parcel out forces among anatomical structures. While sharply limiting its utility for studying the kinematics of the vocal folds, the lumped-element simplicity of the Ishizaka-Flanagan model actually represents an advantage for first-generation studies of laryngeal dynamics. It may profitably be pressed into service for a brief exploration of an important way in which the dynamics of a system may be explored, and for a demonstration of some features that indicate probable laryngeal chaos.

Phase-Space Representation

Examination of vocal data is typically done in the time and frequency domains, using acoustic or glottographic waveforms, F_0 and sound pressure data, or sound spectrograms. Dynamical characteristics, however, are much better evaluated using a phase-space representation. It is created by assigning a different relevant variable to each of n axes of a Cartesian space. The data are plotted as a time series in the n-dimensional region thus created. Geometric and topologic methods can be used to assess the resultant structure.

Behavior of the Two-Mass Model

We are now in a position to explore laryngeal dynamics, as represented by the Ishizaka-Flanagan two-mass model. Normal phonation can be modeled by choosing moderate vocal fold tension (125 kdynes) and subglottal pressure (5 kdynes $\approx$ 5 cm H_2O). The results are illustrated in Figure 2. Plot A is the time-series record of the position of the two vocal fold masses. The regularity of the waveshapes, periods, and amplitudes after the start-up period demonstrate the excellent stability of glottal function under the conditions selected.

Figure 2B is a phase-space representation of the behavior of the model from the time of startup and continuing for approximately 75 cycles after stabilization. The phase space is two-dimensional, the two dimensions being the mediolateral position of the upper (vertical axis) and lower (horizontal axis) masses of the vocal folds. The midline is at 0.0; negative position values represent compressive deformation of the vocal fold mass when it collides with the opposing vocal fold. The line formed by the data is the system's trajectory in this phase space.

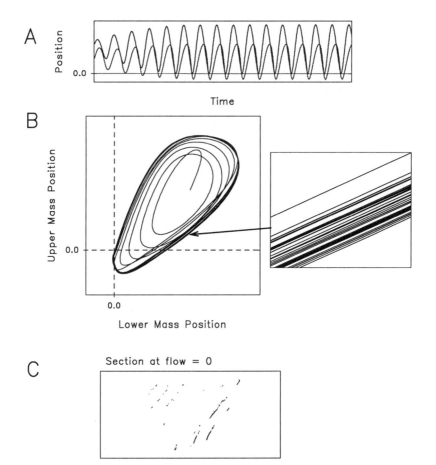

FIGURE 2. Behavior of the two-mass model with "normal" parameters. A: Position of the upper (solid line) and lower (dashed line) masses over time. B: Phase space representation of the motion of the upper and lower masses. A small portion of the trajectory is greatly enlarged to the right. C: Poincaré section (at transglottal flow = 0) through the trajectory.

It begins at an initial rest position in the central portion of the plane, and then spirals outward until it settles down to a roughly-elliptical path, which it will follow as long as the model is allowed to run. The interpretation of the trajectory's behavior is relatively easy: the oscillatory motions of the upper and lower masses of the folds increase until they ultimately reach a stable amplitude and phase relationship. The trajectory seems to be drawn to a final oscillatory condition, represented by the elliptical path. In fact, trajectories starting in a large area of this region of the phase space will all be drawn into the same final elliptical pathway. For this reason it is known as an attractor. (All starting points that produce trajectories that fall into an attractor constitute its basin of attraction.)

The attractors of classical dynamical systems can be of two types. Point attractors are fixed locations in the phase space to which the trajectory is drawn. A point attractor characterizes, for instance, an undriven swinging pendulum. As energy is lost, the magnitude of its oscillatory motion diminishes, until it comes to rest. The rest position is perfectly stable, and is shown by a single point in an appropriate phase space. A limit cycle is the other possible attractor, and is exemplified by the motion of a perfect, driven pendulum, such as might be found on an ideal pendulum clock. Energy losses are made up by a power source, so that the pendulum swings with an invariant pattern, which is shown by a single-lined, invariant orbit in the phase space.

The attractor of Figure 2, of course, represents the orbit onto which the solutions of the model's equations ultimately converge. Given that the two-mass model is a completely deterministic system, one might expect that the attractor would be a limit cycle—a single line in the phase space—like the attractor of a driven pendulum. Peculiarly enough, however, this is not the case. Even at the low resolution of Figure 2B, it is apparent that the attractor is at least a double band in its lower-right quadrant and contains many lines throughout. Enlargement of one of the two bands shows that it is itself formed of many separate lines. In fact, if the model had been allowed to run forever, successive enlargements of the attractor would show ever more individual lines.

The attractor of Figure 2B lies in a two-dimensional phase space. We are, in a sense, looking down on it from our third-dimensional vantage point. Let us assume, for the moment, that there is a third dimension (perpendicular to the page) that represents transglottal flow. The attractor can be assumed to travel toward and away from us in this dimension as the transglottal flow increases and diminishes across the glottal cycle. (We cannot see this motion, because we are looking down from directly overhead.) It is possible to slice across the attractor at any given height above (or below) the plane by selecting those upper-mass/lower-mass positional data that are associated with a given flow value. The result is called a Poincaré section, and it allows us to look end-on at a slice through the attractor.

Figure 2C shows the results of just this process. (The cut was made at flow = 0.) It is clear that the orbits of the mass-position attractor are not randomly dispersed about some mean position, but rather are organized into a layered structure. Clearly, the attractor of the two-mass model is not simply a limit cycle. It is an organized agglomeration of lines, infinite in number, with a definite but extraordinarily complex structure. It is an example of a strange attractor or chaotic attractor. Strange attractors are associated with chaotic systems, and all chaotic systems have strange attractors. The phase-space representation is consistent with the recent analytical demonstration that the Ishizaka-Flanagan two-mass model of the larynx is chaotic (Lucero, 1993). And, to the extent that the model is a valid representation of laryngeal dynamics, vocal fold function is chaotic.

If the vocal fold tension is changed to a quite low 25 kdynes, the two-mass model produces the dichrotic phonation generally associated with glottal fry (Figures 3A and 3B). Decreasing vocal fold tension has resulted in a bifurcation of the model's behavior. It shows in the phase space representation (Figure 3C), where the attractor is now a doubled loop. The spread of trajectory lines makes it clear, however, that the transformed attractor is also chaotic.

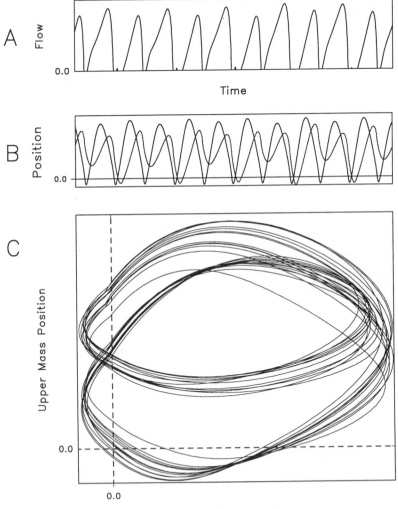

FIGURE 3. Behavior of the two-mass model with moderately low vocal fold tension. A: Transglottal flow over time, showing "dichrotic" waves as often observed in fry phonation. B: Motion of the lower (solid line) and upper (dashed line) vocal fold masses over time. C: Phase space representation of the motion of the upper and lower masses.

One final alteration of the model's parameters proves instructive. Maintaining the low vocal fold tension of Figure 3, let the subglottal pressure now be raised to about 10 cm H_2O (Figure 4). The vocal output becomes very erratic.

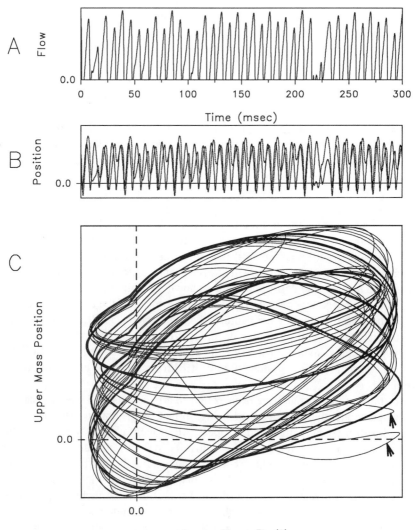

FIGURE 4. Chaotic behavior of the two-mass model. A: Transglottal flow over time. Note the transition from erratic to dichrotic waveforms, followed by a brief phonatory failure and subsequent recovery. B: Motion of the lower (solid line) and upper (dashed line) masses over time. C: Phase plane representation of the motion of the upper and lower masses. Arrows indicate aberrant loops of the trajectory associated with the phonatory failure shown in A.

In the interval from 0 to 100 ms or so glottal flow waves (Figure 4A) vary in shape, period, and amplitude in an apparently random way. But there then ensues a distinct bifurcation, and in the interval from about 125 to about 225 ms there is a train of nicely ordered dichrotic glottal pulses (Figure 4B). At about 225 ms phonation essentially fails for a brief moment (another bifurcation), but then manages to re-establish its dichrotic pattern.

This phonatory record clearly bears several of the hallmarks of serious vocal disorder. If it had been obtained from a patient the examiner might well conclude that there is intermittent instability of phonatory control variables, perhaps vocal fold tension or driving pressure, causing the sporadic appearance of glottal fry and momentary interruption of vocal fold oscillation. Yet, of course, nothing of the sort happened at all: all of the model's physiologic conditions remained constant. The shifts and aberrations of vocal behavior seen in the record are inherent in the original setup of the model. They are not the products of transient causative events.

The phase space portrait (Figure 4C) shows a very complex attractor. (Part of the trajectory has been darkened to make its structure clearer.) It is obviously chaotic, and occupies much more of the area in its region of the plane than the simpler attractors of Figures 2 and 3. Among its interesting features are apparently-aberrant loops (arrows) that constitute unexpected deviations from the general pattern of the attractor. They represent "aberrant" motions of the two vocal fold masses, and correspond to the phonatory "failure" at about 225 ms in the flow record. Note, however, that as the trajectories of the abnormal motions cross the phase plane they are captured by the attractor, and hence oscillation continues. Because the attractor occupies so much of the phase space the probability that a deviant trajectory will encounter the attractor is high. In short, the chaotic nature of the system endows it with an enhanced ability to correct itself.

The three attractors that have been examined are obviously all characteristic of the single model that we have been using. In fact, as the control parameters are altered, still more attractors would appear. All of them obviously coexist in the same phase space. The import of this is that the system can change its behavior to new stable states very quickly as controlling conditions alter by jumping from one attractor to another. Some of those states produce very steady phonation. Some produce interesting patterns of vocal cycles, like the dichrotic phonation of glottal fry. And some produce highly erratic and unstable voice.

Chaos restrained endows the laryngeal system with an enormous richness of phonatory potential. Chaos uncontrolled is dysphonia. Dysphonic voice may be qualitatively different from normal phonation. But it seems likely that in many cases the principles of chaotic dynamics may unify them, perhaps showing abnormality to be less complex and multidimensional than we tend to believe.

It is clear that there are significant qualitative differences that seem to put an abnormal voice in a different functional category from normal phonation. For many disorders, chaos may provide an important unification of the normal and the pathologic.

REFERENCES

Askenfelt, A. G., & Hammarberg, B. (1986). Speech waveform perturbation analysis: A perceptual-acoustical comparison of seven measures. *Journal of Speech and Hearing Research, 29,* 50-64.

Beuter, A., Labrie, C., & Vasilakos, K. (1991). Transient dynamics in motor control of patients with Parkinson's disease. *Chaos, 1,* 279-286.

Coumel, P., & Maison-Blanche, P. (1991). Complex dynamics of cardiac arrhythmias. *Chaos, 1,* 335-342.

Crutchfield, J. P., Farmer, J. D., Packard, N. H., & Shaw, R. S. (1986). Chaos. *Scientific American, 254 (12),* 46-57.

Glass L., & Mackey, M. C. (1988). *From clocks to chaos: The rhythms of life.* Princeton, NJ: Princeton U.

Gleick, J. (1987). *Chaos: Making a new science.* New York: Viking.

Goldberger, A. L., Bhargava, V., West, B. J., & Mandell, A. J. (1985). On a mechanism of cardiac electrical stability. *Biophysical Journal, 48,* 525-528.

Goldberger, A. L., Rigney, D. R., & West, B. J. (1990). Chaos and fractals in human physiology. *Scientific American, 262 (2),* 43-49.

Hammarberg, B., Frizell, B., Gauffin, J., & Sundberg, J. (1986). Acoustic and perceptual analysis of vocal dysfunction. *Journal of Phonetics, 14,* 533-547.

Herzel, H., Steinecke, I., Mende, W., & Wermke, K. (1991). Chaos and bifurcations during voiced speech. In E. Mosekilde (Ed.), *Complexity, chaos and biological evolution* (pp. 41). New York: Plenum.

Holden, A. V. (Ed.) (1986). *Chaos.* Princeton NJ: Princeton.

Hollien, H., & Michel, J. (1968). Vocal fry as a phonational register. *Journal of Speech and Hearing Research, 11,* 600-604.

Ishizaka, K., & Flanagan, J. L. (1972). Synthesis of voiced sounds from a two-mass model of the vocal cords. *Bell System Technical Journal, 51,* 1233-1268.

Ishizaka, K., & Isshiki, N. (1976). Computer simulation of pathological vocal-cord vibration. *Journal of the Acoustical Society of America, 60,* 1193-1198.

Kaplan, D. T., & Talajic, M. (1991). Dynamics of heart rate. *Chaos, 1,* 251-256.

Koike, Y. (1969). Amplitude modulations in patients with laryngeal diseases. *Journal of the Acoustical Society of America, 45,* 839-844.

Kronenberg, F. (1991). Menopausal hot flashes: Randomness or rhythmicity. *Chaos, 1,* 271-278.

Kryger, M. H., & Millar, T. (1991). Cheyne-Stokes respiration: Stability of interacting systems in heart failure. *Chaos, 1,* 265-269.

Lucero, J. C. (1993). Dynamics of the two-mass model of the vocal folds: Equilibria, bifurcations, and oscillation region. *Journal of the Acoustical Society of America, 94,* 3104-3111.

Ludlow, C., Coulter, D., & Gentges, F. (1983). The differential sensitivity of frequency perturbation to laryngeal neoplasms and neuropathologies. In D. Bless, & J. Abbs (Eds.), *Vocal fold physiology: Contemporary research and clinical issues.* San Diego: College-Hill.

Mende, W., Herzel, H., & Wermke, K. (1990). Bifurcations and chaos in newborn infant cries. *Physics Letters A, 145,* 418-424.

Moon, F. C. (1987). *Chaotic vibrations: An introduction for applied scientists and engineers.* New York: Wiley.

Ramig, L. A., Scherer, R. C., Titze, I. R., & Ringel, S. P. (1988). Acoustic analysis of voices of patients with neurologic disease: A rationale and preliminary data. *Annals of Otology, Rhinology and Laryngology, 97,* 164-172.

Rapp, P. E., Bashore, T. R., Martinerie, J. M., Albano, A. M., Zimmerman, I. D., & Mees, A. I. (1989). Dynamics of brain electrical activity. *Brain Topography, 2,* 99-118.

Remacle, M., & Trigoux, I. (1991). Characteristics of nodules through the high-resolution frequency analyzer. *Folia Phoniatrica, 43,* 53-59.

Robb, M., & Saxman, J. (1988). Acoustic observations in young children's noncry vocalizations. *Journal of the Acoustical Society of America, 83,* 1876-1882.

Sheldon, R., & Riff, K. (1991). Changes in heart rate variability during fainting. *Chaos, 1,* 257-264.

Skinner, J. E., Goldberger, A. L., Mayer-Kress, G., & Ideker, R. E. (1990). Chaos in the heart: Implications of clinical cardiology. *Biotechnology, 8,* 1018-1024.

Stewart, I. (1989). *Does God play dice: The mathematics of chaos.* Cambridge, MA: Blackwell.

Teich, M. C., Keilson, S. E., Khanna, S. M., Brundin, L., Ulfendahl, M., & Flock, Å. (1991). Chaos in the cochlea. In D. J. Lim (Ed.), *Abstracts of the Fourteenth Midwinter Meeting: Association for Research in Otolaryngology* (pp. 50).

Teich, M. C., Lowen, S. B., & Turcott, R. G. (1991). On possible peripheral origins of the fractal auditory neural spike train. In D. J. Lim (Ed.), *Abstracts of the Fourteenth Midwinter Meeting: Association for Research in Otolaryngology* (p. 50).

Titze, I. R., Baken, R. J., & Herzel, H. (1992). Evidence of chaos in vocal fold vibration. In I. Titze (Ed.), *Vocal fold physiology: New frontiers in basic science.* San Diego: Singular.

Truby, H. M., & Lind, J. (1965). Cry sounds of the newborn infant. In J. Lind (Ed.), *Newborn infant cry* (pp. 8). Uppsala: Almqvist and Wiksells.

Wasz-Höckert, O., Lind, J., Vuorenkoski, V., Partanen, T., & Valanné, E. (1968). *The infant cry: A spectrographic and auditory analysis.* (Clinics in

Developmental Medicine, no. 29). London: Spastics International Medical Publications.

West, B. J. (1990). *Fractal physiology and chaos in medicine.* Teaneck, NJ: World Scientific.

Wieser, M. (1981). Periodendaueranalyse bei spastischen Dysphonien. *Folia Phoniatrica, 33,* 314-324.

17 Imaging of the Larynx— Past and Present

Hajime Hirose
School of Allied Health Sciences, Kitasato University

Imaging of the larynx is very important for the basic study of the speech production mechanism. However, the technology required to view the larynx, which is situated deeply below the pharynx, was not available until the middle of the 19th century. In 1854, Manuel Garcia (Figure 1), a Spanish music teacher who lived in Paris, invented a method of laryngeal observation using a dental mirror. He reported his technique and the result of his observation in the next year in Proceedings of Royal Society of London (1855), in which he wrote:

> The pages which follow are intended to describe some observations made on the interior of the larynx during the act of singing. The method which I have adopted is very simple. It consists of placing a little mirror, fixed on a long handle suitably bent, in the throat of the person experimented on against the soft palate and uvula. The party ought to turn himself towards the sun, so that the luminous rays falling on the little mirror may be reflected on the larynx. If the observer experiments on himself, he ought, by means of a second mirror, to receive the rays of the sun, and direct them on the mirror, which is placed against the uvula.

Garcia's technique, using artificial light instead of sunlight, was quickly introduced to clinical practice in laryngology. It yielded, and continues to yield, considerable knowledge of both normal and pathologic human larynges.

FIGURE 1. A portrait of Manuel Garcia at the age of 100, a year before his death in London.

Recently, the laryngeal mirror has been supplemented by a rigid laryngeal tele-endoscope. This device, which can easily be combined with the techniques of laryngeal photography and videography, has allowed the recording of laryngeal images to become a routine procedure in the laryngological clinic.

Although the laryngeal mirror and rigid endoscope are very useful for laryngeal imaging, the larynx can be observed only while the subject keeps his mouth open with his tongue stuck out, as shown in Figure 2. Therefore, these methods are not very suitable if we want to observe laryngeal behavior during natural speech production.

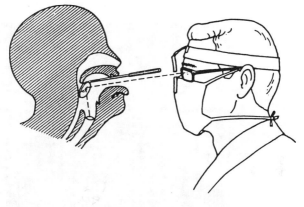

FIGURE 2. Observation of the larynx using (a) a laryngeal mirror, and (b) a rigid endoscope.

In order to overcome this problem, a special instrument, the fiber optic endoscope, was developed in the late 1960s at the Research Institute of Logopedics and Phoniatrics, Faculty of Medicine, the University of Tokyo (Sawashima & Hirose, 1968). The fiberscope consists of a hard tip that houses an objective lens, a flexible optical cable, and a connector to a camera system (Figure 3). The cable is thin enough to be inserted through the nostril into the pharynx. When the tip of the scope is located just above the tip of the epiglottis, the larynx is readily visible.

Since its introduction, the laryngeal fiberscope has been widely used for both clinical and research purposes. The diameter of the cable of the fiberscope has been made much thinner, compared with its prototype, and it is now relatively easy to use, even with young babies.

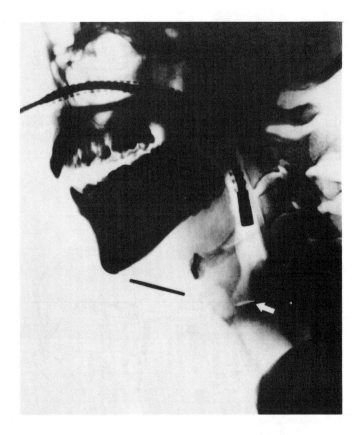

FIGURE 3. The position of the fiberscope during observation of the larynx.

Soon after the introduction of the laryngeal fiberscope to speech research, it was found that the glottis performs an opening gesture associated with the production of voiceless segments. Moreover, experimenters confirmed that various phonetic conditions may be characterized by specific patterns of degree and timing of glottal behavior. Much important research related to this topic was reported from several institutions, including Haskins Laboratories (Sawashima, 1977).

Although the fiberscope is a powerful tool, one disadvantage of ordinary fiberoptic observation is that the resolution of the laryngeal image is limited by the number of image fibers contained in the optical bundle, each of which conveys an element of the image. To meet the need for improved resolution, the video endoscope system has recently been developed. The video endoscope tip contains a high-resolution monochrome CCD chip. The high speed color wheel rotation system delivers color images of very good quality (Figure 4). Although this system is still very expensive and the diameter of the fiber bundle is relatively thick, it can provide laryngeal views with better resolution than the traditional endoscope. This system will be applied more widely for laryngeal research in near future.

In the field of laryngeal research, observation and analysis of vocal fold vibration are quite important. However, ordinary videography or cinematography is inadequate for precise analysis of vocal fold vibration because the frame rate is low. In the past, ultra-high-speed movie systems provided useful data on the nature of vocal fold vibration. Ultra-high-speed movie systems, however, are generally massive and costly, and it is always very time-consuming to obtain final results from the frame-by-frame analysis of the movie film.

Stroboscopy is a very useful tool in clinical situations. However, stroboscopy cannot provide a precise cycle-by-cycle analysis of vocal fold vibration with reference to an acoustic signal.

In recent years, a new method of digitally imaging vocal fold vibration, using a solid-state image sensor attached to a conventional camera system has been developed at The Research Institute of Logopedics and Phoniatrics, University of Tokyo. The method is relatively simple to use and it can provide useful data for both research and clinical purposes (Hirose, 1988; Kiritani, Imagawa, & Hirose, 1986).

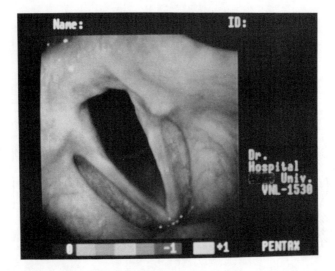

FIGURE 4. The larynx as viewed through a video endoscope.

Figure 5 shows a block-diagram of the system. In this system, a lateral-viewing tele-endoscope is attached to a single-lens reflex camera that has an image sensor at the position of the film plate. In order to obtain a brighter image, a model of the tele-endoscope was constructed, the diameter of which was made larger than that of the ordinary endoscope that is used for clinical purposes. The cross-section of the tube of the scope is elliptic, so that a larger amount of light guide fibers can be contained. The image guide is located at the center of the tube, while the light guides are on both sides of the image guide. The two light-guide cables are connected to separate light sources consisting of two 250 W halogen lamps.

When the shutter is open, an image scan is made and video-image signals are A/D converted and stored in 1-megabyte memory that contains a high-speed, 8-bit A/D converter. The stored data are then reproduced in the form of a slow motion display.

At present, it is possible to obtain image data at a maximum rate of 4,000 frames per second (f/s). For most of the clinical analysis, the frame rate of 2,000 f/s is adopted. One of the great advantages of the system is that the system is relatively free from surrounding noise, so that a simultaneous high-quality recording of the speech signal is possible. A similar laryngeal imaging system was recently reported by Hess and Gross (1993).

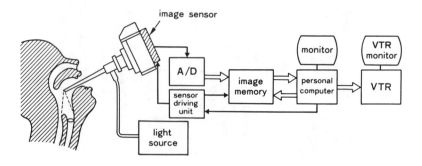

FIGURE 5. Block diagram of the high-speed digital imaging system.

Figure 6 illustrates the vocal fold vibration of a normal male subject taken at a rate of 2,000 f/s. Twenty consecutive frames are displayed from the top-left corner to the bottom-right. The opening and closing phases of the vocal fold vibration are easily identifiable. In this example, the subject produced a sustained vowel /e/ with a fundamental frequency of 200 Hz, so that approximately 2 cycles are displayed in the figure. The two curves shown below the frames are the acoustic signal (the upper curve) and the electroglottographic (EGG) signal (the lower curve).

The present system is very useful for the analysis of the pathological larynx with abnormal voice quality. In particular, asymmetrical or irregular vibratory patterns that are hardly discernible using stroboscopy can be readily identified.

Recently, we treated three cases of diplophonia in our clinic. They consisted of one case of recurrent laryngeal nerve paralysis and two cases of superior laryngeal nerve paralysis following hemithyroidectomy. Definite diagnosis of the involvement of each nerve was made by means of laryngeal electromyographic examinations.

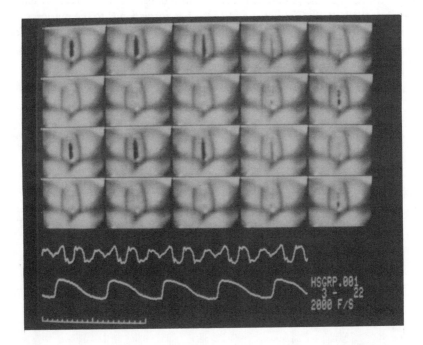

FIGURE 6. An example of images recorded using the high-speed digital imaging system (frame rate: 2000 f/s) of a male subject producing a sustained /e/ at a fundamental frequency of 200 Hz. The two curves shown below the frames are the acoustic signal (the upper curve) and the electroglottographic (EGG) signal (the lower curve).

Diplophonia is defined as the simultaneous production by the voice of two separate tones. Studies have revealed that diplophonia is caused by quasi-periodic variations in vocal fold vibration. For example, in their ultra-high speed movie study on one case of voluntary production of diplophonia, Ward and Moore (1969) reported that the left and right vocal folds vibrated at different frequencies. Previous studies, however, have not provided a detailed description of the relationship between the pattern of the vocal fold vibration and the speech waveform.

Our high-speed digital imaging system was used to analyze the three diplophonic cases mentioned earlier. In order to observe the pattern of vocal fold vibration clearly, the glottal width on the monitor screen was calculated and its temporal course was displayed by a computer together with the speech waveform. The measure of glottal width was defined as shown in Figure 7. Namely, in the glottal image, the area where the brightness curve fell below a selected threshold was taken as the glottal opening. A horizontal scan line at an appropriate position on the glottal opening was then selected and, on this scan line, the width of the glottal opening was measured. In this figure, the glottal width for each frame is consecutively displayed as a series of vertical bars as shown at the bottom. The upper end of the bar corresponds to the margin of the right vocal fold, whereas the lower end corresponds to that of the left vocal fold. The vertical bars are absent when the glottis is completely closed. This figure illustrates an example of analysis of the normal subject shown in Figure 6.

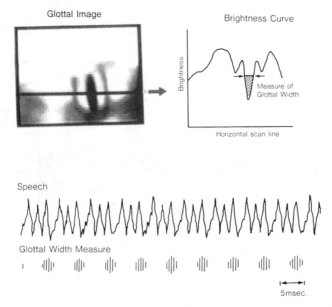

FIGURE 7. An example of the time sequence of the glottal width measure obtained for the normal vocal fold vibration.

The analysis of the diplophonic cases revealed that quasi-periodic variation in the speech waveform, secondary to the phase difference in the vibratory pattern of the left and right vocal folds, was related to diplophonic voice quality.

Figure 8 shows an example of the result of analysis of a case of superior laryngeal nerve paralysis. In this figure, it is apparent that the speech signal shows a pattern of temporal variation over a span of six cycles, while there is an asymmetry in the vibratory frequency of the left and right vocal folds and the degree of phase difference varies with time.

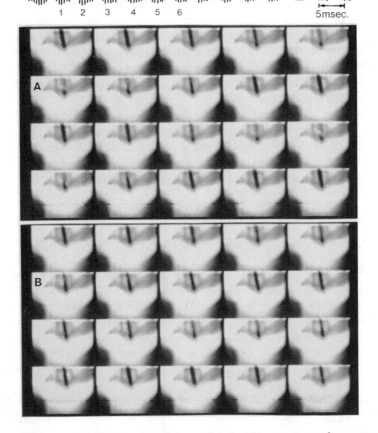

FIGURE 8. Comparison of speech waveform, glottal width measure, and corresponding glottal images in the case with diplophonia.

Near the first cycle, the phase difference is small, and there is a complete glottal closure. However, near the fourth cycle, the phase difference is nearly 180 degrees, and there is no glottal closure. Then, in the fifth cycle the inward movement of the right vocal fold appears as if it is truncated, and in the sixth cycle the left and right vocal folds resume synchrony. When the movements of the vocal folds are in phase, the glottal closure is complete (Series A in the upper part of Figure 8) and the excitation pattern in the speech waveform is strong, whereas when the movements are out of phase, the glottal closure is incomplete (Series B in the lower part of Figure 8), the excitation pattern is weak and high frequency components are lacking in the speech waveform.

Differences between the vibratory frequency of the two vocal folds were also observed in the other two cases of diplophonia (Kiritani, Hirose, & Imagawa, 1991). Thus, it can be assumed that this quasi-periodic pattern of variation in the speech waveform, associated with the phase difference in the vibratory pattern of the two vocal folds, is positively related to diplophonic voice quality. Although the nature of physiological conditions underlying irregular vocal fold vibration has not been completely clarified, it has been suggested the asymmetry in the mass or tension of the left and right vocal folds could be the origin of the irregularity (Isshiki et al., 1977). Further study is necessary to ascertain the precise nature of the asymmetry in the physical properties of the vocal folds in different pathological conditions.

Techniques of laryngeal imaging have improved since the era of Manuel Garcia, and we can expect further developments in the near future. Imaging is one of the powerful techniques for better understanding of the nature of speech production. It is hoped that the future application of different imaging systems will broaden and deepen the scope of research in laryngology.

REFERENCES

Garcia, M. (1855). Observations on the human voice. *Proceedings of Royal Society of London, 7,* 399-410.

Hess, M. M., & Gross, M. G. (1993). High-speed, light-intensified imaging of vocal fold vibrations in high optical resolution via indirect microlaryngoscopy. *Annals of Otology, Rhinology and Laryngology, 102,* 502-507.

Hirose, H. (1988). High-speed digital imaging of vocal fold vibration. *Acta Otolaryngologica (Stockholm), Supplement 458,* 151-153.

Isshiki, N., Tanabe, M., Ishizaka, K., & Broad, D. J. (1977). Clinical significance of asymmetrical vocal cord tension. *Annals of Otology, Rhinology and Laryngology, 86,* 58-67.

Kiritani, S., Imagawa, H., & Hirose, H. (1986). Simultaneous high-speed digital recording of vocal fold vibration, speech and EGG. *Annual Bulletin RILP, 20,* 11-16.

Kiritani, S., Hirose, H., & Imagawa, H. (1991). Vocal fold vibration and the speech waveform in diplophonia. *Annual Bulletin RILP, 25,* 55-62.

Sawashima, M. (1977). Fiberoptic observation of the larynx and other speech organs. In M. Sawashima & F. S. Cooper (Eds.), *Dynamic aspects of speech production* (pp. 31-47), University of Tokyo Press.

Sawashima, M., & Hirose, H. (1968). New laryngoscopic technique by use of fiber-optics. *Journal of Acoustical Society of America, 43,* 168-169.

Ward, P. H., & Moore, G. P. (1969). Diplophonia. *Annals of Otology, Rhinology and Laryngology, 78,* 771-777.

Using the Airway Interruption Method for Aerodynamic Assessment of Voice Disorders

18

Masayuki Sawashima
Yokohama Seamen's Insurance Hospital, Yokohama

Kiyoshi Makiyama and Naoko Shimazaki
Nihon University Surugadai Hospital, Tokyo

TECHNICAL PROBLEMS IN STUDYING THE AERODYNAMICS OF PHONATION

In phonation, expiratory air flow serves as an energy source for generating voiced sound at the glottis. Thus, the evaluation of the aerodynamic conditions at the glottis is important for studying physiology and pathophysiology of voice production. Van den Berg (1956) proposed "glottal efficiency," which is defined as the ratio of the acoustic power of the glottal source sound to the power of the subglottic air flow, as an effective evaluation measure.

The intensity of the voice measured in front of the mouth can be used as a practical substitute for the acoustic power of the glottal source sound. To estimate the subglottic power, which is the product of the mean flow rate and the mean pressure below the glottis, the mean flow rate can be readily measured at the mouth opening with little technical problem. In contrast, there are many technical problems in measuring subglottic pressure.

Several methods have been reported for measuring subglottic pressure during phonation, with tracheal puncture being the most straightforward procedure. This method, however, is not suitable for many experimental sessions or routine clinical examinations because of its potential dangers. The transglottal approach has become viable because of the development of miniaturized pressure transducers (Koike & Perkins, 1968). In this method, direct measurement is made by placing a small pressure transducer through the glottis into the subglottic cavity. The output of the transducer is carried outside the body, by a thin cable passing through the posterior end of the glottis, where it can be recorded. The method is not harmful, but the placement of the transducer involves surface anesthesia of the glottis, and may interfere with the production of natural voicing of the subjects. The measurement of intraesophageal pressure, instead of the subglottic pressure, has also been used in the field of experimental phonetics (Ladefoged, 1960; Ladefoged & McKinney, 1963; van den Berg, 1956). In this method, a balloon attached to a tube is swallowed into the upper part of the esophagus and the pressure inside the balloon is transmitted through the tube to be recorded externally. Although the placement of the balloon may be easier than that of the transducer into the subglottic cavity, this does not necessarily mean that the procedure is easy to use with large numbers of research subjects or clinical cases. Furthermore, subglottic pressure must be calculated from the intraesophageal pressure, and the relationship between esophageal and subglottic pressures varies depending on the air volume in the lungs (Kunze, 1964). Thus, all three methods have some technical problems that limit their practical use, especially for clinical examinations.

AIRWAY INTERRUPTION METHOD

The airway interruption method was originally devised as a technique for measuring alveolar pressure and was used for estimating the resistance of the lower respiratory tract in pulmonary diseases (Neergaad & Wirz, 1927). This technique was later applied to the aerodynamic study of phonation (Nishida, 1967; Nishida & Suwoya, 1964). In this method, respiratory air flow is momentarily interrupted by a shutter attached to a mask covering the face of the subject. Air pressure at the shutter is measured at the moment when it reaches equilibrium with the alveolar pressure. In phonation, this technique provides us with the expiratory lung pressure at the moment of abrupt cessation of voice. Expiratory lung pressure thus obtained, however, is not the same as the subglottic pressure during phonation estimated with the air flow uninterrupted.

A schematic diagram of the airway interruption method in phonation is shown in Figure 1. In the figure, PEX is the expiratory lung pressure obtained through airway interruption; Ps is true subglottic pressure during phonation immediately before the interruption; Rs and RG are the effective resistance of the lower (i.e., subglottic) respiratory tract and the glottis respectively; and U is the flow rate immediately before the interruption. The relation between PEX and Ps can be approximately represented by the following equation:

$$PEX-Ps=U*Rs.$$

According to Nishida and Suwoya (1964), PEX is in good agreement with Ps. In our previous experiment (Sawashima & Honda, 1987), U*Rs was estimated to be less than 5 mm H_2O at a flow rate of 100 ms. The value was less than 10 mm H_2O at a flow rate of 200 ms and exceeded 10 mm H_2O at 400 ms. That is, the increase in U*Rs was roughly proportional to the flow rate.

The airway interruption method is noninvasive and appears to be fairly easy to perform, although continuous recording and dynamic analysis of the expiratory air pressure patterns during speech are not possible. However, in the clinical examination of voice disorders, we are primarily interested in static aerodynamic conditions during sustained phonation, and a momentary sampling during sustained phonation is satisfactory for clinical purposes.

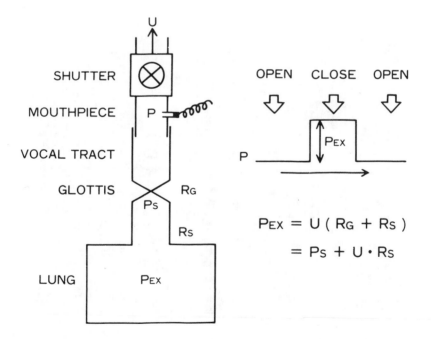

FIGURE 1. Schematic diagram of the airway interruption method in phonation.

For clinical use, we have developed a system using the airway interruption method for measuring the expiratory lung pressure simultaneously with the air flow rate, vocal pitch (i.e., F0), and vocal intensity. The system is shown in Figure 2. A specially designed air shutter with a pressure transducer is connected to the mouth piece. The outlet of the shutter is attached to the flow transducer. Vocal pitch and intensity are measured through a microphone at the distance of 20 cm from the outlet of the flow transducer. The detailed description of the instrumentation system has been reported elsewhere (Sawashima & Honda, 1987), and is commercially available from Nagashima Medical Instrument Co. LTD. in Tokyo.

Measurements are performed in a soundproofed room. The subject produces a sustained phonation while holding the mouthpiece airtight with the lips. The expiratory air flow is momentarily interrupted by the air shutter. The interruption results in a rapid increase of air pressure at the mouthpiece an abrupt cessation of voice. Air pressure at the mouthpiece is measured at the moment when it reaches complete equilibrium with the alveolar pressure. Air flow rate, vocal pitch, and vocal intensity are measured at the moment immediately before the interruption.

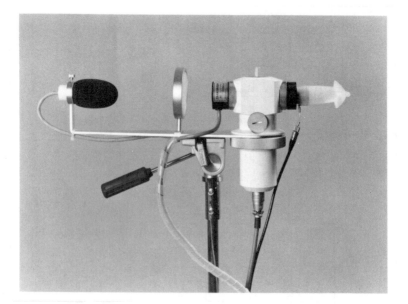

FIGURE 2. Instrumentation of the airway interruption system. (By courtesy of Nagashima Medical Instrument Co.)

A four-channel display of time curves before and after the interruption of a sustained phonation is shown in Figure 3. The uppermost trace is the fundamental frequency of voice, which is 142 Hz at the time moment of the vertical bar immediately before the interruption. The second panel presents the voice intensity trace, 84 dB (SPL) just before the interruption, and the third panel displays the air flow rate, which is 235 ms just before the interruption. The lowermost trace displays the pressure, which reaches equilibrium with the alveolar pressure, indicated by a plateau of the time curve, during the interruption. The pressure in this example is 77 mm H_2O at equilibrium.

The expiratory lung pressure thus obtained can be interpreted as an aerodynamic representation of so-called "expiratory force," or "breath effort," in phonation. If we assume that the flow resistance of the lower respiratory tract is essentially constant, the pressure value, in combination with the flow rate, provides us a good, if not direct, measure for evaluating the aerodynamic conditions at the glottis. Data on normal subjects collected by this technique (Sawashima, Niimi, Horiguchi, & Yamaguchi, 1988) indicated that the expiratory lung pressure increased systematically with increases in vocal intensity, and that the efficiency, defined as the ratio of the vocal intensity to the expiratory power (PEX*U), also increased with the increases in the vocal intensity. We are now collecting clinical data on various organic voice disorders, data that are presented in the following section.

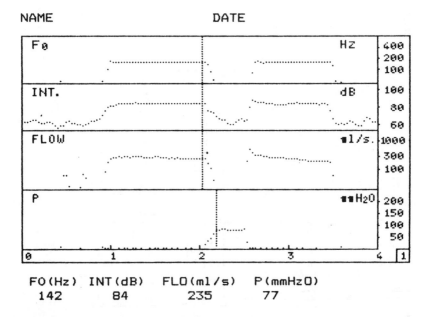

FIGURE 3. Four-channel display of time curves before and after the interruption.

CLINICAL DATA

In the voice clinic of Nihon University Surugadai Hospital, we collected aerodynamic data using the airway interruption method on several organic voice disorders. The subjects were 44 normal speakers (25 males, 19 females); 72 patients with vocal fold polyps (33 males, 39 females); 25 patients with polypoid vocal folds (12 males, 13 females); 22 patients with vocal fold nodules (all females); 18 patients with recurrent nerve paralysis (10 males, 8 females); and 25 patients with vocal fold tumors, which include carcinoma, papilloma, and leucoplakia (all males). Measurements were made during sustained phonation at the habitual pitch and intensity of each subject.

Data for the subjects in each group are summarized in Figure 4A (males) and Figure 4B (females). In the figures, pressure values are plotted in relation to flow rate values. Vocal intensity was roughly in the range of 65 dB to 80 dB for the normal and all pathologic groups except for the female polypoid group, where the intensity range is 60 dB to 75 dB.

In Figure 4A, flow rate in normal male subjects was essentially distributed between 150 ms to 250 ms. A vertical line at 205 ms is the sample value that bounds the upper quarter of normal samples. We have tentatively used this line as a reference for observing flow data in the pathologic groups.

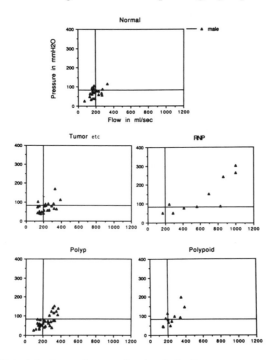

FIGURE 4A. Clinical data on male normal and pathologic cases.

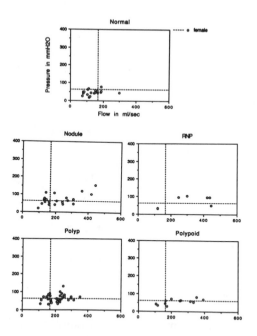

FIGURE 4B. Clinical data on female normal and pathologic cases.

In the male tumor group, flow rate values are within the normal range, although a few samples show somewhat greater flow rate than normal. The same pattern is observed in both polyp and polypoid groups of male patients. In the male group of recurrent nerve paralysis, however, we see samples with extremely high flow rates.

Most of the pressure values in the normal male subjects are in the range of 50-to-100 mm H_2O. The reference level here, indicated by a horizontal line, is at 81 mm H_2O. Most of the tumor samples are in the normal range, while some samples in the polyp and polypoid groups show higher pressure than normal. Some of the recurrent nerve paralysis patients show extremely high pressure.

In Figure 4B, data of female normal and disease groups are shown. We see that flow rates in the female normal group were essentially distributed within the range of 100-to-200 ms. The reference line here is at 166 ms. It is noted that most of pathologic cases had greater flow rates than the normal subjects, with some cases in the nodule, polypoid and recurrent nerve paralysis groups having flow rates as high as 400 ms.

Pressure data for the normal female group show that pressure was distributed mostly in the range of 40-to-60 mm H_2O; the reference line is at 55 mm H_2O. Polypoid cases show pressure levels within the range of normal subjects, but individuals in the nodule, recurrent nerve paralysis, and polyp groups had increased pressure values.

As mentioned before, expiratory lung pressure is interpreted as an aerodynamic representation of the "breath effort" in phonation. For a given expiratory lung pressure, the air flow rate varies with varying flow resistance of the respiratory tract, which includes the lower respiratory tract and the glottis. If we assume that there is little difference in the resistance of the lower respiratory tracts of normal subjects and persons with organic voice disorders, variation in the flow rate in the pathologic cases in this study may indicate variations in resistance at the glottis during phonation.

Thus, an increase in the flow rate that is greater than the increase in pressure indicates a decrease in glottal resistance. Pathophysiologically, this condition corresponds to insufficient glottal closure in phonation. Such a decrease in glottal resistance is typically seen in some cases of the recurrent nerve paralysis in both male and female patients, and in some female patients with polypoid vocal folds and vocal fold nodules.

An increase in the lung pressure greater than that in flow rate may indicate an increase in glottal resistance. A slight increase in the glottal resistance is seen in some samples in male polyp and polypoid groups, as well as in female nodule and polyp groups.

The distribution of the data points tells us that there is considerable variation in the aerodynamic conditions within the same disease group. Furthermore, there is considerable overlap in the values of different groups, including normal subjects. Thus, the data must be studied carefully in relation to other pathophysiological conditions of the larynx, for each case, to explore more fully the nature of the relations between aerodynamic conditions and vocal fold pathology.

REFERENCES

Koike, Y., & Perkins W. (1968). Applications of miniaturized pressure transducer for experimental speech research. *Folia Phoniatrica, 20,* 360-368.

Kunze, L. E. (1964). Evaluation of methods of estimating subglottal air pressure. *Journal of Speech and Hearing Research, 7,* 151-164.

Ladefoged, P. (1960). The regulation of subglottal pressure. *Folia Phoniatrica, 12,* 169-175.

Ladefoged, P., & McKinney, N. P. (1963). Loudness, sound pressure, and subglottal pressure in speech. *Journal of the Acoustic Society of America, 35,* 454-460.

Neergaad, V. K., & Wirz, K. (1927). Die Messung der Stromungswiderstande in den Atemwegen des Menschen, insbesondere bei Asthoma und Emphysem. *Zeitschrift fur Klinische Medizin, 105,* 51-82.

Nishida, Y. (1967). Aerodynamic studies on voice regulation. *Otologia Fukuoka, 13, (Supplement)*, 44-66.

Nishida, Y., & Suwoya, H. (1964). Indirect measuring method of the subglottic pressure in the voice productions; intraesophageal method and interruption method. *Otologia Fukuoka, 10*, 264-270.

Sawashima, M., & Honda, K. (1987). An airway interruption method for estimating expiratory air pressure during phonation. In T. Bear, C. Sasaki, & K. S. Harris (Eds.), *Laryngeal function in phonation and respiration* (pp. 439-447). Boston: College-Hill Press.

Sawashima, M., Niimi, S., Horiguchi, S., & Yamaguchi, H. (1988). Expiratory lung pressure, air flow rate, and vocal intensity: data on normal subjects. In O. Fujimura (Ed.), *Vocal physiology: Voice production, mechanisms and functions* (pp. 415-422). New York: Raven Press.

Van den Berg, Jw. (1956). Direct and indirect determination of the mean subglottic pressure. *Folia Phoniatrica, 8*, 1-24.

19 Examination of the Laryngeal Adduction Measure EGGW

R. C. Scherer, V. J. Vail, and B. Rockwell
Wilbur James Gould Voice Research Center
The Denver Center for the Performing Arts

INTRODUCTION

Adduction of the larynx at the level of the vocal folds and arytenoid cartilages is a primary peripheral and mechanical control variable in phonation. Laryngeal qualities from breathy to constricted phonation are dependent on glottal adduction (e.g., Scherer, Gould, Titze, Meyers, & Sataloff, 1988). A noninvasive measure of laryngeal adduction is of importance for both theoretical and applied purposes.

The electroglottograph (EGG) is a noninvasive instrument that provides a signal related to glottal kinematics (see Baken, 1987, 1992; Colton & Conture, 1990; and Orlikoff, 1991, for a relatively thorough review of principles, history, pitfalls, and relationships to laryngeal function). Values of the EGG waveform function correspond to the amount of contact area between the two vocal folds, but not in ways that are completely understood or straightforward (Anastaplo & Karnell, 1988; Childers, Alsaka, Hicks, & Moore, 1987; Childers, Hicks, Moore, Eskenazi, & Lalwani, 1990; Childers & Krishnamurthy, 1985; Scherer, Druker, & Titze, 1988; Titze, 1990). However, the shape of the EGG waveform may be related to specific configurations and motions of the vocal folds relevant to normal, abnormal, and trained voices (e.g., Baken, 1987; Brown & Scherer, 1992; Childers, Alsaka, Hicks, & Moore, 1986; Dejonckere & Lebacq, 1985; Fourcin, 1974; Gerratt, Hanson, & Berke, 1987; Motta, Cesari, Iengo, & Motta, 1990; Painter, 1988; Scherer & Titze, 1987; Titze, 1984, 1989).

EGGW: A PROPOSED MEASURE OF VOCAL FOLD ADDUCTION

This study examines a simple measure of the waveform of the EGG, called the EGGW measure, to determine its relationship to other measures of adduction, including a direct measure of the gap between the vocal processes of the arytenoid cartilages. The results will show that EGGW appears to be a significant measure of adduction, at least for the limited number of subjects and phonatory conditions reported in this study.

EGGW: Definition

Figure 1 illustrates an EGG waveform during normal phonation and the definition of the simple measure EGGW (the W stands for width). At the 25% level (Orlikoff, 1991; cf. Higgins & Saxman, 1993, who used 40%, and Rothenberg & Mahshie, 1988, who used 35%), the duration A corresponds to an approximate time of glottal closure, and B to an approximate time of glottal opening. EGGW, the ratio of A to A+B, is an estimate of the portion of the cycle the glottis is closed, and could be called a glottal closed quotient. EGGW is obtained for each cycle of phonation. The type of EGG used throughout this study was the Research Laryngograph produced by Dale Teaney (1987).

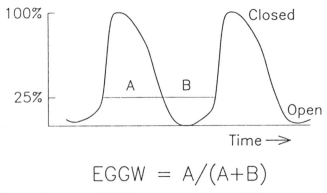

$$EGGW = A/(A+B)$$

FIGURE 1. Definition of the EGGW measure taken at the 25% height location on the electroglottograph waveform. 100% represents the portion of the waveform corresponding to maximum glottal closure (or maximum glottal contact area); 0% represents maximum glottal opening (or minimum glottal contact area).

Comparison Measures

Titze (1984) suggested the Abduction Quotient, Qa, a ratio of the width of the glottis (essentially between the vocal processes) to twice the amplitude of motion of a vocal fold as an index of laryngeal adduction. Qa is one of a number of measures given by Titze (1984) obtained from a theoretical approach to the mechanics of motion of the vocal folds. The Qa measure was obtained from the software analysis and synthesis program GLIMPES. Qa tends to decrease as vocal quality changes from breathy to normal to pressed, with values above approximately 0.5 associated with hypoadduction, and those below −1.0 with hyperadduction (Scherer, Gould, Titze, Meyers, & Sataloff, 1988).

The derivative of the EGG signal may give prominent positive and negative peaks that can be used to approximate the glottal open quotient (e.g., Childers et al., 1990). As Figure 2 illustrates, the positive and negative peaks of the EGG derivative refer to locations near glottal closure and opening, respectively, during which the EGG signal changes (increases and decreases, respectively) most rapidly. The ratio of the interval B divided by A+B is designated Qodegg, and is an approximation to the glottal open quotient. If the glottis is viewed with stroboscopy and recorded onto video tape, the glottal open quotient can be estimated by noting the number of frames in which the glottis appears to be partly-to-most-fully open as a proportion of the total number of frames in the complete glottal phonatory cycle; this is designated Qostrb.

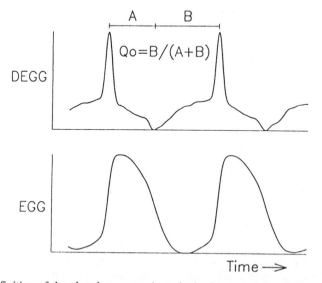

FIGURE 2. Definition of the glottal open quotient obtained by using the differentiated EGG waveform. The upper trace is the differentiation of the lower EGG waveform. Markers on the upper trace are taken at the maximum and minimum values of the differentiated EGG waveform.

Relation between EGGW and Qa

Seven normal adult community actors (4 males and 3 females, age range of 23-35 years, with no reported history of nontransient vocal problems), were asked to produce three prolonged /a/s in a steady manner at comfortable pitch, loudness, and equal effort levels for each of three vocal qualities, breathy, normal, and pressed (or constricted). The middle 1 sec of each EGG recording was digitized using a 16- (effective 15-) bit analog-to-digital system (Digital Sound Corporation 200 Audio Data Conversion System) at 20,000 samples per second, and stored in a VAX 11/750 computer. Analyses were performed on three consecutive cycles near the beginning, middle, and end of each digitized file. About 25% of the data was discarded because of the inability of GLIMPES to run successfully (Scherer, Gould, Titze, Meyers, & Sataloff, 1988; see also Scherer & Titze, 1987).

Figure 3 shows the data comparing EGGW and Qa fit to a cubic equation with an R^2 of 0.877. This suggests a reasonably strong relationship between these two variables. On a retest (Figure 4) with a professional tenor (BB) and a non-professional bass-baritone (RS), again using comfortable pitch for sustained /a/ but over a wide range of adduction intentions, the relationship shown in Figure 3 was supported (average difference of 11.9%, sd = 18.6%, between the data of Figure 4 and the cubic fit of Figure 3).

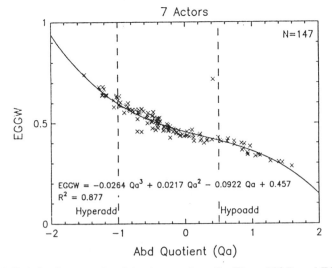

FIGURE 3. Relation between the abduction quotient, Qa (Titze, 1984), and EGGW (at 25% height of the EGG waveform) for seven community actors with normal voices. The subjects prolonged the vowel /a/ over a wide range of intended voice qualities, from very breathy to very pressed. The vertical dashed lines mark expected regions of hyperadduction and hypoadduction determined from Scherer, Gould, Titze, Meyers, & Sataloff, 1988.

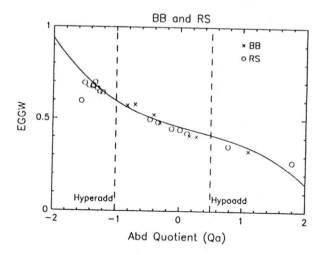

FIGURE 4. Relation between the abduction quotient and EGGW for two normal adult male subjects, BB and RS.

The vertical dashed lines on Figure 3 and Figure 4 represent approximate markers for perceived hyperadduction and hypoadduction (Scherer, Gould, Titze, Meyers, & Sataloff, 1988). The corresponding range between these conditions (the "normal" range) involves values of EGGW between 0.4 and 0.6. This would suggest that for normal larynxes, values of EGGW below 0.4 may correspond to the perceptual label of hypoadduction and above 0.6 to hyperadduction.

Relations among EGGW25, EGGW50, EGGW75, and Qa

The measure EGGW discussed above (and throughout this report) was taken at the 25% amplitude location from the baseline of the EGG waveform. The measure can be taken at any reasonable height, however. Using the data associated with Figure 3, Table 1 shows the correlations among EGGW taken at the 25, 50, and 75% height levels, and Qa. The correlations are reasonably high (minimum = 0.711). EGGW at the 50% level may be the measure of choice when there appears to be too much noise or waveform distortion on the lower portions (open glottis region) of the EGG waveform. A reasonable correspondence between the EGGW50 and EGGW25 values is y = 1.067x + 0.081, where x is the EGGW50 value and y is the associated predicted EGGW25 value. The highest correlation is between the measures EGGW50 and EGGW75, suggesting that these measures are essentially redundant. The table also indicates that EGGW25 is the measure most highly correlated with Qa (r = −0.912) of the three EGGW measures.

TABLE 1. *Pearson product moment correlation values for EGGW and Qa measures. N for Qa correlations is 147, and 189 for the other correlations.*

	EGGW50	EGGW75	Qa
EGGW25	0.820	0.711	−0.912
EGGW50		0.966	−0.854
EGGW75			−0.715

Relation between EGGW and Qostrb

A Wolf stroboscopic system was used with subject BB to determine the open quotient of the larynx by counting the number of frames during which the glottis was partly to fully open, and the number of frames in the complete cycle. The EGG signal was recorded simultaneously with the stroboscopic image. Comfortable pitch and loudness levels were used. Figure 5 shows the data for corresponding Qostrb and EGGW measures (the quantity 1–Qostrb, the equivalent to a closed quotient, is used in the figure). The figure suggests that, for a wide range of adduction, video frame counting and the EGGW measure not only are strongly related ($r = 0.93$), but that the values of 1–Qostrb and EGGW at the 25% level are nearly the same if the corresponding linear fit line is considered. The relation between EGGW and 1–Qostrb is given by 1–Qostrb = 0.999EGGW + 0.015 in Figure 5.

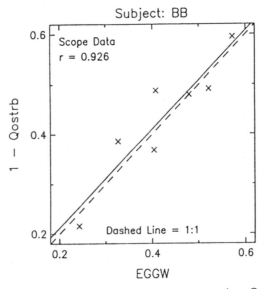

FIGURE 5. Relation between EGGW and the glottal open quotient, Qostrb, obtained by counting frames of stroboscopic images of the larynx of subject BB.

Relation between EGGW and Qodegg

EGGW obtained from the nondifferentiated EGG signal, and the open quotient from the differentiated EGG signal, were obtained for both subjects BB and RS. The results are shown in Figure 6. The figure indicates that the EGGW measure yielded larger values than did 1–Qodegg, but with a strong relationship (r=0.982 for the two subjects combined). There appears to be a greater difference between values for EGGW and Qodegg as adduction increases. EGGW is related to 1–Qodegg in Figure 6 by the linear equation 1–Qodegg = 0.902EGGW–0.0139.

Figure 7 helps to explain the discrepancy between EGGW and 1–Qodegg: the peaks of the derivative of the EGG waveform tend to fall in a narrower range (d_1) than the 25% height markings (d_2) for the EGGW measure. As adduction decreases, the discrepancy may decrease, as suggested by Figure 6.

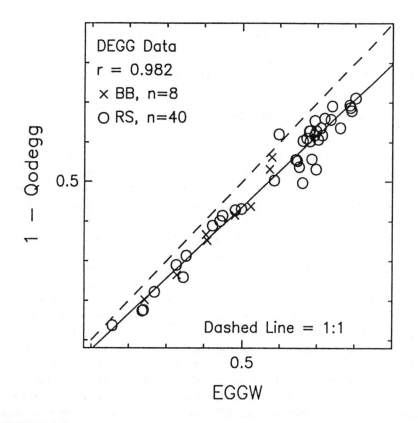

FIGURE 6. Relation between EGGW and the glottal open quotient, Qodegg, obtained from the differentiated EGG waveform for subjects BB and RS.

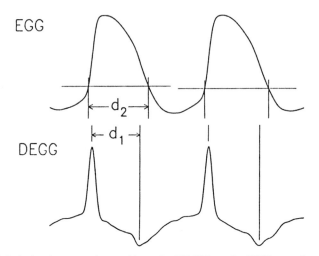

EGG

DECG

FIGURE 7. Relation between the markings for EGGW on the EGG waveform (above), and the markings for the closed quotient on the derivative of the EGG waveform (below). The markings on the EGG derivative (giving distance d_1) are closer together than are those for the 25% cut for EGGW (giving distance d_2), showing the reason for the larger EGGW measures of glottal closed quotient in Figure 6.

Relation between EGGW and Glottal Distance Measures

The larynx of subject RS was viewed using a Wolf rigid laryngoscopic system and recorded onto video tape during a wide range of adduction conditions as the subject produced an open hypopharynx vowel at comfortable pitch (one octave above the lowest pitch of the chest register) and comfortable loudness. Under visual observation and video recording, a length of cleaned soldering wire with a turned tip was passed through the vocal tract airway. The end was placed on top of the left cuneiform cartilage. This permitted the estimation of the superior oblique width of the left cuneiform cartilage (LAW as shown in Figure 8), and, thus, also an approximation of the width of the gap between the vocal processes (VPG). The vocal process gap and the width of the oblique diameter of the left cuneiform cartilage were measured directly on the video monitor. Measurements were made by two persons; the calculated maximum error expected for measurements of the vocal process gap, in cm, was +/– 13.4%. This error was calculated using measurement variabilities for the VPG monitor measurement, the LAW monitor measurement, the estimate of the actual LAW measure determined from the width of the tip of the soldering wire (including the variability for the solder width measurement), and the height discrepancy between the level of the true vocal folds and the top of the cuneiform cartilage (assumed to be 1 cm). Figure 8 is a tracing and schematic of the glottal measures for subjects RS and BB, respectively.

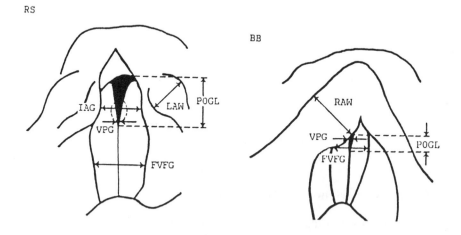

FIGURE 8. Tracings of laryngeal images for subjects RS and BB indicating the various measures. VPG = vocal process gap; IAG = interarytenoid gap; LAW = oblique width of the left cuneiform cartilage of subject RS; RAW = oblique width of the right cuneiform cartilage of subject BB; POGL = posterior glottal length; FVFG = false vocal fold gap.

Figure 9 shows the relationship between EGGW and the vocal process gap VPG for subject RS. The figure strongly suggests a reduction in the space between the vocal processes (greater adduction) as EGGW increases. The data suggest that the vocal processes touch when EGGW is between 0.60 and 0.65. For the nonlinear relationship shown in Figure 9, VPG (cm) = $1.205EGGW^2 - 1.571EGGW + 0.511$, for $0.2 \leq EGGW \leq 0.65$. The data suggest, for example, that for a vocal process gap of 0.1 cm, EGGW equals approximately 0.36 for this subject. It is also noted that a value of EGGW = 0.6, the value near which the vocal processes touch, corresponds to the perceptual boundary of hyperadduction (see discussion above for Figure 3).

A similar experiment was performed with subject BB, although without absolute measures of the vocal process gap. The larynx of subject BB was video taped with the Wolf system and the laryngeal images seen on the video monitor were copied to a Tektronix 4632 hard copy unit. The VPG measure was made at the visually consistent region where the viewed right cuneiform border intersected the vocal process border (see Figure 8). The value of the VPG was normalized using the oblique diameter of the right cuneiform cartilage (VPG/RAW). Actual gap values were not obtained.

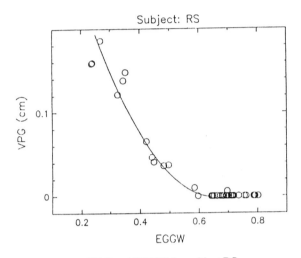

FIGURE 9. Relation between VPG and EGGW for subject RS.

Figure 10 illustrates that the relationship between EGGW and VPG/RAW for subject BB appears to be linear. Again, the data suggest that the vocal processes touch when the EGGW value is between 0.60 and 0.65 (the best fit line suggests 0.64, whereas there is one data point near VPG=0 at approximately 0.575).

After the vocal processes touch, greater adductory forces can further approximate the arytenoid cartilages. For subject RS, the medial boundaries of the arytenoids (cartilagenous glottis) could be viewed. The interarytenoidal gap (IAG, Figure 8) was approximated by measuring the video monitor distance between the bilateral supero-medial arytenoid cartilage eminences, and normalized by the left oblique cuneiform diameter. Figure 11 shows the IAG measure (maximum measurement error of +/–11.2%) vs. the EGGW measure for a wide range of adduction. This figure strongly suggests a change in the relationship near a value of EGGW = 0.65, the approximate value corresponding to a vocal process gap of zero. With greater adduction, IAG decreases rapidly as EGGW increases slowly. Near the lowest values of IAG, the scatter of EGGW is relatively high. The data suggest that values of EGGW greater than 0.65 correspond to forceful adduction of the arytenoid cartilages. Smaller IAG values and EGGW values greater than 0.65 suggest effective compression at the vocal processes and "closer" vocal folds. The scatter of EGGW data corresponding to values of IAG below about 0.3 cm suggests adjustments of the thyroarytenoid muscles, interarytenoid muscles, and perhaps subglottal pressure, resulting in a variety of widths of the closed glottis portion of the EGG waveform. It is noted that the reported values of the IAG measure may be unique to subject RS because of individual differences of the structure of the arytenoid cartilages and adductory function across individuals.

Data shown in Figure 11 (and Figure 6) suggest that the total range of expected EGGW values may be 0.15 to 0.80.

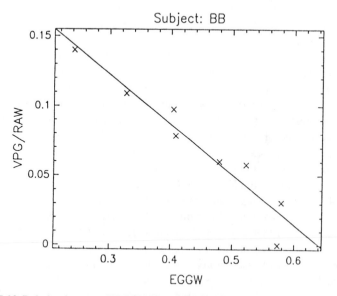

FIGURE 10. Relation between VPG/RAW and EGGW for subject BB.

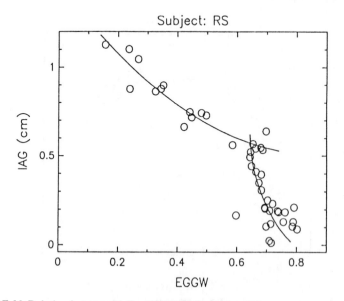

FIGURE 11. Relation between IAG and EGGW for subject RS.

Also examined (see Figure 8) were the distance changes between the ventricular folds (FVFG, the false vocal fold gap) and the anterior-posterior distance of the visible cartilaginous glottis (POGL, the posterior glottal length). Figure 12 shows the data for FVFG vs. EGGW for subject RS (the estimated maximum error for the FVFG measures for RS was +/–11.1%). Although there is some scatter of data, there was apparently little change in the distance between the medial edges of the ventricular folds until the EGGW values reached approximately 0.65, consistent with the IAG measure; beyond this point, there was a sharp change in FVFG, corresponding to the inferred hyperadduction. Figure 13 shows the corresponding measure, FVFG/RAW, for subject BB. Here the data suggest that the distance between the false vocal folds begins to decrease at a value of EGGW of approximately 0.41, a smaller value than for subject RS. The decrease in FVFG may suggest greater contraction of the superior portions of the thyroarytenoid muscle lateral to the ventricular folds.

Data for posterior glottal length (POGL, Figure 8) for subject RS are shown in Figure 14 (estimated maximum error for POGL data was +/–10.8%). POGL values decreased linearly as adduction (EGGW) increased to (once again) about 0.65, beyond which POGL values dropped sharply. The distance decreased as a result of the bunching of the soft tissue on the posterior wall and greater posterior contact of the medial arytenoid surfaces at and posterior to the vocal processes. The POGL/RAW vs. EGGW values for subject BB (Figure 15) show a similar trend as for RS; that is, BB showed a relatively linear decrease in posterior glottal length with increasing adduction over the same range of EGGW values as subject RS.

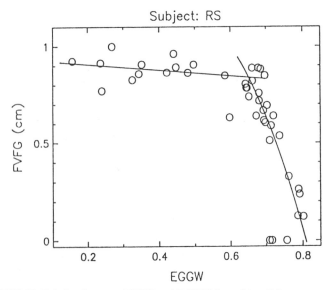

FIGURE 12. Relation between FVFG and EGGW for subject RS.

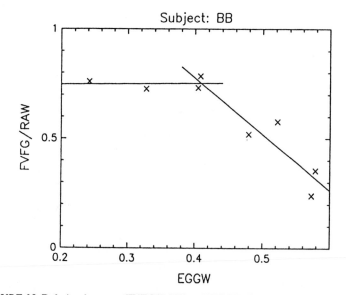

FIGURE 13. Relation between FVFG/RAW and EGGW for subject BB.

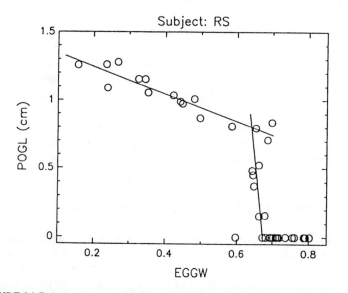

FIGURE 14. Relation between POGL and EGGW for subject RS.

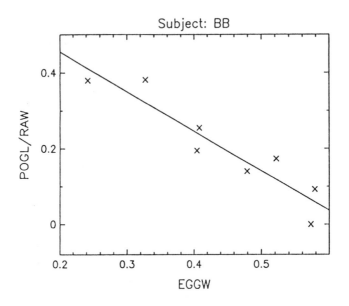

FIGURE 15. Relation between POGL/RAW and EGGW for subject BB.

Relation of EGGW to Theoretical Adduction Measures

Under the useful assumption that vocal fold tissue moves in a sinusoidal manner or that glottal area can be modelled by a truncated sinusoid, Titze (1988) and Rothenberg and Mahshie (1988) describe the abduction quotient Qa and abduction measure D, respectively, with respect to a diagram similar to Figure 16. Tissue movement or glottal area is represented by the sinusoidal waveform, tissue contact by the baseline value zero, the distance of the vocal process of one arytenoid from the midline by W/2, and the amplitude of motion of the vocal fold by A.

Using the quantities represented in Figure 16, Titze's abduction quotient is given by $Qa = W/(2A) = -\cos(\pi Qo)$, where $Qo = To/T$, To is the time the glottis is open, and T is the period of the cycle. Rearranging this statement yields

$$1 - Qo = 1 - (1/\pi)\cos^{-1}(-Qa) \qquad \text{[Equation 1]}.$$

This nonlinear relation between Qa and 1–Qo is shown in Figure 17. Rothenberg and Mahshie (1988) define their abduction measure $D = (1/2)(1 - \cos(\pi Qo))$ which also equals $(1/2)(1+Qa)$. Using the first expression for D with appropriate substitution of Qa leads to Equation 1. A more direct comparison of D and Qa is $1-D = (1/2) (1-Qa)$, and this is also shown in Figure 17. The cubic fit to the Qa values obtained by applying GLIMPES (Titze, 1984) to EGG recordings from humans, and shown above in Figure 3, is also shown in Figure 17.

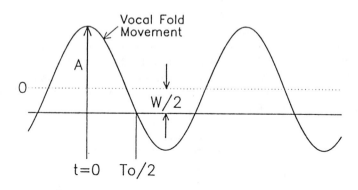

FIGURE 16. Sinusoidal representation of vocal fold movement. "A" is the amplitude of motion of one vocal fold. The dotted zero line represents the medial glottal closure location. W/2 represents half of the prephonatory glottal width at the vocal processes. "To" is the time the glottis is open. The figure is after Titze (1988) and Rothenberg and Mahshie (1988).

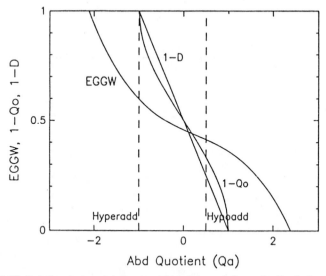

FIGURE 17. Relation between human subject data and theoretically derived functions. The function between the abduction quotient Qa and EGGW is empirical (Figure 3). The function relating 1–Qo (one minus the open quotient) and Qa is theoretically derived from Titze (1988); the function relating 1–D (one minus the abduction measure) and Qa is theoretically derived from Rothenberg and Mahshie (1988).

The range of Qa values is beyond the theoretically expected values; that is, the theoretical values of Qa range from −1 to +1, whereas human data values range from about −1.5 to +1.5 as shown in Figure 3.[1] The form of the theoretical and actual data curves, however, is not dissimilar in shape.

An abduction quotient of Qa=0 would imply that the vocal processes just touch. Figures 18(a) and 18(b), which show the data for subjects RS and BB, respectively, suggest that the vocal processes were still separated when GLIMPES gave a value of Qa=0. For subject RS, the vocal process gap was approximately 0.04 to 0.06 cm when Qa=0.

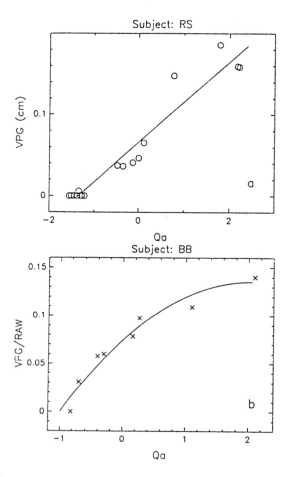

FIGURE 18. (a) Relation between VPG and Qa for subject RS; (b) relation between VPG/RAW and Qa for subject BB.

---

[1]A later application of Qa by Titze (1990) shows wider ranges of Qa than did his 1984 paper.

DISCUSSION AND CONCLUSIONS

A reliable and straightforward measure of glottal adduction is required to evaluate and establish adequate phonation within a wide variety of communication requirements, and to determine the most acoustically and physiologically efficient glottal configuration (e.g., Titze, 1988; also see Scherer, 1991). This study examined the simple glottal adduction measure EGGW. It is derived from the EGG waveform by taking a ratio of durations (or distances) obtained using an intersection line through the signal waveform at the 25% height location.

EGGW was shown to be strongly related (via a cubic equation) to Titze's (1984) abduction quotient Qa (which had been related to visual judgments of adduction in Scherer, Gould, Titze, Meyers, & Sataloff, 1988). EGGW was also shown to be strongly related to (and nearly equal to) measures of the glottal closed quotient (that is, one minus the value of the glottal open quotient, 1–Qo) using frame counting from stroboscopic views, although these data were not extensive. Values of EGGW were greater than 1–Qo obtained by the EGG derivative method, but varied from 1–Qo in a consistent manner. Since the EGG derivative method is troublesome for EGG waveforms without clear derivative peaks, EGGW may be a more reliable method.

The most significant result of this study may be the relation revealed between EGGW and the actual distance between the vocal processes of the arytenoid cartilages. The results suggested that EGGW monotonically increased as the vocal process gap decreased, at least for the comfortable pitch and loudness levels used by the two normal subjects studied. EGGW reached a value between approximately 0.60 and 0.65 when the vocal processes just touched. EGGW tended to increase as adduction was increased after the vocal processes touched, suggesting additional compression of the vocal processes of the arytenoid cartilages. This study suggests that EGGW may eventually be useful in inferring actual glottal adduction distances in subjects or patients.

Other measures of tissue approximation, such as the interarytenoid gap, the distance between the ventricular folds, and the length of the open posterior glottis, also appear to be viable corresponding measures of glottal adduction. The degree of closure of the posterior glottis is important because it may relate to the degree of hyperadduction (as suggested here), dynamic stability of arytenoid movement, and interarytenoid pressures (Scherer, Cooper, Alipour-Haghighi, & Titze, 1985), and aeroacoustic influence on the glottal volume velocity signal affecting vocal tract excitation (Cranen & Schroeter, 1992). However, the length of the open posterior glottis is not easily seen in many persons because of the "overhang" of the cuneiform and corniculate cartilages. The distance between the ventricular folds may also be a relevant measure, especially when it begins to decrease during phonation, suggesting the inclusion of additional muscle forces.

In addition, this study suggests that the range of values for the EGGW measure for normal phonation (neither hypoadducted nor hyperadducted) is between about 0.4 and 0.6. However, this conclusion drawn from a study for normal speakers may not hold (for example) for classically trained male operatic voices during singing, where full glottal closure might be the normal expectation (e.g., Scherer & Titze, 1987; also cf. Howard, Lindsey, & Allen, 1990) and phonation would not be labelled as hyperadduction with the connotation of abnormal function.

It is expected that EGGW should be useful as a glottal adduction measure for comfortable ranges of pitch and loudness for a subject over time. However, caution must be exercised in interpreting the results of this study until more extensive studies of EGGW are carried out. Thus it will be necessary to evaluate the relation between EGGW and vocal fold length (decreasing with greater length as the vertical glottal depth decreases) and subglottal air pressure (increasing with greater pressure as larger collision forces and greater contact area are expected; Orlikoff, 1991, demonstrated a significant increase in the EGGW measure with intensity increase; see also Kempster, Preston, Mack, & Larson, 1987, and Dromey, Stathopoulos, & Sapienza, 1992). Furthermore, EGGW will probably vary with any vocal fold abnormality (e.g., increasing with edema, decreasing with bowing—see, e.g., Kitzing, 1990; abnormalities associated with neurological diseases should also be explored—see, e.g., Countryman & Ramig, 1993; Ramig, Scherer, Winholtz, Benjamin, Lane, & Countryman, 1992). It also seems reasonable to expect EGGW to be affected by larynx height (if a lowered larynx tends to lengthen the vertical glottis dimension) and by vocal tract distortions (in the sense that simultaneous tilting of the head or protrusion of the mandible, as was performed in this study with subjects BB and RS for laryngeal visualization, may place the glottis in an atypical posture). Clearly, then, the relations among EGGW and independent variables of phonation (vocal fold length, subglottal pressure, and arytenoid adduction), oscillatory dependencies on normal biomechanical changes of the vocal fold (e.g., degree of contraction of the vocalis muscle), and vocal folds abnormalities, need to be mapped out—this study was performed at comfortable pitch and loudness levels only. Finally, and obviously, it is critical that only valid EGG recordings be used in evaluating the usefulness of the EGGW measure (e.g., Colton & Conture, 1990; Houben, Buekers, & Kingma, 1992), and also that comparisons of EGGW values across subjects be made with great care.

ACKNOWLEDGMENTS

Data analysis and manuscript preparation support has been provided by grant number 1 P60 DC00976 from the National Institute on Deafness and Other Communication Disorders. The authors greatly appreciate the analysis and graphics help given by Chwen-Geng Guo. An earlier version of this paper was presented at the 116th Meeting of the Acoustical Society of America, Honolulu, November, 1988. Thanks are extended to Dr. David Kuehn for a discussion of a part of this paper.

REFERENCES

Anastaplo, S., & Karnell, M. P. (1988). Synchronized videostroboscopic and electroglottographic examination of glottal opening. *Journal of the Acoustical Society of America, 83,* 1883-90.

Baken, R. J. (1987). *Clinical measurement of speech and voice.* Boston: College-Hill Press.

Baken, R. J. (1992). Electroglottography. *Journal of Voice, 6/2,* 98-110.

Brown, L. R., & Scherer, R. C. (1992). Laryngeal adduction in trillo. *Journal of Voice, 6,* 27-35.

Childers, D. G., Alsaka, Y. A., Hicks, D. M., & Moore, G. P. (1986). Vocal fold vibrations in dysphonia: Model vs. measurement. *Journal of Phonetics 14,* 429-434.

Childers, D. G., Alsaka, Y. A., Hicks, D. M., & Moore, G. P. (1987). Vocal fold vibrations: An EGG model. In T. Baer, C. Sasaki, & K. Harris (Eds.), *Laryngeal function in phonation and respiration* (pp. 181-202). Boston: Little, Brown and Company, Inc.

Childers, D. G., Hicks, D. M., Moore, G. P., Eskenazi, L., & Lalwani, A. L. (1990). Electroglottography and vocal fold physiology. *Journal of Speech and Hearing Research, 33,* 245-254.

Childers, D. G., & Krishnamurthy, A. K. (1985). A critical review of electroglottography. *CRC Critical Review Biomedical Engineering, 12,* 131-161.

Colton, R. H., & Conture, E.G. (1990). Problems and pitfalls of electroglottography. *Journal of Voice, 4,* 10-24.

Countryman, S., & Ramig, L. O. (1993). Effects of intensive voice therapy on voice deficits associated with bilateral thalamotomy in Parkinson disease: A case study. *Journal of Medical Speech Language Pathology, 1,* 233-250.

Cranen, B., & Schroeter, J. (1992). Modeling a leaky glottis. *Journal of the Acoustical Society of America, 91,* 2420(A).

Dejonckere, P. H., & Lebacq, J. (1985). Electroglottography and vocal nodules: An attempt to quantify the shape of the signal. *Folia Phoniatrica, 37,* 195-200.

Dromey, C., Stathopoulos, E. T., & Sapienza, C. M. (1992). Glottal airflow and electroglottographic measures of vocal function at multiple intensities. *Journal of Voice, 6,* 44-54.

Fourcin, A. J. (1974). Laryngographic examination of vocal fold vibration. In B. Wyke (Ed.), *Ventilatory and phonatory control systems* (pp. 315-333). New York: Oxford University Press.

Gerratt, B. R., Hanson, D. G., & Berke, G. S. (1987). Glottographic measures of laryngeal function in individuals with abnormal motor control. In T. Baer, C. Sasaki, & K. Harris (Eds.), *Laryngeal function in phonation and respiration* (pp. 521-532). Boston: Little, Brown and Company, Inc.

Higgins, M. B., & Saxman, J. H. (1993). Inverse-filtered airflow and EGG measures for sustained vowels and syllables. *Journal of Voice, 7,* 47-53.

Houben, G. B., Buekers, R., & Kingma, H. (1992). Characterization of the electroglottographic waveform: A primary study to investigate vocal fold functioning. *Folia Phoniatrica, 44,* 269-281.

Howard, D. M., Lindsey, G. A., & Allen, B. (1990). Toward the quantification of vocal efficiency. *Journal of Voice, 4,* 205-212.

Kempster, G., Preston, J., Mack, R., & Larson, C. (1987). Preliminary investigation relating laryngeal muscle activity to changes in EGG waveforms. In T. Baer, C. Sasaki, & K. Harris (Eds.), *Laryngeal function in phonation and respiration* (pp. 339-348). Boston: Little, Brown and Company. Inc.

Kitzing, P. (1990). Clinical applications of electroglottography. *Journal of Voice, 4,* 238-249.

Motta, G., Cesari, U., Iengo, M., & Motta Jr., G. (1990). Clinical application of electroglottography. *Folia Phoniatrica, 42,* 111-117.

Orlikoff, R. F. (1991). Assessment of the dynamics of vocal fold contact from the electroglottogram: Data from normal male subjects. *Journal of Speech and Hearing Research, 34,* 1066-1072.

Painter, C. (1988). Electroglottogram waveform types. *Archives of Oto-Rhino-Laryngology, 245,* 116-121.

Ramig, L. O., Scherer, R. C., Winholtz, W., Benjamin, P., Lane, K., & Countryman, S. (1992). Voice treatment and Parkinson's disease: Impact on respiratory, laryngeal and articulatory function. Paper presented to the Conference on Motor Speech Disorders, Boulder, April.

Rothenberg, M., & Mahshie, J. (1988). Monitoring vocal fold abduction through vocal fold contact area. *Journal of Speech and Hearing Research, 31,* 338-351.

Scherer, R. C. (1991). Physiology of phonation: A review of basic mechanics. In C. N. Ford, & D. M. Bless (Eds.), *Phonosurgery: Assessment and surgical management of voice disorders* (pp. 77-93). New York: Raven Press.

Scherer, R. C., Cooper, D., Alipour-Haghighi, F., & Titze, I. R. (1985). Contact pressure between the vocal processes of an excised bovine larynx. In I. R. Titze & R. C. Scherer (Eds.), *Vocal fold physiology: Biomechanics, acoustics and phonatory control* (pp. 292-303). Denver, CO: The Denver Center for the Performing Arts.

Scherer, R. C., Druker, D. G., & Titze, I. R. (1988). Electroglottography and direct measurement of vocal fold contact area. In O. Fujimura (Ed.), *Vocal physiology: Voice production, mechanisms and functions* (pp. 279-291). New York: Raven Press, Ltd.

Scherer, R. C., Gould, W. J., Titze, I. R., Meyers, A. D., & Sataloff, R. T. (1988). Preliminary evaluation of selected acoustic and glottographic measures for clinical phonatory function analysis. *Journal of Voice, 2*, 230-244.

Scherer, R. C., & Titze, I. R. (1987). The abduction quotient related to vocal quality. *Journal of Voice, 1*, 246-251.

Teaney, D. T. (1987). Fabre-Wheatstone electroglottograph—A precision R-F plethysmograph. *Proceedings of the Ninth Annual Conference of the IEEE Engineering in Medicine and Biology Society*, Boston, November, 13-16.

Titze, I. R. (1984). Parameterization of the glottal area, glottal flow, and vocal fold contact area. *Journal of the Acoustical Society of America, 75*, 570-80.

Titze, I. R. (1988). Regulation of vocal power and efficiency by subglottal pressure and glottal width. In O. Fujimura (Ed.), *Vocal physiology: Voice production, mechanisms and function* (pp. 227-238). New York: Raven Press, Ltd.

Titze, I. R. (1989). A four parameter model of the glottis and vocal fold contact area. *Speech Communication, 8*, 191-201.

Titze, I. R. (1990). Interpretation of the electroglottographic signal. *Journal of Voice 4*, 1-9.

20 Pathophysiology of the Spasmodic Dysphonias

Christy L. Ludlow
Voice and Speech Section, Division of Intramural Research, National Institute on Deafness and Other Communication Disorders

INTRODUCTION: A CURIOUS SYMPTOM COMPLEX

The spasmodic dysphonias (SD) are a group of laryngeal motor control disorders which have mystified all those who have studied them. They are both task specific, that is specific to speech, and focal, primarily affecting the functioning of the laryngeal musculature (Ludlow, 1991; Rosenbaum & Jankovic, 1988). Because of the former attribute, at first patients were thought to have an hysterical disorder. They were able to cough, clear their throat, and yell and sometimes even sing normally, but had a great deal of difficulty speaking (Aronson, 1980; Aronson, Brown, Litin, & Pearson, 1968a; Aronson, Brown, Litin, & Pearson, 1968b). Their whispered speech characteristics seemed to be within the normal range (Bloch, Hirano, & Gould, 1985).

The symptoms are acquired, often appearing in middle age, more frequently in women, and usually after a stressful event, an upper respiratory infection, or an injury (Izdebski, Dedo, & Boles, 1984). The onset may be gradual or immediate, but after an initial developmental period of two years or less, the symptoms remain chronic for life if untreated. Spontaneous remission occurs in some cases, but these are assumed to be psychogenic (Chevrie-Muller, Arabia-Guidet, & Pfauwadel, 1987).

The types of SD include: adductor SD; abductor SD; and essential voice tremor with or without either adductor or abductor SD (Aronson et al., 1968b; Aronson & Hartman, 1981; Hartman, Abbs, & Vishwanat, 1988; Hartman, 1984). The symptoms are most evident during speech, and, even if present at rest, become exacerbated with continued speaking. At present, diagnosis is solely on the basis of speech symptoms (Gates, 1992). Certain symptoms make the SDs distinguishable from other voice disorders, although some symptoms are difficult to distinguish from other voice disorders such as early Parkinsonian dysarthria, muscular tension dysphonia, and psychogenic dysphonia (Aronson et al., 1968b; Hartman et al., 1988; Hartman & Vishwanat, 1984; Hartman & Aronson, 1981; Morrison, Nichol, & Rammage, 1986).

In adductor SD, the major symptom is voice breaks occurring during vowels when quick adductory movements of the vocal folds produce an involuntary glottal stop. In abductor SD, voiceless consonants are prolonged when the vocal folds remain abducted and are slow to close for the following vowel. In voice tremor, the vocal folds may move normally until voicing begins and then they begin to oscillate at about 5 Hz with the onset of voicing. In some patients, tremor can also be seen at rest, particularly during exhalation. All three disorders are laryngeal motor control disorders specific to speech production; when the patients are not producing speech, their vocal fold movements are usually normal in range, speed, and symmetry. Thus, mechanisms involved in the neural, kinematic, kinetic and/or airflow components particular to speech must be involved in the generation of symptoms in these disorders.

Electromyographic Studies of the Spasmodic Dysphonias

Clinically, spasmodic bursts can sometimes be seen in the electromyographic (EMG) recordings of the thyroarytenoid (TA) muscles in patients with adductor SD (Shipp, Izdebski, Reed, & Morrissey, 1985). These spasmodic bursts are sporadic and can be seen in muscles other than the TA, like the cricothyroid (CT) in adductor SD (Ludlow, Hallett, Sedory, Fujita, & Naunton, 1990).

Also, high levels of muscle activity can be seen clinically in the TA muscles in some adductor SD patients. When five adductor SD patients were compared with controls, the level of TA muscle activity (in microvolts, μV) was higher in the adductor SD patients (Ludlow, Baker, Naunton, & Hallett, 1988). This measure, however, cannot be used for such comparisons because the amplitudes (in μV) depend upon the proximity of the recording electrode to a large number of motor units. In clinical neurophysiology, electromyographic recordings are normalized as a percentage of the maximal level of muscle recruitment either during supramaximal nerve stimulation or maximal task recruitment. Because of the inaccessibility of the Recurrent Laryngeal Nerve for electrical stimulation, nonspeech gestures are used to measure maximal recruitment: effort closure (for the TA), high pitch /i/ during a glide (for the CT), head raise against resistance (for the thyrohyoid, TH, and sternothyroid, ST), and sniffing (for the posterior cricoarytenoid, PCA). However, SD patients vary considerably in their abilities to perform these tasks, possibly because of their laryngeal motor control disorder. During the valsalva maneuver (effort closure), some patients and controls fail to produce a sustained TA activation, because they use other vocal tract structures to close the airway. During swallowing, laryngeal elevation and depression produce large DC shifts in EMG recordings that are movement artifacts. Also, in SD the levels of muscle activation during swallowing and valsalva are often lower than during speech, resulting in speech measures being two or three times the 'maximum level' (Watson et al., 1991; Schaefer et al., 1992). This may reflect subjects' difficulties performing these relatively unfamiliar reference tasks in comparison with speech, a highly familiar task. Alternatively, SD patients may have higher levels of muscle activation during speech than during nonspeech tasks, because they have a speech motor control disorder.

Because of these procedural difficulties, few objective comparisons have been conducted between laryngeal muscle activation levels in patients with adductor or abductor SD and normal speakers. One study (Schaefer et al., 1992) found greater variation of muscle activity in adductor SD patients than in controls, but no clear pattern of activation abnormality. These results differed across speech tasks, with the greatest variation in TA levels found in the patients on repeated word and sentence tasks. Another study, investigating physiological differences in laryngeal muscle activity in adductor and abductor SD (Watson et al., 1991), examined TA and PCA muscles during sustained /i/ and /s/. When the percent changes in muscle level were computed relative to levels during the valsalva gesture (for the TA) and forced inhalation (for the PCA), the expected patient differences were not found. Some adductor SD patients had normal levels of TA activity while some abductor SD patients had higher levels of TA activity than the adductor SD patients or the controls. The results differ somewhat from the findings of Schaefer et al. (1992) and pose the dilemma of accounting for different voice symptoms in adductor and abductor SD patients when muscle activation does not differ among patient groups and normals.

In a study of untreated adductor and abductor SD patients (Van Pelt, Ludlow, & Smith, 1994), the levels of muscle activity were compared with those of controls during syllable repetition. The syllables selected were those particularly difficult for SD patients. Clinically, glottal stops and vowels are particularly difficult for adductor SD patients, while repetition of the syllable "see" is particularly difficult for abductor SD patients. The activity of intrinsic (TA, CT, and PCA) and extrinsic (TH and ST) laryngeal muscles was measured during two adduction gestures (for phonation and for a glottal stop), and during two abduction gestures (after a vowel and after a glottal stop). Muscle activity was measured while the patients attempted these four speech gestures, albeit with considerable difficulty, but not during complete voice breaks. Resting muscle activity, activity increases for speech, and percent changes for adduction and abduction, did not differ from normal in either patient group. Thus, normal muscle activation levels and normal patterns of muscle activity occurred during speech in the patients.

In a preliminary EMG study of five adductor SD patients before and after botulinum toxin injection (Ludlow et al., 1990), the spasmodic bursts between 100 and 250 ms in duration that were 50% greater than surrounding EMG activity were counted. Two examiners counted bursts from unidentified EMG recordings of TA and CT muscles in normal controls and patients before treatment. Bursts were more frequent in the patients than in the controls in both the TA and CT muscles. After botulinum toxin injection into either one or both TA muscles, the number of bursts was reduced in both the treated and untreated muscles. Therefore, the number of bursts may be better related to symptom manifestation than muscle activation levels in SD.

To date, the results suggest that voice breaks in SD may be due to an *intrusion* of spasmodic bursts upon an otherwise normal muscle activation pattern. To better understand the pathophysiology of SD, therefore, the mechanism responsible for the generation of these spasmodic bursts must be identified.

Differences between the Pathophysiology of Vocal Tremor and Adductor and Abductor Spasmodic Dysphonia

Clinically, some mechanisms that can reduce or heighten symptoms of adductor or abductor SD do not seem to affect voice tremor, and vice versa. Following recurrent laryngeal nerve section (Izdebski, Shipp, & Dedo, 1979), patients' voice breaks improved, while the tremulous modulations in the voice did not. This was also noted during a lidocaine recurrent nerve block (Ludlow, Naunton, & Bassich, 1984).

Voice symptoms are often reduced during fiberoptic laryngoscopy, in many adductor and abductor SD patients. Certain components of the fiberoptic exam might account for this phenomenon: (a) the mucosal anesthetic usually sprayed in the hypopharynx before the exam, and/or (b) the unusual sensation caused by the insertion of the scope into the oropharynx. Surface anesthesia will interfere with the mucosal afferents in the supraglottic region, reducing afferent feedback to the brain stem. Similarly, some patients report that their symptoms improve when they suck benzocaine cough drops. Sensory tricks are also known to modify symptoms in patients with other types of focal dystonia (Marsden & Fahn, 1987). Patients with oromandibular dyskinesia, or Meige syndrome, often report that when another person places a hand on their face, their dyskinetic movements abate. The insertion of the fiberscope into the hypopharynx may be a similar "sensory trick." Vocal tremor, however, does not seem to be affected during fiberoptic laryngoscopy.

Family history also tends to differ between patient groups. Patients with vocal tremor alone are more frequently women who report that a similar voice disorder affected women in previous generations in their family.

Further, we have noted that during administration of sodium amytal, a barbiturate, symptoms remain the same or become worsened in adductor and abductor SD, but are significantly improved in patients with vocal tremor. Therefore, some differences are apparent in the pathophysiology underlying vocal tremor and adductor and abductor SD.

Differences between the Pathophysiology of Adductor and Abductor Spasmodic Dysphonia

Although the speech symptoms differ between adductor and abductor SD, it is not known if the two disorders have similar pathophysiological mechanisms underlying their motor control. Only a few studies have examined the pathophysiology of laryngeal muscle activation in abductor SD, the rarer type of SD. In a clinical report on 10 abductor SD patients (Ludlow, Naunton, Terada, & Anderson, 1991), muscle bursts were associated with prolonged voice offsets in a variety of muscles. Six patients had CT muscle bursts, while only four had bursts in the PCA muscle. Many had bursts in several muscles simultaneously: the TA, PCA, and CT. One patient did not have spasmodic bursts in any muscles; rather a lack of TA activity was noted during phonation. In another study, however, an abductor patient had higher levels of TA activation than either the controls or the adductor SD patients (Watson et al., 1991).

Treatment results using botulinum toxin injection have also suggested differences between adductor and abductor SD. Botulinum toxin injections in the TA muscles either unilaterally (Miller, Woodson, & Jankovic, 1987; Ludlow, Naunton, Sedory, Schulz, & Hallet, 1988) or bilaterally (Blitzer & Brin, 1991; Brin et al., 1987; Truong, Rontal, Rolnick, Aronson, & Mistura, 1991) have significant benefits and will reduce or eliminate symptoms in adductor SD. Fewer benefits and greater side effects follow botulinum toxin injections in abductor SD (Ludlow, Bagley, Yin, & Koda, 1992). In one study, ten abductor patients with muscle bursts observed clinically in the CT muscle, were injected in that muscle bilaterally (Ludlow et al., 1991). Only six of the ten were improved based on objective speech measures and patient reports. In two reports, the PCA muscle was injected in abductor patients (Blitzer, Brin, Stewart, Aviv, & Fahn, 1992; Rontal et al., 1991). In the larger series (Blitzer et al., 1992), 32 abductor SD patients were treated using a percutaneous approach to the PCA. None of the patients' speech reached the normal range; the greatest improvement was to 70% of normal, with an average percent improvement of 39%. Of the 32 patients, 20 improved with either unilateral or bilateral PCA injections, 9 required both bilateral PCA and bilateral CT injections, and 3 required a Type I Thyroplasty and PCA injections.

To date, no evidence is available to suggest whether or not abductor SD is a different type of disorder from adductor SD. Abductor SD patients have long been assumed to be similar to those with adductor SD but with hyperactivity or spasmodic bursts in the PCA muscle. The physiological study and treatment results thus far do not support this assumption. In fact, in some patients, problems may not be due to muscle hyperactivity or spasmodic muscle bursts, but rather to reductions in muscle activation for speech. The variety of muscles found active during voice breaks in abductor SD could be due to patients attempting to use other muscles to achieve phonation when they cannot normally produce phonation.

Neurological Abnormalities Found in the Spasmodic Dysphonias

Several indications of brain stem abnormalities have been found in adductor SD patients (Schaefer, 1983). In a study of the middle ear acoustic reflex (Hall & Jerger, 1976), the median contraction time of the stapedial muscle was lengthened by 100 ms in some SD patients in comparison with normal. A study of auditory brain stem responses reported prolongation of the inter-peak latency in a significant proportion of the SD patients studied (Finitzo-Hieber, Freeman, Gerling, Dobson, & Schaefer, 1981). Another study examined both auditory brain stem responses and the cephalic gastric reflex in a cohort of SD patients. Abnormalities were found in some SD patients on both tests: a wave V shift in latency was found in 57% of patients on ABR and reduced acid secretions in 11 of 12 patients. These findings suggest evidence of pathology in at least two pathways at the brain stem level in these patients. The abnormalities were greatest in patients with the longest duration of symptoms and in those patients who had vocal tremor in addition to adductor SD (Schaefer, 1983).

Hyperactive blink reflex responses have been found in some adductor SD patients in two studies (Cohen et al., 1989; Tolosa, Montserrat, & Bayes, 1988). Abnormal blink reflex recovery curves suggested a reduction in reflex inhibition in the affected adductor SD patients. Blink reflex abnormalities were not found in all patients, however, which may suggest that these are associated disorders occurring in some patients rather than part of the pathophysiology of SD.

Studies of cortical structure have employed magnetic resonance imaging while brain electrical activity mapping (BEAM) and single positron emission tomography (SPECT) were employed as measures of cortical physiology. In a study of 26 SD patients without controls (Devous et al., 1990), brain activity was studied at rest and during the elicitation of visual and auditory evoked potentials. Regional cerebral blood values from SPECT were abnormally low in the left and right hemispheres in 35% of patients while 57.6% had bilateral abnormalities on some of the BEAM measures. MRI abnormalities, suggesting unilateral periventricular white matter lesions, were also reported in some SD patients by the same research team (Finitzo et al., 1987; Schaefer et al., 1985). However, the cortical physiological abnormalities did not relate to the MRI findings and no physiological abnormalities were found in the same regions as the MRI abnormalities (Devous et al., 1990). Only in a few instances did the patients' results on SPECT or BEAM suggest unilateral dysfunction; these were equally in the right and left hemispheres and most abnormalities were suggestive of bilateral dysfunction (Devous et al., 1990). These findings need to be substantiated by studies of patients and controls with blind readings of the results.

None of these findings, however, are suggestive of the type of pathology that might be involved in the generation of the laryngeal motor control abnormalities seen in the SDs. Some possible mechanisms are suggested from conditioning studies in other focal dystonias.

Conditioning Abnormalities Found in Neurological Disorders

Mechanisms that could be responsible for the generation of spasmodic muscle bursts have been studied in other focal dystonias. The blink reflex is elicited either by mechanical or electrical stimulation of the supraorbital branch of the trigeminal nerve. An ipsilateral R1 response in the obicularis oculi is followed by a later bilateral response, the R2. This reflex can be elicited by a single stimulus, but when stimuli are presented in pairs with short intervals between them, responses to the second stimulus are reduced in amplitude. This is known as a conditioning effect, and demonstrates the presence of inhibitory mechanisms normally responsible for the control of these reflex responses (Kimura, 1983; Kimura, Powers, & Allen, 1969). Studies of blepharospasm, torticollis, and SD (mentioned above) have demonstrated that the conditioning effect is absent or reduced in many patients (Berardelli, Rothwell, Day, & Marsden, 1988; Berardelli, Rothwell, & Marsden, 1985; Tolosa et al., 1988). Although these studies demonstrated a reduction in normal inhibitory mechanisms in the blink reflex in various focal dystonias, not all these patients' had blepharospasm. Therefore, the reductions in conditioning effects for the blink reflex are subclinical findings. Similar conditioning abnormalities have been reported in parkinsonism suggesting that basal ganglia disease can alter the inhibitory mechanisms (Agostino, Berardelli, Cruccu, Stocchi, & Manfredi, 1987; Estaban & Gimenez-Roldan, 1975; Iriarte, Chacon, Madrozo, & Chaparro, 1989; Iriarte, Chacon, Madrazo, Chaparro, & Vadillo, 1989). Perhaps inhibitory mechanisms are affected in the focal dystonias, but to varying degrees in different cranial systems, with symptoms of a focal dystonia appearing in the system most affected.

Conditioning of Laryngeal Sensori-Motor Responses in SD

The laryngeal adductor reflex is similar to the blink reflex in physiological characteristics (Ludlow, VanPelt, & Koda, 1992). A single electrical stimulus to the superior laryngeal nerve produces an ipsilateral R1 and bilateral R2 responses in the TA muscles. The R1 response occurs at about 17 ms with a duration of approximately 15 to 20 ms. The R2 responses occur between 65 and 70 ms, with a duration of 40 to 80 ms. The R1 pathway in the cat is thought to involve afferents contained in the internal branch of the superior laryngeal nerve whose cell bodies are contained in the nodose ganglion and terminate in the nucleus tractus solitarius. Interneurons are then thought to be involved in firing both ipsilateral and contralateral motor neurons in the nucleus ambiguus (Anonsen, Lalakea, & Hanley, 1989; Cohen, Esclamado, Telian, Aloe, & Kileny, 1992; Isogai, Suzuki, & Saito, 1987). No information is currently available on the R2 pathway in the cat, although it may be transcortical (Mochida, 1990).

The laryngeal adductor reflex is usually inhibited in normal speakers (Ludlow, Schulz, Yamashita, & Koda, 1992). As with the blink reflex, when two stimuli are presented in rapid succession, the response to the second stimulus may show conditioning effects, that is, a reduction in response amplitude. This indicates that inhibitory mechanisms are evoked by the first stimulus and modify the response to the second stimulus. In normal awake humans preliminary results (Ludlow, Schulz, Yamashita, & Koda, 1992) show a linear reduction in R1 responses with reductions in interstimulus intervals down to 50 ms. R2 responses seem to be markedly reduced at interstimulus intervals below 750 ms. These conditioning effects suggest that inhibitory mechanisms normally control these responses. Further, when studied during speech and tasks like effort closure, R2 responses are reduced in normal speakers.

Currently, we are investigating these conditioning and task modulations in patients with spasmodic dysphonia with the expectations that the inhibitory modulation may be altered in SDs. As mentioned earlier, clinical observations suggest that these patients' symptoms are altered by changes in laryngeal sensation. We hope to determine if the lack of inhibition of the laryngeal adductor reflex is related to the types of spasmodic bursts that occur in the TA muscle in adductor SD (Ludlow, Koda, & Schulz, 1992). We are now studying these conditioning and task modulation effects in patients with vocal tremor, adductor SD, and abductor SD. There may not be a direct relationship, however, between the conditioning abnormalities and spasmodic bursts in SD. For example, it has been suggested that the increased eye blink rates and reduced reflex conditioning in blepharospasm are both due to abnormal excitatory drive from the basal ganglia (Jankovic, 1988). Thus, the conditioning abnormalities and spasmodic bursts may both result from a disruption of the central pathways affecting inhibitory mechanisms. Recent brain activation studies are suggestive of disinhibition mechanisms that may present in some of the focal dystonias.

Positron Emission Tomography Studies of Brain Activation Abnormalities in the Idiopathic Dystonias

Anatomical brain imaging studies of patients with *acquired* dystonias following brain injury or disease have found striatal lesions more often involving the putamen than the caudate (Fross et al., 1987). Positron emission tomography (PET) studies in *idiopathic* dystonias have found abnormalities in the basal ganglia, most often showing hypometabolism in the caudate and lentiform and the frontal projection field of the mediodorsal thalamic nucleus (Karbe, Holthoff, Rudolf, Herholz, & Heiss, 1992). Others have suggested that the relationships between the basal ganglia and the thalamus via the pallido-thalamic projection system are affected (Stoessel et al., 1986; Brooks, 1989). When patients were studied at rest, no cortical abnormalities were found (Stoessel et al., 1986).

Two recent studies of dopamine metabolism have found increased dopamine labeling in the striatum when compared with controls (Otsuka et al., 1992; Leenders et al., 1993). One study found increased striatal uptake of 18F-dopa in patients with idiopathic dystonia, but no striatal differences from normal in glucose metabolism (Otsuka et al., 1992). Another study of specific binding of striatal dopamine D2 receptors reported a trend towards higher striatal tracer uptake on the side contralateral to the affected side in torticollis patients (Leenders et al., 1993). This may suggest that the pathogenic mechanism in dystonia involves increased pre-synaptic activity of the dopaminergic system of the striatum (Otsuka et al., 1992).

Two studies have evaluated patients with idiopathic dystonia during vibrotactile stimulation, a method of eliciting a cortical response in both patients and controls, that would not reflect the patients' abnormal motor task performance. Cortical hypometabolism was found in hemispheres contralateral to the affected and unaffected sides in the primary sensori-motor and supplementary motor areas in patients with torticollis (Tempel & Perlmutter, 1990) and in patients with writer's cramp (Tempel & Perlmutter, 1993). The authors interpreted the results as suggesting that subcortical-cortical connections were disrupted. However, the cortical deficits could be a reflection of either interneuronal cortical dysfunction, or altered cortical activity due to increased inhibitory projections from the globus pallidus and/or substantia nigra via the pallido-thalamic pathway, thus reducing thalamic excitation of the supplementary motor area (Tempel & Perlmutter, 1993).

None of these studies of cortical and subcortical metabolism or neurotransmitter uptake has been conducted using SD patients. Therefore, although they do suggest that a possible basal ganglia dysfunction may alter cortical function, it is not yet known whether or not they are relevant to SD.

Hypotheses Regarding the Etiology of the Spasmodic Dysphonias

Many studies have been aimed at identifying the abnormal neurophysiological processes underlying SD. An understanding of the abnormal processes is essential to a consideration of the possible etiologies of SD. The following is a discussion of some of them.

Peripheral Nerve Injury

The possibility that a peripheral nerve injury might predispose patients to develop a focal dystonia has been suggested by Jankovic and Linden (1988). It is well known that hand cramps in professional musicians can follow a hand injury. Injury to the recurrent laryngeal nerve has also been suggested as a possible cause of SD (Dedo, Townsend, & Izdebski, 1978). One clinical report of EMG findings in SD patients with similar voice symptoms reported that some had larger motor units than normal (Blitzer, Lovelace, Brin, Fahn, & Fink, 1985), suggesting that some peripheral nerve injury with subsequent reinnervation may have taken place in these patients. A case study also reported symptoms of SD following denervation (Lieberman & Reife, 1989). However, morphologic studies of the recurrent laryngeal nerve found no evidence of recurrent nerve abnormalities in SD (Ravits, Aronson, Desanto, & Dyck, 1979).

Another possibility might be a peripheral injury to the internal branch of the superior laryngeal nerve. Because this nerve carries afferent fibers only, it might produce a "silent" peripheral injury. Basic studies of afferent discharge following peripheral injury have demonstrated that there is heightened afferent discharge following peripheral injury that produces reorganization of central circuits (Chi, Levine, & Basbaum, 1993a). The massive release of glutamate following nerve transection can have neurotoxic effects on some central cells. If the cell death is concentrated in inhibitory interneurons, it could result in increased firing of disinhibited neurons (Chi, Levine, & Basbaum, 1993b). Such a mechanism could account for changes in central inhibition following peripheral injury to afferent fibers.

Neurotoxicity

Neurotoxins can target particular cells causing selective cell death, as occurs with 1-methyl-4-phenyl-1,2,3,6-tetrahydropyridine (MPTP). This neurotoxin causes selective cell death of more than 80% of nigrostriatal neurons, producing clinical symptoms similar to Parkinsonism (Chiueh, Burns, Markey, Jacobowitz, & Kopin, 1985). The possibility of an unknown neurotoxin being specific to targeting inhibitory neurons in the brain stem must be considered.

Degenerative Neurological Disease

Several degenerative diseases are well known to affect laryngeal motor control specifically, either early in the disease, as in Parkinsonism (Logemann, Fisher, Boshes, & Blonsky, 1977); or affecting only some motor neurons, as in abduction paralysis in Multiple Systems Atrophy (also known as Shy Drager Syndrome) (Bassich, Ludlow, & Polinsky, 1984; Hanson, Ludlow, & Bassich, 1983); or causing a selective unilateral paralysis and spasticity, as in pseudobulbar palsy (Hartman, 1984). Given these highly specific laryngeal motor control disorders in known neurologic degenerative diseases, it seems possible that SD could also be a specific degenerative disease. One pathology study reported moderate-to-severe neuronal loss in several brainstem nuclei (substantia nigra compacta, locus coruleus, raphe nuclei, and pedunculopontine nucleus) in a 68 year old man with Meige syndrome (Zweig et al., 1988). However, in most focal dystonias the symptoms rarely progress, suggesting that this is not a degenerative disease. Rather, the problems in SD seem to appear at one time or progress for two years, and then remain chronic.

Central Injury

Three cases of SD appearing following a head injury have been reported (Finitzo et al., 1987). All were closed-head injuries, however, and no specific lesion or site could be identified. Because SD is not usually seen in head-injured patients, and patients with SD rarely report head trauma, this etiology is unlikely.

Genetic Predisposition

A genetic basis for idiopathic torsion dystonia in the Ashkenazi Jewish population has been identified (Kramer et al., 1990). In addition, some report a high frequency of SD patients in families with dystonia (Blitzer, Brin, Fahn, & Lovelace, 1988). When patients with only SD are studied, a very small proportion has first or second degree relatives with other forms of dystonia (Ludlow, Bless, Sedory, & Bishop, 1991; Izdebski et al., 1984).

Viral

Retrograde transport, such as occurs with neurotropic viruses like herpes simplex and rabies viruses, might account for selective central cell death as a result of a viral infection (Schwartz, 1985). Many patients report that their symptoms began immediately following an upper respiratory infection. The infection may involve an unknown virus. Further, changes in the airway mucosa during an upper respiratory infection may alter the absorption of the mucosa, making entry of viruses to the afferent nerve terminals more likely. Transsynaptic transport to higher levels in the central nervous system might account for disturbances in the pathways involving the cortex and basal ganglia.

Discussion

This review has highlighted some of the suggestions in the literature regarding the type of pathophysiology that may produce the speech symptoms seen in the SDs. Some hypotheses seem more attractive than others for investigation of this disorder. First, the disorder seems to be attributable to a loss of inhibition, perhaps due to dysfunction of brain stem inhibitory circuits and/or interruptions in the pathways between the striatum and regions normally inhibited by the striatum.

The concept of the SDs as disorders of involuntary spasmodic activity interfering in an otherwise normal motor control system needs to be investigated further. One possibility is that sensory feedback may elicit muscle bursts that are normally inhibited.

Two of the possible reasons symptoms appear only during speech and not during other laryngeal motor control activities, are: (a) speech requires much finer control of muscle activation and any loss of control would therefore become more evident in speech, or (b) the pathophysiology affects pathways in the brainstem and basal ganglia that are particular to speech.

The onset of the SDs in midlife, at times of increased stress, illness, or injury may be because of changes in the immune system that may make individuals more susceptible to a disease process. The SDs are quite rare disorders and may depend upon the coincidence of two or more factors for the pathophysiology to occur.

Finally, we do not yet know if the different disorders are the result of the same type of pathophysiology or are very different disorders. As this review indicated, vocal tremor seems independent from adductor and abductor SD and patients with SD and tremor may have both disorders. However, the fact that it frequently co-occurs with adductor or abductor SD suggests that the types of pathophysiology are closely related. Finally, although adductor and abductor SD certainly have different speech symptoms, it is not yet known if the pathophysiologies differ in the two disorders. The different results of treatment with botulinum toxin suggest that they differ, but this question needs to be addressed directly.

ACKNOWLEDGMENT

In my case, Katherine Harris has been a mentor without knowing it. At every opportunity she has always been a guiding light. She asks the critical questions, and is always open to new ideas to advance our thinking. I appreciate all she has done for so many of us.

REFERENCES

Agostino, R., Berardelli, A., Cruccu, G., Stocchi, F., & Manfredi, M. (1987). Corneal and blink reflexes in Parkinson's disease with "on-off" fluctuations. *Movement Disorders, 2,* 227-235.

Anonsen, C. K., Lalakea, M. L., & Hanley, M. (1989). Laryngeal brain stem evoked response. *Annals of Otology, Rhinology, and Laryngology, 98,* 677-683.

Aronson, A. E. (1980). *Clinical voice disorders: An interdisciplinary approach.* New York: Thieme-Stratton.

Aronson, A. E., Brown, J. R., Litin, E. M., & Pearson, J. S. (1968a). Spastic dysphonia. I. Voice, neurologic, and psychiatric aspects. *Journal of Speech and Hearing Disorders, 33,* 203-218.

Aronson, A. E., Brown, J. R., Litin, E. M., & Pearson, J. S. (1968b). Spastic dysphonia. II. Comparison with essential (voice) tremor and other neurologic and psychogenic dysphonias. *Journal of Speech and Hearing Disorders, 33,* 219-231.

Aronson, A. E., & Hartman, D. E. (1981). Adductor spastic dysphonia as a sign of essential (voice) tremor. *Journal of Speech and Hearing Disorders, 46,* 52-58.

Bassich, C. J., Ludlow, C. L., & Polinsky, R. J. (1984). Speech symptoms associated with early signs of Shy Drager syndrome. *Journal of Neurology, Neurosurgery, and Psychiatry, 47,* 995-1001.

Berardelli, A., Rothwell, J. C., Day, B. L., & Marsden, C. D. (1988). The pathophysiology of cranial dystonia. In S. Fahn (Ed.), *Advances in neurology: Dystonia 2* (50th ed.) (pp. 525-535). New York: Raven Press.

Berardelli, A., Rothwell, J., & Marsden, C. D. (1985). Pathophysiology of blepharospasm and oromandibular dystonia. *Brain, 108,* 593-608.

Blitzer, A., & Brin, M. F. (1991). Laryngeal dystonia: A series with botulinum toxin therapy. *Annals of Otology, Rhinology, and Laryngology, 100,* 85-89.

Blitzer, A., Brin, M. F., Fahn, S., & Lovelace, R. E. (1988). Clinical and laboratory characteristics of focal laryngeal dystonia: Study of 110 cases. *Laryngoscope, 98,* 636-640.

Blitzer, A., Brin, M., Stewart, C., Aviv, J. E., & Fahn, S. (1992). Abductor laryngeal dystonia: A series treated with botulinum toxin. *Laryngoscope, 102,* 163-167.

Blitzer, A., Lovelace, R. E., Brin, M. F., Fahn, S., & Fink, M. E. (1985). Electromyographic findings in focal laryngeal dystonia (spastic dysphonia). *Annals of Otology, Rhinology, and Laryngology, 94,* 591-594.

Bloch, C. S., Hirano, M., & Gould, W. J. (1985). Symptom improvement of spastic dysphonia in response to phonatory tasks. *Annals of Otology, Rhinology, and Laryngology, 94,* 51-54.

Brin, M. F., Fahn, S., Moskowitz, C., Friedman, A., Shale, H. M., Greene, P. E., Blitzer, A., List, T., Lange, D., Lovelace, R. E., & McMahon, D. (1987). Localized injections of Botulinum toxin for the treatment of focal dystonia and hemifacial spasm. *Movement Disorders, 2,* 237-254.

Brooks, D. J. (1989). Positron emission tomographic studies of the subcortical degenerations and dystonia. *Seminars in Neurology, 9,* 351-359.

Chevrie-Muller, C., Arabia-Guidet, C., & Pfauwadel, M. C. (1987). Can one recover from spasmodic dysphonia? *British Journal of Disordered Communication, 22:2,* 117-128.

Chi, S.-I., Levine, J. D., & Basbaum, A. I. (1993a). Peripheral and central contributions to the persistent expression of spinal cord fos-like immunoreactivity produced by sciatic nerve transection in the rat. *Brain Research, 617,* 225-237.

Chi, S. -I., Levine, J. D., & Basbaum, A. I. (1993b). Effects of injury discharge on the persistent expression of spinal cord fos-like immunoreactivity produced by sciatic nerve transection in the rat. *Brain Research, 617,* 220-224.

Chiueh, C. C., Burns, R. S., Markey, S. P., Jacobowitz, D. M., & Kopin, I. J. (1985). Primate model of Parkinsonism: Selective lesion of nigrostriatal neurons by 1-methyl-4-phenyl-1,2,3,6-tetrahydropyridine produces an extrapyramidal syndrome of rhesus monkeys. *Life Sciences, 36,* 213-218.

Cohen, L. G., Ludlow, C. L., Warden, M., Estegui, M. D., Agostino, R., Sedory, S. E., Holloway, E., Dambrosia, J. A., & Hallett, M. (1989). Blink reflex curves in patients with spasmodic dysphonia. *Neurology, 39,* 572-577.

Cohen, S., Esclamado, R. M., Telian, S., Aloe, L., & Kileny, P. (1992). Laryngeal brain stem evoked response in the porcine model. *Annals of Otology, Rhinology, and Laryngology.*

Dedo, H. H., Townsend, J. J., & Izdebski, K. (1978). Current evidence for the organic etiology of spastic dysphonia. *Otolaryngology-Head and Neck Surgery, 86,* 875-880.

Devous, M. D., Pool, K. D., Finitzo, T., Freeman, F. J., Schaefer, S. D., Watson, B. C., Kondraske, G. V., & Chapman, S. B. (1990). Evidence for cortical dysfunction in spasmodic dysphonia: Regional cerebral blood flow and quantitative electrophysiology. *Brain and Language, 39,* 331-344.

Estaban, A., & Gimenez-Roldan, S. (1975). Blink reflex in Huntington's chorea and in Parkinson's disease. *Acta Neurologica Scandinavia, 52,* 145-157.

Finitzo, T., Pool, K. D., Freeman, F. J., Cannito, M., Schaefer, S. D., Ross, E. D., & Devous, M. D. (1987). Spasmodic dysphonia subsequent to head trauma. *Archives of Otolaryngology-Head and Neck Surgery, 113,* 1107-1110.

Finitzo-Hieber, T., Freeman, F. J., Gerling, I., Dobson, L., & Schaefer, S. (1981). Auditory brainstem response abnormalities in adductor spasmodic dysphonia. *American Journal of Otolaryngology, 3,* 26-30.

Fross, R. D., Martin, W. R., Li, D., Stoessel, A. J., Adam, M. J., Ruth, T. J., Pate, B. D., Burton, K., & Calne, D. B. (1987). Lesions of the putamen: Their relevance to dystonia. *Neurology, 37,* 1125-1129.

Gates, G. (1992). Introduction. *Journal of Voice, 6,* 293

Hall, J. W., & Jerger, J. (1976). Acoustic reflex characteristics in spastic dysphonia. *Archives of Otolaryngology, 102,* 411-415.

Hanson, D. G., Ludlow, C. L., & Bassich, C. J. (1983). Vocal fold paresis in Shy-Drager syndrome. *Annals of Otology, Rhinology,and Laryngology, 92,* 85-90.

Hartman, D. E. (1984). Neurogenic dysphonia. *Annals of Otology, Rhinology, and Laryngology, 93,* 57-64.

Hartman, D. E., Abbs, J. H., & Vishwanat, B. (1988). Clinical investigations of adductor spastic dysphonia. *Annals of Otology, Rhinology, and Laryngology, 97,* 247-252.

Hartman, D. E., & Aronson, A. E. (1981). Clinical investigations of intermittent breathy dysphonia. *Journal of Speech and Hearing Disorders, 46,* 428-432.

Hartman, D. E., & Vishwanat, B. (1984). Spastic dysphonia and essential (voice) tremor treated with primidone. *Archives of Otolaryngology, 110,* 394-397.

Iriarte, L. M., Chacon, J., Madrozo, J., & Chaparro, P. (1989). Blink reflex in Parkinson's disease with levodopa-induced dyskinesia. *Functional Neurology, 4,* 257-251.

Iriarte, L. M., Chacon, J., Madrazo, J., Chaparro, P., & Vadillo, J. (1989). Blink reflex in dyskinetic and nondyskinetic patients with Parkinson's disease. *European Neurology, 29,* 67-70.

Isogai, Y. M., Suzuki, M., & Saito, S. (1987). Brainstem response evoked by the laryngeal reflex. In M. Hirano, J. A. Kirchner & D. M. Bless (Eds.), *Neurolaryngology: Recent advances* (pp. 167-183). Boston: College-Hill Press.

Izdebski, K., Dedo, H. H., & Boles, L. (1984). Spastic dysphonia: A patient profile of 200 cases. *American Journal of Otolaryngology, 5,* 7-14.

Izdebski, K., Shipp, T., & Dedo, H. H. (1979). Predicting postoperative voice characteristics of spastic dysphonia patients. *Otolaryngology-Head and Neck Surgery, 87,* 428-434.

Jankovic, J. (1988). Etiology and differential diagnosis of blepharospasm and oromandibular dystonia. In J. Jankovic & E. Tolosa (Eds.), *Advances in neurology: Facial dyskinesias* (49th ed.) (pp. 103-117). New York: Raven Press.

Jankovic, J., & Linden, C. V. (1988). Dystonia and tremor induced by peripheral trauma: Predisposing factors. *Journal of Neurology, Neurosurgery and Psychiatry, 51,* 1512-1519.

Karbe, H., Holthoff, V. A., Rudolf, J., Herholz, K., & Heiss, W. D. (1992). Positron emission tomography demonstrates frontal cortex and basal ganglia hypometabolism in dystonia. *Neurology, 42,* 1540-1544.

Kimura, J. (1983). *Electrodiagnosis in diseases of nerve and muscle: Principles and practice* (pp. 323-351). Philadelphia: F. A. Davis.

Kimura, J., Powers, M., & Allen, M. W. V. (1969). Reflex response of orbicularis oculi muscle to supraorbital nerve stimulation. *Archives of Neurology, 21*, 193-199.

Kramer, P. L., deLeon, D., Ozelius, L., Risch, N., Bressman, S. B., Brin, M. F., Schuback, D. E., Burke, R. E., Kwiatkowski, D. J., Shale, H. G., Gusella, J. F., Breakefield, X. O., & Fahn, S. (1990). Dystonia gene in Ashkenazi Jewish population is located on chromosome 9q32-34. *Annals of Neurology, 27*, 114-120.

Leenders, K., Hartvig, P., Forsgren, L., Holmgren, G., Almay, B., Eckernas, S. A., Lundquist, H., & Langstrom, B. (1993). Striatal [11C]-N-methyl-spirone binding in patients with focal dystonias (torticollis) using positron emission tomography. *Journal of Neural Transmission: Parkinson's Disease and Dementia Section, 5*, 79-87.

Lieberman, J. A., & Reife, R. (1989). Spastic dysphonia and denervation signs in a young man with tardive dyskinesia. *British Journal of Psychiatry, 154*, 105-109.

Logemann, J. A., Fisher, H. B., Boshes, B., & Blonsky, R. (1977). Frequency and co-occurrences of vocal tract dysfunctions in the speech of a large sample of Parkinson patients. *Journal of Speech and Hearing Disorders, 42*, 47-57.

Ludlow, C. L. (1991). Characteristics and treatment of laryngeal dystonias affecting voice users. *Medical Problems of Performing Artists, 16*, 128-131.

Ludlow, C. L., Bagley, J. A., Yin, S. G., & Koda, J. (1992). A comparison of different injection techniques in the treatment of spasmodic dysphonia with botulinum toxin. *Journal of Voice, 6*, 380-386.

Ludlow, C. L., Baker, M., Naunton, R. F., & Hallett, M. (1988). Intrinsic laryngeal muscle activation in spasmodic dysphonia. In R. Benecke, B. Conrad, & C. D. Marsden (Eds.), *Motor disturbances* (pp. 119-130). Orlando: Academic Press.

Ludlow, C. L., Bless, D. M., Sedory, S. E., & Bishop, S. G. (1991). Factors associated with spasmodic dysphonia and other disorders. *Asha, 33*, 162.

Ludlow, C. L., Hallett, M., Sedory, S. E., Fujita, M., & Naunton, R. F. (1990). The pathophysiology of spasmodic dysphonia and its modification by botulinum toxin. In A. Berardelli, R. Benecke, M. Manfredi, & C. D. Marsden, (Eds.), *Motor disturbances* (2nd ed.) (pp. 274-288). Orlando: Academic Press.

Ludlow, C. L., Koda, J., & Schulz, G. M. (1992). Laryngeal muscle responses to afferent stimulation in spasmodic dysphonia. *Movement Disorders, 7*, 127.

Ludlow, C. L., Naunton, R. F., & Bassich, C. J. (1984). Procedures for the selection of spastic dysphonic patients for recurrent laryngeal nerve section. *Otolaryngology-Head and Neck Surgery, 92*, 24-31.

Ludlow, C. L., Naunton, R. F., Sedory, S. E., Schulz, G. M., & Hallet, M. (1988). Effects of botulinum toxin injections on speech in adductor spasmodic dysphonia. *Neurology, 38,* 1220-1225.

Ludlow, C. L., Naunton, R. F., Terada, S., & Anderson, B. J. (1991). Successful treatment of selected cases of abductor spasmodic dysphonia using botulinum toxin injection. *Otolaryngol-Head Neck Surgery, 104,* 849-855.

Ludlow, C. L., Schulz, G. M., Yamashita, T., & Koda, J. (1992). Modulation of laryngeal reflexes in humans during repeated stimulation and volitional tasks. *Society for Neuroscience Abstracts, 18,* 1402.

Ludlow, C. L., VanPelt, F., & Koda, J. (1992). Characteristics of late responses to superior laryngeal nerve stimulation in humans. *Annals of Otology, Rhinology, and Laryngology, 101,* 127-134.

Marsden, C. D., & Fahn, S. (1987). Problems in the dyskinesias. In C. D. Marsden & S. Fahn (Eds.), *Movement disorders* (2nd ed.) (pp. 305-312). London: Butterworths.

Miller, R. H., Woodson, G. E., & Jankovic, J. (1987). Botulinum toxin injection of the vocal fold for spasmodic dysphonia. *Archives of Otolaryngology-Head and Neck Surgery, 113,* 603-605.

Mochida, A. (1990). Reflex control on laryngeal functions: Vibration effect of the laryngeal mucosa on recurrent laryngeal nerve reflexes. *Journal of Otolaryngology of Japan, 93,* 938-947.

Morrison, M. D., Nichol, H., & Rammage, R. A. (1986). Diagnostic criteria in functional dysphonia. *Laryngoscope, 94,* 1-8.

Otsuka, M., Ichiya, Y., Shima, F., Kuwabara, Y., Sasaki, M., Fukumura, T., Kato, M., Masuda, K., & Goto, I. (1992). Increased striatal 18F-dopa uptake and normal glucose metabolism in idiopathic dystonia syndrome. *Journal of the Neurological Sciences, 111,* 195-199.

Ravits, J. M., Aronson, A. E., Desanto, L. W., & Dyck, P. J. (1979). No morphometric abnormality recurrent laryngeal nerve in spastic dysphonia. *Neurology, 29,* 1376-1382.

Rontal, M., Rontal, E., Rolnick, M., Merson, R., Silverman, B., & Truong, D. D. (1991). A method for the treatment of abductor spasmodic dysphonia with botulionum toxin injections: A preliminary report. *Laryngoscope, 101,* 911-914.

Rosenbaum, F., & Jankovic, J. (1988). Task specific focal tremor and dystonia: Categorization of occupational movement disorders. *Neurology, 38,* 522-527.

Schaefer, S. D. (1983). Neuropathology of spasmodic dysphonia. *Laryngoscope, 93,* 1183-1204.

Schaefer, S. D., Freeman, F., Finitzo, T., Close, L., Cannito, M., Ross, E., Reisch, J., & Maravilla, K. (1985). Magnetic resonance imaging findings and correlations in spasmodic dysphonia patients. *Annals of Otology, Rhinology, and Laryngology, 94,* 595-601.

Schaefer, S. D., Roark, R. M., Watson, B. C., Kondraske, G. V., Freeman, F. J., Butsch, R. W., & Pohl, J. (1992). Multichannel electromyographic observations in spasmodic dysphonia patients and normal control subjects. *Annals of Otology, Rhinology, and Laryngology*, *101*, 67-75.

Schwartz, J. H. (1985). Synthesis and distribution of neuronal protein. In E. R. Kandel & J. H. Schwartz, (Eds.), *Principles of neural science* (2nd ed.) (pp. 37-48). New York: Elsevier.

Shipp, T., Izdebski, K., Reed, C., & Morrissey, P. (1985). Intrinsic laryngeal muscle activity in a spastic dysphonic patient. *Journal of Speech and Hearing Disorders*, *50*, 54-59.

Stoessel, A. J., Martin, W. R., Clark, C., Adam, M. J., Ammann, W., Beckman, J. H., Bergstrom, M., Harrop, R., Rogers, J. G., & Ruth, T. J. (1986). PET studies of cerebral glucose metabolism in idiopathic torticollis. *Neurology*, *36*, 653-657.

Tempel, L. W., & Perlmutter, J. S. (1990). Abnormal vibration-induced cerebral blood flow responses in idiopathic dystonia. *Brain*, *113*, 691-707.

Tempel, L. W., & Perlmutter, J. S. (1993). Abnormal cortical responses in patients with writer's cramp. *Neurology*, *43*, 2252-2257.

Tolosa, E., Montserrat, L., & Bayes, A. (1988). Blink reflex studies in focal dystonias: Enhanced excitability of brainstem interneurons in cranial dystonia and spasmodic torticollis. *Movement Disorders*, *3*, 61-69. (A)

Truong, D. D., Rontal, M., Rolnick, M., Aronson, A. E., & Mistura, K. (1991). Double-blind controlled study of botulinum toxin in adductor spasmodic dysphonia. *Laryngoscope*, *101*, 630-634.

Van Pelt, F., Ludlow, C. L., & Smith, P. J. (1994). A comparison of muscle activation patterns in adductor and abductor spasmodic dysphonia. *Annals of Otology, Rhinology, and Laryngology, 103,* 192-200.

Watson, B. C., Schaefer, S. D., Freeman, F. J., Dembroski, J., Kondraske, G., & Roark, R. (1991). Laryngeal electromyographic activity in adductor and abductor spasmodic dysphonia. *Journal of Speech and Hearing Research*, *34*, 473-482.

Zweig, R. M., Hedreen, J. C., Jankel, W. R., Cassanova, M. F., Whitehouse, P. J., & Price, D. L. (1988). Pathology in brainstem regions of individuals with primary dystonia. *Neurology*, *38*, 702-706.

Motor and Sensory Components of a Feedback-Control Model of Fundamental Frequency

21

Ingo R. Titze
National Center for Voice and Speech
The University of Iowa

INTRODUCTION

It is generally understood that the fundamental frequency of oscillation (F_0) of the vocal folds is controlled primarily by the cricothyroid muscles, secondarily by the thyroarytenoid muscles, and tertiarily by subglottal pressure. Increased activity in the cricothyroid (CT) muscles and increased subglottal pressure (P_S) usually raise F_0, whereas increased activity in the thyroarytenoid (TA) muscles may either raise or lower F_0 (Larson & Kempster, 1983; Titze, Luschei, & Hirano, 1989). The bidirectional changes of F_0 with TA activity result from the complex way in which the effective length, tension, and mass of the vocal folds change when laryngeal muscles contract. Stiffness in the vocal fold *cover*, for example, can vary in a direction opposite to the stiffness in the *body*, particularly at high F_0. The quantitative nature of these biomechanical interactions between tissue stiffness and intrinsic laryngeal muscle activity have recently been modeled (Titze, 1991) and will be reviewed later.

In this paper we begin to investigate the contributions of the sensory system to F_0 control. In particular, we wonder if the glottal closure reflex (Sasaki & Suzuki, 1976) or any other mechanical reflex from laryngeal afferents (Ludlow,

Pelt, & Koda, 1992) can play a significant role in F_0 control. It will be shown that a reflex with an 80-90 ms latency can provide the delayed negative feedback for spontaneous tremor and vibrato in the laryngeal structures (Stein & Oguztoreli, 1976). We also wonder about the importance of auditory feedback in stabilizing F_0 (Sapir, McClean, & Larson, 1983; Sapir, McClean, & Luschei, 1983). Recent findings suggest that the auditory feedback loop is corrective (Kawahara, 1993); that is, an error signal between the target F_0 and the perceived F_0 is minimized by muscle response.

OVERVIEW OF THE F_0 CONTROL SYSTEM

The F_0 control mechanisms are shown in a flow-chart (Figure 1). Three levels of activity are indicated, the cortical level, the midbrain and spinal level, and the peripheral mechanical level. At the cortical level, an intended pitch is mapped by memory into target muscle activities. In general, the target muscle activities include all respiratory and phonatory components for laryngeal posturing, but only three primary control variables are indicated here: cricothyroid target muscle activity (A_{CT}), thyroarytenoid target muscle activity (A_{TA}), and lung pressure (P_L).

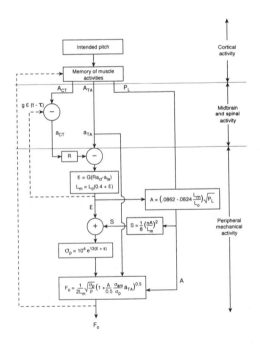

FIGURE 1. Block diagram of a fundamental frequency control system involving motor and sensory pathways.

At the midbrain and spinal level, mechanical reflexes are integrated with the target muscle activities to produce the modulated muscle activities a_{CT} and a_{TA}. (For simplicity, only one CT loop is shown, dashed lines.) This modulation is in part governed by the feedback gain, g, and by the feedback delay, τ, that exists between the vocal strain, ϵ, and muscle activity. The modulated a_{CT} is then the target A_{CT} minus the correction $g\epsilon\,(t-\tau)$.

At the mechanical level, the combined modulated muscle activities a_{CT} and a_{TA} produce a vocal fold strain, ϵ, and membranous vocal fold length L_M. We assume that these mechanical variables are sensed by mechanoreceptors (muscle spindles, pressure receptors, joint receptors, etc.) so that the state of elongation is fed back to the nervous system as mentioned above. Meanwhile, a dynamic vocal fold stretch, s, is added to the vocal fold strain, ϵ, to determine the passive elastic vocal fold stress, σ_P. This dynamic stretch s depends on the amplitude of vibration, A, which in turn depends on lung pressure, P_L, and the length of the membranous vocal fold, L_M. Finally, the passive stress, σ_P, is combined with the active stress, σ_{aM}, and a_{TA}, where σ_{aM} is the maximum active stress in the TA muscle and a_{TA} is the normalized muscle activity (ranging from 0 to 1), to produce the fundamental frequency F_0. Details of the empirical equations for the peripheral mechanical system have been reported elsewhere (Titze, Jiang, & Druker, 1988; Titze et al., 1989; Titze, 1991).

The auditory feedback system senses F_0 and modulates the target muscle activities A_{CT}, A_{TA}, and P_L to retain the target pitch. Kawahara (1993) states that the latency in this auditory loop is between 100 - 200 ms, and typically about 130 ms.

PERFORMANCE OF THE MODEL

It is difficult to show the dependence of F_0 on three independent variables (a_{CT}, a_{Ta}, and P_L) in a single diagram. Much effort has gone into constructing such a diagram, however. It is labeled a Muscle Activation Plot (MAP) and is shown in Figure 2. Normalized cricothyroid muscle activity a_{CT} is shown along the ordinate and normalized thyroarytenoid muscle activity a_{TA} is shown along the abscissa. On each axis, the value 1.0 represents maximum (100%) muscle activity. Lung pressure, P_L, and F_0 are shown as parameters. Note that the dependent variable F_0 displays a banded structure, with constant fundamental frequency bands shown in steps of 100 Hz, from 100 to 500 Hz. Owing to the width of the bands, it is clear that a large combination of a_{CT}, a_{TA}, and P_L can produce the same fundamental frequency F_0. We will now choose some specific regions within the MAP to show how F_0 varies with any single variable.

Variation of F_0 with Cricothyroid Muscle Activity

As a first example, assume that we have a low a_{TA} (0.05) and a moderate P_L (0.6 kPa). Now let us vary a_{CT} from 0% to 100%. This variation follows a vertical path from the bottom to the top of the MAP, very close to the vertical axis in Figure 2. It is clear that for a such a path F_0 increases monotonically as various bands are reached between 100 Hz and 500 Hz. This increase is shown more explicitly in Figure 3(a) (curve labeled $a_{TA} = 0.05$). Note that for this level of TA activity and lung pressure, F_0 shows a consistently positive slope, but the slope increases with higher CT activity. The summation of the strain and stretch, on the other hand, is a nearly linear variation, as shown in Figure 3(b).

The picture changes somewhat when the TA activity is increased to 50%. On the curve labeled $a_{TA} = 0.5$, we see that F_0 first decreases and then increases again. Looking back to Figure 2, an imaginative vertical line drawn from the middle of the horizontal axis ($a_{TA} = 0.5$) will first intersect the 300 Hz band, then touch the 200 Hz band, then intersect the 300 Hz band again, and finally progress towards 400 Hz. Thus, the change with cricothyroid activity is not monotonic, but shows both a decrease and an increase, depending on the region of MAP. The vocal fold stretch and strain combination, on the other hand, maintains a monotonic increase with a_{CT}, as seen in Figure 3(b).

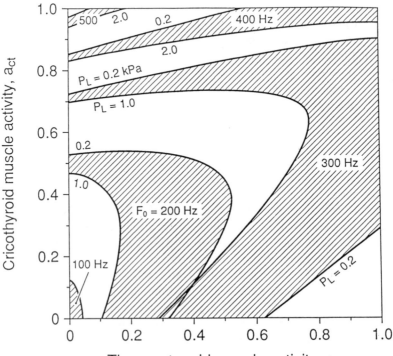

FIGURE 2. A muscle activation plot (MAP) in which cricothyroid activity (A_{CT}) is shown as a function of thyroarytenoid activity, with lung pressure and fundamental frequency shown as parameters.

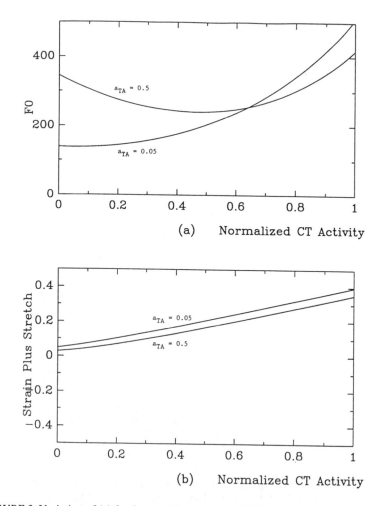

FIGURE 3. Variation of (a) fundamental frequency, and (b) strain ϵ plus dynamic stretch s as a function of cricothyroid activity. Lung pressure is constant at 0.6 kPa.

Change in F_0 with Thyroarytenoid Muscle Activity

Assume now an example where the cricothyroid activity is low, around 5%. Again, assume a moderate lung pressure of 0.6 kPa. A path of constant increase of thyroarytenoid muscle activity would be a horizontal line parallel to the abscissa in the MAP of Figure 2. It is clear that for such a path, the increase in F_0 is once again monotonic as the 100 Hz, 200 Hz, and 300 Hz bands are approached sequentially. This is shown explicitly in Figure 4(a), where we see a nearly linear increase in F_0 for $a_{CT} = 0.05$.

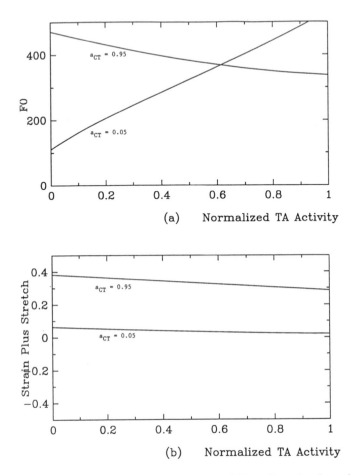

FIGURE 4. Variation of (a) fundamental frequency, and (b) strain ∈ plus dynamic stretch *s* as a function of normalized TA activity. Lung pressure is constant at 0.6 kPa.

When the cricothyroid activity is large ($a_{CT} = 0.95$), however, an increase in TA activity systematically lowers F_0, as seen by the overall negative slope in Figure 4(a). In Figure 2, this is obtained by drawing an imaginary horizontal line near the top of the MAP. It will first intersect the 500 Hz band, then the 400 Hz band, and so forth. It is clear, therefore, that the TA muscle can be used for both F_0 raising and F_0 lowering.

Change in F_0 with Lung Pressure

An increase in lung pressure is generally a move from the outer edge of the F_0 band to the inner edge in the MAP of Figure 2. That is, to keep fundamental frequency constant, some muscle activities have to decrease as lung pressure increases. On the other hand, if we do not constrain the fundamental frequency, but allow it to change with constant muscle activities, the picture shown in Figure 5 emerges. For $a_{CT} = 0.95$ and $a_{TA} = 0.05$, a highly strained vocal fold, we see only a gradual increase in F_0. The rate of change is about 20 Hz/kPa or 2 Hz/cm H_2O, which is the low end reported in the literature (Baer, 1975; Hixon, Klatt, & Mead, 1971; Lieberman, Knudson, & Mead, 1969; Titze, 1989). A much different picture emerges, however, when the $a_{CT} = 0.05$ and $a_{TA} = 0.50$.

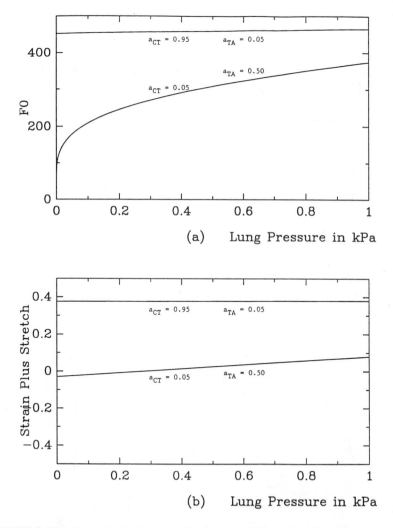

FIGURE 5. Variation of (a) fundamental frequency, and (b) strain ϵ plus dynamic stretch s as a function of lung pressure. Muscle activities are indicated.

Now, the vocal fold is shortened and the cover is made lax in relation to the body. At that point, the dynamic stretch s can be very significant and a large slope in the F_0 with lung pressure is realized (more than 200 Hz/kPa or 20 Hz/cm H_2O). Very large increases or decreases in F_0 can therefore be obtained with relatively small changes in lung pressure, which provides an ideal mechanism for frequency lowering at the end of sentences and in tonal languages. The claim made by Honda (this volume) and Maeda (1976) that the rate of change of F_0 with lung pressure is insufficient to account for the declinations measured in speech (Gelfer, Harris, Collier, & Baer, 1983), is perhaps a bit premature.

Variation of F_0 with Delayed Feedback

Figure 6 shows how modulations of F_0 and vocal fold elongation can be obtained by a feedback loop that involves a time delay, τ, and a gain, g.

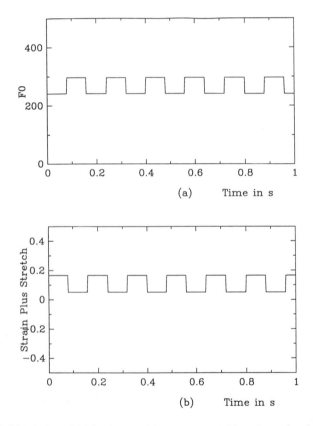

FIGURE 6. Variation of (a) fundamental frequency, and (b) strain ϵ plus dynamic stretch s as a function of time when the reflex gain is 2.5 and the reflex delay is 80 ms.

Referring back to the flow chart of Figure 1, a time delay of 80 ms and a loop gain of 2.5 was introduced. Oscillatory behavior in fundamental frequency and vocal fold elongation is seen in Figure 6. The oscillations are square instead of sinusoidal because there is as yet no inertia built into the model. Response to feedback is instantaneous. The frequency of the oscillation is approximately 6 Hz, which is remarkably close to the tremor or vibrato frequencies observed in humans. Thus, we establish the preliminary hypothesis that negative (corrective) feedback with a latency of 80 to 90 ms can be responsible for self-sustaining oscillations in F_0, that is, vibrato or tremor.

CONCLUSION

A skeletal outline of a control systems theory of fundamental frequency has been presented. Many of the building blocks of the system are yet to be quantified. In particular, a number of experiments need to be performed in which certain feedback paths are perturbed or certain blocks are disengaged, so that the individual blocks can be defined more specifically. At this point, the peripheral mechanical mechanisms are specified in greater detail than the feedback components. Furthermore, very little is quantitative at this point about the auditory reflex, but it is believed that this reflex can also play a substantial role in the control of F_0.

ACKNOWLEDGMENT

This work was supported by grant P60 DC00976 from the National Institutes on Deafness and Other Communication Disorders.

REFERENCES

Baer, T. (1975). *Investigation of phonation using excised larynxes.* Unpublished doctoral dissertation, MIT, Cambridge, MA.

Gelfer, C. E., Harris, K. S., Collier, R., & Baer, T. (1983). Is declination actively controlled? In I. Titze & R. Scherer (Eds.), *Vocal fold physiology: Biomechanics, acoustics, and phonatory control* (pp. 113-126). Denver, CO: The Denver Center for the Performing Arts.

Hixon, T. J., Klatt, D. H., & Mead, J. (1971). Influence of forced transglottal pressure on fundamental frequency. *Journal of the Acoustical Society of America, 49,* 105(A).

Kawahara, H. (1993). Transformed auditory feedback: Effects of fundamental frequency perturbation. *ATR Technical Report,* 1-13, Kyoto, Japan.

Larson, C. R., & Kempster, G. B. (1983). Voice fundamental frequency changes following discharge of laryngeal motor units. In I. R. Titze & R. C. Scherer, (Eds.), *Vocal fold physiology: Biology, acoustics, and phonatory control* (pp. 91-104). Denver CO: The Denver Center for the Performing Arts.

Lieberman, P., Knudson, R., & Mead, J. (1969). Determination of the rate of change of fundamental frequency with respect to subglottal air pressure during sustained phonation. *Journal of the Acoustical Society of America, 45*, 1537-1543.

Ludlow, C., Pelt, F., & Koda, J. (1992). Characteristics of late responses to superior laryngeal nerve stimulation in humans. *Annals of Otology, Rhinology, and Laryngology, 101*, 127-134.

Maeda, S. (1976). *A characterization of American English intonation.* Unpublished doctoral dissertation, MIT, Cambridge, MA.

Sapir, S., McClean, M. D., & Larson, C. R. (1983). Human laryngeal response to auditory stimulation. *Journal of the Acoustical Society of America, 73*, 315-321.

Sapir, S., McLean, M. D., & Luschei, E. S. (1983). Effects of frequency-modulated auditory tones on the voice fundamental frequency in humans. *Journal of the Acoustical Society of America, 73*, 1070-1073.

Sasaki, C., & Suzuki, M. (1976). Laryngeal reflexes in cat, dog, and man. *Archives of Otolaryngology, 102*, 400-402.

Stein, R., & Oguztoreli, M. (1976). Does the velocity sensitivity of muscle spindles stability the stretch reflex? *Biological Cybernetics, 23*, 219-228.

Titze, I. R. (1989). On the relation between subglottal pressure and fundamental frequency in phonation. *Journal of the Acoustical Society of America, 85*, 901-906.

Titze, I. R. (1991). Mechanisms underlying the control of fundamental frequency. In Gauffin, J., & Hammergerg, B. (Eds.), *Vocal fold physiology: Acoustic perceptual and physiological aspects of voice mechanisms* (pp. 129-138). San Diego: Singular Publishing Group, Inc.

Titze, I. R., Jiang, J., Druker, D. (1988). Preliminaries to the body-cover theory of pitch control. *Journal of Voice, 1*, 314-319.

Titze, I. R., Luschei, E. S., & Hirano, M. (1989). The role of the thyroarytenoid muscle in regulation of fundamental frequency. *Journal of Voice, 3*, 213-224.

Producing Speech: The Utterance

22 Prosodic Phrasing at the Sentence Level

Angelien Sanderman and René Collier
Institute for Perception Research/IPO

Prosody can play various roles in speech communication. One possible function of pitch contour choice, temporal variation, or pause duration is to give phonetic support to the information structure of the message. In particular, prosody is capable of demarcating the syntactic-semantic units that make up the utterance. This paper examines how a professional speaker behaves in that respect.

We dedicate our study to Kathy Harris, whose stimulating work in speech production has covered much more ground than the mechanics of articulation or phonation, but also addressed the issue of how a speaker's behavior reflects the higher-order, linguistic, organization of the message.

INTRODUCTION

Prosody can be defined most easily as the ensemble of phonetic properties that do not enter into the definition of individual speech sounds. When phonemes are turned into actual phones, the perceived identities of the vowels and consonants are not basically affected by their relative pitch or intensity, nor by the accentual or rhythmical structure of the utterance. Speech melody, loudness, prominence, temporal variation, and voice quality are suprasegmental features: they encompass higher-order units in the flow of speech, such as syllables, words, phrases, sentences, and paragraphs. Their communicative function is to carry information that is not usually expressed by the lexical or syntactic make-up of an utterance. This extra information may relate to the identity of the speaker or to his or her emotional or attitudinal state. It may also serve to indicate the relative importance of old and new elements in the flow of information. Especially in longer utterances and, *a fortiori,* in spoken texts, prosody can highlight the information structure of the message: it can mark the beginning and ending of information units, give an indication about the relative position of utterances in the discourse, or mark the boundaries between the constituents (phrases) within a sentence.

The phenomenon of prosodic phrasing is interesting from a communicative point of view to the extent that it is based on an agreement between speaker and listener as to the cue value of melodic, durational, and other variables. Therefore, the study of the speaker's behavior has to go hand in hand with observations of the effect it produces in the listener. More specifically, the perceptual prosodic boundaries should direct the investigator to those locations in the speech signal where interesting production facts are likely to be observed. Consequently, our study first looks at the speech data from a perceptual angle, using listeners' responses as a pointer to where the speaker may have deliberately inserted one or more phrasing cues. The assumption that prosodic phrasing is communicatively functional implies that the word groups that are set apart build elements of linguistic structure: they should correspond to pieces of information that can be processed as a whole. Therefore, our study also investigates whether the prosodic boundaries that turn out to be salient in perception and production terms can be motivated independently as constituents of syntactic (surface) structure.

In the prosodic domain, not much seems to be obligatory: speakers have a great deal of freedom as to whether and how they use prosody in order to transmit information. In particular, in reading aloud and in other forms of planned speech, professional speakers tend to produce richer prosodic structures than less experienced talkers. The latter do not use other, but rather fewer cues (Collier, de Pijper, & Sanderman, 1993; de Pijper & Sanderman, 1994). For this reason our study concentrates on the performance of a single professional speaker who exhibits the whole gamut of prosodic variation that one is likely to find in the linguistic population at large.

The contents of the paper are organized as follows: part 1 sketches the general research procedure; part 2 presents the analysis of the phonetic variables that mark perceptual prosodic boundaries; part 3 discusses the correspondence between prosodic boundaries and elements of linguistic structure; and part 4 offers a general discussion and some conclusions.

1. EXPERIMENTAL PROCEDURE

1.1. Speech Materials

In a pilot investigation on the same topic (Collier et al., 1993; de Pijper & Sanderman, 1994), the database was limited to only twenty (Dutch) utterances. They were rather short and differed from each other randomly. In the present investigation the materials were constructed in a more systematic way, so that a wider range of variability in length and syntactic structure was obtained, one which would elicit a reasonable amount of variation in the prosodic behavior of the speaker. The selection of materials also took into account the possibility of making comparisons between pairs or groups of utterances that differ minimally in syntactic or phonetic composition.

The number of sentences in the database was 114. The syntactic structure of the majority of them can be represented by the following set of rewrite rules:

(1) S -> NP VP

(2) NP -> Det ((AdvP) AdjP) N {(PP)}
$\qquad\qquad\qquad\qquad$ {(S)}
(3) VP -> V Det ((AdvP)AdjP) N (PP) {(PP)}
$\qquad\qquad\qquad\qquad\qquad\quad$ {(S)}

Most of these sentences also appeared as elements in conjoined or subordinate structures and parentheses were inserted into some of them. Surface structure variation was created mainly by interchanging the order of certain constituents ('fronting').

As to the phonetic make-up, we ensured that preboundary lengthening could be measured by having identical vowel-consonant sequences preceding different types of syntactic boundaries. In particular, the sequences '-aas' en '-af' occurred in most of the critical positions. The words following a phrase boundary often started with a vowel, so that the occurrence of a glottal stop as an additional boundary marker could be examined. Syntactic variability and phonetic constraints combined into sentences of the following type:

(i)   De huisbaas introduceerde de geitekaas
      (The landlord introduced the goat cheese)
(ii) Op de grote bijeenkomst 'De Kaasbaas' introduceerde de huisbaas de geitekaas en in het mooie hotel 'Bergaf' informeerde de personeelstaf de verkoopstaf.
      (At the big meeting 'The Cheese Boss,' the landlord, introduced the goat cheese and in the beautiful hotel 'Downhill' the personnel staff informed the sales staff.)

1.2. The Speaker

One professional male speaker was asked to read the sentences aloud in a relaxed and non-emphatic way. The texts of the sentences appeared in a random order on a monitor, one after the other. The pace of presentation was under the control of the experimenter, who also monitored disfluencies. The recordings were made in a sound-treated room; the speech was digitized at a 16 kHz sampling frequency.

1.3. The Listeners

Of all word boundaries in the database, 494 were selected to be evaluated for the strength of the concomitant prosodic boundary. Eighteen untrained listeners were asked to express that perceptual boundary strength (PBS) on a ten-point scale. To that end they were allowed to listen to the utterance as often as necessary. A previous investigation (Collier et al., 1993; de Pijper & Sanderman, 1994) had indicated that, in making these prosodic judgments, listeners were not biased by the fact that they could understand the sentences. The correlation between the PBS values of intelligible and unintelligible versions of the same utterance was very high ($r = 0.92$, $p < 0.01$). Therefore, it was decided this time not to present the listeners with unintelligible versions of the utterances (a procedure that greatly complicates stimulus preparation and the execution of the experimental task). The 114 randomized utterances were presented in four sessions of approximately 30 minutes. Two utterances appeared in all sessions, so that we were able to check the consistency of the subjects. Both the within-subjects and the between-subjects agreement turned out to be very high: the correlation coefficients reached significance at the 0.0001 level.

1.4. Phonetic Analyses

As stated in the Introduction, the listeners' PBS values served as a guide for looking at the speaker's performance in a selective and focused way. In particular, it was our primary aim to establish some of the phonetic bases of the perceived variation in prosodic boundary strength. This amounts to an analysis of which prosodic variables the speaker controlled in which way, whenever he produced a conspicuous boundary. The following parameters were analyzed in greater detail:

a. *Pause duration*. This was measured, to the nearest 10 ms, in the digitized speech wave. The mean duration of the few silent intervals associated with plosives (which were all < 100 ms) was subtracted from the total pause duration. When interpreting these data, the durations were categorized in classes 100 ms wide.

b. *Pitch contour characteristics both before and after the boundary*. These were established by having an expert intonologist transcribe the course of the pitch in terms of the notation conventions applicable to Dutch ('t Hart, Collier, & Cohen, 1990). The auditory transcription was checked jointly by both authors, who compared it to pitch measurements of the utterances (Hermes, 1988) and in case of uncertainty made 'close copy stylizations' of the measured curves. The auditory and acoustic analyses also paid attention to the occurrence of declination resets at the boundaries. These could be either an upward shift of the baseline or (particularly in case of parentheses) a downward one.

   c. *Preboundary lengthening.* The amount of lengthening of the phonetic segments was established with reference to the duration of the corresponding phones when averaged over all their occurrences, and using the standard deviation as a unit to express the distance to the mean. The data that pertain to this variable did not show any systematic relation to the PBS values, nor did they correlate with other prosodic variables in a significant way. Therefore they will not be included in the further presentation and discussion of the results.

   d. *Glottal stops.* The occurrence of a glottal stop at the onset of a post-boundary vowel was checked auditorily, in combination with a visual inspection of the waveform. It appeared that the speaker had produced a glottal stop in all cases, which means that he does not use this variable as a prosodic cue.

2. RESULTS

In this section we will take the perception data, in particular the PBS values assigned by the listeners, as our starting point and examine how the speaker's prosodic behavior may have cued them. Figure 1 presents a general overview of the data. It plots the average PBS values against the ten combinations of three prosodic variables that were observed. The most conspicuous trend is that the speaker induces the perception of increasingly stronger boundaries by combining a larger number of cues. A general linear model was applied to the ensemble data, using a method of least squares regression and a stepwise fitting procedure (Cohen, 1982; Keren, 1982; Stevens, 1992). First, the model comprised the addition of three main factors, each represented by a binary value: the presence or absence of a pause P, the presence or absence of a melodic cue M, and the presence or absence of a declination reset R (in an upward or downward direction). This model, expressed as PBS = P + M + R, accounts for 82% of the observed variance. If, secondly, the value of P is quantized in six categories (each 100 ms wide) and M is differentiated into six contour types (one of them being mere declination), then the explained variance rises to 89%. Finally, if the significant interactions between the variables are included in the model (PBS = P + M + R + P*M + P*R + M*R) it explains 91% of the variance. The fact that nearly all of the variance can be explained is in line with our observation that prosodic duration, in particular preboundary lengthening, is not really used (by this speaker) as a major cue.

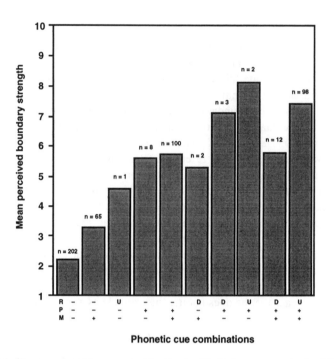

Phonetic cue combinations

FIGURE 1. The ten phonetic cue combinations with their mean perceptual boundary strength. The phonetic cue R (reset) could either be absent (–) or present with a shift in upward direction (U) or downwards direction (D). The phonetic cues P (pause) and M (melodic contour) could be absent (–) or present (+).

Let us now examine in greater detail how the speaker uses the variation in P, M, and R to mark prosodic boundaries of varying strength.

2.1. Pause and Pitch Contour

Table 1 shows the average PBS value (and the number of observations on which each is based) for specific combinations of pause-duration category and pitch contour type. If there is no pause, some pitch contour types seem to produce a higher PBS value than others. However, this small effect of pitch-contour type completely disappears when pause duration increases. Basically, the speaker lengthens the duration of his pauses to produce an increase in PBS. The net effect of pause, in the absence of any melodic cue, can be gauged in the top row of Table 1: the increasing length of the pause results in correspondingly higher PBS scores. This tendency persists in combination with pitch contours of different sorts (as can be seen in rows 2 to 6). It can also be noted that the speaker has a mild preference for particular co-occurrences of pitch contour type and pause duration: instances of pause without a melodic cue are very rare, contour types 1Ø, 1E, and 1E2 typically occur with no pause or a short one, while the remaining types seem to go with pauses of any duration.

TABLE 1. *Mean perceptual boundary strength for each combination of pause-duration category (column headings) and pitch contour type (row headings). (Number of observations in parentheses.)*

		0 ms	<100 ms	100-199ms	200-299ms	300-399ms	>399ms
0 (Declination)		2.2 (203)	5.0 (8)	7.8 (1)	8.2 (3)	-	9.7 (1)
1Ø	⟋	2.7 (12)	4.9 (5)	6.6 (2)	7.4 (3)	7.9 (5)	8.2 (1)
1E2	⟋⌢	3.0 (4)	4.4 (4)	6.4 (1)	-	-	-
1E	⟋‾	3.5 (23)	5.0 (25)	7.0 (3)	7.3 (15)	8.3 (1)	-
1A2	⟋⟍	3.6 (25)	5.2 (51)	7.2 (17)	7.4 (28)	7.5 (21)	8.4 (13)
1Ø2	⟋	3.8 (3)	6.4 (4)	-	7.3 (4)	7.7 (6)	8.3 (1)

2.2. Pitch Before and After the Boundary

The different types of pitch contour that are observed before a prosodic boundary basically derive their cue value from the fact that they end in nonlow pitch. This suggests 'continuation' or, more specifically, the end of one information unit and the beginning of the next one (Swerts, 1994). The boundary impression can be reinforced if, after the juncture, there is an audible discontinuity in the course of the pitch. To that effect, the speaker can suddenly lower the pitch immediately after the boundary. In combination with the nonlow pitch before the juncture, this results in a stronger melodic break than if the pitch were lowered more gradually. Table 2 allows a comparison of both situations. It can be observed that the PBS scores in the 'sudden lowering' (left-hand) column are systematically higher than in the 'gradual lowering' (middle) column.

TABLE 2. *Mean perceptual boundary strength for each type of contour. Row headings indicate the type of contour before the boundary. Column headings indicate the pitch value after the boundary. (Number of observations in parentheses.)*

	Declination	Sudden lowering	Gradual lowering	High declination
Declination	2.4 (216)			
1E		5.2 (41)	4.0 (12)	5.8 (14)
1E2		4.1 (6)	3.5 (2)	4.3 (1)
1Ø		5.1 (23)	4.5 (5)	-
1A2		6.4 (102)	5.5 (9)	5.7 (44)
1Ø2		6.7 (13)	4.7 (2)	7.8 (3)

After the boundary, the course of the pitch can continue in yet another way, viz. by remaining high on the upper declination line. In that case, the perceptual effect is diverse. If the pitch is high before the boundary, there is no melodic discontinuity and the PBS values in the 'high declination' (right-hand) column should resemble those in the 'gradual lowering' (middle) column. This appears to hold in the case of contour 1A2 only. In the case of contour 1Ø2, the high PBS value can be explained by the co-occurrence of a very long pause which overrides all other effects. With contour 1E, the PBS value is also rather high, but this can be understood as well: the contour before the boundary ends in mid-level pitch, so that the high pitch after the break actually creates a melodic discontinuity. The only counterexample is contour 1E2, but this single exemplar may not be representative.

2.3. Pause and Declination Reset

The speaker can also induce the impression of a major pitch discontinuity, mentioned in the previous section, by producing a declination reset after the boundary. Table 3 presents the PBS scores that result from particular combinations of pause duration and reset type (upward or downward). Resets are very unlikely if there is no pause. Upward resets predominantly co-occur with pauses longer than 200 ms, while downward resets seem to combine with pauses of less than 100 ms (the number of observations in each pause-duration group is given in parentheses). An upward reset mainly contributes to the PBS score if the pause is absent or very short. As soon as pause duration exceeds 100 ms, the presence of an upward reset raises the PBS score only little. Due to lack of sufficient data, the picture is less clear for downward resets.

In summary, this speaker produces prosodic boundaries of variable strength. In doing so, he primarily adapts the duration of the pauses; secondarily, and depending to some extent on that duration, he produces a particular type of melodic discontinuity by choosing an appropriate pitch contour and/or by resetting the declination baseline. Preboundary lengthening does not appear to be part of his strategy.

TABLE 3. *Mean perceptual boundary strength for each combination of pause-duration category (columns) and reset (rows). (Number of observations in parentheses.)*

	0 ms	<100 ms	100-199ms	200-299ms	300-399ms	>399 ms
no reset	2.4 (267)	5.1 (79)	7.1 (14)	7.7 (12)	8.4 (1)	8.5 (3)
upward R	4.6 (1)	6.0 (8)	7.0 (6)	7.3 (42)	7.6 (32)	8.3 (12)
downward R	5.2 (2)	5.0 (10)	7.8 (4)	-	-	9.7 (1)

3. PROSODIC PHRASING AND SYNTACTIC CONSTITUENCY

Apparently, our speaker takes care to use prosodic means in order to mark certain chunks in the flow of speech differentially. These cues are interpreted by our listeners as prosodic boundaries of varying strength. This systematic behavior of both partners in the communication chain makes it likely that the prosodic chunking is meaningful from the perspective of information transfer. To that effect, the units that are delineated by the speaker should correspond to elements of linguistic structure that play a role in the syntactic and semantic make-up of the message. In particular, the locations of the prosodic boundaries should coincide with those of phrase boundaries, and their strength should reflect the depth of those phrase boundaries.

We have tested this hypothesis selectively, primarily by looking at the prosodic marking of the NP-VP boundary and by comparing it to the boundary that occurs before a PP in two syntactically different contexts. Also, some observations about sentence boundaries have been included. One interesting fact is that the NP-VP boundary is accompanied by PBS scores that vary between 2.8 and 8.9, rather than having a nearly invariable marking that would correspond to its fixed status in a syntactic sense. The variability is far from random, however: the PBS score increases with the growing length (and complexity) of the NP (see Table 4A). In a sentence of the type [Det N][V Det N] the average PBS value is 5, whereas it rises to 6.5 when the NP is a [Det Adv Adj N] sequence and increases to 8.0 when a relative clause follows the head noun of the NP ([Det N S]). The length and complexity of the VP, on the other hand, play no major role. Whether the VP is as simple as [V Det Adj N] or whether it contains an additional adverbial clause is of no consequence to the strength of the prosodic boundary that precedes it. Apparently, the speaker pays more attention to properties of the sentence part that has been produced than to what is still to follow (even if he is actually able to anticipate because he is reading a printed text).

TABLE 4. *Mean perceptual boundary strength in different syntactic structures. '#' indicates which boundary was judged.*

A Syntactic structure	PBS	B Syntactic structure	PBS
[Det N] $_{NP}$ # [V Det N] $_{VP}$	5.0 (4)	[Det N # PP] $_{NP\ or\ VP}$	3.0 (17)
[Det A N] $_{NP}$ # [V Det N] $_{VP}$	5.3 (2)	[V NP] $_{VP}$ # PP	5.7 (10)
[Det Adv Adj N] $_{NP}$ # [V Det N] $_{VP}$	6.5 (2)	S $_{<main>}$ # S $_{<sub>}$	7.4 (22)
[Det N PP] $_{NP}$ # [V Det N] $_{VP}$	6.9 (2)		
[Det A N PP] $_{NP}$ # [V Det N] $_{VP}$	7.0 (2)		
[Det Adv Adj N PP] $_{NP}$ # [V Det N] $_{VP}$	7.2 (2)		
[Det N Srel] $_{NP}$ # [V Det N] $_{VP}$	8.1 (8)		

As to the boundary that precedes a PP constituent, two cases can be distinguished: one in which the PP is a nominal adjunct in a NP, and one in which it serves as an adverbial adjunct inside or outside the VP. Supposedly, the former syntactic boundary requires a weaker prosodic demarcation. Our data show that this is indeed the case: in the nominal adjunct case the PBS score is 3.0, while in the adverbial adjunct case it is 5.7 (see Table 4B). In the former case, the strength of the boundary seems to decrease as a function of the increasing length of the part of the NP that has already been uttered. The PBS score in the adverbial adjunct case can be compared to the boundary between a main clause and the subclause that follows. The latter PBS value appears to be 7.4 (Table 4B). Exactly the same score is obtained if the subclause opens the sentence and therefore precedes the main clause (averaged over six instances). Whether a full sentence or a major phrasal constituent is in front position makes no difference. If they are of comparable length, they receive the same PBS score: 7.4 and 7.2, respectively (averaged over six observations in each case). This generalization can be taken one step further: if the sentence starts with the subject NP and if that constituent has roughly the same length as the PP or S in the two cases just mentioned, again a similar score results (7.2, average of two observations). In summary, the strength of the boundary appears to depend on the length (or complexity) of the preceding constituent; this applies both to initial and final constituent positions. Secondly, the syntactic nature of the preceding constituent plays a role: a boundary inside a major constituent (a PP in an NP) is weaker than that between major constituents (a PP outside a NP or VP; a S next to another S).

From this overview it can be concluded that the positions of the prosodic boundaries and their strengths are well motivated from a syntactic-semantic point of view. Clearly, this kind of prosodic demarcation can facilitate and support the listener's processing of the incoming flow of speech, or rather, flow of information. Let us now return briefly to the speaker's perspective.

The wide range of PBS values that marks the boundaries corresponds to the different ways in which the speaker combines his prosodic cues. As we have seen, low PBS values usually go hand in hand with a short pause, a contour of type 1E, and the absence of a declination reset. High PBS values frequently imply a long pause, a contour of the type 1A2, and an (upward) declination reset. It is important to emphasize that there is a great deal of freedom and variability in our speaker's prosodic behavior. For one thing, equivalent PBS values may be obtained by different combinations of cues: for instance, a shorter pause may be combined with a larger reset, or vice versa, in order to produce the same perceptual effect. On the other hand, despite all the regularities in the data, it remains true that the same syntactic boundary may not always be marked equally strongly in all its occurrences. For instance, a particular NP-VP boundary received a PBS score of 3.6 in one version and 6.1 in another (mainly as a function of pause duration). Finally, it should be kept in mind that the freedom of the speaker may lead to a complete absence of any prosodic markers.

4. GENERAL DISCUSSION AND CONCLUSION

Our investigation set out to address the question of whether a speaker can make the linguistic and informational structure of his utterances transparent by prosodic means. In the case of our one subject, a professional speaker reading sentences from a screen, the answer is clearly affirmative. It could be observed that variable combinations of up to three prosodic cues resulted in systematic differences in the perception of prosodic boundaries by a group of listeners. In our speaker's strategy, pause duration is the one variable that correlates in greatest detail with elements of linguistic structure: the length of the pause is determined by the length of the major syntactic constituent that precedes it, irrespective of whether that constituent is a phrase or a full sentence, and irrespective of its position or syntactic function in the utterance. Longer pauses result in higher PBS values and their effect is often intensified by the presence of a melodic discontinuity and (or) a declination reset.

From the speaker's point of view, this strategy is easy to implement. It requires keeping track of the amount of linguistic material that has been produced since the beginning of the utterance or since the previous boundary. But the length of a constituent is independent of its linguistic complexity. In a previous study, based on the same speaker's performance, Terken and Collier (1992) manipulated both factors and found that they had a separate and significant influence on pause duration. Furthermore, their effects turned out to be additive. This means that the speaker also has to be aware of the surface syntactic make-up of what he has already produced, so that a long but simple constituent gets a different prosodic treatment than a long and complex one. This awareness is also required in order to differentiate between minor and major syntactic boundaries. In the former case, when the boundary occurs inside a larger component (as with a PP in an NP), a shallow prosodic break is required and pause duration has to be short. Our data show that the speaker's sentence-internal prosody is motivated by certain linguistic properties of the sentence. On the other hand, we have seen that the listeners derive perceptual boundaries of variable strength from the speaker's prosodic cues. Therefore, it seems highly plausible that listeners use that prosodic information to compute the informational structure of the input speech. Whether they actually do remains to be investigated separately. A simple test would involve the ability of listeners to disambiguate sentences by relying on the speaker's prosodic indications. A more challenging task would be to verify the hypothesis that, at least under difficult listening conditions, a prosodically well-marked text is easier to comprehend.

REFERENCES

Cohen, J. (1982). "New Look" multiple regression/correlation analysis and the analysis of variance/covariance. In G. Keren (Ed.), *Statistical and methodological issues in psychology and social research* (pp. 41-69). Hillsdale, NJ: Lawrence Erlbaum Associates.

Collier, R., de Pijper, J. R., & Sanderman, A. A. (1993). Perceived prosodic boundaries and their phonetic correlates. *Human language technology: Proceedings of a workshop held at Plainsboro* (pp. 341-345). San Francisco: Morgan Kaufmann Publishers.

de Pijper, J. R., & Sanderman, A. A. (1994). On the perceptual strength of prosodic boundaries and its relation to suprasegmental cues. *Journal of the Acoustical Society of America, 96,* 2037-2047.

Hermes, D. J. (1988). Measurement of pitch by subharmonic summation. *Journal of the Acoustical Society of America, 83,* 257-264.

Keren, G. (1982). A balanced approach to unbalanced designs. In G. Keren (Ed.), *Statistical and methodological issues in psychology and social research* (pp. 155-186). Hillsdale, NJ: Lawrence Erlbaum Associates.

Stevens, J. (1992). *Applied multivariate statistics for the social sciences.* Hillsdale, NJ: Lawrence Erlbaum Associates.

Swerts, M. G. J. (1994). *Prosodic features of discourse units.* Unpublished doctoral dissertation, University of Eindhoven.

Terken, J., & Collier, R. (1992). Syntactic influences on prosody. In Y. Tokhura, E. Vatikiotis-Bateson, & Y. Sagisaki (Eds.), *Speech perception, production and linguistic structure* (pp. 427-438). Amsterdam, Washington, Oxford: IOS Press and Tokyo, Osaka, Kyoto: Ohmsha.

't Hart, J., Collier, R., & Cohen, A. (1990). *A perceptual study of intonation.* Cambridge University Press.

23 Supralaryngeal Declination: Evidence from the Velum

Rena A. Krakow
Temple University
Haskins Laboratories

Fredericka Bell-Berti
St. John's University
Haskins Laboratories

Q. Emily Wang
University of Connecticut
Haskins Laboratories

INTRODUCTION

In several recent papers, Fowler and her colleagues (Fowler, 1988; Vatikiotis-Bateson & Fowler, 1988; Vayra & Fowler, 1992) have advanced an intriguing hypothesis—that there is a general "winding down" in speech that affects the upper articulators (e.g., the jaw and lips), in addition to fundamental frequency, acoustic amplitude, and subglottal pressure. The present study extends to another articulatory subsystem, the velum, the hypothesis that declination is a general phenomenon of speech production.

In contrast to supralaryngeal declination, fundamental frequency declination (the tendency of F_0 to fall over the course of a major syntactic unit or breath group) has received considerable attention in the phonetics and phonology literature (e.g., Breckenridge, 1977; Cohen, Collier, & 't Hart, 1982; Cooper & Sorenson, 1981; Pierrehumbert, 1979). The question of whether F_0 declination is actively controlled, long debated in the literature (see e.g., Lieberman, 1967; Cooper & Sorenson, 1981), was reconsidered by Gelfer (1987). Speech production necessitates regulation of muscular and recoil forces to control sub-glottal pressure and to extend expiration. Gelfer's results show that declining subglottal pressure over the course of a sentence can account, in large measure, for both a decline in F_0 and a concomitant decline in acoustic amplitude (see also 't Hart, Collier, & Cohen, 1990). Hence, declining F_0 and acoustic amplitude can be described as the passive consequences of the regulation of respiratory activity for speech.

To test the hypothesis that declination in speech affects supralaryngeal behavior, Fowler and her colleagues examined a number of articulatory and acoustic measures of vowels in English and Italian. Vayra and Fowler (1992) examined F_1 and F_2 frequency measures for vowels in the first and last syllables of bi- and tri-syllabic Italian pseudo-words produced by three speakers of Standard Italian. They found declination in the form of centralization for vowels at the extremes of the vowel triangle. That is, the open vowel /a/ became less open (F_1 decreased) and the close vowels, /i/ and /u/, became less close (F_1 increased). The F_2 measures indicated that there was also centralization along the front-back dimension: F_2 of front vowels decreased, and F_2 of back vowels increased. These results are consistent with the hypothesis that the latter part of an utterance is produced with increased relaxation or reduced energy.

Vayra and Fowler (1992) examined jaw opening as well as F_1 for the vowel /a/, produced by two other speakers of Standard Italian, in sequences of the form: 'BAbaba,' 'baBAba,' 'babaBA.' Their results showed a monotonic decrease in jaw opening and a corresponding decrease in F_1 for stressed vowels across the initial, medial, and final syllables of the sequences. The unstressed vowels, on the other hand, showed a V-shaped pattern, in which the least jaw opening and lowest values of F_1 occurred in the medial syllables. To determine whether jaw and formant changes over the course of the isolated trisyllables could be viewed as phrase- rather than word-level in origin, Vayra and Fowler compared the measures of the isolated trisyllables to measures of corresponding trisyllables embedded in a carrier phrase. They explained that word-level effects ought to be retained when the words are embedded in a larger context, while phrase-level effects ought to disappear or be considerably reduced. Their results were largely consistent with the interpretation that the original effects were phrasal in nature. That is, for one subject, F_1 and jaw declination occurred only in the trisyllables produced in isolation. For the second subject, F_1 declination occurred only in the isolated trisyllables, whereas jaw declination occurred in both kinds of productions, but was reduced in magnitude in the sentence as compared with the isolated word condition.

A separate study by Vatikiotis-Bateson and Fowler (1988) focused entirely on sentence-level effects; in this case, measures were made of changes over the entire span of a sentence, rather than of changes across a word produced in isolation or embedded in a sentence. This study examined a variety of kinematic measures of lip opening in reiterant sentences (CVCVCV...) produced by speakers of English. They observed weak, but consistent, declination in the extent, peak velocity, and duration of opening and closing gestures over the course of the utterances.

Unfortunately, this sentence-level study did not include measures of corresponding F_0 data. While Gelfer's study showed parallel functions for subglottal pressure and F_0, no previous study of sentence-level declination has compared measures of F_0 and articulation or formant frequencies. But F_0 and acoustic amplitude data were obtained by Vayra and Fowler (1992) for the Italian trisyllables, described above, in which jaw and formant frequency declination were observed. Of interest was their finding of a divergence between the patterns of declination for the supralaryngeal measures, on the one hand (i.e., jaw and formant frequency), and the respiratory-laryngeal measures, on the other (i.e., F_0 and acoustic amplitude). That is, whereas declination of F_1 and jaw opening had been largely restricted to stressed vowels, declination of F_0 and acoustic amplitude occurred for both stressed and unstressed vowels. This suggests that while declination may occur across different components of the speech production mechanism, its occurrence may be manifested differently from one component to the other.

The present work reflects our interest in extending the hypothesis that declination is a general phenomenon in speech production to another articulatory subsystem, the velum. Finding declination for the velum would further support earlier studies indicating that the velum participates in the prosodic organization of speech. Previous studies have shown, for example, robust effects of stress (Krakow, 1989, 1993; Vaissière, 1988) and speaking rate (Bell-Berti & Krakow, 1991; Kent, Carney, & Severied, 1974; Krakow, 1993; Kuehn, 1976) on the velum. Data collected for other purposes (Bell-Berti & Krakow, 1991) revealed what appeared to be declination in peak velum positions for oral consonants over the course of individual sentences, but the sequences were not appropriately designed to test for declination systematically. Hence, we decided to explore this issue in more detail with more suitably designed sequences. Guided by the studies described above, we investigated a number of specific issues.

We compared initial with final velar peaks in both natural and reiterant speech, to answer the question of whether there was declination from the first to the last syllables of the sentences, as has been reported for fundamental frequency. Since, in measuring F_0 early and late in a sentence, researchers have sought stability in one or the other endpoint and/or in the total amount of declination in sentences of varying length, we looked for such stability in our velar measures. We asked whether initial or final velar peaks, or the difference between them, was influenced by sentence length. Conceivably, examination of initial and final measures might provide little insight regarding declination throughout the utterance. We therefore compared measures of medial velar peaks (from the reiterant speech samples) to determine whether declination occurred throughout individual sentences, and whether the nature of the decline, if observed, changed over the time course of a sentence. Measurement of medial peaks also provided information on the relation between stress and declination in a manner that initial and final peaks did not. That is, the initial and final peaks occurred in stressed syllables whereas the medial peaks occurred in both stressed and unstressed syllables. Looking across medial sentence positions, we were able to compare declination in stressed and unstressed syllables, and to determine whether the contrast between the stressed and unstressed syllables in a word changed as a function of the word's position in a sentence. Finally, to determine whether the relation between stress and position in sequence might distinguish velar (i.e., supralaryngeal) declination from F_0 declination (as Vayra and Fowler observed for formant and jaw declination vs. F_0 declination), we compared our measures of velar peaks to corresponding measures of F_0 peaks in stressed and unstressed syllables variously positioned in our sentences.

Methods

Data Collection

Subjects. Three native speakers of American English served as subjects. One of the subjects (S1) was a co-author. No subject had a history of speech or hearing impairment.

Stimuli. The utterance list was devised to enable us to measure velar peaks over the course of individual sentences ranging in length from three to nine syllables. We used both natural and reiterant sentence stimuli (Table 1), making most of our measures on the latter, to avoid the confounding of segmental and prosodic effects on the velum. To ensure that the reiterant sentence data at least resembled the natural sentence data, we did make a number of measures on both. We also attempted, to the extent possible, to have our natural sentences make sense and yet vary minimally with respect to consonant segments. In this experiment, we chose to look at velar peaks. Among speech sounds, obstruents require the highest positions of the velum. Thus, all of the syllable-initial consonants in our study were obstruents. In our natural sentences, most were /s/. In our reiterant sentences, all were /t/ (the reiterant syllable was "ten").

TABLE 1. *Stimuli.*

Natural speech sentences	# of syllables
Sue saw Sid.	3
Suzy saw Sid.	4
Suzy saw sad Sid.	5
Suzy saw sexy Sid.	6
Suzy saw sad sexy Sid.	7
Suzy saw sexy sassy Sid.	8
Suzy saw sad sexy sassy Sid.	9
reiterant syllable: "ten"	

Instrumentation. The time-varying vertical position of the velum was tracked with the Velotrace (Horiguchi & Bell-Berti, 1987). The device, which is inserted through the nose, consists of three major parts: a curved internal lever that rests on the nasal surface of the velum, an external lever that remains in full view outside of the nose, and a push rod (carried on a thin support rod) that connects the internal and external levers. Movements of the velum result in changes in the angle of the internal lever with respect to its fulcrum that are reflected in corresponding angular movements of the external lever; the levers are connected so that when the internal lever is raised (by a raising movement of the velum), the external lever moves toward the subject. Hence, measurement of the movement of the external lever in the x-dimension provides information on the y-dimension of velum displacement. The external lever of the Velotrace is considerably longer than the internal lever, so the obtained displacements are larger (about twice as large) than the actual displacements.

An optoelectronic tracking system (Kay, Munhall, Vatikiotis-Bateson, & Kelso, 1985) was used to track the movements of light-emitting diodes (LEDs) mounted on the external lever and on the fulcrum of the Velotrace (the latter, for reference). The positions of the LEDs in the midsagittal plane were tracked by a position-sensitive detector, and the reference signal was subtracted from the data signal in order to factor out head movement. Acoustic recordings were obtained simultaneously with the movement recordings. All signals were recorded onto a 14-channel FM data recorder.

Procedure. The natural sentences were printed on individual file cards and shown to subjects one at a time. Each reiterant sentence was produced with the same number of syllables as its natural speech model (see Liberman & Streeter, 1978). Subjects were instructed to produce the natural sentence first, followed by the matched reiterant sentence. Subjects produced 12 randomized repetitions of each of the seven natural-reiterant sentence pairs. A small number of tokens had to be deleted from analysis because they were truncated during recording.

Data Analysis

Measurement. We measured the syllable-initial velar peaks for the first and last syllables of both the natural and reiterant sentences. For the natural sentences, we measured the velar peaks for /s/ in "Sue" or "Suzy" and "Sid." For the reiterant sentences, we measured the velar peaks for /t/ in all of the syllables. Measurements of the first and last syllable-initial peaks in each sentence enabled us to see how well the reiterant speech was modeled on the natural speech. One of the measures used to compare the two was the interval between the first and last peaks, as a function of the number of syllables in each sentence. Figure 1 reveals that the natural and reiterant sentences were of similar duration.

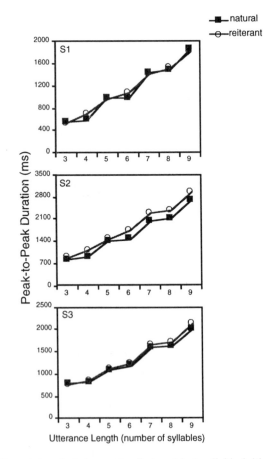

FIGURE 1. Mean intervals between the first and last syllable-initial velar peaks as a function of the number of syllables in each sentence. Natural sentences are plotted with solid squares, and the corresponding reiterant sentences, with open circles. Data for S1, S2, and S3, are shown at the top, middle, and bottom, respectively.

The other comparisons between natural and reiterant sequences involved the declination measures, which are described in the results section, below. In order to examine the interaction of stress with position in a sentence on velar vs. F_0 data, we also measured peak F_0 for the reiterant stressed and unstressed syllables of "Suzy," "sexy," and "sassy."

Results

Initial vs. Final Positions: Velar Data

Figure 2 shows initial and final peak values for natural and reiterant sentences. (We remind you that our figures are derived from velar movements as monitored with the Velotrace, which magnifies those movements about two-fold.) Testing for declination from the beginning to the end of the sentences, we first ran a three-way ANOVA that examined the variables: sentence *position* (initial vs. final peak), sentence *length* (3,4,5,6,7,8,9 syllables), and sentence *type* (natural vs. reiterant). The results of this ANOVA are shown in Table 2.

The results revealed a significant main effect of *type* for each of the three subjects reflecting the fact that natural peaks were consistently higher than reiterant peaks. This finding seems largely to be a function of the fact that the reiterant speech contained nasal segments (which display lower velar peaks because of coarticulation) while the natural speech did not. The effect may also reflect other segmental differences between the two types of sentences (e.g., syllable-initial /s/ vs. /t/). The results also showed a main effect of *position* (i.e., initial vs. final peaks) for the three subjects, supporting the notion that there is declination of velar movements from the beginning to the end of sentences.

For S2, the main effect of *length* was not significant and there were no significant two- or three-way interactions of factors. Hence, only *position* and *type* showed a significant influence on peak velum position. For S1 and S3, there was a main effect of *length*, suggesting that velar peaks were affected by the number of syllables in the sentence. However, both of these subjects also showed significant interactions that likely bear on the nature of this main effect. A significant *position* × *length* interaction for both S1 and S3 meant that we needed to examine the effect of *length* on first vs. last peaks. For S3, the difference between first vs. last measures, examined in an analysis of simple main effects, showed significance for sentences of each *length* ($p < 0.01$), except those of five syllables, which approached significance ($p = 0.06$). For S1, the effect was significant for all but the shortest *length* (3 syllables: $p > 0.1$; 4 - 5 syllables: $p < 0.05$; 6 - 9 syllables: $p < 0.01$—see Table 3, for the detailed results). These results suggest that declination, as assessed by comparisons between the first and last syllable in a sentence, is observed across a range of sentence lengths for all three subjects.

To clarify the nature of the interaction between *length* and *position*, we also ran an analysis of simple main effects comparing first vs. last peaks separately for the reiterant and natural speech sentences. The effect of sentence *length* on initial measures was not significant for either S1 or S3 (S1: [$F(6,298) < 1.0$, $p = 0.7893$]; s3: [$F(6,308) = 1.02$, $p = 0.4127$]), while the effect of sentence length on final measures was significant for both (S1: [$F(6,298) = 5.20$, $p < 0.0001$]; S3: [$F(6,308) = 5.87$, $p < 0.0001$]). This means that for these subjects, the main effect of *length* stems from effects on final measures alone.

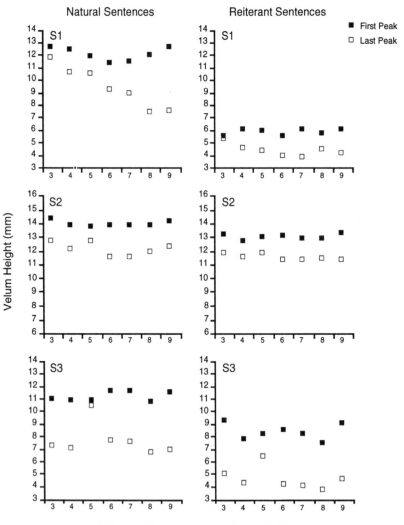

FIGURE 2. Mean values of velum height at the first (solid squares) and last (open squares) syllable-initial peaks in the natural (left) and reiterant (right) sentences as a function of the number of syllables in each sentence. Data for S1, S2, and S3, are shown at the top, middle, and bottom, respectively.

TABLE 2. *Results of three-way ANOVA on the effects of position, length, and type on peak velar positions.*

Subject	S1			S2			S3		
variable	**F**	**(df)**	**p**	**F**	**(df)**	**p**	**F**	**(df)**	**p**
position	76.9	(1,298)	<0.0001	31.18	(1,303)	<0.0001	245.81	(1,308)	<0.0001
length	3.2	(6,298)	<0.01	0.32	(6,303)	NS	3.48	(6,308)	<0.01
type	586.72	(1,298)	<0.0001	6.92	(1,303)	<0.01	163.34	(1,308)	<0.0001
position × type	6.12	(1,298)	<0.05	0.43	(1,303)	NS	0.34	(1,308)	NS
length × type	1.55	(6,298)	NS	0.05	(6,303)	NS	0.82	(6,308)	NS
position × length	2.52	(6,298)	<0.05	0.18	(6,303)	NS	3.4	(6,308)	<0.01
position × type × length	1.42	(6,298)	NS	0.06	(6,303)	NS	0.26	(6,308)	NS

Because there was no significant interaction of *length* with *type* for either subject, we collapsed the measures for natural and reiterant sentences at the final position to look at the nature of *length* effects in final positions for the two subjects. We observed a negative correlation between sentence *length* and the final peak that was significant for S1 [$r = -0.264$, $p < 0.001$] but not for S3 [$r = -0.126$, $p > 0.1$]. Thus, for S1, increasing *length* resulted in increasingly lower final velar peaks. For S3, who showed an effect of *length* on final consonants, the nature of that effect is not clear.

Only S1 showed another significant interaction, that of *position × type*, meaning that the first vs. last peak difference was influenced by their occurring in natural or reiterant sentences. Testing the simple main effect of *position* (i.e., first vs. last) for the natural [$F(1,298) = 61.89$, $p < 0.01$] and then for the reiterant sentences [$F(1,298) = 20.24$, $p < 0.01$], we found that while both showed higher initial than final peaks, the difference was larger for the natural sentences (mean 12 mm vs. 9.43 mm) than for the reiterant sentences (mean 5.82 mm vs. 4.36 mm).

In studies investigating the decline of F_0, researchers have measured endpoint values, as we have just described for the velum; they have also examined the magnitude of the difference between the two endpoint values. Therefore, to complete our examination of the relation between sentence length and velum

height, we ran a two-way ANOVA on the effects of sentence *length* (3-9 syllables) and sentence *type* (natural/reiterant) on the difference in velum height from
the first to the last syllable-initial peak (Table 4).

The effect of sentence *type* on the difference in velum position was significant
only for S1 and S2, but the effect was not consistent across subjects: on average,
S1 showed greater differences for natural than reiterant sentences, whereas S2
showed greater differences for reiterant than natural sentences. The effect of
length was significant for all three subjects. To determine whether this latter
effect reflected greater declination for longer sentences, we tested the correlation
between sentence *length* and the measured difference between the first and last
peaks.

TABLE 3. *Results of an analysis of the simple main effect of measurement
position on peak velar height at each sentence length.*

Subject:		S1			S3	
number of syllables	*F*	(df)	*p*	*F*	(df)	*p*
3	0.89	(1,298)	NS	42.61	(1,308)	<0.0001
4	6.64	(1,298)	<0.05	35.08	(1,308)	<0.0001
5	5.82	(1,298)	<0.05	3.44	(1,308)	NS
6	9.01	(1,298)	<0.01	45.91	(1,308)	<0.0001
7	15.10	(1,298)	<0.001	45.15	(1,308)	<0.0001
8	22.69	(1,298)	<0.0001	40.14	(1,308)	<0.0001
9	30.36	(1,298)	<0.0001	53.63	(1,308)	<0.0001

TABLE 4. *Results of a two-way ANOVA on the effects of sentence length and
type on the change in velar position from the earliest to the latest
measurement point.*

Subject:	S1			S2			S3		
variable	*F*	(df)	*p*	*F*	(df)	*p*	*F*	(df)	*p*
length	9.98	(6,149)	<0.0001	3.60	(6,152)	<0.01	7.02	(6,154)	<0.0001
type	24.01	(1,149)	<0.0001	88.11	(1,152)	<0.0001	0.69	(1,154)	NS
length × type	5.55	(6,149)	<0.0001	0.42	(6,152)	NS	0.53	(6,154)	NS

For S2 and S3, we collapsed measures from the reiterant and natural sentences, while for S1 (who showed an interaction between the two factors), we did not. (For S1 natural, r = –0.684, p < 0.001; for S1 reiterant, r = –0.229, p < 0.05; for S2, r = –0.189, p < 0.05; for S3, r = –0.156, p < 0.05). Clearly, the difference measures captured something that the measures of first and last peaks taken separately did not. Nonetheless, these correlations, while significant, were weak, with the exception of S1's natural sentence data.

Medial Positions: Velar Data

A reduction in peak velum height from the beginning to the end of a sentence is consistent with the notion of decreasing energy from start to finish in a sentence, as higher velum positions are normally associated with increased activity of the levator palatini (see Bell-Berti, 1993). However, the notion that the phenomenon of declination occurs throughout the sentence requires measurements of medial sentence positions. Figure 3 (which displays sample velum movement patterns for reiterant sentences) shows that while there was a general decline in the velar peaks over the course of sentences, there were also some local increases and decreases, similar to patterns observed for F_0 and acoustic amplitude. Such local effects have been attributed, at least in part, to stress effects (see Gelfer, 1987).

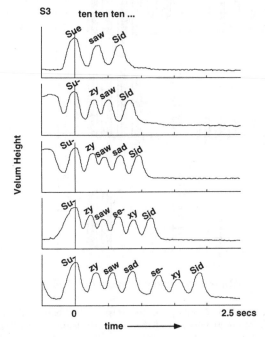

FIGURE 3. Sample velum movement patterns for reiterant sentences at each of five different syllable lengths, produced by Subject 3. The corresponding natural sentences are shown above the traces.

To determine whether there were declination-type effects on medial as well as marginal positions, and to examine the nature of other co-occurring effects, we examined the peaks in the reiterant syllables corresponding to the bisyllabic words, "Suzy," "sexy" and "sassy," in the natural sentences. In this way, it was possible to compare velar peaks for both stressed and unstressed syllables occurring earlier and later in the same sentence. It was, of course, also possible to compare stressed vs. unstressed syllables in the same word.

Note that not all bisyllabic words occurred in all of our sentences. That is, "Suzy" occurred in six of the sentence types, "sexy" in four, and "sassy," in only two (see Table 1). Our first analysis including medial sentence positions examined the effect of stress for each subject's productions of each bisyllabic word. In almost every case, peaks for stressed syllables were higher than peaks for unstressed syllables in the same word (Figure 4). We ran a two-way ANOVA on the effects of stress and sentence length on each of the bisyllables (Table 5).

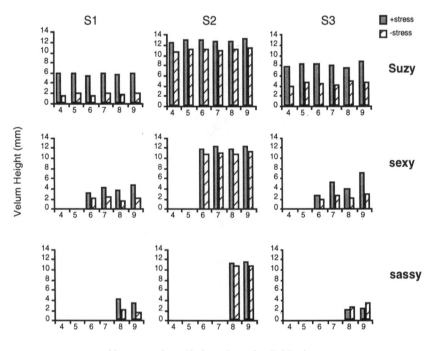

FIGURE 4. Mean values of velar peaks for stressed (dark bars) and unstressed (light bars) syllables in the reiterant versions of the three bisyllabic words, "Suzy" (top), "sexy" (middle), "sassy" (bottom). Data for S1, S2, and S3 are shown at the left, middle, and right, respectively, and are plotted as a function of the number of syllables in the sentences from which the bisyllables were measured.

TABLE 5. *Results of a two-way ANOVA on the effects of stress and sentence length on velar peaks in bisyllabic words.*

Subject:	S1			S2			S3		
SUZY	*F*	**(df)**	*p*	*F*	**(df)**	*p*	*F*	**(df)**	*p*
stress	195.29	1,132	<0.0001	16.16	1,130	<0.0001	91.98	1,132	<0.0001
length	0.38	5,132	NS	0.13	5,130	NS	0.46	5,132	NS
stress									
× length	0.13	5,132	NS	0.00	5,130	NS	0.42	5,132	NS
SEXY	*F*	**(df)**	*p*	*F*	**(df)**	*p*	*F*	**(df)**	*p*
stress	26.82	1,88	<0.0001	3.57	1,86	NS	38.89	1,88	<0.0001
length	1.36	3,88	NS	0.16	3,86	NS	9.03	3,88	<0.0001
stress *× length*	0.43	3,88	NS	0.01	3,86	NS	3.31	3,88	<.05
SASSY	*F*	**(df)**	*p*	*F*	**(df)**	*p*	*F*	**(df)**	*p*
stress	12.78	1,44	<.001	0.68	1,42	NS	1.12	1,44	NS
length	1.01	1,44	NS	0.01	1,42	NS	0.59	1,44	NS
stress									
× length	0.01	1,44	NS	0.03	1,42	NS	0.17	1,44	NS

For the reiterant version of "Suzy," the effect of *stress* was significant for all three subjects, but neither the effect of sentence *length* nor the interaction between the two variables was significant for any subject. That is, stressed syllables consistently had higher peaks than unstressed syllables regardless of sentence length. For "sexy," the effect of *stress* was significant for S1 and S3, and approached significance for S2 ($p = 0.0623$). In all cases, the stressed syllables had the higher peaks. The effect of sentence *length* and the *stress* × *length* interaction on "sexy" was not significant for S1 or S2. However, both were significant for S3, apparently because the *stress* effect was simply stronger for some sentence lengths than others. For "sassy," there was only one significant effect for any subject, and that was the effect of *stress* for S1, where once again the stressed syllable had a higher peak than the unstressed syllable.

While these analyses examined all stressed and unstressed syllables in bisyllabic words in the study, a separate analysis was undertaken to focus on the two sentences in which all three words occurred (those 8 and 9 syllables in length); they were always in the same order ("Suzy" first, "sexy" next, "sassy" last). This analysis enabled us to determine whether stressed and/or unstressed syllables declined in peak velum height throughout the sentence and whether the difference between stressed and unstressed syllables was affected by sentence position (i.e., earlier vs. later). The data are plotted in Figure 5.

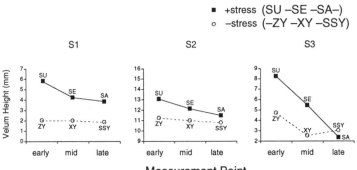

FIGURE 5. Mean values of velar peaks for stressed (filled squares) and unstressed (open circles) syllables in the reiterant speech corresponding to the three bisyllabic words as a function of their relative positions in the sentences ("Suzy" = early; "sexy" = mid; "sassy" = late). Data were obtained from the two longest sentences (8 and 9 syllables in length) as those contained all three bisyllabic words. Data for S1, S2, and S3 are shown at the left, middle, and right, respectively.

We ran two-way ANOVAs on the effects of *word* (first, second, third) and *stress* (stressed/unstressed syllable) on the measures of the reiterant sentences (Table 6). All subjects showed a main effect of *stress*, reflecting the fact that, in all but one case for one speaker, stressed syllables had higher velar peaks than unstressed syllables; the only exception was the reversal of that pattern for S3's last bisyllabic word, in which the unstressed peak was higher than the stressed peak. S1 and S3 also showed significant effects of *word* and significant interactions between *word* and *stress*. The main effect of *word* reflected the fact that overall, earlier words had higher velar peaks than later words; however the interaction and the pattern shown in the graphs meant that stress and word effects had to be untangled.

We examined the effect of *stress* for each of the three words, in an analysis of simple main effects. For S1, the effect of *stress* was significant for 'Suzy,' 'sexy,' and 'sassy' (all $p < 0.01$), although the F value declined across the three words (F = 58.09, 20.18, 15.64, respectively), indicating a reduction in the magnitude of the stress difference over the time course of the sentence.

TABLE 6. *Results of two-way ANOVA on the effects of word position and stress on velar peaks.*

Subject:	S1			S2			S3		
variable	**F**	**(df)**	**p**	**F**	**(df)**	**p**	**F**	**(df)**	**p**
word	5.10	(2,138)	<0.01	1.58	(2,132)	NS	34.47	(2,138)	<0.0001
stress	86.07	(1,138)	<0.0001	6.62	(1,132)	<.05	25.71	(1,138)	<0.0001
word × stress	3.92	(2,138)	<0.05	0.49	(2,132)	NS	12.00	(2,138)	<0.0001

For S3, the effect of *stress* was significant for 'Suzy' and 'sexy' (both p<.01), with a reduction in F values (F = 28.87, 19.77, respectively), but the effect was not significant for 'sassy,' again consistent with the notion that stress effects are reduced later in a sentence.

Next, we examined the effect of *word* separately for stressed and unstressed syllables. The effect of *word* was significant for stressed (F = 8.97, $p < 0.01$), but not unstressed syllables (F<1.0, $p = 0.9441$) for S1. The effect of *word* was significant for both stressed and unstressed syllables for S3, with a decline in F values (40.35 and 6.12, respectively) and significance levels ($p < 0.0001$ and $p < 0.01$, respectively). All subjects, including S2, showed a flatter function in the unstressed than stressed syllables, although the difference between stressed and unstressed syllables as a function of position was not significant for S2.

A shortcoming of our study is that *stress* is confounded here with syllable position in a word; stressed syllables were always first syllables and unstressed syllables were always second syllables in these bisyllabic words. Hence, it is unclear whether the effects described are due to stress and/or to within-word declination. However, there are data that indicate that stress probably played an important role. For example, Vaissière (1988) showed that velar peaks for the oral consonants in CVN sequences were higher in stressed than in unstressed syllables—just the pattern that was observed in these data. And, as mentioned in the Introduction, Vayra and Fowler (1992) showed that declination patterns of the jaw and formant frequencies observed in isolated trisyllables are weakened or disappear when the trisyllables are embedded in carrier phrases. Thus, if velar declination is like jaw and formant declination, then there should be relatively little, if any, effect of consonant position in a word when that word is embedded in a sentence (Vayra & Fowler, 1992) assuming, of course, that the consonant occurs in the same syllable position (Krakow, 1993), which is the case here. In light of Vayra and Fowler's study and those of Vaissière and of Krakow, the present data suggest that, in general, velar declination (in the form of increasing relaxation, i.e., decreasing height of velar peaks) is largely restricted to stressed syllables, and that this results in a reduction in the contrast between velar peaks in stressed and unstressed syllables as they occur later in a sentence.

F_0 Data

Figure 6 shows data that we obtained on peak F_0 in the stressed and unstressed syllables of reiterant versions of the three bisyllabic words, "Suzy," "sexy," and "sassy" in the two longest sentences in our study. (These measures correspond to the velar measures presented in Figure 5). The figure shows that F_0 declination occurs in stressed and unstressed syllables, whereas velar declination is limited to stressed syllables. We ran the same statistical analyses on the F_0 data as on the corresponding velar data beginning with a two-way ANOVA on the effects of word (first, second, third) and stress (stressed/unstressed syllable) (Table 7).

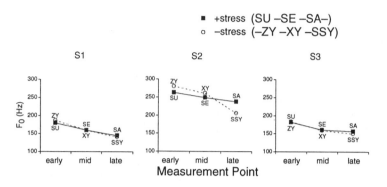

FIGURE 6. Mean values of F_0 peaks for stressed (filled squares) and unstressed (open circles) syllables in the reiterant speech corresponding to the three bisyllabic words as a function of their relative positions in the sentences ("Suzy" = early; "sexy" = mid; "sassy" = late). Data were obtained from the two longest sentences (8 and 9 syllables in length) as those contained all three bisyllabic words. Data for S1, S2, and S3 are shown at the left, middle, and right, respectively.

TABLE 7. Results of two-way ANOVA on the effects of word position and stress on F_0 peaks.

Subject:	S1			S2			S3		
variable	**F**	**(df)**	**p**	**F**	**(df)**	**p**	**F**	**(df)**	**p**
word	307.19	(2,138)	<0.0001	25.67	(2,132)	<0.0001	183.30	(2,138)	<0.0001
stress	0.19	(1,138)	NS	0.02	(1,132)	NS	.27	(1,138)	NS
word × stress	6.82	(2,138)	<0.01	7.42	(2,132)	<0.001	1.42	(2,138)	NS

In contrast to the velar data, where the main effect of *stress* was significant for all three subjects, the main effect of *stress* on F_0 peaks was not significant for any of the subjects. In the velar data, only two of the three subjects showed a main effect of *word*; here, all three subjects showed a *word* effect, reflecting the fact that F_0 peaks declined from early to late in the sentences. Two subjects (S1 and S2) showed a *word × stress* interaction on F_0. The nature of this interaction was examined in analyses of simple main effects. For S2, the effect of *word* was weaker (but still significant) in the stressed ($p < 0.05$), than in the unstressed syllables ($p < 0.0001$), the opposite of what was generally observed for the velum; for S1, the effect of *word* was significant for both stressed and unstressed syllables with no difference in the level of significance ($p < 0.0001$). Testing the effect of *stress* for each of the three words for the two subjects who showed an interaction, we observed the following: For S2, *stress* was significant only for "sassy" (where the stressed syllable had a higher F_0 peak than the unstressed syllable; p <0.01); for S1, *stress* was significant for "Suzy" (where the unstressed syllable had a higher F_0 peak than the stressed syllable; $p < 0.01$) and "sassy" (where the stressed syllable had a higher F_0 peak than the unstressed syllable; $p < 0.05$). Overall, then, stress did not appear to play an important or consistent role in declination of fundamental frequency for these data; stressed and unstressed syllables alike showed declination. This result is consistent with what Vayra and Fowler observed for F_0 and acoustic amplitude.

DISCUSSION

We can conclude that declination occurs across the three major physiological systems important in speech—respiratory, laryngeal, and supralaryngeal—based on measures of subglottal pressure, F_0, acoustic amplitude, formant frequencies, and movements of the jaw, lips, and velum. In this study, we observed declination in the velum, thus supplementing the literature on supralaryngeal declination, which previously had been limited to articulators affecting (or acoustic measures reflecting) oral tract shape. We found that peaks of velar elevation for syllable-initial obstruents were consistently lower at the end of a sentence than at the beginning. In addition, the difference between initial and final velar peaks was largest for the longest sentences and smallest for the shortest ones.

Although declination can be described as a general phenomenon in speech, we must conclude that there are differences in the nature of declination as a function of the part of the speech production mechanism involved. Our study, taken with that of Vayra and Fowler (1992), shows a contrast between patterns of the jaw and velum, on the one hand, and F_0 and acoustic amplitude, on the other. That is, the patterns of declination for the supralaryngeal articulators appear more similar to each other than to those of the laryngeal-respiratory systems, which likewise resemble each other. It is important to note, also, that across the studies of declination, causative relations are only evident in the effects of respiration (sub-glottal pressure) on F_0 and acoustic amplitude (e.g., Gelfer, 1987; Vayra & Fowler, 1992). And while the supralaryngeal measures of jaw, velum and formant frequencies have in common the fact that declination seems largely restricted to stressed syllables, other aspects of the patterns render them dissimilar.

To aid in our discussion of "types" of declination patterns, we have plotted, in Figure 7, data from Vayra and Fowler (1992), (Table 4, p. 55) in a manner parallel to our plots of velar and F_0 data in the sentences with the three bisyllabic words (Figures 5 and 6). Vayra and Fowler's data include F_1, jaw opening, F_0, and acoustic amplitude measures for the vowel /a/ in bVbVbV sequences. Note that while both studies used measures at early, mid, and late positions, Vayra and Fowler's utterances were trisyllabic words, while ours were sentences of eight and nine syllables. In addition, Vayra and Fowler's subjects were Italian speakers while ours were English speakers.

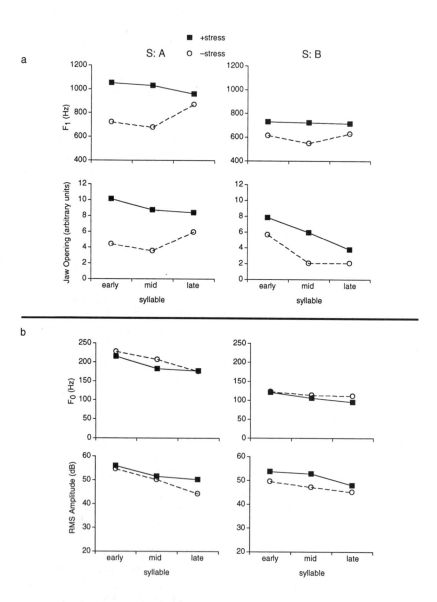

FIGURE 7. Mean values of (a) F_1 and jaw opening and (b) F_0 and acoustic amplitude in stressed (filled squares) and unstressed (open circle) /a/'s in Italian trisyllables, as a function of their relative positions in the sequences (early = vowel 1; mid = vowel 2; late = vowel 3). Data for subject A are at the left, and for B, at the right. (Adapted from Vayra & Fowler, 1992, Table 4, p. 55.)

Figure 7(a) shows their F_1 and jaw opening measures for stressed and unstressed /a/ in each of the three syllables in their trisyllabic words, here labeled 'early,' 'mid,' and 'late.' Consistent with our velar data, Vayra and Fowler reported declination across the three measurement points only for the stressed syllables. Jaw opening decreases and F_1 decreases from early to late, consistent with relaxation of the extreme opening and F_1 measures ordinarily associated with the vowel /a/. In the case of the velar data, increased relaxation was observed in the form of decreasing velar height for the stressed syllables across the three measurement positions. Examination of the F_1 and opening measures in the unstressed syllables show the V-shaped pattern described by the authors, with the smallest jaw opening and the lowest F_1's associated with the middle, rather than the final, syllable. This contrasts with most of the velar data, in which the unstressed syllables showed a rather flat function, with little or no effect of position. However, one of our subjects (3), showed a V-shaped pattern in her unstressed velar peaks, that is, the velar peak for /t/ was lowest in the middle, rather than the final, unstressed measurement position. For both sets of supralaryngeal data, we note that the combined patterns of the stressed and unstressed syllables are such that the contrast between stressed and unstressed syllables is most reduced in the final position.

Figure 7(b) shows Vayra and Fowler's F_0 and acoustic amplitude measures for stressed and unstressed syllables in the three syllables of the test words. Consistent with our F_0 data, these measures show declination effects running from early to late in the sequence, in both stressed and unstressed syllables, with lower F_0's and reduced amplitude at each successive measurement point. The results for the respiratory-laryngeal measures, then, are highly similar across the two studies, whether obtained in isolated trisyllables or in sentences, supporting Vayra and Fowler's notion that isolated trisyllables take on sentence intonation. And, as noted in both studies, the supralaryngeal measures differ from the F_0 and/or acoustic amplitude measures, in that the former show continuous declination effects only for stressed syllables. The lack of a parallel pattern for velar and F_0 data supports Vayra and Fowler's (1992) contention, based on the difference in their jaw opening and F_1 data vs. F_0 and acoustic amplitude data, that supralaryngeal declination cannot be said to be the result of respiratory declination in the way that the decline in F_0 and acoustic amplitude is a result of subglottal pressure declination. As noted by Fowler (1988), the exclusive focus of earlier declination studies on F_0 was misguided not only because the effects are more general, but because F_0 effects in particular are largely the result of declining subglottal pressure.

In summary, our findings of velar declination expand the investigations of supralaryngeal declination begun by Fowler and her colleagues that included a variety of measures of the jaw, lips, and formant frequencies (Fowler, 1988; Vatikiotis-Bateson & Fowler, 1988; Vayra & Fowler, 1992). Our findings also add to a growing body of literature that highlights the importance of velar movement patterns at different levels of speech organization, from the segmental to the syllabic and prosodic (for reviews, see Bell-Berti, 1993; Krakow, 1993). Previously reported prosodic effects on the velum include those of stress (Vaissière, 1988; Krakow, 1993) and speaking rate (e.g., Bell-Berti & Krakow, 1991; Kuehn, 1976), and the present work highlights an interaction between position in a sentence and stress. An interaction between these two variables, it is suggested here, may turn out to be a hallmark of supralaryngeal declination. How other non-segmental variables, such as speaking rate and position in a syllable interact with velar declination, or, indeed, how different segment types affect the pattern, have yet to be investigated.

ACKNOWLEDGMENTS

We owe a tremendous debt to Carole Gelfer and to Carol Fowler, whose pioneering work on respiratory-laryngeal and supralaryngeal declination, respectively, stimulated our interest in velar declination. We also thank Ignatius Mattingly and Carol Fowler for providing helpful comments on an earlier version of this paper and Dorothy Ross for assisting with data analysis. This work was supported by NIH grant DC-00121 to Haskins Laboratories.

REFERENCES

Bell-Berti, F. (1993). Understanding velic motor control: Studies of segmental context. In M. K. Huffman & R. A. Krakow (Eds.), *Nasals, nasalization, and the velum. (Phonetics & Phonology V)* New York: Academic Press, 63-86.

Bell-Berti, F., & Krakow, R. (1991). Anticipatory velum lowering: A coproduction account. *Journal of the Acoustical Society of America, 90,* 112-123.

Breckenridge, J. (1977). Declination as a phonological process. (Bell System Manuscripts, Murray Hill, NJ)

Cohen, A., Collier, R., & 't Hart, J. (1982). Declination: Construct or intrinsic feature of pitch? *Phonetica, 39,* 254-273.

Cooper, W., & Sorenson, J. (1981). *Fundamental frequency in sentence production.* New York: Springer Verlag.

Fowler, C. (1988). Periodic dwindling of acoustic and articulatory variables in speech production. *Paw Review, 3,* 10-13.

Gelfer, C. (1987). *A simultaneous physiological and acoustic study of fundamental frequency declination.* Unpublished doctoral dissertation. New York: City University of New York.

Horiguchi, S., & Bell-Berti, F. (1987). The Velotrace: A device for monitoring velar position. *Cleft Palate Journal, 24,* 104-111.

Kay, B. A., Munhall, K. G., Vatikiotis-Bateson, E. V. , & Kelso, J. A. S. (1985). A note on processing kinematic data: Sampling, filtering, and differentiation. *Haskins Laboratories Status Reports on Speech Research, SR 81*, 291-303.

Kent, R. D., Carney, P. J., & Severeid, L. R. (1974). Velar movement and timing: Evaluation of a model for binary control. *Journal of Speech and Hearing Research, 17*, 470-488.

Krakow, R. A. (1989). *The articulatory organization of syllables: A kinematic analysis of labial and velar gestures.* Unpublished doctoral dissertation, Yale University.

Krakow, R. A. (1993). Nonsegmental influences on velum movement patterns: Syllables, sentences, stress, and speaking rate. In Huffman, M. K. & Krakow, R. A., (Eds.) *Nasals, nasalization, and the velum.* (*Phonetics & Phonology V*) New York: Academic Press, 87-118.

Kuehn, D. P. (1976). A cineradiographic investigation of velar movement in two normals. *Cleft Palate Journal, 13*, 88-103.

Lieberman, P. (1967). *Intonation, perception and language.* Research Monograph No. 38. Cambridge: MIT Press.

Liberman, M., & Streeter, L. A. (1978). Use of nonsense-syllable mimicry in the study of prosodic phenomena. *Journal of the Acoustical Society of America, 63*, 231-233.

Pierrehumbert, J. (1979). The perception of fundamental frequency declination. *Journal of the Acoustical Society of America, 66*, 363-369.

't Hart, J., Collier, R., & Cohen, A. (1990). *A perceptual study of intonation: An experimental-phonetic approach to speech melody.* Cambridge, New York: Cambridge University Press.

Vaissière, J. (1988). Prediction of articulatory movement of the velum from phonetic input. *Phonetica, 45*, 122-139.

Vatikiotis-Bateson, E., & Fowler, C. (1988). Kinematic analysis of articulatory declination. *Journal of the Acoustical Society of America, 84*, S128 (A).

Vayra, M., & Fowler, C. A. (1992). Declination of supralaryngeal gestures in spoken Italian. *Phonetica, 49*, 48-60.

24 Acoustic and Kinematic Correlates of Contrastive Stress Accent in Spoken English

Carol A. Fowler
Haskins Laboratories
University of Connecticut
Yale University

INTRODUCTION

Many studies have examined correlates of metrical prominence or "stress." A remarkable observation, evident across the studies, is that most variables that have been and imaginably could be examined show systematic change under stress variation. The following is a brief survey of acoustic and articulatory measures that have been found to index stress.

Acoustic Variables

As compared to unstressed vowels, stressed vowels are longer in their steady-state durations (e.g., Summers, 1987), and less centralized in their formant values (Harris, 1978; Summers, 1987; Tiffany, 1959). Formant transitions into and out of the steady-state are steeper for stressed than unstressed vowels (Summers, 1987). The stressed vowel's average fundamental frequency is generally found to be higher and to show more change (Lehiste, 1970; Lieberman, 1960; Summers, 1987; Tiffany, 1959) than the fundamental frequency of unstressed vowels. With effects of vowel quality controlled, stressed vowels are more intense than unstressed vowels (Lehiste, 1970). Finally, the slope of a "locus equation" relating F_2 at the midpoint of a vowel to F_2 at the transition onset of a stressed CV is shallower than the slope of the equation for an unstressed syllable (Krull, 1989), indexing a smaller coarticulatory influence of the stressed than the unstressed vowel on the preceding consonant; that is, the "coarticulation resistance" (Bladon & Al-Bamerni, 1976) of stressed segments is greater than that of unstressed segments.

Articulatory Variables

Measures of electromyographic activity for muscles implementing stressed consonants and vowels consistently show higher activity than for unstressed segments (Harris, 1978; Tuller, Kelso, & Harris, 1982). In addition, stressed syllables may be associated with increased activity of the internal intercostal muscles, which increase expiratory effort, as compared to unstressed syllables (Ladefoged, 1967).

As for kinematic measures, actions of articulators that implement gestures of stressed segments are generally longer in duration, show greater displacement and greater peak velocities than gestures for unstressed syllables (tongue dorsum for velar consonants and back vowels: Ostry, Keller, & Parush, 1983; jaw and lower lip for reiterant /ba/: Kelso, Vatikiotis-Bateson, Saltzman, & Kay, 1985; jaw for contrastively stressed CVCs: Summers, 1987). In Summers' research, in addition, low vowels were associated with lower positions of the jaw when stressed than when unstressed, and their "steady-state" positions were held longer.

All of these findings appear consistent with the proposal (Öhman, 1967; Lehiste, 1970) that stressing consists of a global increase in production effort. The globality of the increase in effort, if that is what it is, is underscored by the remarkable findings of Kelso, Tuller, and Harris (1983) that movements of the finger that accompany stressed syllable production are greater in amplitude than those accompanying unstressed syllable production, even though actors are instructed to keep movement amplitude constant.

New findings on stress are reported by investigators adopting a "dynamical systems" perspective on speech production. (See Turvey, 1990, for a discussion of the perspective itself applied to intentional action generally, and see Saltzman & Kelso, 1987, for a model. For applications to stress, see e.g., Ostry et al., 1983; Kelso et al., 1985; Vatikiotis-Bateson, 1987; Beckman & Edwards, 1992.) In general, speech gestures are modeled as varieties of oscillatory systems. In a simple mass spring system, the spring displacement over time, x, is equal to $A\cos\omega t$, where A is the maximum displacement of the spring during an oscillatory cycle, ω is angular velocity, and t is time. The velocity of the change in displacement is $\dot{x} = \omega A\sin\omega t$, where ωA is the peak velocity of the displacement change. A characteristic of this simple oscillatory system that is seen repeatedly in speech gestures (e.g., Kuehn & Moll, 1976) and other movements (e.g., Cooke, 1980) is a positive covariation between maximum displacement and peak velocity—that is, bigger displacements are associated with higher peak velocities. The slope of the line relating maximum displacement (A) to peak velocity (ωA) is ω.

Some investigators have found a difference in the slope of the displacement/peak velocity relationship for stressed and unstressed syllables, with unstressed syllables showing a steeper slope (Ostry et al., 1983; Kelso et al., 1985). That is, in unstressed gestures (associated, as summarized earlier, with smaller displacements and lower velocities than stressed gestures), greater changes in peak velocity accompany given changes in displacements than for stressed syllables. The slope of the line relating displacement to peak velocity is $\omega = (k/m)1/2$, where k is spring stiffness and m is mass. Given that articulator masses do not change under variation in stress, the change in slope can be ascribed to changes in k or stiffness. From this perspective, the findings suggest that unstressed syllables are produced by a stiffer oscillatory system than stressed syllables.[1]

It is worth noting in this regard that not all investigators have found the slope difference between stressed and unstressed syllables. Vatikiotis-Bateson (1987) did find the difference in closing gestures (of reiterant /ma/ and /ba/ productions of the Rainbow passage), but he did not generally find it for opening gestures. Beckman and Edwards (1992) did not find a slope difference. They speculate that the reason for the difference between their findings and those of Ostry et al. and of Kelso et al. was that, whereas the latter investigators looked at lexical stress, they looked at accented and unaccented lexically stressed words. This account does not explain Vatikiotis-Bateson's inconsistent findings, however.

Beckman and Edwards (1992) proposed two other ways in which stressing might be achieved simply in a dynamical system. Stressed gestures may increase overall in amplitude relative to unstressed gestures, or sequential gestures may decrease in overlap. They contrasted the three hypotheses: a difference between stressed and unstressed gestures in stiffness, in amplitude or in phasing as shown in Figure 1 (following Beckman & Edwards, 1992).

---

[1]Kelso et al. (1985) entertained, but rejected, a hypothesis that the stiffness difference they found between stressed and unstressed reiterant syllables was the product of a spring that was nonlinear such that net stiffness decreased with movement displacement. In such a system, stiffness would be greater for unstressed syllables not because talkers set stiffness to a higher value, but rather because unstressed gestures are smaller in magnitude than are stressed gestures. Making the assumption that each displacement midpoint approximated the equilibrium position of the oscillatory system responsible for the gesture, the investigators examined stiffness near the midpoint. If the data from stressed and unstressed gestures reflected a single nonlinear spring, then stiffness values near the equilibrium position should have been the same for stressed and unstressed gestures. Instead, they differed systematically. Accordingly, the investigators concluded that different stress levels were achieved using different settings of a stiffness parameter.

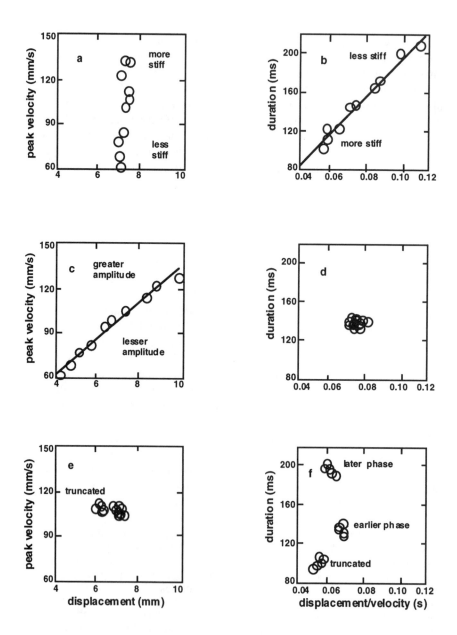

FIGURE 1. Kinematic manifestations of accenting if accenting is achieved by variation in (a,b) stiffness of an oscillatory system, (c,d) in amplitude, or (e,f) in phasing of gestures. (Adapted from Beckman & Edwards, 1992. See text for further explanation.)

According to the stiffness hypothesis, stressed and unstressed gestures should differ in peak velocity, but not in displacement (Figure 1a). This is because a change in k changes the value of ω but not of A. (Notice that this is not exactly the finding of Ostry et al. or of Kelso et al. described above and ascribed to a change in stiffness.) Further, because the peak velocity of the gesture will increase with an increase in k, with displacement held constant, unstressed gestures will be associated with smaller displacement/velocity ratios and shorter durations than stressed gestures (Figure 1b). If, instead, talkers stress gestures by increasing only their amplitudes, peak velocity (ωA) and displacement (A) will change proportionally. Therefore, gestures for stressed and unstressed syllables will fall on a common line relating displacement to peak velocity with slope ω (Figure 1c). Because the two measures change proportionally, the ratio of the one to the other will not change with stress variation, and gesture duration will not change (Figure 1d). If, finally, talkers change the phasing of sequential stressed gestures by reducing overlap, gestures will not change in velocity or displacement (unless highly overlapped unstressed gestures are truncated by following ones; Figure 1e). Because displacement and velocity do not change, their ratio does not change either (except, again, with truncation), but the measured duration of the stressed gestures will be longer (Figure 1f).

Beckman and Edwards examined kinematic measures of productions of the target syllable "pop" in the utterances below:

1. Pop, opposing the question strongly, refused to answer it.
2. Poppa, posing the question loudly, refused to answer it.
3. Poppa posed the question loudly and then refused to answer it.

This gave them a comparison of an accented, intonational phrase final syllable (sentence 1), an accented nonfinal syllable (sentence 2) and an unaccented syllable preceding a nuclear accent on "posed" (sentence 3). Overall, their kinematic (jaw-movement) data most closely favored the third, phasing hypothesis—a result consistent with Krull's (1989) findings, summarized earlier, of shallower locus equation slopes for stressed than unstressed syllables. However, the data were not fully consistent with any of the three hypotheses. In particular, whereas a change in phasing should not be associated with a change in movement velocity, jaw opening movements did show differences as a function of accent. The investigators judged them small relative to displacement changes, and ascribed the displacement changes, therefore, to truncation due to a change in phasing rather than to a true change in gesture amplitude.

A subsequent report by Beckman and Edwards (in press) examines utterances 4b and 5b below, produced in answer to questions 4a and 5a, respectively.

4a. Was her *mama* a problem about the wedding?
  b. Her POPPA posed a problem
5a. Did *his* dad pose a problem as far as their getting married?
  b. HER poppa posed a problem.

In these utterances, both syllables of "poppa" were examined. The final syllable is unaccented and lexically unstressed. The first syllable is lexically stressed
in both utterances, but accented only in 4b. In these sentences, the accents express a contrast with information in the preceding question. In the findings on
these utterances (with the same speakers as in the earlier study), there generally
was overlap in measures of lip displacement and velocity for the two lexically
stressed syllables. Striking kinematic differences were seen only comparing
these two syllables to the unaccented, lexically unstressed final syllable of
"poppa." The conclusion in this paper is that, whereas kinematic differences
may be seen between accented and unaccented lexically stressed syllables, the
differences are not consistent across talkers and utterances. The most consistent
marker of accented words is the pitch accent on the word. Kinematic differences
are "ancillary" to the pitch accent difference; perhaps kinematic differences arise
sometimes to accommodate the pitch contour. Large consistent kinematic differences do differentiate lexically stressed and unstressed words; accordingly,
stress is marked differently at different levels of a prosodic hierarchy. If a syllable is stressed lexically—that is, if it is the head of a metrical stress foot (see,
e.g., Selkirk, 1980)—its stress will be marked kinematically. (For the two talkers
whose data are discussed, the conclusion was that lexical stress was achieved by
a change in stiffness.) If a lexically stressed word receives a sentence accent, the
accent will be implemented by a pitch accent on the word, and, if necessary,
something will be done (a change in stiffness for one of the two talkers, a change
in phasing for the other) to lengthen the gestures to accommodate the accent.

This account is very neat and very interesting. It is not entirely clear, however,
that it is consistent with other findings. In particular, Summers' (1987) findings,
summarized above, of consistent kinematic correlates of stress were obtained
from three talkers who produced contrastive stress either on the target CVC utterance or on the word before it in a carrier sentence. These words would have
been heads of stress feet that were accented when contrastively stressed. In unpublished research, Fowler, Gracco, Vatikiotis-Bateson, and Romero (in press)
find large and consistent kinematic differences among productions of a target
word "pop" that is either accented, because it provides the new information in an
answer to a question or, unaccented, when another word in the sentence, either
before the target word or after it, is accented. For these sentences, "pop" always
is stressed lexically. Of course, it remains possible that these kinematic correlates of sentence accents are "ancillary" to production of pitch accents; however,
they were found to be consistent and reliable across talkers and utterances.

METHOD

Subjects and Speech Materials

The present report provides another data set from three native speakers of
English, the same three talkers as in Fowler et al. (in preparation). In the data set

reported on here, utterances are the six listed below. In one set, the target word (underlined) appears early in the utterances, and it is either contrastively stressed (capitalized) or it precedes or follows contrastive stress on another word. In the second set, the target word appears late in the sentence and it is either contrastively stressed or it precedes or follows contrastive stress on another word.

6a. He's <u>POPE</u> John not VICAR John.
 b. He's <u>pope</u> JOHN not pope FRED.
 c. HE's <u>pope</u> John the man in blue's pope Fred.
7a. Her CIGAR'S in the ashtray but her <u>PIPE</u>'S in her pocket.
 b. HIS pipe's in the ashtray but HER <u>pipe</u>'s in her pocket.
 c. Her pipe's OUT of the ashtray but her <u>pipe</u>'s IN her pocket.

Measurements

We measured various acoustic variables: vowel duration, measured as the voiced interval in each word, and average fundamental frequency (F_0), and average F_1, F_2, and RMS amplitude during the vowel of the target syllables. In addition, we measured numerous kinematic variables. Discussion here will be limited to measures of lip-aperture opening duration, displacement, and peak velocity.

Movements of the upper lip, lower lip, and jaw for one subject (VG) were obtained using strain gauge transducers on a head-mounted frame (see Barlow, Cole, & Abbs, 1983). For the other subjects, movements were obtained from light emitting diodes (subjects EB and JS). Signals were recorded on FM tape and were digitized off line with 12 bit resolution. The sampling rate for the movement signals was 500 Hz. Audio signals were filtered at 4.8 kHz and sampled at 10 kHz. Following digitization, movement signals were low passed filtered at 25 Hz using a procedure that effectively eliminates time delays. Finally, the lip and jaw signals were combined in software to obtain a signal reflecting lip aperture. Durations of opening and closing movements, displacements, and peak velocities served as the data to be described below.

RESULTS

Acoustic Measures of Contrastive Stress

Figure 2 shows the fundamental frequency contour of typical productions of sentences 6a-to-6c. The accents on contrastively stressed syllables are clearly visible. The second contrastively stressed syllable in each sentence ("vic," "Fred," and "blue," in 6a-c, respectively) has a lower peak than the first. The fundamental frequency patterns for typical productions of the "pipe" sentences were similar to those in Figure 2.

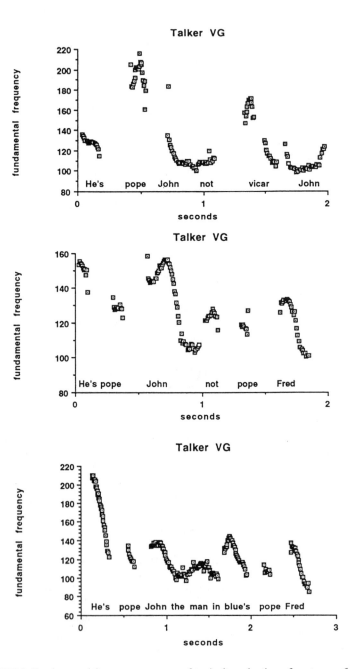

FIGURE 2. Fundamental frequency contour of typical production of sentences 6a-c.

Early Contrastive Stress

By far, the most salient acoustic marker of early contrastive stress was fundamental frequency. In analyses of variance, the F value comparing the average fundamental frequencies during the vowels of the three sets of productions of "pope" exceeded 200 for all three talkers. In each data set, average fundamental frequency on the accented word was higher than that in either unaccented word. No other acoustic or kinematic measure was as reliable. The only other acoustic measure that was significant for all three talkers was RMS amplitude; further, it showed the expected pattern of a higher value for the accented word for just two of the three talkers. The patterning of F_1 (lower for unaccented "pope") and F_2 (higher for unaccented "pope") may reflect centralization of the /o/ component of the diphthongized vowel of the word. This pattern was present for two of the three talkers, but the pattern on F_1 differs from that reported by Tiffany (1959), who took measures of stressed and unstressed vowels one quarter of the way through each vowel.

Late Contrastive Stress

In the "pipe" sentences, in which the target word was late and was the second of two contrastively stressed syllables, although fundamental frequency differed reliably as a function of accent for all three talkers, and differed such that the accented word had the highest average fundamental frequency, it was not the most reliable acoustic marker of contrastive stress (Talker JS: $F(2,27) = 10.89$; VG: $F(2,27) = 37.10$; EB: $F(2,27) = 15.75$, $ps < 0.001$). The remaining acoustic measures are significant for all three talkers, but not always for the same reason. For two of the talkers, F_1 was lowest for the accented syllables, and for all three talkers, F_2 was lowest. This is the same pattern of changes found by Tiffany (1959). Duration and RMS amplitude showed different patterns across the three talkers. If we ask what variable both patterned such that the accented word was distinguished from the unaccented productions and was the most reliably distinct, that variable is F_2 for JS ($F(2,27) = 15.75$, $p < 0.001$), RMS amplitude for VG ($F(2,27) = 55.57$, $p < 0.001$), and fundamental frequency for EB ($F(2,27) = 15.75$, $p < 0.001$).

Kinematic Measures of Contrastive Stress

Early Contrastive Stress

Opening duration, displacement, and peak velocity pattern in the same way for the three talkers. Opening durations are longest for the accented syllable (JS: $F(2,27) = 142.75$; VG: $F(2,27) = 48.4$; EB: $F(2,27) = 19.02$), displacements are largest for the accented syllable (JS: $F(2,27) = 171.03$; VG: $F(2,27) = 56.45$; EB: $F(2,27) = 34.4$), and velocities are highest (JS: $F(2,27) = 82.63$; VG: $F(2,27) = 10.4$; EB: $F(2,27) = 6.81$). There are, for these speakers, then, highly reliable kinematic correlates of accent. Too, somewhat surprisingly, these did not, for any of the talkers, give rise to an acoustically longer duration vowel. Accordingly, if the reason for the larger accented than unaccented gestures was to better accommodate the pitch accents of Figure 2, the effort was not successful.

Turning to the dynamical means by which stress accent might have been achieved by the talkers, as depicted in Figure 1, I can rule out a change in stiffness as an important mechanism. Figure 3 shows regression lines fit to sentences 6a-c. Clearly the data do not resemble the idealized picture of Figure 1a. (Separate regression lines fit to each sentence separately (Table 1) do not show a shallower slope for accented words for any of the three talkers. This is consistent with the suggestion of Beckman and Edwards (1992, in press) that accenting lexically stressed words is not like the comparison of lexically stressed to unstressed syllables in the research of Ostry et al. and Kelso et al.)

Talker EB (Figure 3b) shows a pattern similar to that of Beckman and Edwards' (in press) talkers in which points for accented and unaccented syllables overlap considerably. However, JS and VG (Figures 3a and c) show a clean and nearly clean separation, respectively. The patterning in their data is most consistent with a hypothesis that the amplitude of gestures (Figure 1c) is increased under contrastive stress. However, neither VG nor EB shows a consistent relation between variation in the displacement/velocity ratio and duration, a finding also predicted by a controlled increase in gesture amplitude (Figure 1d). JS does show a consistent relation however ($R^2 = 0.65$), with a steep line relating a clump of points for the unaccented words to a clump for the accented words (Figure 4). This pattern looks most consistent with an interpretation of a change in stiffness (Figure 1b), an interpretation, however, ruled out above on other grounds.

Late Contrastive Stress

As for the measure of fundamental frequency, kinematic differences between accented and unaccented productions of "pipe" are generally less reliable than those for "pope" in the early sentence position. Measures of opening duration (JS: $F(2,27) = 5.39$; VG: $F(2,27) = 18.49$; EB: $F(2,27) = 15.27$) are uniformly significant, but only EB and VG show the expected pattern with accented productions longest. Measures of opening displacement are significant only for JS and EB (JS: $F(2,27) = 64.33$; VG: $F(2,27) = 2.78$, $p = 0.08$; EB: $F(2,27) = 21.45$), and the measure patterns numerically as expected only for JS and VG. Opening velocity is uniformly significant and it patterns as expected for all three talkers (JS: $F(2,27) = 116.65$; VG: $F(2,27) = 8.29$; EB: $F(2,27) = 21.12$).

Displacement-peak velocity relations are shown in Figure 5. Again no subject shows the pattern expected if stiffness were reduced under constrastive stress.

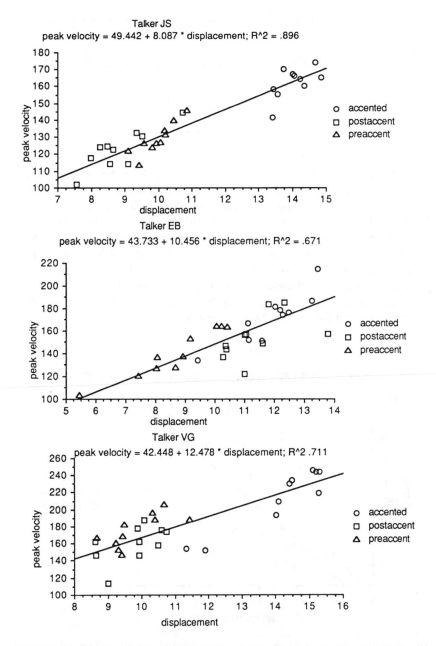

FIGURE 3. Scattergrams and regression analyses of displacement predicting peak velocity for the three talkers' productions of sentences 6a-c (early contrastive stress).

TABLE 1. *Slopes of regression lines relating displacement and peak velocity. Nonsignificant values (p > .0 05) are in parentheses.*

	Pope sentences			Pipe sentences		
	Accented	Post-accented	Pre-accented	Accented	Post-accented	Pre-accented
JS	11.6	11.3	16.0	(8.1)	8.6	(5.1)
VG	23.8	(15.0)	17.5	21.2	14.9	(10.4)
EB	17.2	21.9	13.2	14.4	10.8	15.4

As for the "pope" sentences, JS shows a close fit of points to the line and no overlap between accented and unaccented words. Both EB and VG show overlap, here perhaps an indication that there is little separation kinematically between contrastively stressed and unstressed words. (JS is the only one of the three subjects to show overall reliable differences in both velocity and displacement favoring the accented word in the analyses above.) Only JS shows a pattern that is interpretable in the context of the mechanisms depicted in Figure 1. His pattern conforms most closely to that expected if the major dynamical parameter that is changed under stress is amplitude. This interpretation does not fit the data for EB and VG, but then they distinguish the syllables only weakly or not at all on one of the relevant kinematic measures.

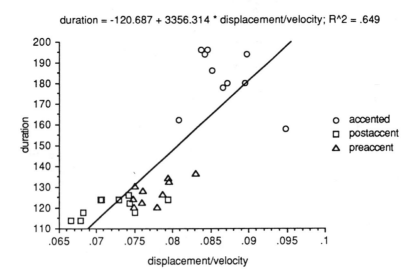

Talker JS

duration = -120.687 + 3356.314 * displacement/velocity; R^2 = .649

FIGURE 4. Duration predicted by the displacement/peak velocity ratio of talker JS on sentences 6a-c.

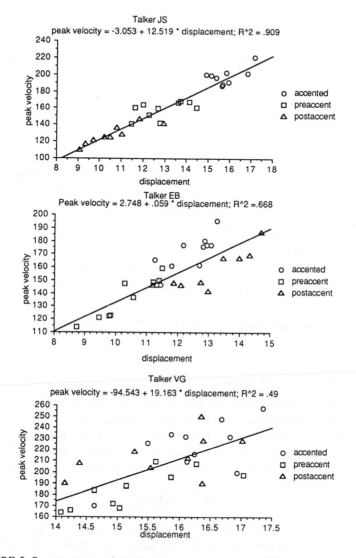

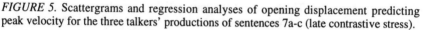

FIGURE 5. Scattergrams and regression analyses of opening displacement predicting peak velocity for the three talkers' productions of sentences 7a-c (late contrastive stress).

The relationship between the displacement/velocity ratio and the duration of the opening gesture is significant only for JS (Figure 6). However, the scatterplot is not consistent with any of the patterns in Figure 1. There is considerable overlap in points for accented and unaccented words.

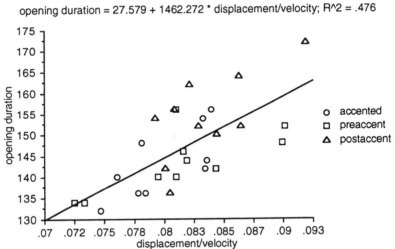

Talker JS

opening duration = 27.579 + 1462.272 * displacement/velocity; R^2 = .476

FIGURE 6. Duration predicted by the displacement/peak velocity ratio of talker JS on sentences 7a-c.

Pre- and Post-accented Words

We had included sentences 6b and 6c and 7b and 7c both to provide comparisons with the accented versions of each target but also to see whether unaccented words preceding and following a major stress or accent were produced differently. In brief, we found no systematic differences between unaccented words preceding or following an accented word in either acoustic or kinematic measures.

DISCUSSION

The data permit a few conclusions. A positive conclusion is that contrastive stress accents are reliably marked by pitch accents in both early and late positions in the sentence. These are not the only reliable correlates of accentuation, however, especially for early accents. There are also highly reliable kinematic correlates of accents. Whereas it may be the case, as Beckman and Edwards (in press) suggest, that these are subsidiary to the pitch accent, for a few reasons, I am not convinced that this is the most plausible interpretation. First, the present data do not conform to a hypothesis that, kinematically, stressing is achieved in ways that increase acoustic syllable durations in the most straightforward way possible. As Figure 1 shows, changes in stiffness and phasing should increase the acoustic duration of a syllable. But our data do not fit those patterns well. This is not to say that gestures were not larger for accented than unaccented words; they were, reliably so. Rather it is that the kinematic consequences were not confined to amplitude changes and their kinematic correlates that might be expected on the assumption that speakers are intending only to lengthen a syllable. Another point that perhaps not too much should be made of, because it is unusual, is that, for these subjects, acoustic duration was not a reliable indicator of stress accent. Related to these considerations is that we took between nine and eleven kinematic measures from our subjects, the lip aperture opening measures already described, compatible measures of closing, measures of respiratory activity using Respitrace and, for two subjects, measures of intraoral pressure during /p/. JS showed significant effects of accent with the expected patterning on all but two of the kinematic measures on the early accented sentences, VG on all but three, and EB on six of ten measures. A final point is that Kelso et al. (1983) found larger amplitude movements of the finger accompanying accented as compared to unaccented productions of the strong syllable "stock." Together these findings appear most consistent with the older "global effort" hypothesis. Perhaps it is not that kinematic adjustments occur to accommodate a pitch accent, but rather that the perceptually salient pitch accent highlights global effort increases for the listener.

What should be made of the observation that the data did not conform to predictions of any of the three ways, illustrated in Figure 1, in which talkers might produce stress accents simply from a dynamical systems perspective? Incompatibility with predictions can have several origins. One likely reason is that the simple oscillatory system used to generate the predictions of kinematic correlates of stress accent is not realistic. As noted earlier (see, especially, footnote 1), this possibility has been considered also by Kelso et al. (1985). Another reason, however, is that the model oscillator represents a presumed control system for vocal tract actions and does predict the effects on action of the biomechanics of the vocal tract itself. These effects, with an origin peripheral to any oscillatory control system, may make model testing difficult.

Differences in the Magnitude of Accentuation in the Sentence Sets

Correlates of stress accent were generally weaker for the later accented word "pipe" than for the earlier word "pope." Our sentence sets 6 and 7 do not permit a clear interpretation of this difference. It may be a consequence of general articulatory dwindling early-to-late in an utterance for which weak evidence has been reported in the literature (Bell-Berti & Krakow, 1991; Fowler, 1988; Krakow, 1993; Vatikiotis-Bateson & Fowler, 1988; Vayra & Fowler, 1992). However, other interpretations are possible. The utterance sets under examination here both consisted of two intonational phrases, and the late accented word was always in the second phrase. Perhaps a nuclear accented word in an intonational phrase following one in an earlier phrase is reduced in magnitude. Third, the second accented words in both sentence sets are redundant as compared to the first accented word, and the redundancy may underlie the reduction (cf. Fowler & Housum, 1987; Lieberman, 1963). That is, in a sentence like "He's POPE John not VICAR John," the listener knows, after hearing "not," that an accented word will be produced that contrasts in some way with the first accented word. Further, it is very likely that the contrasting word will be in the same broad semantic category as the first one.

Two observations help to distinguish among the interpretations. First, the later of two nuclear accented words need not be smaller in pitch accent than the first. In an utterance in which a talker produces a word substitution error and then corrects it (e.g., "It's a pen holder...I mean a pipe holder"), the correction is accented, and its pitch accent has a higher peak and a higher average fundamental frequency than the accented word that it corrects. This outcome is not consistent with a hypothesis that the second of two nuclear accented words in sequentially produced intonational phrases *must* be reduced. A second observation may help to distinguish these remaining hypotheses. This observation derives from a data set described in detail in Fowler et al. (1993). In this study, the same talkers as in the present study produced answers to questions in which the target word "pop" either provided the new information in the answer or provided given information. Across sentence sets, "pop" appeared in four positions in the answers, early (initial) to late (final). In these sentences, accented "pop" did not provide redundant information; further, when "pop" was accented, it was the nuclear accent in the answers, which were produced on just one intonational phrase. These utterances should provide a clean test of the hypothesis that there is a global dwindling of acoustic and kinematic correlates of accenting early-to-late in an utterance, or perhaps in an intonational phrase. In brief, however, there was no convincing evidence that the difference in acoustic or kinematic correlates of accented and unaccented, productions of "pop" decreased early-to-late in an utterance. Nor was there evidence that utterance final accented "pop" was less distinct from unaccented "pop" than were target words elsewhere in the utterance. The most viable interpretation of the difference in magnitude of markers of accenting in the "pope" sentences as compared to the "pipe" sentences, therefore, is that it reflects a difference in the redundancy of the accented word in the two sentences.

SUMMARY AND CONCLUSIONS

I interpret the present findings as showing that lexically stressed, accented words are distinguished from stressed but unaccented words in multiple ways. A salient acoustic marker of stress accent is the occurrence of a pitch accent on an accented word. However, accented words are also distinguished by supralaryngeal kinematic markers that, presumably, index the dynamical systems that are used to achieve and release constrictions for the component consonants and vowels of the accented words. These kinematic differences between accented and unaccented words themselves have acoustic consequences.

I found no simple way to characterize the kinematic differences in terms of a change in a single dynamical parameter. In this respect, findings are most consistent with suggestions that accenting reflects an overall increase in global articulatory effort. These conclusions are qualified, however, by recognition that kinematic measures may reflect operation of dynamical control systems as their effects are mediated by the biomechanics of the vocal tract anatomy and physiology.

No evidence was obtained in the data for a global dwindling in accentuation early to late in an utterance. The smaller effects of late than early accenting most likely reflect the redundancy of the later, as compared to the earlier, accented words in the particular utterances we asked talkers to produce.

ACKNOWLEDGMENTS

Preparation of this manuscript was supported by NICHD Grant HD-01994 and by NIDCD Grant DC-00121 to Haskins Laboratories. The data reported on here were collected in collaboration with Vincent Gracco and Eric Vatikiotis-Bateson, who I thank for their contributions to the research project.

REFERENCES

Barlow, S., Cole, K., & Abbs, J. (1983). A new headmounted lip-jaw movement transduction system for the study of speech-motor disorders. *Journal of Speech and Hearing Research, 26,* 283-288.

Beckman, M., & Edwards, J. (in press). Articulatory evidence for differentiating stress categories. In P. Keating (Ed.), *Papers in laboratory phonology, III: Phonological structure and phonetic form* (pp. 7-33). Cambridge University Press.

Beckman, M., & Edwards, J. (1992). Intonational categories and the articulatory control of duration. In Y. Tohkura, E. Vatikiotis-Bateson, & Y. Sagisaka (Eds.), *Speech perception, production and linguistic structure* (pp. 359-376). Tokyo: IOS Press.

Bell-Berti, F., & Krakow, R. (1991). Velar height and sentence length: declination? *Journal of the Acoustical Society of America, 89,* 1916(A).

Bladon, A., & Al-Bamerni, A. (1976). Coarticulation resistance in English /l/. *Journal of Phonetics, 4,* 137-150.

Cooke, J. D. (1980). The organization of simple skilled movements. In G. Stelmach & J. Requin (Eds.), *Tutorials in motor behavior* (pp. 199-212). Amsterdam: North-Holland.

Fowler, C. (1988). Periodic dwindling of acoustic and articulatory variables in speech production. *Perceiving-Acting Workshop, 3,* 10-13.

Fowler, C., Gracco, V., Vatikiotis-Bateson, E., & Romero, J. (1993). *Global correlates of stress accent in spoken sentences.* Unpublished manuscript.

Fowler, C., & Housum, J. (1987). Talkers' signalling of "new" and "old" words in speech and listeners' perception and use of the distinction. *Journal of Memory and Language, 26,* 489-504.

Harris, K. (1978). Vowel duration change and its underlying physiological mechanisms. *Language and Speech, 21,* 354-361.

Kelso, J. A. S., Tuller, B., & Harris, K. (1983). A "dynamic pattern" perspective on the control and coordination of movement. In P. MacNeilage (Eds.), *The production of speech* (pp. 137-173). New York: Springer-Verlag.

Kelso, J. A. S., Vatikiotis-Bateson, E., Saltzman, E. L., & Kay, B. (1985). A qualitative dynamic analysis of reiterant speech production: Phase portraits, kinematics, and dynamic modeling. *Journal of the Acoustical Society of America, 77,* 266-280.

Krakow, R. (1993). Nonsegmental influences on velum movement patterns: Syllables, segments, stress and speaking rate. In M. Huffman & R. Krakow (Eds.), *Phonetics and phonology, 5: Nasals, nasalization and the velum* (pp. 87-116). New York: Academic Press.

Krull, D. (1989). Consonant-vowel coarticulation in spontaneous speech and in reference words. *PERILUS, 10,* 101-105.

Kuehn, D., & Moll, K. (1976). A cineradiographic study of VC and CV articulatory velocities. *Journal of Phonetics, 4,* 303-320.

Ladefoged, P. (1967). *Three areas of experimental phonetics.* London: Oxford University Press.

Lehiste, I. (1970). *Suprasegmentals.* Cambridge, MA: MIT Press.

Lieberman, P. (1960). Some acoustic correlates of word stress in American English. *Journal of the Acoustical Society of America, 32,* 451-460.

Lieberman, P. (1963). Some effects of semantic and grammatical context on the production and perception of speech. *Journal of the Acoustical Society of America, 6,* 172-187.

Öhman, S. (1967). Word and sentence intonation: A quantitative model. *Speech Transmission Laboratory: QPSR (Royal Institute of Technology), 2-3,* 20-54.

Ostry, D., Keller, E., & Parush, A. (1983). Similarities in the control of the speech articulators and the limbs: Kinematics of tongue dorsum movement in speech. *Journal of Experimental Psychology: Human Perception and Performance, 9,* 637-651.

Saltzman, E., & Kelso, J. A. S. (1987). Skilled action: A task-dynamic approach. *Psychological Review, 94,* 84-106.

Selkirk, E. (1980). The role of prosodic categories in English word stress. *Linguistic Inquiry, 11,* 563-605.

Summers, W. V. (1987). Effects of stress and final consonant voicing in vowel production: Articulatory and acoustic analyses. *Journal of the Acoustical Society of America, 82,* 847-863.

Tiffany, W. (1959). Nonrandom sources of variation in vowel quality. *Journal of Speech and Hearing Research, 2,* 305-317.

Tuller, B., Harris, K., & Kelso, J. A. S. (1982). Differential transformations of articulation. *Journal of the Acoustical Society of America, 71,* 1534-1543.

Turvey, M. T. (1990). Coordination. *American Psychologist, 45,* 938-953.

Vatikiotis-Bateson, E. (1987). *Linguistic structure and articulatory dynamics: A cross-language study.* Unpublished doctoral dissertation, Indiana University.

Vatikiotis-Bateson, E., & Fowler, C. (1988). Kinematic analysis of articulatory declination. *Journal of the Acoustical Society of America, 84,* S128(A).

Vayra, M., & Fowler, C. (1992). Declination of supralaryngeal gestures in spoken Italian. *Phonetica, 49,* 48-60.

25 Apraxia of Speech Reconsidered

Paula A. Square
University of Toronto

INTRODUCTION

The role that the dominant hemisphere plays in the regulation of linguistic function is indisputable. There has, however, been continuing debate regarding left hemisphere specialization for the control of speech motor behavior, and this debate has spawned confusion regarding the nature of speech errors that occur in nonfluent aphasia. Two diametrically opposed philosophical camps emerged in the 1970s—one suggesting that aphasic speech errors could be explained largely by linguistic theory (e.g., Blumstein, 1973), and the other proposing that the speech errors had a motoric base and were specifically apractic in nature (e.g., Johns & Darley, 1970). A more recent and moderate position is that the basis of the articulatory-prosodic abnormalities in nonfluent aphasia is largely motoric and the errors are most likely both 'apractic' and 'dysarthric' in nature (Square & Martin, 1993). Further, the motor speech errors in left hemisphere damage (LHD) will rarely be strictly 'apractic' nor strictly 'dysarthric.' In addition, nonfluent aphasic patients will also likely present with phonemic paraphasic errors, but linguistic errors in the absence of a motor speech impairment will be fairly infrequent. Finally, among nonfluent aphasic patients there will be

375

variable combinations of 'apraxic,' 'dysarthric,' and/or 'phonological' disruptions.

The purposes of this chapter are to: (1) review the evidence that highlights the significant role that the left hemisphere plays in the control of motor behavior; (2) review the range of motor disturbances that may result from LHD; (3) explore the implications of this broader base of information regarding left cortical and subcortical motor control as it applies to the reconsideration of the disorder, 'apraxia of speech;' and (4) argue for studies that correlate motor speech behavior as a result of brain damage with neuro-imaging and neurometabolic information rather than studying patients who have been classified *a priori* as having 'apraxia of speech.'

LEFT (DOMINANT[1]) HEMISPHERE SPECIALIZATION FOR MOTOR CONTROL

There are several converging lines of evidence that indicate that the left (or dominant) hemisphere is specialized for the control of motor behavior across several modalities, including manual, orofacial nonverbal, and orofacial verbal. The first line of evidence is that the disorders of ideomotor limb apraxia (e.g., Faglioni & Basso, 1985; Nass & Gazzaniga, 1977) and bucco-facial apraxia (e.g., Mateer & Kimura, 1977) are frequently associated with LHD to frontal and parietal lobes but rarely occur in right hemisphere damage (RHD).[2]

The second line of evidence is that LHD results in specific deficits of motor control for the production of single nonverbal oral, speech, and hand *postures* (Kimura & Archibald, 1976; Mateer, 1978; Mateer & Kimura, 1977). Work undertaken in our laboratory has indicated that similar types of errors occur across the three modalities—limb, oral-nonverbal, and speech movement—in subjects with LHD (Roy & Square-Storer, 1990; Square-Storer & Roy, 1989). These include: disturbances in the initiation of movement, errors of spatial targeting, discoordination of subcomponents of movement, disturbances in rate of movement, coexisting additive or augmentative movements, omitted components of movements, sequencing disturbances and perseverations of components of movements or whole movements (Square-Storer & Roy, 1989).

Despite the similarities of error types across modalities, the neuroanatomical regions that control the production of single manual and oral gestures, the latter of which include both verbal and nonverbal gestures, may differ. While disorders in the production of single oral postures, both speech and nonspeech,

---

[1]For convenience the notation LHD is used rather than dominant hemisphere damage. Unlike the more extensive literature on aphasia, there have been few reports in the literature of the right hemisphere being responsible for the control of speech, and oral nonverbal and limb movements.

[2]The definition of apraxia being used here is that originally proposed by Liepmann—a reduced ability to perform skilled learned movements (Liepmann, 1908).

have been reported to result principally from left frontal lesions, and disorders of single manual postures principally from left parietal lobe lesions, the oral and manual systems appear more separable in the frontal than in the parietal lobe, suggesting that the parietal lobe '...may play a general programming role that is enacted through the left frontal region' (Kimura, 1982, p. 135).

Reduced abilities to produce *sequences* of several postures, whether manual, nonspeech oral, or speech sequences, also result from both frontal (premotor) and posterior (parietal) left hemisphere lesions (Kimura, 1982). Recent work from our laboratory (Roy & Square, 1992; Roy et al., 1993) as well as by others (Jason, 1983, 1985) has indicated that the sequencing deficit likely has as its basis a memory disorder. While able to select *an* appropriate response from memory, i.e., one posture, LHD subjects are unable to generate sequences of responses from memory (Roy & Square, 1992). Nonetheless, sequence generation is relatively unimpaired in LHD subjects if they are not required to produce the sequence from memory (as occurs, for example, under pictorial representation of manual or oral nonverbal sequences as opposed to when they are, in both verbal command and visual imitative conditions (Roy & Square, 1992)).

Acquisition of newly learned limb (Roy, 1981), oral nonverbal (Mateer, 1978) and verbal sequences (Mateer & Kimura, 1977) also has been reported to be inferior in subjects with LHD, regardless of whether aphasia is present or not. Memory deficits have been implicated as the factor underlying the inability to learn novel motor sequences in all modalities (Mateer, 1978; Roy 1981).

Fine motor control for the production of elemental movements versus gross abilities to generate sequences appear to be two distinct aspects of motor control governed by the left hemisphere. To elaborate further on the former, fine-grained analyses of the production of single nonverbal oral postures has revealed that the productions of LHD subjects, compared to those of normal subjects, were marked by a greater frequency and severity of movement augmentation and spatial alignment (Square-Storer, Qualizza, & Roy, 1989). These error types increased significantly in productions of sequences (as compared to production of isolated postures) by the LHD, but not by normal subjects. In addition, the LHD subjects revealed a disability for phasing subcomponents of movements, reflected by a dramatic increase of complex errors on sequenced items, even when subjects were *not* clinically diagnosed as having buccofacial apraxia (Square-Storer et al., 1989). These results from a study of the production of oral nonverbal movements indicated that LHD subjects may have deficits for both *fine motor control* and *sequencing*.

The effects of lateralized brain damage on one of the most fundamental levels of fine motor control, repetitive movement, has also been studied in our laboratory (Roy, Clark, Aigbogun, & Square-Storer, 1992). Evidence of a bilateral regulatory effect for fine motor control by the left hemisphere was dramatic. Specifically, fine motor control impairments affecting both right (contralateral) and left (ipsilateral) hands were found to accompany LHD when subjects attempted to tap regularly with their index fingers. Both a decreased rate and greater variability of tapping, as measured by intertap interval and amount of time in the depressed and raised postures, were found. While LHD resulted in finger tapping that was disrupted bilaterally, right hemisphere damage resulted in a similar fine motor control disruption, but for the contralateral (left) hand only (Roy et al., 1992). Surprisingly, the fine motor control deficit was unrelated to performances on traditional tasks of limb praxis, complex visual motor behavior as measured by the Purdue Pegboard task (Lezak, 1976), and joint movement sensitivity.

In my opinion, the most intriguing finding from this study was the bilateral disruption of the *rhythm* of repeated movements subsequent to LHD. The significance of the rhythmic disruption is largely speculative, but if the hypothesis of Grillner (1982) is correct, i.e., that rhythm underlies the regulation of all motor behavior including speech, the disruptive effect of LHD on the organization of motor behavior would be profound.

The last line of evidence regarding the significant role that the left hemisphere plays in the regulation of motor control for speech has been highlighted in a recent case study of the loss of speech in a five-year-old child following acquired brain damage that affected primarily, but possibly not exclusively, the left fronto-parietal region (Square, Aronson, & Hyman, 1994). Acoustic analyses of the re-emergence of motor speech control over a 46-week period revealed that the child first reacquired the ability to produce gross speech postures. Although, relatively undifferentiated in the early stages (11 days to approximately 17 weeks post onset) of motor speech recovery, vocal tract postures for the production of vowels, measured acoustically, became progressively more differentiated throughout the recovery period studied. Short functional utterances (sequenced speech movements) emerged about the 26th week post insult. These ranged in length from one-word utterances at 26 weeks to 15-word utterances at 46 weeks. Throughout the recovery period, however, intelligibility was compromised as utterance length (sequencing demands) increased. Last to emerge was the control of prosody, which remained distorted because of intersyllabic dissociation throughout the 46 weeks studied. The hypothesis put forward by our group based upon this case study was as follows: regulation of motor control for speech is a dominant hemisphere function that appears to lateralize in early childhood (Square et al., 1994).

In conclusion, there are several substantive lines of converging evidence that indicate that the left hemisphere is a supramodal regulator of motor behavior. There are many motor behaviors that appear to be controlled and it remains to be seen whether these behaviors are independent of one another, interdependent, or hierarchical in their emergence and control. Nonetheless, the motor behaviors that appear to be governed by the left hemisphere include: (1) posture generation; (2) fine motor control; (3) governance over the learning of new skilled motor behaviors; and (4) seriation and/or melodic flow of movements.

The Traditional Definition of Apraxia of Speech (AOS)

The term "apraxia of speech" (AOS) was popularized by Darley (1968) to describe the motor speech impairment that typically co-occurred with aphasia. AOS was described as an articulatory-prosodic speech disorder caused by "...an impaired ability to program the positioning of the speech musculature...and the sequencing of muscle movements..." (Darley, Aronson, & Brown, 1975a). This definition included two key elements of left hemisphere motor control as described above: posture generation and seriation of gestures. AOS was described as different from the dysarthrias in that examination of the speech musculature revealed insignificant impairments of muscle tone, strength, or range of movement. Further, abnormal oral reflexes were not present. Nonetheless, speech was disorganized. While the disorganization was less evident on automatic speech tasks (e.g., counting), self-formulated speech, and especially attempts at longer utterances, were disrupted with regard to initiation, kinetic melody, and articulatory accuracy.

Kent and Rosenbek (1983) elaborated upon the symptomatology of apraxia of speech based upon acoustic analyses of the speech of seven patients diagnosed as having AOS and minimal aphasia; each also had either transient or chronic hemiplegia. They described AOS as a central disruption of motor speech control in which '...errors in sequencing, timing, coordination, initiation and vocal tract shaping...' (p. 245) occur. Indeed, their description of the speech symptomatology associated with AOS parallels the numerous levels of motor control that were described above as being governed by the left hemisphere. The question arises, however, whether all of these heterogeneous symptoms should be considered to be elements of 'apraxia,' per se.

Dysarthria Resulting from LHD

There is increasing evidence that suggests that many studies concerning the nature of the disorder 'apraxia of speech' are questionable because subjects also had confounding dysarthria. Indeed, motor strip involvement that probably disrupts the execution of fractional movements (Freund, 1987) or the use of sets of muscles for executing certain movements (Evarts, 1986) frequently coexist with premotor and parietal lobe damage. Lesions also often extend into subcortical motor areas. Such lesion extension is likely to contaminate the neuromotor speech symptoms that arise from premotor and/or parietal lobe damage in the left hemisphere. Evidence that suggests that a left-hemisphere 'dysarthria' may exist has been reported by Duffy and Folger (1986) and Hartman and Abbs (1992), who refer to the 'dysarthria' as unilateral upper motor neuron (UUMN) dysarthria. Associated with UUMN dysarthria are phonatory, articulatory, and rate deviancies similar to those associated with pseudobulbar dysarthria, the latter of which is caused by bilateral motor strip damage (Darley et al., 1975a).

Another line of evidence suggesting that 'dysarthria' may have contaminated the speech output of many of the subjects used in studies of 'apraxia of speech' has its basis in combined information from two fields: neuroradiologic imaging and motor neurophysiology. Studies of apraxia of speech have typically used patients who have suffered a single thrombo-embolic incident involving the distribution of the left middle cerebral artery. Lesions resulting from middle cerebral artery infarction can affect the premotor cortex as well as the motor cortex, the first and second temporal gyri, the primary and secondary sensory facial regions in the parietal lobe, and the angular and supra-marginal gyri. Frontal damage may extend to the lateral cortex, the mesial premotor area (i.e., the supplementary motor area) and the cingulate gyrus and insula. More frequently, however, subcortical extension of these lesions occurs involving the white matter through which motor-sensory tracks course, the left basal ganglia (BG), the internal capsule, and sometimes even the thalamus. Each of these subcortical areas has either basic motor functions or is a conduit through which pyramidal and extrapyramidal motor tracts travel (see Square & Martin, 1993, for a review). To deny that a 'dysarthric' motor impairment would result from damage to the BG, thalamus, internal capsule or white matter would be to deny basic principles of motor physiology. In addition, there may be subtle influences on symptomatology caused by hypometabolism of contralateral cerebellum as a secondary result of frontal damage (see, for example, Metter et al., 1987). Therefore, ataxic-like symptoms could further complicate the clinical presentation of the speech dysfunction typically diagnosed as 'apraxia of speech.' Indeed, a case of ataxico-apraxic speech has been reported (Rosenbek, McNeil, Teetson, Odell, & Collins, 1981).

The Neuromotor Speech Disorder Associated with Left-Hemisphere Damage

Aphasic speech errors (phonemic paraphasias) are also likely to co-occur with neuromotor speech errors caused by LHD. One way to circumvent confusion between motor-based and phonemic selection errors is to study speech-impaired LHD subjects with no clinically demonstrable aphasia. The first to describe the acoustic and perceptual speech characteristics of LHD speakers with minimal aphasia were Alajouanine, Ombredane, and Durand (1939). They concluded that the motor speech disorder that results from LHD is a variable combination of apraxia, paresis, and dystonia. I concur with the conclusion of Alajouanine et al. (1939) that heterogeneity prevails in the range and possible combinations of parameters of the neuromotor speech impairment associated with LHD.

In our lab, perceptual and acoustic studies of LHD patients without aphasia and with varying sites of lesions resulted in the identification of dramatically different speech patterns (Square, Darley, & Sommers, 1982; Square & Mlcoch, 1983; Square-Storer & Apeldoorn, 1991). One of these patterns was found in two non-hemiparetic patients with left parietal lobe lesions, identified by CT-scans, whose speech was characterized by marked difficulties with initiation, audible and inaudible postural groping generally preceding but also within utterances, and numerous off-target approximations of phonemes with repeated attempts to self-correct (Square, 1981; Square et al., 1982). Absolute and relative durations of phonemes, as well as F_0 contours, were within normal limits, except in some cases of self-correcting reattempts (Square-Storer & Apeldoorn, 1991). Also, there were few instances of syllable dissociation but some occurrences of syllable segregation, especially on reattempted initiating syllables (Kent & Rosenbek, 1983, for definitions; Square-Storer & Apeldoorn, 1991).

The role of the dominant parietal lobe in speech motor control, specifically *speech sequencing*, has been noted historically (Canter, 1969; Deutsch, 1984; Kimura, 1982; Luria, 1966). Indeed, the apractic speaker who has come to be known as the "Tornado man" (Darley, Aronson, & Brown, 1975b) demonstrated speech patterns that were perceptually similar to those of our patients with parietal lobe lesions. Examination of the medical records of the 'Tornado man' revealed a left parietal lobe thrombo-embolic lesion.

It is purely speculative that Broca's area lesions that do not extend subcortically may result in speech patterns that are perceptually and acoustically similar to subjects with left cortical midparietal lesions (see, for example, Mohr et al., 1975). In our lab, however, a "pure apractic speaker" with a strictly Broca's area lesion has never been studied. Further, models based upon results from positron emission tomography (PET) studies predict that Broca's area has no direct influence on fluency (Metter et al., 1990) and the parietal lobe lesion subjects reported in the literature have been decidedly dysfluent due to trial and error groping and sound syllable and word repetitions (Square et al., 1982).

A somewhat different pattern was found in our third patient, a hemiplegic patient with a lesion to the left BG caused by thrombo-embolic insult, who also displayed the speech symptoms described above. In addition, there was a salient and overriding 'slow and effortful' quality. This perceived slowness was acoustically verified by long absolute durations of phonemes and syllables, relatively equal durations of syllables, significant syllable dissociation, abnormal F_0 contours and numerous vowel distortions (Square-Storer & Apeldoorn, 1991). All of the later deviancies were pervasive, i.e., unrelenting. Speech was marked by severe effortfulness and some explosiveness (Square et al., 1982; Square & Mlcoch, 1983). This lead to the hypothesis that LHD, especially damage to the left BG, may result in a motor speech syndrome that is a combination of initiating and sequencing deviancies (as those observed in patients with midparietal lesions) coupled with 'paretic-dystonic' features (Square & Apeldoorn, 1991).

Another pattern variation was found in a fourth subject, whose data have not been published, and who had a fronto-parietal lesion with its greatest volume being subcortical. She also had a right hemiparesis. Her speech was marked by all the characteristics present in the parietal lesion subjects. She, too, had 'paretic' features but her speech was not nearly as effortful and explosive as that of the BG-lesion subject. Both the third and fourth patients also had mild phonatory harshness and extremely mild yet pervasive hypernasality; these characteristics also occur in pseudobulbar dysarthria (Darley et al., 1975a). This led to a more recent and here-to-fore unpublished hypothesis: 'dysarthric' qualities, including slow, effortful speech, may result from damage to the motor tracts coursing through the white matter underlying motor and/or sensory facial regions in the left hemisphere. In conclusion, we hypothesize that symptoms typically thought to be 'apractic,' including initiation difficulties marked by groping, off target productions and attempts to self-correct, and a disability for producing longer sequences of speech, are most likely consistent with cortical damage to the dominant parietal lobe. Other 'dysarthric-like' qualities such as slow effortful speech may or may not compound this motor sequencing disorder.

CONCLUSIONS

Among patients clinically diagnosed as having 'apraxia of speech' (Wertz, LaPointe, & Rosenbek, 1984), there is heterogenity regarding speech symptomotology identified perceptually and acoustically. We hypothesize that these differences relate to damage to different functional units of the motor control symptoms of the dominant hemisphere. The symptoms may result from direct damage to structures or from deferred hypometabolism (see Metter et al., 1987; Metter, Riege, Hanson, Jackson, Kempler, & Van Lancker, 1988; Metter, Riege, Hanson, Phelps, & Kuhl, 1988; Illes et al., 1989). Square and Martin (1993) have summarized the probable motor deviancies that result from damage to each of the following cortical areas: primary cortex, lateral premotor cortex, mesial premotor cortex (supplementary motor area), and parietal cortex. The motor sequalae are qualitatively different. It would, therefore, appear prudent for speech scientists to begin to explore the effects of specific sites of lesions in the left hemisphere on the motor control of speech in nonaphasic subjects. It is proposed that *a priori* classification of subjects as 'apractic' or 'dysarthric' has served only to increase the confusion that has existed regarding the enigmatic disorder, apraxia of speech. Acoustic and kinematic speech analyses, coupled with neuroanatomical information derived from MRI, and neurometabolic information derived from positron emission tomography, should illuminate our understanding of the parameters of speech motor control governed by the left hemisphere and clarify the neuromotor speech impairment(s) that may result from LHD. Such studies should resolve the confusion regarding the nature of the neuromotor speech disorder that often, but not always, accompanies nonfluent aphasia (Darley, 1968).

REFERENCES

Alajouanine, T., Ombredane, A., & Durand, M. (1939). *Le syndrome de disintegration phonetique dans l'aphasie.* Paris: Masson.

Blumstein, S. (1973). Some phonological implications of aphasic speech. In H. Goodglass & S. Blumstein (Eds.), *Psycholinguistics and aphasia* (pp. 123-137). Baltimore, MD: The Johns Hopkins University Press.

Canter, G. J. (1969). The influence of primary and secondary verbal apraxia on output disturbances in aphasic syndromes. Paper presented to the American Speech and Hearing Association, Chicago (unpublished).

Darley, F. L. (1968). Apraxia of speech: 107 years of terminological confusion. Paper presented to the American Speech and Hearing Association, Denver (unpublished).

Darley, F. L., Aronson, A. E., & Brown, J. (1975a). *Motor speech disorders.* Philadelphia: Saunders.

Darley, F. L., Aronson, A. E., & Brown, J. (1975b). *Motor speech disorders: Audio seminars in speech pathology.* Philadelphia: Saunders.

Deutsch, S. E. (1984). Prediction of site of lesion from speech apraxic error patterns. In J. C. Rosenbek, M. R. McNeil, & A. E. Aronson (Eds.), *Apraxia of speech: Physiology, acoustics, linguistics, and management* (pp. 113-134). San Diego: College Hill Press.

Duffy, J. R., & Folger, W. N. (1986). Dysarthria in unilateral central nervous system lesion: a retrospective study. Paper presented at the annual convention of the American-Speech-Language and Hearing Association, Detroit.

Evarts, E. V. (1986). Motor cortex outputs in primates (Chapter 6). *Cerebral cortex, Vol. 5, Sensory motor areas and aspects of cortical connectivity* (pp. 217-241).

Faglioni, P., & Basso, A. (1985). Historical perspectives on neuroanatomical correlates of limb apraxia. In E. A. Roy (Ed.), *Neuropsychological studies of apraxia and related disorders* (pp. 3-45). Amsterdam: North-Holland Press.

Freund, H. J. (1987). Abnormalities of motor behavior after cortical lesions in humans. In S. Geiger, F. Plum & V. Mountcastle (Eds.), *Handbook of physiology, Vol. 5, The nervous system* (pp. 763-810). Bethesda: American Physiological Society.

Grillner, S. (1982). Possible analogies in the control of innate motor acts and the production of sound in speech. In S. Grillner, B. Lindblom, J. Labker, & A. Persson (Eds.), *Speech motor control* (pp. 217-230). New York: Pergamon Press.

Hartman, D. E., & Abbs, J. H. (1992). Dysarthria associated with focal unilateral upper motor neuron lesion. *European Journal of Disorders of Communication, 27,* 187-196.

Illes, J., Metter, E. J., Dennings, R., Jackson, C., Kempler, D., & Hanson, W. (1989). Spontaneous language production in mild aphasia: relationship to left prefrontal glucose hypometabolism. *Aphasiology, 3,* 527-537.

Jason, G. (1983). Hemispheric asymmetries in motor function: 1. Left hemisphere specialization for memory but not performance. *Neuropsychologia, 21,* 35-46.

Jason, G. (1985). Manual sequence learning after focal cortical lesions. *Neuropsychologia, 23,* 35-46.

Johns, D. F., & Darley, F. L. (1970). Phonemic variability in apraxia of speech. *Journal of Speech and Hearing Research, 13,* 556-583.

Kent, R. D., & Rosenbek, J. C. (1983). Acoustic patterns of apraxia of speech. *Journal of Speech and Hearing Research, 26,* 231-249.

Kimura, D. (1982). Left-hemisphere control of oral and brachial movements and their relation to communication. *Philosophical Transactions of the Royal Society of London, 298,* 135-149.

Kimura, D., & Archibald, Y. (1976). Motor functions of the left hemisphere. *Brain, 97,* 337-350.

Lezak, M. D. (1976). *Neuropsychological assessment.* New York: Oxford University Press.

Liepmann, H. (1908). *Drei Aufsatze aus dem Apraxiegebiet,* Volume 1. Berlin: Karger.

Luria, A. R. (1966). *Higher cortical functions in man.* New York: Basic Books.

Mateer, C. (1978). Impairments of nonverbal oral movements after left-hemisphere damage: A follow-up analysis of errors. *Brain and Language, 6,* 334-341.

Mateer, C., & Kimura, D. (1977). Impairment of nonverbal oral movements in aphasia. *Brain and Language, 4,* 262-276.

Metter, E. J., Kempler, D., Jackson, C., Hanson, W., Riege, W., Camras, L., Mazziota, J., & Phelps, M. E. (1987). Cerebellar glucose metabolism in chronic aphasia. *Neurology, 37,* 1599-1606.

Metter, E. J., Riege, W. R., Hanson, W. R., Jackson, C., Kempler, D., & Van Lancker, D. (1988). Subcortical structures in aphasia: Analysis based on FBG, PET, and CT. *Archives of Neurology, 45,* 1229-1234.

Metter, E. J., Riege, W. R., Hanson, W. R., Phelps, M. E., & Kuhl, D. E. (1988). Evidence for caudate role in aphasia from FBG positron computed tomography. *Aphasiology, 2,* 33-43.

Metter, E. J., Hanson, W. R., Jackson, C. A., Kempler, D., & van Lancker, D. (1990). Brain glucose metabolism in aphasia: A model of the interrelationship of frontal lobe regions on fluency. *CAC,* 69-75.

Mohr, J. P., Funkenstein, H. H., Finkelstein, S., Ressin, M. S., Duncan, G. W., & Davis, K. R. (1975). Broca's area infarction versus Broca's aphasia. *Neurology, 25,* 349.

Nass, R. D., & Gazzaniga, M. S. (1977). Cerebral lateralization and specialization in human central nervous system. *The Handbook of Physiology: The Nervous System,* 701-761.

Rosenbek, J., McNeil, M., Teetson, M., Odell, K., & Collins, M. (1981). A syndrome of neuromotor speech deficit and dysgraphia? In R. Brookshire (Ed.), *Clinical aphasiology: Conference proceedings* (pp. 309-315). Minneapolis: BRK Publishers.

Roy, E. A. (1981). Action sequencing and lateralized cerebral damage: Evidence for metries in control. In J. Long & A. Baddeley (Eds.), *Attention and performance* (pp. 487-498). Hillsdale, NJ: Erlbaum.

Roy, E. A., Brown, L., Winchester, T., Square, P. A., Hall, C., & Black, S. (1993). Memory processes and gestural performance in apraxia. *Adapted Physical Activity Quarterly, 10,* 293-311.

Roy, E. A., Clark, P. C., Aigbogun, S., & Square-Storer, P. A. (1992). Ipsilesional disruptions to reciprocal finger tapping. *Archives of Clinical Neuropsychology, 7,* 213-219.

Roy, E. A., & Square-Storer, P.A. (1990). Evidence for common expressions of apraxia. In G. Hammond (Ed.), *Cerebral control of speech and limb movements* (pp. 477-502). Amsterdam: Elsevier Science.

Roy, E. A., & Square, P.A. (1992).Impairments to limb and oral sequencing: Effects of memory demands and sequence length. *Journal of Clinical and Experimental Neuropsychology, 14,* 63.

Square, P. A. (1981). *Apraxia of speech in adults: Speech perception and production.* Unpublished doctoral dissertation, Kent State University.

Square, P. A., Aronson, A. E., & Hyman, E. (1994). An acoustic study of the redevelopment of motor speech control subsequent to acquired left hemisphere damage in early childhood: A case study. *American Journal of Speech-Language Pathology, 3,* 67-80.

Square, P. A., Darley, F. L., & Sommers, R. K. (1982). An analysis of the productive errors made by pure apractic speakers with differing loci of lesions. In R. Brookshire (Ed.), *Clinical aphasiology: Conference proceedings* (pp. 245-250). Minneapolis: BRK Publishers.

Square, P. A., & Mlcoch, A. G. (1983). The syndrome of subcortical apraxia of speech: An acoustic analysis. In Brookshire, R. (Ed.), *Clinical aphasiology: Conference proceedings* (pp. 239-243). Minneapolis: BRK Publishers.

Square, P. A., & Martin, R. E. (1993). The nature and treatment of neuromotor speech disorders in aphasia. In R. Chapey (Ed.), *Language intervention strategies in adult aphasia* (3rd ed.) (pp. 467-499). Baltimore, MD: Williams & Wilkins.

Square-Storer, P. A., & Apeldoorn, S. (1991). An acoustic study of apraxia of speech in patients with different lesion loci. In C. A. Moore, K. M. Yorkston, & D. R. Beukelman (Eds.), *Dysarthria and apraxia of speech: Perspectives on management* (pp. 271-286). Baltimore, Maryland: Paul H. Brookes Publishing Co.

Square-Storer, P.A., & Roy, E. A. (1989). The apraxias: Commonalities and distinctions. In P. Square-Storer (Ed.), *Acquired apraxia of speech in aphasic adults* (pp. 20-63). London: Taylor & Francis.

Square-Storer, P. A., Qualizza, L., & Roy, E. A. (1989). Isolated and sequenced oral motor posture production under different input modalities by left-hemisphere damaged adults. *Cortex, 25,* 371-386.

Wertz, R. T., LaPointe, L. L., Rosenbek, J. C. (1984). *Apraxia of speech in adults: The disorder and its management.* New York: Grune and Stratton.

26

A Case Study of a Child with Neurologically-Based Dysfluency: From Semantic Representation to Word Production

Claude Chevrie-Muller
Institut National de la Santé et de la Recherche Médicale, Paris

INTRODUCTION

The most frequently documented disorder affecting speech fluency is stuttering, and of the types of stuttering, primarily developmental stuttering. Acquired neurogenic dysfluencies (AND), in contrast, have been much less frequently reported, although well documented case studies are currently available and have aroused interest in the hope that they may offer an explanation of pathophysiological mechanisms of developmental stuttering that have remained unknown.

In 1975, Farmer quoted 21 reports of AND between 1943 and 1970, and Mazzucchi, Moretti, Carpegiani, Parma, and Paini (1981) added six more for the period 1899 to 1970. During the 1970s and 1980s cases were reported using various labels like "acquired stuttering" (Bhatnagar & Andy, 1989; Cipolotti, Bisacchi, Denes, & Gallo, 1988; Fleet & Heilman, 1985; Helm, Butler, & Benson, 1978; Lebrun & Leleux, 1985; Mazzucchi et al., 1981), "neurogenic stuttering," "neurological stuttering" (Canter, 1971; Helm, Butler, & Canter, 1980; Helm-Estabrooks, 1986; Lebrun, Leleux, & Retif, 1987; Quinn & Andrew, 1977), and "adult-onset stuttering" (Baratz & Mesulam, 1981; Nowack, 1986). The more general label, "neurogenic dysfluency," was used by Horner and Massey (1983) and by Koller (1983).

387

Whatever the label used, it appears from the literature that AND is not a unitary disorder. Attempts have been made to relate different types of dysfluencies to the neurological cause and localization of brain damage. Helm-Estabrook (1986) summarized the characteristics of stroke-induced stuttering, and of stutterers with head-trauma or extra-pyramidal disease. Horner and Massey (1983) compared cortical stuttering with palilalia, as did Koller (1983) describing vascular/trauma dysfluencies versus extrapyramidal ones. As for localization in the CNS, Lebrun et al. (1987) reported right- and left-sided and bilateral cortical lesions, diffuse brain damage or dysfunction, and extrapyramidal disease. Differences in the clinical aspects of AND can be in the modality of onset (abrupt or progressive), in its course (transient or persistent dysfluency), and in the absence (Helm et al., 1978; Rosenbek, Messert, Collins, & Wertz, 1978; Rosenfield, 1972) or presence of associated symptoms, such as dysarthria, dyspraxia, or dysnomia (Canter, 1971; Farmer, 1975) (in cases of anterior or posterior left-hemispheric damage or subcortical dysfunction), as well as (nonverbal) symptoms of right-hemisphere lesions .

To further knowledge of the pathophysiological mechanisms of developmental stuttering (DS), one challenge has been to define similarities and differences between this disorder and AND (Canter, 1971; Koller, 1983). Nine characteristics, generally assumed to differentiate these disorders, were discussed by Lebrun (unpublished lecture, European Postgraduate Course in Language Pathology, Brussels, 1989), who concluded that "if the time of onset is disregarded, acquired and developmental stuttering cannot always be safely distinguished" (on the basis of their clinical symptoms).

The time of the dysfluency onset was considered by Lebrun to be one of the main points in diagnosis. In fact, the youngest patient with AND (subsequent to head trauma followed by a coma) reported in the literature was 19 years old (Mazzucchi et al., 1981), possibly because following an acute encephalopathy in childhood, some acquired dysfluencies are transient (Roulet et al., 1993), just as some symptoms of acquired aphasia were similarly underestimated in children because of their transitory nature (Van Hout, Evrard, & Lyon, 1985). This report is of a case of childhood acquired neurogenic dysfluency that persisted into adulthood.

CASE REPORT

T. E. was the elder male child of two siblings. He was born preterm (seven-months, birth weight 2400 g) and suffered from a benign neonatal icterus. His early motor and language development were normal. Neither stuttering nor "stuttering-like" behavior was described by his parents. He entered primary school at 6 years 5 months.

The initial pathologic event was an acute encephalopathy with a sudden onset at age 6 years 11 months. The first symptom was a rhinopharyngitis with hyperpyrexia and vomiting. After a few days the child went into a coma (stage

III). Agitation, acidosis and a high level of propionic acid were noted. Despite further thorough enzymatic analyses it was impossible to identify the cause of this encephalopathy. After four days in a coma the child completely regained consciousness but did not speak at all. He demonstrated bilateral pyramidal symptoms, a pseudobulbar palsy (tongue, pharynx, and lips) and poor coordination between head and eye movements.

After one month he returned home, having produced no speech during the month in the hospital. The first day the child was at home he produced a few words, and the next day he uttered intelligible and grammatically correct sentences. Speech and language were first assessed, by the author, a few days after this speech recovery.

Initial Assessments

T.E. was seven years old at this first assessment. He looked well and was alert. Oropharyngeal motor function was impaired by a paresis of the tongue, a moderate velopharyngeal paresis (slight nasal air escape during phonation), and a weakness of orbicularis oris contraction. The child occasionally drooled.

At each re-evaluation, speech was recorded during tasks like picture naming and story telling (from pictures). At those times, speech sound articulation was unremarkable except for some misarticulation of /s/, /ʃ/, /z/, /ʒ/ and a moderate overall nasal emission that was the consequence of the moderate pseudobulbar palsy. Speech was quite intelligible and perfectly fluent (especially in a Story Telling from Pictures task), and its rate rather rapid. Phonologically, syntactically, and lexically there was no impairment in language production.

Follow-up Assessments

Detailed neurological and speech and language follow-up assessments were performed at 7.4, 8.8, 9.7, and 10.4 years. Between the initial evaluation at 7 years and the first re-evaluation at 7.4 years, there was no change in voluntary oropharyngeal function, and despite a worsening in articulation (in addition to the four consonants mentioned above, /t/, /d/, and /n/ were impaired), the patient's speech was quite intelligible. There was, however, a dramatic change in speech fluency. In a standardized story-telling task (Chevrie-Muller, Simon, & Decante, 1981) the following difficulties were observed: severe difficulty in initiating speech, inhibition, need for stimulation, prolonged pauses, filled pauses, changes in speech rate from slow scanned to rapid speech, and odd intonation. The narration was, nevertheless, coherent, with no grammatical disorder. A similar fluency disorder was observed when the child had to answer questions (which had to be asked several times). Word-finding problems were confirmed in a picture-naming task (long pauses before naming some pictures and failure in word-retrieval for others that were, nevertheless, correctly pointed to in the receptive part of the vocabulary task). The parents of the child could not say precisely when the fluency disorders had begun.

Between 7.4 years and 21 years, no dramatic change was observed in the type of speech disorders, but there was, on the whole, a worsening of the symptoms, especially of the dysarthria. Indeed, at the final assessment the intelligibility was lower than at the second assessment (at 7.4 years). The ages at which some changes were observed during the follow-up period are reported below. Figure 1 shows performance on the same verbal task that was used at each re-evaluation (telling a story from a series of pictures, using a standardized material, Chevrie-Muller et al., 1981).

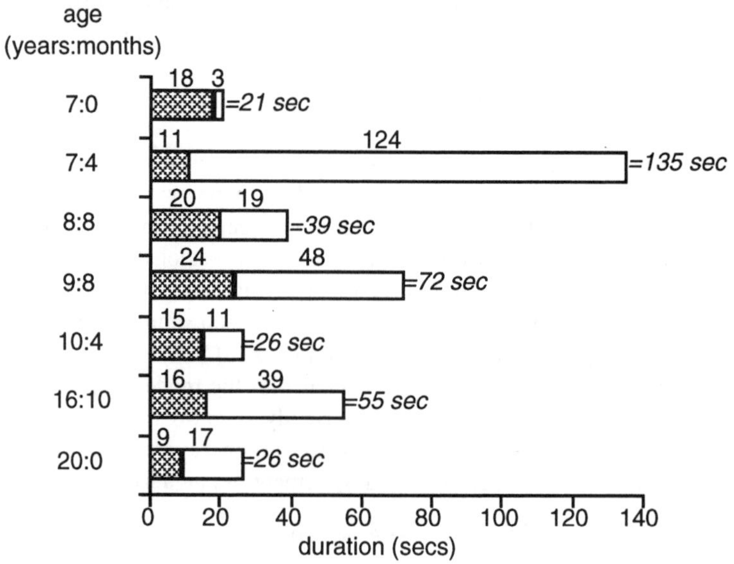

age
(years:months)

7:0	18 3 =21 sec							
7:4	11		124				=135 sec	
8:8	20 19 =39 sec							
9:8	24	48	=72 sec					
10:4	15 11 =26 sec							
16:10	16	39	=55 sec					
20:0	9 17 =26 sec							

0   20   40   60   80   100   120   140
duration (secs)

"Speech" Content:

Age at Evaluation	total words	verbs	repetition episodes
7 years	w=55	v= 8	r= 1
7 years 4 mos*	w= 20	v=3	r= 0
8 years 8 mos	w= 43	v=9	r= 5
9 years 8 mos	w= 28	v=7	r= 4
10 years 4 mos	w= 38	v=8	r= 2
16 years 10 mos	w= 42	v=8	r= 5
20 years	w= 30	v=5	r= 2

*no continuous speech, answers to questions only

FIGURE 1. Duration of speech and pauses in a story-telling task (the same picture materials were used at each recording session). The "speech" and "pauses" durations are represented by hatched and blank bars, respectively. "Speech" duration includes normal continuous speech and repetitions; "pauses" duration includes fillers (e.g., vocalizations, throat clearing) in pauses, except when these were words. Total duration is shown in italics.

Final Assessments

Detailed neurological and speech and language follow-up assessments were performed at 16.10 years, and less complete re-evaluations were made at 20 and 21 years. The two symptoms of primary interest at 16.10 years (final assessments) were a dystonia and the persistent speech dysfluency with an apparent word-finding impairment.

The neurological assessment revealed a moderate dystonia at the cranial and upper limb levels. At the cranial level, grimacing during pathological speech arrests (first noted at 9.7 years) and a slight dystonia of peri-oral muscles in non-speech lip movements (first noted at 16.10 years) were present, as well as upward and downward movements of the larynx before and after phonation. But there were no glottic spasms like those of laryngeal dystonia. The worsening of tongue movements in speech might have been the result of a progressive dystonia following the incomplete tongue paresis noted at 7.0 years (velopharyngeal and lip motor function had been normal since 9.7 years).

At the level of the upper-limbs, a slight postural hand dystonia and a moderate writing cramp were diagnosed at 16.10 years. On the whole, except for the tongue, the dystonia was moderate. Cerebral imaging did not show any structural abnormality, either on CT scan or with Magnetic Resonance Imaging. A Positron Emission Tomogram was not completely interpreted because a partial loss of data in the computer, but there was neither hemispheric asymmetry in blood flow nor hypoperfusion in basal ganglia.

The neuropsychological tests showed cognitive and spatial skills to be at average levels (WAIS, Wechsler, 1955; Rey-Osterrieth Complex Figure Copy, Rey, 1959), but auditory and visual memory skills were rather poor (WAIS Memory of digits; Rey-Osterrieth Complex Figure Memory). The patient completed his academic studies at age 17 (10th grade) and obtained a moderate qualification. He was rather impulsive and irritable, and had infrequent temper tantrums, but had no specific psychiatric disorder. He was quite cooperative for long and relatively frequent assessments.

The neurolinguistic assessment allowed the analysis of three kinds of symptoms: dysarthria, dysfluency, and word-finding problems.

The dysarthria consisted of a general "blurred" articulation, the primary impairment being on the anterior consonants (/t, d, n, s, z/) and an inconstant slightly nasal emission. Words, and sometimes even phrases, were unintelligible.

The dysfluency was severe and caused a handicap in social communication. The speech modalities in which dysfluencies occurred were: continuous self-formulated speech (telling a story); answering questions; repetition of meaningful and meaningless words and or sentences; and reading sentences.

The types of dysfluency were: (1) latencies in initiating phonation and answering questions; (2) pauses between words and phrases, often filled by *"comment?"* ("what?") or, exceptionally, by the whole sentence *"comment ça s'appelle?"* ("what's that called?") or by swearwords or repeated throat clearing;

(3) word and phrase repetitions, or occasional syllable repetitions; there were, however, no isolated segment repetitions (see Tables 1 and 2). Utterances, which occurred between pauses of different durations, were frequently produced at a rapid rate.

For the most part, repetitions were composed of words or groups of words, and included function words more often than content words (e.g., *"de le"*—"of the;" *"avec un"*—"with a;" *"j'ai un"*—"I have a;" *"c'est un"*—"that's a"). Often, the repetition seemed to suggest a progressive control of the whole phrase (see Table 3).

TABLE 1. *Distribution of episodes of dysfluency types—Conversational speech (16.10 years).*

Total number of sentences (*)   = 154   Total number of words   = 455
(*) 30 "one-word" answers: "oui" or "non" (*"yes" or "no"*)

Total number of repetition episodes   = 54   Total number of "fillers"   = 35

Number of episodes

content of repetitions: number (*) of repetitions	segment (part of word)	syllable (part of word)	1 or 2 words	3 to 5 words	7 words
1	0	3	22	5	1
2-3	0	0	14	8	0
4	0	0	1	0	0
6	0	0	1	0	0

(*) number of repetitions per episode

TABLE 2. *Distribution of episodes of dysfluency types—Reading short sentences.*

Total number of sentences (*)   = 10   Total number of words   = 40

Total number of repetition episodes   = 10   Total number of "fillers"   = 22

Number of episodes

content of repetitions: number (*) of repetitions	segment (part of word)	syllable (part of word)	1 or 2 words	3 to 5 words
1	0	1	6	1
2-3	0	2	3	0

(*) number of repetitions per episode

TABLE 3. *Samples of progressively longer repetition dysfluencies.*

(Spontaneous speech)	(Spontaneous speech—in answer to a question)
1. j'ai vu	1. oui parceque
2. j'ai vu ce matin	2. parceque, *comment(*)*,
3. ce matin	3. parce que je suis dans
4. j'ai vu ce matin	4. parceque je suis dans un centre de cure
5. ce matin	5. parce que je suis dans un centre de cure
6. j'ai vu ce matin sur la "5"	*" yes, because I'm in a rehabilitation center"*
"I saw this morning on "5" (TV channel)"	(*) *"what"*

Problems in word recall were obvious. In self-formulated speech or when answering questions the patient was struggling for the right word, he was helped in two ways: phonemic cueing by the interviewer and spelling the letters of the word by himself.

Word finding was also disturbed in two other tasks. First, in naming pictures (and objects) latencies might occur both for frequent and infrequent words, but they occurred for a smaller number of words at adolescent age than in childhood; there were no paraphasias. Furthermore, although it happened very rarely, when the expected word was not produced it was replaced by another semantically close word. In a verbal fluency test (Thurstone test, Lezak, 1983) the scores were very low, from 1 to 5 words per 60 seconds depending on the subtests (lowest scores were for words beginning with the letters C, P, F, L). These low scores were consistent on repeated tests. The same slight problems in naming tasks and the severe ones in fluency tasks also occurred in the written modality: the score on a written word-fluency task was as low as in the oral tests.

Latencies in producing words also occurred in repetition tasks. That is, latencies occurred for two of 15 meaningful three-syllable words that the patient had to repeat (the latencies were 6 seconds and 17 seconds). In repeating 15 three- and four-syllable nonsense words, a latency of 15 seconds was observed for one word.

DISCUSSION

It has been shown here that a dysfluency had become firmly established in a child at age 7, within four months following an acute encephalopathy (of unknown origin). Despite the patient's age at onset, it was obvious that the disorder was not a developmental dysfluency (developmental stuttering—DS) because the child never experienced speech or language disorders prior to the encephalopathy and, moreover, because he demonstrated fluent speech at the first evaluation session just after the acute disease episode.

Descriptively, this acquired dysfluency shared some characteristics with DS, including word repetitions (more often in function words), phrase repetitions, and prolonged phonatory arrests (> 5 seconds). But even these features, according to the literature (Riley & Riley, 1983), are not sufficient for the diagnosis of (developmental) stuttering. Sound repetitions were never observed and syllable repetitions or part word repetitions were very rare. The typical "fragmentation" of words (cf. Bijleved, 1989, unpublished lecture - European Postgraduate Course in Language Pathology, Brussels, 1989) was, therefore, an exceptional feature, and when present such part word repetitions never occurred more than twice. Other "core behaviors" of stuttering were also absent: there were no sound prolongations, no secondary motor manifestations (except, beginning in adolescence, repeated throat clearing), no effort to overcome a possible blockage, no severe behavioral responses, and no marked anxiety.

While the dysfluency was obviously acquired and neurogenic, it is interesting to note that it has some similarities with the disorder described as palilalia (Horner & Massey, 1983; Lebrun et al., 1987), especially because in palilalia— as in the dysfluency case reported here—the repetitions may occur on any part of the utterance and not just at its beginning, and are most often repetitions of words and phrases (rather than being predominantly of segments and syllables). Other similarities between T. E.'s case and extrapyramidal syndromes (Canter, 1971) were the very long pauses and the difficulties in initiating speech, especially at the time when the child began to have problems with speech fluency (at 7.4 years). But the impediment in speech production for our patient does not fit, for example, with the classical description of Parkinson's palilalia, which includes repetitions mostly of the latter part of the phrase that are accompanied by decreasing loudness and increasing rate of speech. Finally, as no hemispheric lesion was demonstrated on cerebral brain imaging, the mechanism of the dysfluency cannot be easily compared to the acquired cortical stuttering described in adult hemispheric or bihemispheric cerebral lesions (Ardila & Lopez, 1986; Horner & Massey, 1983; Jones, 1966; Lebrun & Leleux, 1985; Lebrun et al., 1987; Mazzucchi et al., 1981; Rosenbek et al., 1978).

In an attempt to understand the mechanism of the dysfluency reported here, we must consider two of the neurological symptoms, the dystonia and the impaired articulation (i.e., the dysarthria), and one neuropsychological symptom, the word retrieval impairment.

The facial dystonia during speech was noted at age 9.6 years, and later (at 16 years), peribuccal dystonia was noted in nonspeech lip movements and there was also a dystonia of the upper limbs. The progressive worsening of speech articulation from 7 (pseudobulbar palsy with slight dysarthria) to 20 years (frequently unintelligible speech) may be the consequence of a tongue dystonia. Even if the early sequelae of the acute encephalopathy were mainly of the pyramidal type (paralysis), a secondary degeneration of the subcortical projection system in response to cortical damage may be hypothesized as a possible mechanism of the tongue dystonia. But while a lingual dystonia may be the cause of the articulatory deficit and of the decrease in speech intelligibility, it seems very unlikely that it was the origin of the prolonged pauses during which there was neither effort in articulation nor symptoms of any laryngeal block, like laryngealization or the strained/strangled voice characteristic of spastic dysphonia.

The last point to be discussed is the relationship between, on the one hand, the word recall impairment observed in naming and repetition tasks and the very low score at the Thurstone test and, on the other hand, the dysfluency. From a psycholinguistic point of view, the impairment was obviously not at the level of conceptual or semantic representation (or semantic access) of the word (Dunn, Russel, & Drummon, 1989; Garret, 1982; Kay & Ellis, 1987), since one way for the patient to produce a "difficult" word was for him to spell the word letter-by-letter (sometimes the first letter was sufficient); that is, the production was helped by access to the "alphabetic" form of the word.

Since it was obvious that at the moment when the subject had retrieved a word he could produce it (most of the time in its entirety and fluently), the level of impairment appears to be neither that of motor programming nor of motor control. That is, articulatory motor disturbance interpreted as a secondary lingual dystonia could not be the cause of repetition of words and groups of words; moreover, one must remember that the fluency disturbance was at least as severe at the earliest follow-up (at age 7.4), when the tongue impairment was just a mild paresis, as later, when the tongue dystonia had developed. It can be noted that neither in pseudobulbar dysarthria nor in localized lingual dystonias nor in generalized ones are such word and word-group repetitions a clinical feature of the patients' speech impairment.

One hypothesis that can be put forth is that the problem may be at an intermediate level, between the semantic representation (or access) and the motor level; that is, at the level of activation of spoken word forms in a phonological (output) lexicon, as has been described in a "phonologically based anomia" by Kay and Ellis (1987). These authors gave examples of a phonological anomic patient's production, including "five in it" (i.e., letters), and "begins with 't', table"; the manner in which their patient succeeded in producing the words may be comparable to the letter-by-letter spelling of words by our patient.

When one considers cerebral damage that might be the cause of such a disruption in the phonological word-form activation, a possible site would be the prefrontal area. The two most notable modifications in verbal behavior with cerebral lesions (or dysfunctions) in this area are: (i) a decrease of the verbal fluency in tasks like the Thurstone test; and (ii) a particular form of anomia characterized by a greater disturbance in nonspecific situations (dialogs, narratives) than in naming tasks, the presence of a cueing effect, a delay or absence of production, or of answers with neither paraphasias, nor circumlocutions or commentaries (Botez, 1987). Palilalias have also been described in frontal patients (Valenstein, 1975).

It is obvious that our patient shared some of the verbal behavior features of patients with prefrontal lesions. The assumption that a frontal dysfunction may be the cause of the speech dysfluency in the case reported here nevertheless deserves comment, because no other feature of the frontal syndrome, except the moderate attentional and behavioral disturbances, was observed in our patient.

CONCLUSION

Observation of pathological cases with specific communicative disturbances is one way to identify the different processing levels in speech production. This case of acquired dysfluency with a dysnomic component has allowed us to identify specific difficulties at the level of activation of phonological word forms as opposed to the lower level at which disturbances appear to occur in developmental stuttering (i.e., at the motor programming level).

ACKNOWLEDGMENTS

The author thanks Marie-Thérèse Le Normand, Marie-Claire Goldblum, Marise Forgue, and Anne-Marie Simon who, at different stages of the follow-up, performed neuropsychological and language assessments of the patient, Françoise Gouttières and Professor Pierre Rondot for the data concerning the medical neurologic examination. She is grateful to Marie-Thérèse Rigoard for her technical assistance and Suzan Orsoni for the revision of the manuscript.

REFERENCES

Ardila, A., & Lopez, M. V. (1986). Severe stuttering associated with right hemisphere lesion. *Brain and Language, 27*, 239-246.

Baratz, R., & Mesulam, M. M. (1981). Adult-onset stuttering treated with anticonvulsivants. *Archives of Neurology, 38*, 132.

Bhatnagar, S., & Andy, O. J. (1989). Alleviation of acquired stuttering with human centromedian thalamic stimulation. *Journal of Neurology, Neurosurgery, and Psychiatry, 52*, 1182-1184.

Botez, M. J. (1987). *Neuropsychologie clinique et neurologie du comportement.* (pp. 124-125). Montréal: PUM, & Paris: Masson.

Canter, G. J. (1971). Observations on neurogenic stuttering: A contribution to differential diagnosis. *British Journal of Disorders of Communication, 6*, 139-145.

Chevrie-Muller, C., Simon, A. M., & Decante, P. (1981). *Epreuves pour l'examen du langage.* Paris: Centre de Psychologie Appliquée.

Cipolotti, L., Bisacchi, P. S., Denes, G., & Gallo, A. (1988). Acquired stuttering: A motor programming disorder. *European Neurology,* 28, 321-327.

Dunn, N. D., Russel, S. S., & Drummon, S. S. (1989). Effect of stimulus context and response coding variables on word-retrieval performances in dysphasia. *Journal of Communication Disorders,* 22, 209-223.

Farmer, A. (1975). Stuttering repetitions in aphasic brain damaged adults. *Cortex,* 11, 391-396.

Fleet, W. S., & Heilman, K. M. (1985). Acquired stuttering from a right-hemisphere lesion in a right-hander. *Neurology,* 35, 1343-1346.

Garrett, M. F. (1982). Production of speech: Observation from normal and pathological use. In A. Ellis (Ed.), *Normality and pathology in cognitive functions.* London/New York: Academic Press.

Helm, N. A., Butler, R. B., & Benson, D. F. (1978). Acquired stuttering. *Neurology,* 28, 1159-1165.

Helm, N. A., Butler, R. B., & Canter, G. J. (1980). Neurogenic acquired stuttering. *Journal of Fluency Disorders,* 5, 269-279.

Helm-Estabrooks, N. (1986). Diagnosis and management of neurogenic stuttering in adults. In K. St. Louis (Ed.), *The atypical stutterer* (pp. 193-217). London/New York: Academic Press.

Horner, J., & Massey, E. W. (1983). Progressive dysfluency with right-hemisphere disease. *Brain and Language,* 18, 71-85.

Jones, R. K. (1966). Observation on stammering after localized cerebral injury. *Journal of Neurology, Neurosurgery, and Psychiatry,* 29, 192.

Kay, J., & Ellis, A. (1987). A cognitive neuropsychological case study of anomia. Implications for psychological models of word-retrieval. *Brain,* 110, 613-629.

Koller, W. C. (1983). Dysfluency (stuttering) in extrapyramidal disease. *Archives of Neurology,* 40, 175-177.

Lebrun, Y., & Leleux, C. (1985). Acquired stuttering following right brain damage in dextrals. *Journal of Fluency Disorders,* 10, 137-141.

Lebrun, Y., Leleux, C., & Retif, J. (1987). Neurogenic stuttering. *Acta Neurochirurgica,* 85, 103-109.

Lezak, M. D. (1983). *Neuropsychological assessment.* New York: Oxford University Press.

Mazzucchi, A., Moretti, G., Carpegiani, P., Parma, M., & Paini, P. (1981). Clinical observation on acquired stuttering. *British Journal of Disorders of Communication,* 16, 19-30.

Nowack, W. J. (1986). Adult onset stuttering and seizures. *Clinical Electroencephalography,* 17, 142-145.

Quinn, P. T., & Andrews, G. (1977). Neurological stuttering—A clinical entity? *Journal of Neurology, Neurosurgery, and Psychiatry,* 40, 699-701.

Rey, A. (1959). *Test de Copie d'une Figure Complexe.* Paris: Centre de Psychologie Appliquée.

Riley, G. D. & Riley, J. (1983). Evaluation of a basis for intervention. In D. Prins & R. J. Ingham (Eds.), *Treatment of stuttering in early childhood: Treatment methods and issues*. San Diego: College Hill Press.

Rosenbek, J., Messert, B., Collins, M., & Wertz, R. T. (1978). Stuttering following brain damage. *Brain and Language, 6*, 82-96.

Rosenfield, D. B. (1972). Stuttering and cerebral ischemia. *New England Journal of Medicine, 287*, 991.

Valenstein, E. (1975). Non-language disorders of speech reflect complex neurologic apparatus. *Geriatrics, 30*, 117-121.

Van Hout, A., Evrard, P., & Lyon, G. (1985). On the positive semiology of acquired aphasia in children. *Developmental Medicine and Child Neurology, 27*, 231-241.

Wechsler, D. (1955). *Wechsler adult intelligence scale*. New York: Psychological Corporation.

REFERENCE NOTES

Bijleved, H. (1989). Unpublished lecture—European Postgraduate Course in Language Pathology, Brussels, 1989

Lebrun, Y. (1989). Unpublished lecture—European Postgraduate Course in Language Pathology, Brussels, 1989

Roulet, E., Davidoff, V., Buschini, M. F., & Deonna, T. (1993). Bégaiement acquis transitoire d'origine neurologique. Paper presented at the Vème Congrès de la Société Européene de Neurologie Pédiatrique, Strasbourg, France.

27

Fluency, Disfluency, Dysfluency, Nonfluency, Stuttering: Integrating Theories

Frances Jackson Freeman
University of Texas at Dallas
Callier Center for Communication Disorders

Each time I approached the writing of this manuscript, I attempted to begin with a statement of respect, admiration, and affection for Dr. Katherine S. Harris. Each time I failed, paralyzed by the task. I have never found adequate words. I can only say that the honor of being her student is a blessing for which I am eternally grateful.

This chapter is based on studies of adult developmental stutterers. At the core of this research are studies of brain metabolic activity measured through regional cerebral blood flow (rCBF) (Pool, Devous, Freeman, Watson, & Finitzo, 1991). Brain electrical activity, in resting EEG spectra and auditory evoked potentials, was analyzed to support or refute the blood flow findings (Finitzo, Pool, Freeman, Devous, & Watson, 1991). Finally, behavioral performance results from tests of speech production and language were related to cortical function patterns (Watson, Freeman, Chapman et al., 1991; Watson, Pool, Devous, Freeman, & Finitzo, 1992; Watson et al., 1994). While findings have been reported in five journal articles and two conference proceedings, this chapter is the first comprehensive summary of the work.

Equally important, this report offers an interpretation of results based on the deliberations of our multidisciplinary research team. While I accept responsibility for this presentation (including any unintentional distortions or oversimplifications), I fully acknowledge that ideas, theories, approach, and sometimes the words are the product of the group (see Acknowledgment). Specifically, the data and discussions shifted our focus from the disorder of stuttering to the broader context of fluency. The purpose of this paper is to persuade, intrigue, and challenge others to make a similar shift.

REVIEW OF EXPERIMENTAL FINDINGS

Analyses of cortical blood flow revealed global absolute flow reductions in stutterers, at a $p < 0.00001$ (Pool et al., 1991). While hypoperfusion can be a consequence of vascular disease, it can also be related to autoregulatory phenomena at the neuronal level since blood flow is related to the activity and metabolic demands of the neurons involved. This finding of hypoperfusion in stutterers as compared with normal controls was tested through analyses of quantitative EEG spectra. Absolute spectra in the eyes-closed condition were analyzed for five frequency bands: Delta (0.5-3.5 Hz), Theta (4-7.5 Hz); Alpha 8-11.5 Hz), Beta 1 (12-15.5 Hz), and Beta 2 (16-19.5Hz). When stutterers were compared with normal controls, significant reductions were observed in Beta 1 and Beta 2, but not in other frequency bands (Finitzo et al., 1991). Reduction of Beta is congruent with hypoperfusion because Beta is the first EEG frequency to be affected in mild ischemia (Faught, Mitchem, Conger, Garcia, & Halsey, 1988).

In addition to global flow reductions, relative blood flow asymmetries were identified in the stutterers. Flow asymmetries, with left hemisphere flows reduced significantly relative to right hemisphere flows, were found in three regions: anterior cingulate, superior temporal, and middle temporal gyri. Less pronounced asymmetries were identified in the inferior frontal region adjacent to the cingulate (Pool et al., 1991).

Examination of individual asymmetry values revealed that all of the 20 stuttering subjects had below normal median flows for at least one of the three implicated cortical regions. Further, all stuttering subjects demonstrated below median asymmetry values for at least one of the temporal regions (middle and/or superior) (Pool et al., 1991). Of ten subjects classified as severe, only one had normal median asymmetry values for the anterior cingulate region. The six stutterers with the most pronounced asymmetry in the anterior cingulate region were all severe. When the ten severe stutterers were compared to the ten mild/moderate stutterers, the severe stutterers demonstrated significantly greater flow asymmetries in the anterior cingulate region. Thus, blood flow asymmetries in the anterior cingulate region, which includes portions of the supplementary motor area and limbic projections, were associated with severe stuttering.

However, it should be understood that, within this subject population, severe stutterers demonstrated anterior cingulate flow asymmetries *in addition to* asymmetries in one or both temporal regions. Blood flow asymmetries *only* in the superior and/or middle temporal regions were associated with mild or moderate stuttering (Pool et al., 1991).

The rCBF findings of flow asymmetries in the superior and middle temporal regions were tested through analyses of auditory evoked potentials (AEPs). Three major components of the auditory evoked potential (P1, N1, and P2) were selected because their generators arise, at least in part, from the temporal cortex. Significantly lower amplitudes for all three AEPs were found for stutterers as compared to normals (Finitzo et al., 1991). Thus, findings of temporal blood flow asymmetries in stutterers were supported by reduced amplitudes of AEPs generated (at least in part) in the temporal cortex.

The subgrouping of stutterers on the basis of regional blood flow asymmetries was further explored through quantitative EEG. When the ten severe stutterers were compared to the ten mild/moderate stutterers, significant reduction in Beta was observed in the severe stutterers. Conversely, the mild/moderate stutterers showed significantly greater reduction in AEP amplitudes (P1 & N1, generated at least in part from the temporal regions) than did the severe stutterers (Finitzo et al., 1991). Thus, rCBF and quantitative EEG findings were congruent with respect to classifying stuttering severity sub-groups.

In addition to the work reported above, we collected and analyzed results of performance measures of language and speech motor control. Linguistic performance was evaluated using tasks that assess relatively high-level production and comprehension processes (Bond, 1986; Cannito, Hayashi, & Ulatowska, 1988; Ulatowska, Allard, & Chapman, 1989). Results identified a subgroup of 11 stutterers who differed significantly from normal controls and from 8 "linguistically normal" stutterers. This linguistic impairment was unrelated to age, educational level, or stuttering severity (Watson, Freeman, Chapman et al., 1991; Watson et al., 1994).

Next, we explored relation(s) between performance on linguistic tasks and performance on a speech task sensitive to linguistic/motoric complexity (Watson, Freeman, & Dembowski, 1991). Our laryngeal reaction tasks require responses of increasing complexity, from a simple /ɑ/ vowel, to a bisyllabic word "Oscar," to a sentence, "Oscar took Pete's cat." All three subject groups (controls, linguistically normal stutterers, and linguistically impaired stutterers) showed longer laryngeal response times (LRTs) with increasing complexity of the response (Watson, Freeman, & Dembowski, 1991). However, the magnitude of the increase in LRT, as a function of response complexity, was greatest for the linguistically impaired stutterers. They differed significantly from normal controls on both the word and sentence tasks. Neither measure (linguistic performance or LRT) was systematically related to stuttering severity (Watson, Freeman, Chapman et al., 1991).

Given these results, we explored the potential relation(s) between LRTs and rCBF (Watson et al., 1992). We found that increased latencies in LRT for complex utterances were systematically related to rCBF asymmetry in the left middle and superior temporal regions. Stutterers with normal relative flows in *one* of these temporal regions did not differ significantly from controls on LRT latencies. Stutterers with reduced relative rCBF to *both* temporal regions had significantly delayed LRTs for both the word and sentence conditions. No systematic relation was found between LRT and flow asymmetries in the anterior cingulate.

Finally, we explored relation(s) between linguistic performance and asymmetries in regional cerebral blood flow (Watson et al., 1994). Between group comparisons (controls vs. "linguistically normal" stutterers, vs. "linguistically impaired" stutterers) of rCBF data were performed for the four previously identified cortical regions (superior temporal, medial temporal, anterior cingulate, and inferior frontal). Controls and "linguistically normal" stutterers did not differ for any rCBF comparison. "Linguistically impaired" stutterers differed significantly from controls on the medial temporal and inferior frontal rCBF comparisons. "Linguistically impaired" and "linguistically normal" stutterers differed significantly on middle temporal rCBF comparisons (Watson, Freeman, Chapman, Devous, Finitzo, & Pool, 1994). Thus, linguistic performance and LRTs for simple and complex responses were related to each other, and to patterns of regional cerebral blood flow in the superior and middle temporal lobes.

These investigations are, admittedly, preliminary and controversial (Viswanath, Rosenfield, & Nudelman, 1992; Pool, Devous, Watson, Finitzo, & Freeman, 1992; Fox, Lancaster, & Ingham, 1993; Pool, Devous, Watson, Finitzo, & Freeman, 1993). Our desire to be conservative—to double check validity through a multiple internal tests—dictated the design. The design tests the results of one approach to the assessment of brain function (i.e., rCBF by SPECT) against the results of a second measure of brain function, quantitative EEG. Finally, both of these measures are compared to behavioral findings. It is the internal consistency of the results that most persuasively supports their validity.

By adopting conservative statistical approaches, we sought to avoid Type I errors, but readily admit to possible Type II errors. The most severely limiting factor, however, is the number of subjects. These limitations are most apparent when subgroups of stutterers are compared. Thus, although replication and expansion are absolutely essential, consideration of theoretical implications is appropriate.

THEORETICAL IMPLICATIONS

Given these empirical findings, we considered the implicated cortical regions. First, dysfunction in the left cingulate and left inferior frontal gyri is consistent with our current understanding of motor control for speech. Motor initiation deficits have been localized to the mesial frontal cortex in the region of the cingulate or supplementary motor area. Motor programming for speech has been localized to the left perisylvian (inferior frontal) cortex.

In contrast, the implicated temporal regions are associated with linguistic processing and performance. Critical components in the processing of complex temporal patterns, including speech and speech-like stimuli, have been localized to these regions. In several articles and chapters, Kent has described aspects of performance related to function of the temporal lobe, and has reminded us that the name "temporal" was meaningfully assigned to this region (see Kent, 1983; Kent, 1984; and Kent & Rosenbek, 1983).

One approach to the data is to create diagnostic sub-groups consisting of: 1) stutterers with a "motor" deficit, 2) stutterers with a temporal processing deficit, and 3) stutterers with both. The data suggest that severe stutterers are most likely to have a "motor" component. Mild and moderate stutterers have, primarily or exclusively, a temporal deficit. Stutterers with temporal dysfunction demonstrate greater difficulties on more complex speech tasks, and deficits on measures of receptive and expressive language.

The idea that stuttering is sometimes an expression of a language disorder and sometimes of a motor speech disorder was advanced in explaining within-group differences among young stuttering children (Meyers & Freeman 1985a, 1985b). In phonological studies of young stuttering children, Wolk, Edwards, and Conture (1991) have also proposed two types of stuttering, one occurring with and one without disordered phonology.

This interpretation, however, suffers from several limitations. Most notably, it lures us toward the naive assumption that the processes of language and speech generation are resolvable into separate and distinct motor and linguistic components. To conclude simplistically that some stutterers are "dyspraxic" while others are "dysphasic" does little to advance our understanding of the neurophysiologic system that subserves these theoretically distinct constructs. It appears more productive and more accurate to interpret these findings as *supporting an integrative relation between linguistic and speech motor processes in the production of fluency.*

INTEGRATION OF THEORIES OF STUTTERING

The results of our investigations do not dictate the creation of a new theory of stuttering. Rather, they offer experimental evidence supporting several theories. For example, theories describing stuttering as a cognitive or linguistic disorder (e.g., Hamre, 1984; Wingate, 1988) are supported. On the other hand, theories of stuttering as a motor-control, sequencing and timing, or temporal programming disorder (e.g., MacKay & MacDonald, 1984; Kent, 1984) are also supported. Differential deficit patterns (or subgroups) within the stuttering population offer an explanation for conflicting findings with respect to neuropathophysiology, auditory function, articulatory and phonatory characteristics (e.g., Rosenbek, 1984; Moore, 1984; Rosenfield & Jerger, 1984; Zimmerman, 1984).

Unfortunately, the "umbrella" utility of our findings with respect to theories can be problematic. For example, if stuttering is a linguistic disorder, a temporal programming disorder, and a motor-control disorder, how is it related to, or differentiated from, other disorders of fluency? Clearly there is a blurring of the distinctions our professions have established between dysfluency (as applied to stuttering) and nonfluency (as applied to aphasia).

DISFLUENCY, DYSFLUENCY, AND NONFLUENCY

General lexicology designates both *dis-* and *non-* as meaning the *"opposite of"* or *"lack of"* the quality or attribute denoted. *Dys-* is a synonym of the other two, with the negative connotation of *abnormal* or *bad*. However, as all speech/language pathologists know, normals have disfluencies (or sometimes normal nonfluencies); stutterers have disfluencies and Dysfluencies; and aphasics are either fluent or non-fluent.[1]

Fortunately, our research team included an audiologist uninitiated in these intradisciplinary semantic mysteries. Terese Finitzo, a key investigator in the stuttering research, was also involved in studies of fluent and non-fluent aphasics. She perceived a potential relation based on the commonality of fluency disruption. Since she had been analyzing EEG spectral data from both populations, she plotted Beta-1 for our 20 stuttering subjects, for 24 mildly non-

---

[1]While disfluent and dysfluent are frequently considered synonyms for each other, and for the more generally used term nonfluency, within the stuttering literature some authorities have made a distinction. The term *dys* takes the connotation of pathological or abnormal. Thus, *dysfluency* has been recommended for use in denoting abnormal instances of fluency failure, i.e., those most closely associated with stuttering. Dysfluencies have been broadly defined as including all instances of broken words (i.e., part-word, syllable, or sound repetitions or prolongations) and/or fluency disruptions associated with excess tension or struggle. In contrast, *disfluency* has been proposed as the appropriate term for normal nonfluencies. *Disfluencies* or normal *nonfluencies* are considered to include silent, nontense pauses, interjections, revisions, word or phrase repetitions without indications of stress or struggle. Although professional dictionaries define *dys*fluency more frequently than *dis*fluency, the latter term has come to be more widely used in professional journals and texts. Both terms are often used as synonyms for stuttering. This euphemism originated through the influence of the general semanticist, Wendell Johnson. Johnson believed that the application of the label "stuttering" to a child was a primary cause of the disorder. Disfluency was used to avoid the label stuttering.

fluent aphasics, and for 30 severely nonfluent aphasics. The rational was unorthodox, but the results were striking (see Figure 1).

As measured at the T3 electrode, a continuum of Beta reduction is observed over the patient groups. The stuttering and mildly nonfluent aphasic groups do not differ significantly from one another, but both differ significantly from the control group and from the severely non-fluent aphasic group. A similar pattern was observed when amplitudes of auditory evoked potentials (P1 and N1) were plotted and compared for these subject groups. These data suggest a continuum of electrophysiologic dysfunction that appears to relate to a continuum of fluency impairment. These results (in combination with findings of linguistic performance deficits in a subgroup of stutterers) led to further consideration of potential relations between stuttering and the aphasias.

Authors who have observed relations between stuttering and aphasia may be divided into those who view stuttering as related primarily or exclusively to the sensory-motor components of aphasia (i.e., cortical dysarthria or apraxia of speech), and those who view stuttering as related more broadly to the linguistic components of aphasia. Kussmaul (1887) described stuttering as an aphasia that produced a syllabic dysarthria characterized by a lack of coordination of voice, respiration, and articulation. Shtremel (1963) defined stuttering as an apraxia. Caplan (1972) compared stutterers with subjects having predominantly expressive aphasia with coexisting anomia and apraxia, described mutual patterns of fluency disruption, and concluded that stuttering is a form of apraxia.

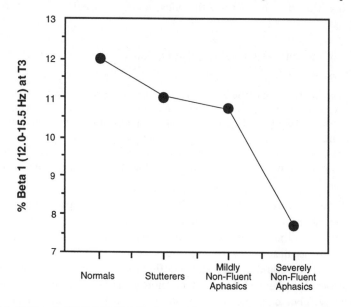

FIGURE 1. % Beta 1 (12.0-15.5 Hz) for four subject groups: normals, stutterers, mildly nonfluent aphasics, and severely nonfluent aphasics.

Other investigators have not equated the disorders of stuttering and apraxia, but have explored intriguing similarities. Schuell, Jenkins, and Jimenez-Pabon (1964) concluded that every aphasic patient with sensorimotor involvement passes through a period of stuttering during recovery. Johns and Darley (1970) and Trost (1971) described the perceived similarity between fluency disruptions of stutterers and apraxics.

Farmer (1975) made a broader comparison between the aphasias and stuttering. She studied stuttering repetitions in the speech of aphasics with diagnoses of Wernicke's aphasia (3 patients), conduction aphasia (2 patients), Broca's aphasia (4 patients), and anomic aphasia (3 patients). Wernicke's aphasics produced the most disfluencies, conduction aphasics were second, followed by Broca's aphasics, and finally by anomic aphasics. The Wernicke's aphasics were significantly more disfluent than the Broca's aphasics and the anomic aphasics. This is an apparently paradoxical finding since Wernicke's aphasia is widely considered to be synonymous with the behavioral category of "fluent aphasia," while Broca's aphasia is generally considered "nonfluent" aphasia.

Farmer interpreted her findings as indicating that "lesions to the left hemisphere resulting in inefficient language performance may reflect temporal disorganization in the form of stuttering repetitions" (p. 395). Rosenbek (1980) reviewed the literature on stuttering and apraxia of speech, and reached a parallel conclusion: "stuttering is a frequent speech response to a number of different abnormalities in the neural substrates serving speech-language function" (p. 247).

Most recently Wingate (1988) has devoted a volume to reviewing the stuttering literature as related to emerging models of linguistic performance. In the synopsis of this review, Wingate states:

> As for stuttering per se, the converging lines of evidence indicate that the disorder reflects a special kind of neurologic dysfunction involving, at least as the principal focus, neuronal systems of the left prefrontal cortex and related subcortical structures. There are many parallels between stuttering and aphasia, ordinarily not clearly evident, that become discernible once the parameters of comparison are brought into focus. (p. 266)

As Wingate perceived, "bringing the parameters of comparison into focus" (p. 267) lies at the heart of our quandary. *Fluent aphasic, nonfluent aphasic, stutterer, nonstutterer* are useful clinical diagnostic categories. *Disfluency* and *dysfluency* as descriptors of specific speech events have practical utility. However, their acquired, specialized definitions, as part of the "jargon" of our professions, have robbed them of their more general definition as the "opposite of fluency." For example, fluent/nonfluent as applied in aphasia is a binary distinction derived from a global assessment of oral language production. In order to be classified as "fluent," the aphasic patient must produce a flow of words with relatively normal prosody and grammatical structure. An expansion of this definition would classify the vast majority of stutterers as "fluent." The difficulty lies in the fact that the presence of numerous disfluencies in the speech of "fluent" aphasics is not relevant to this clinical distinction.

Similarly, the counting of dys/disfluencies does not encompass the negative of fluency. These terms define speech events (hesitations, repetitions, revisions, prolongations, interjections, etc.), but fail to capture the global percept of stuttering or fluency. Abnormal fluency can be identified by listeners in utterances that do not contain dys/disfluencies (or from which all dys/disfluencies have been excised). Non-fluent aphasics demonstrate fewer dys/disfluencies than fluent aphasics or stutterers, yet there can be little doubt that as a group these patients represent the most pronounced degree of fluency disruption.

These negative terms have appropriate clinical, diagnostic application. The limitations and paradoxes do not impair their usefulness. The danger arises when we mistakenly assume that *fluency* can be defined or understood by reference to them. For example, the phrase "fluency evaluation" is used when the process is actually the differential diagnosis of stuttering; or "measurement of fluency" is used to describe the process of counting dys/disfluencies, and/or computation of speech rates. These fallacies are dangerous because they disguise our basic lack of knowledge or understanding of the construct of *fluency* as applied to the production of language and speech.

Two decades have passed since I first heard Martin Adams describe the inadequacy of negative definitions of fluency, and the limitations our lack of knowledge of fluency imposes on our conceptualization of stuttering (Adams, 1974). The only notable change over the past 20 years appears to be an expanding appreciation of the relevance and importance of these observations. Most importantly, we now recognize that these limitations have implications and ramifications that extend beyond the traditional disorders of fluency.[2]

---

[2]At the beginning of this decade, the NIH sponsored a conference of experts in stuttering and published the results of their deliberations (Cooper, 1990). "The experts" agreed that no clear, precise definition of stuttering exists, and that there is no objective measure of its severity. Although Cantor (1971) was referring to "neurogenic" stuttering, it can be argued that his comment is equally applicable to "developmental" stuttering. He observed that too many kinds of disfluency have been called by one name (see comments by Rosenbek, 1984). In this context, the observations of Zimmermann are extremely cogent:

It is not playing semantic games to point out how viewing stuttering as a discrete entity has very different implications than viewing stuttering only as a perceptually reliable perceptual category. In fact, the case has been made that stuttering does not warrant definition other than as a diagnostic category (Zimmermann & Kelso, 1982), which is important only because it demarcates a significant difference from normal which can impede communication and which is held in low esteem by the culture. This view leads

MODELING FLUENCY

Intra- and intersubject variations in fluency are recognized in disordered and normal speakers. Conditions as diverse as dementia, psychosis, deafness, developmental disorders of language or intelligence, neuroses of public speaking (e.g., "stage-fright"), Parkinson's disease, dysarthria, laryngectomy, or spasmodic dysphonia demonstrate variations and differing patterns of impaired or significantly degraded fluency. However, our ability to quantify or otherwise reliably measure degrees of fluency is virtually nonexistent. Not only do we lack measurement tools and behavioral definitions of the characteristics of fluency, we lack a theoretical model. The search for perceptual and production variables underlying the construct of fluency can best be focused and directed through reference to a general model of fluency. In the absence of such a model, experimental exploration of the basic constructs of fluency through hypothesis-driven research is impeded.

Current models of the normal processes of language and speech generation appear to offer the best potential for development of a general model of fluency. Models of language and speech production vary from relatively basic from (Borden, Harris, & Raphael, 1994) to elaborate and detailed (e.g., Levelt, 1989), but they share several critical elements. These shared components include: 1) a cognitive or conceptualizer level involved in selection and organization of the thoughts, ideas, and emotions that underlie communicative intent; 2) a linguistic or formulator level at which lexical, grammatic, semantic, morphologic, syntactic, and phonologic encoding occurs; and 3) a neuromotor or articulator level for execution of the phonetic plan as a series of neuromuscular instructions for sequences of muscle contractions and structural movements. Further, most models of speech/language production explicitly acknowledge efficient integration of cognitive, linguistic, and motoric processes as critical to successful oral/verbal expression.

Thus, optimal levels of fluency are the product of efficient functioning, appropriate synchronization, and full integration of the processes of speech/language generation. If any component process is delayed or if the integration or synchrony between components is disrupted, some degree of degradation in fluency will result. Fluency can thus be viewed as a reflection of both the integrity and coordination of the components of the system. Impaired fluency reflects inefficiency or dysfunction in some component process, disruption in the integration of components, or both. Disorders of fluency arise from differing forms of disruption at multiple sites within the neurological system subserving these processes. Patterns of fluency failure offer insight into normal processes, disorders of fluency, and the neurophysiological subsystem.

---

us away from seeking simplistic, univariate causes of "stuttering..." (Zimmermann, 1984, p. 133).

In this context, fluency can be viewed as a sensitive, analog index of the overall efficiency of speech/language production.

Many of the constructs of this model have been proposed by others, or are congruent with views expressed by colleagues. For example, in the course of his monograph on speaking, Levelt (1989) uses the term "fluency" 28 times to describe the results of appropriate functioning of elements of the speech/language production system. Conversely, he uses "disfluency" five times to describe the product of functional failures within the system. Further, we find parallels between this integrative approach and the "systems analysis" model described by Nudelman, Herbrich, Hoyt, and Rosenfield (1991).

In their multidimensional formulation of a theory of neurophyscholinguistic function in stuttering, Perkins, Kent, and Curlee (1991) include consideration of the production of fluency. They state:

> Central to the theory is the idea that speech involves linguistic and paralinguistic components, each of which is processed by different neural systems that converge on a common output system. Fluent speech requires that these components be integrated in synchrony. (p. 734)

Finally, while he limits his observations to stuttering, Wingate (1988) approaches the broader construct of fluency when he observes:

> Stuttering is not simply a problem of words per se, but of words as pivotal elements in a system that can transduce ideas and thoughts into an audible code—speech. When functioning properly, as it does in the normal speaker, the system produces and transmits the code with a form and degree of continuity identifiable as having a normal flow. In stutterers the system is not functioning properly. (p. 267)

CONCLUSION

The ultimate value of a model is not its validity, but rather its usefulness as a basis for hypothesis formulation and testing. We have kept the constructs of the model broad and relatively general, while retaining sufficient specificity for hypothesis formulation. This approach was adopted to promote usefulness and adaptability to the interests and paradigms of other investigators.

With respect to stuttering, we conclude with this statement from Watson et al. (1994):

> At a very basic level, we are reiterating a well-accepted fact—stuttering is appropriately classified as a "fluency" disorder. However, this "fact" has some consequences. First, stuttering should be studied and understood within the broad context of fluency. This means we must expand our presently limited understanding of fluency. Second, stuttering will be best understood when investigated within the context of other disorders of fluency. (p. 1126)

ACKNOWLEDGMENTS

The research team responsible for the studies reported in this Chapter was composed of Terese Finitzo, Kenneth Pool, Ben Watson, Michael Devous, Steven Schaefer, and Sandy Bond Chapman. Others who made major contributions include Diane Mendelssohn, George Kondraske, Rick Roark, Susan Miller, Jim Dembowski, and Mari Hayashi. Scott Self, Georgeanne Self, Denise Wadsworth, and Shannon Beasley assisted in preparation of the manuscript. Support was provided by grant NS18276-05 from the National Institute of Neurologic Communicative Disorders and Stroke.

REFERENCES

Adams, M. R. (1974). Clinical interpretations and applications, in L. M. Webster & L. Furst, (Eds.), *Vocal tract dynamics and dysfluency* (pp. 196-228). Speech and Hearing Institute of New York.

Bond, S. (1986). *Reference: Processing of nouns and pronouns in narrative discourse.* Unpublished doctoral dissertation, The University of Texas at Dallas, Dallas.

Borden, G., Harris, K. S., & Raphael, L. J. (1994). *Speech science primer: Physiology, acoustics, and perception of speech* (3rd ed.). Baltimore MD: Williams & Wilkins.

Cannito, M., Hayashi, M., & Ulatowska, H. (1988). Discourse in normal and pathologic aging: Background and assessment strategies. In H. Ulatowska (Ed.), *Seminars in speech and language: Aging and communication* (pp. 117-134). New York: Thieme, Inc.

Cantor, G. (1971). Observations on neurogenic stuttering: A contribution to differential diagnosis. *British Journal Disorders of Communication, 6,* 139-143.

Caplan, L. (1972). An investigation of some aspects of stuttering-like speech in adult aphasic subjects. *Journal of the South African Speech and Hearing Association 19,* 52-66.

Cooper, J. (1990). Research directions in stuttering: Consensus and conflict. In J. Cooper (Ed.), *Research needs in stuttering: Roadblocks and future directions* (pp. 98-100). Rockville, MD: ASHA.

Farmer, A. (1975). Stuttering repetitions in aphasic and nonaphasic brain damaged adults. *Cortex, 11,* 391-396.

Faught, E., Mitchem, H. L., Conger, K. A., Garcia, J. H., & Halsey, J. H. (1988). Patterns of EEG frequency content during experimental transient ischemia in subhuman primates *Neurological Research, 10,* 184-191.

Finitzo, T., Pool, K., Freeman, F. J., Devous, M., & Watson, B. (1991). Cortical dysfunction in developmental stutterers. In H. F. M. Peters, W. Hulstijm, & C.

W. Starkweather (Eds.), *Speech motor control and stuttering* (pp. 251-262). Amsterdam: Elsevier.

Fox, P., Lancaster, J., & Ingham, R. (1993). On stuttering and global ischemia. Letters to the editor, *Archives of Neurology, 50*, 1287-1288.

Hamre, K. (1984). Stuttering as a cognitive-linguistic disorder. In R. F. Curlee & W. H. Perkins (Eds.), *Nature and treatment of stuttering: New directions* (pp. 237-257). College-Hill Press, San Diego.

Johns, D. F., & Darley, F. L. (1970). Phonemic variability in apraxia of speech. *Journal of Speech and Hearing Research, 13*, 556-583.

Kent, R. D. (1984). Stuttering as a temporal programming disorder. In R. F. Curlee & W. H. Perkins (Eds.), *Nature and treatment of stuttering: New directions* (pp. 283-301). College-Hill Press, San Diego.

Kent, R. D., & Rosenbek, J. (1983). Acoustic patterns of apraxia of speech. *Journal of Speech and Hearing Disorders, 26*, 231-248.

Kent, R. D. (1983). Facts about stuttering: Neuropsychologic perspectives. *Journal of Speech and Hearing Disorders, 48*, 249-255.

Kussmaul, A. (1887). Die Storungen der sprache. In H. Ziemssen (Ed.), *Cyclopaedia medica.* Reported in Van Riper, C. (1982). *The nature of stuttering* (2 ed.). Englewood Cliffs, NJ: Prentice Hall.

Levelt, W. J. M. (1989). *Speaking: from intention to articulation.* Cambridge: MIT Press.

MacKay D., & MacDonald, M. (1984). Stuttering as a sequencing and timing disorder. In R. F. Curlee & W. H. Perkins (Eds.), *Nature and treatment of stuttering: New directions* (pp. 261-282). College-Hill Press, San Diego.

Meyers, S., & Freeman, F. J. (1985a). Mother and child speech rates as a variable in stuttering and disfluency. *Journal of Speech and Hearing Research, 28:* 436-444 (1985).

Meyers, S., & Freeman, F. J. (1985b). Interruptions as a variable in stuttering and disfluency. *J. Speech and Hearing Research, 28:* 428-435.

Moore, W. H. (1984). Central nervous system characteristics of stutterers. In F. Curlee & W. H. Perkins (Eds.), *Nature and treatment of stuttering: New directions* (pp. 49-72). College-Hill Press, San Diego.

Nudelman, H. B., Herbrich, K. E., Hoyt, B. D., & Rosenfield, D. B. (1991). A neuroscience approach to stuttering. In H. F. M. Peters, W. Hulstijm, & C. W. Starkweather (Eds.), *Speech motor control and stuttering* (pp. 157-162). Amsterdam: Elsevier.

Perkins, W., Kent, R., & Curlee, R. F. (1991). A theory of neurolinguistic function in stuttering. *Journal of Speech and Hearing Research, 34*, 734-752.

Pool, K. D., Devous, M., Freeman, F. J., Watson, B. C., & Finitzo, T. (1991). Regional cerebral blood flow in developmental stutterers. *Archives of Neurology, 48*, 509-512.

Pool, K. D., Devous, M., Watson, B., Finitzo, T., & Freeman, R. J. (1992). Reply to Viswanath, Rosenfield, & Nudelman. Letters to the Editor, *Archives of Neurology, 49.*

Pool, K., Devous, M., Watson, B. Finitzo, T., & Freeman, F. (1993). Reply to Fox, Lancaster, and Ingham. Letters to the Editor, *Archives of Neurology, 50,* 1289-1290.

Rosenbek, J. (1980). Apraxia of speech—relationship to stuttering. *Journal of Fluency Disorders, 5,* 233-235.

Rosenbek, J. (1984). Stuttering secondary to nervous system damage. F. Curlee & W. H. Perkins (Eds.), *Nature and treatment of stuttering: New directions* (pp. 31-48). College-Hill Press, San Diego.

Rosenfield, D. & Jerger, J. (1984). Stuttering and auditory function. In F. Curlee & W. H. Perkins (Eds.), *Nature and treatment of stuttering: New directions* (pp. 73-88). College-Hill Press, San Diego.

Schuell, H., Jenkins, J. J., & Jimenez-Pabon, E. (1964). *Aphasia in adults.* New York: Harper & Row.

Shtremel, A. K. (1963). Stuttering in left parietal lobe syndrome. *Zhurnal Nevropatologii i Psikhaitrii imeni S.S. Korsakove, 63,* 828-832.

Trost, J. E. (1971). Apraxic dysfluency in patients with Broca's aphasia. Paper presented at the American Speech and Hearing Association Convention, Chicago. Reported in J. Rosenbek (1984). Stuttering secondary to nervous system damage. In F. Curlee & W. H. Perkins (Eds.), *Nature and treatment of stuttering: New directions* (pp. 31-48). College-Hill Press, San Diego.

Ulatowska, H., Allard, L., & Chapman, S. (1989). Narrative and procedural discourse in aphasia. In Y. Joanette & H. Brownell (Eds.), *Discourse ability and brain damage.* New York: Springer-Verlag.

Viswanath, H., Rosenfield, D., & Nudelman, H. (1992). Response to Pool, Devous, Watson, Finitzo, & Freeman. Letters to the Editor, *Archives of Neurology, 49.*

Watson, B. C., Freeman, F. J., Chapman, S., Miller, S., Finitzo, T., & Pool, K. (1991). Linguistic performance deficits in stutterers: Relation to laryngeal reaction time profiles. *Journal of Fluency Disorders 16,* 85-100.

Watson, B. C., Freeman, F. J., Devous, M., Chapman, S., Finitzo, T., & Pool, K. (1994). Linguistic performance and regional cerebral blood flow in persons who stutter. *Journal of Speech and Hearing Research,* 37, 1221-1228).

Watson, B. C., Pool, K., Devous, M., Freeman, F. J., & Finitzo, T. (1992). Brain blood flow related to acoustic laryngeal reaction time in adult developmental stutterers. *Journal of Speech and Hearing Research, 35,* 555-561.

Watson, B. C ., Freeman, F. J., & Dembowski, J. (1991). Respiratory/laryngeal coupling and complexity effects on acoustic LRT in normal speakers. *Journal of Voice, 5,* 18-28.

Wingate, M. E. (1988). *The structure of stuttering: A psycholinguistic analysis.* New York, Springer-Verlag.

Wolk, L., Edwards, M. L., & Conture, E. (1991). Coexistence of stuttering and disordered phonology in young children. *Journal of Speech and Hearing Research, 36*, 906-917.

Zimmermann, G. (1984). Articulatory dynamics of stutterers. In F. Curlee & W. H. Perkins (Eds.), *Nature and treatment of stuttering: New directions* (pp. 131-147). College-Hill Press, San Diego.

Zimmermann, G., & Kelso, J. A. S. (1982). Remarks on the "causal" basis of stuttering. Paper presented at the Van Riper Lecture Series, Western Michigan University, Kalamazoo, MI. April 1-2.

SECTION 4

Producing Speech: Feedback

28 Central and Peripheral Components in the Control of Speech Movements

Vincent L. Gracco
Haskins Laboratories

In the late 1950s and 1960s much of the work at Haskins Laboratories focused on issues related to the representation of the phoneme—the presumed unit of speech. Attempts to find evidence for invariant signal properties in the acoustic stream representing the phoneme had been generally unsuccessful because of contextual variation, that is, coarticulation. As suggested by MacNeilage (1970) this led to two related positions regarding the nature of phonemic invariance. One position, posited by Lindblom (1963), Ladefoged (1967), and Stevens and House (1963), was that invariance is in the vocal tract targets rather than in the peripheral manifestations at any observable level. The Haskins position was "that the EMG [electromyographic] correlates of the phoneme will prove to be invariant in some significant sense" (Liberman, Cooper, Harris, MacNeilage, & Studdert-Kennedy, 1967, p. 84). The motivation for this search was that if there were phonemic units of speech, and the evidence from speech perception studies suggested that there were, they should be invariantly present in the muscular activity output. They reasoned that the acoustic signal was too far removed from the source of the invariance, since the invariance was a reflection of characteristics internal to the organism, in the central motor commands. Therefore, the best place to search for invariance was in the peripheral manifestation of the central motor commands, which can be examined in the

417

activity of the muscles of the vocal tract using electromyography. The earliest studies of this kind where conducted by Katherine Harris and her colleagues at Haskins and resulted in a number of interesting and important findings that have had both theoretical and technical impact on the field of speech science (Harris, Lysaught, & Schvey, 1965; Harris, Schvey, & Lysaught, 1962; Lysaught, Rosov, & Harris, 1961; MacNeilage, 1963). However, even at this level of observation, numerous attempts by various investigators failed to find an acceptable degree of invariance in the peripheral manifestation of the central motor commands, and the search was abandoned. One of the limiting factors in the earlier research into invariance in speech production resulted from the limited consideration of the sensorimotor nature of the behavior. As will be suggested below, speech, as all behavior, is built from and maintained by an integration of sensory and motor signals operating at different functional levels and on different time scales. Attempting to identify the underlying processes and components of speech production without explicit consideration of the sensorimotor character of the behavior can only result in incomplete models of limited relevance to understanding speech motor behavior.

Some History

Perhaps the longest-standing issue in motor control has been the role of sensory feedback in the control of behavior. This issue has generated considerable controversy and dichotomous theoretical positions among students of both speech production and general motor control. Perhaps the most prevalent position is one in which voluntary movement is viewed as being built from explicit sensory-mediated consequences, but that once the movement pattern is acquired, sensory information is no longer necessary. In the field of speech, such a position seems to have originated from the apparent resistance of various methods of sensory reduction to have a significantly degrading effect on speech motor performance in adults (cf. Borden, 1979; Lane & Tranel, 1971). For example, following experimentally induced reduction in oral kinesthesia, global measures of speech production have been found to be minimally disrupted (Gammon, Smith, Daniloff, & Kim, 1971; Ringel & Steer, 1963; Scott & Ringel, 1971). Additionally, reduced or distorted auditory information has resulted in only mildly distorted or essentially normal speech motor output (Kelso & Tuller, 1983; Lane & Tranel, 1971). Other considerations, like neural transport delays involving afferent-to-efferent loops (Kent & Moll, 1975) and the apparent ballistic nature and short duration of many speech movements, have led to the position that speech movements are exclusively preprogrammed with sensory information that is used only in long-term adaptation or speech skill acquisition (Borden, 1979). From this perspective speech movements would be generated from preset motor patterns and executed independently of any afferent-dependent actions. Similar theoretical positions have been postulated based, in part, on limb studies showing that functionally deafferented animals (Fentress, 1973; Polit & Bizzi, 1979; Taub & Berman, 1968) and humans (Rothwell et al., 1982) are capable of executing learned and novel motor tasks (Kelso & Stelmach, 1976 for review). These results indicate that motor tasks can be carried out with reduced or absent afferent input, apparently relying on some stored motor commands. However, it is also true that motor acts executed in the absence of afferent information are often only grossly normal; that is, they often lack their normal precision (Sanes & Evarts, 1983).

An alternative perspective can be generated from consideration of the numerous studies demonstrating movement changes following dynamic mechanical perturbation applied to a moving articulator. It has been observed that loads applied to the lips or jaw result in changes in articulatory movement. Perturbation of a moving articulator results in patterns of compensation: disruptions to the lips result in compensatory changes in the lips and jaw (Abbs & Gracco, 1984; Gracco & Abbs, 1985, 1988) and the larynx (Löfqvist & Gracco, 1991; Munhall, Löfqvist, & Kelso, 1994); jaw loads result in compensatory changes in the tongue (Kelso, Tuller, Vatikiotis-Bateson, & Fowler, 1984), lips (Folkins & Abbs, 1975; Shaiman, 1989), and velum (Kollia, Gracco, & Harris, 1992). Task-specific responses are observed when an articulator is actively involved in the sound segment being produced, but not when an articulator is inactive (Kelso et al., 1984; Shaiman, 1989). These studies suggest that speech is always produced with at least some consideration of ongoing sensory information. Experimental interference with hearing one's own speech (using distorted auditory feedback or masking) results in changes in a number of speech output variables including fundamental frequency, vocal intensity and to a lesser extent, speech movements (Siegel & Pick, 1974; Lane & Tranel, 1971; Forrest, Abbas, & Zimmermann, 1986). Delayed auditory feedback results in a slowing of speaking rate that can result in a breakdown in fluency (Fairbanks, 1955; Howell, El Yaniv, & Powell, 1987; Lee, 1950), whereas low pass filtering results in changes in nasal resonance (Garber & Moller, 1979) and both increases and decreases in lip, jaw, and tongue movement (Forrest et al., 1986). Recently, a number of investigations have used perturbations of the frequency content of the auditory feedback signal to examine changes in fundamental frequency and intonation (Elman, 1981; Kawahara, 1994).

The apparent inconsistency in speech production theories regarding the sensory contribution to speech motor control is one of interpretation. The results of all studies support a general intuitive position that all behaviors involve, to some extent, the integration of sensory information. A simple reflex, of any origin and in any organism, requires sensory stimulation. Human thought, often assumed to represent the highest level of ontogenetic advancement and to set humans apart from the nonhumans, is stimulated by the organisms' environment. However, although thoughts and actions may be conceptualized as emergent properties, they emerge in relation to some antecedent event or series of events. The registration of events can only come through sensory channels. Given both the density of sensory receptors found within the vocal tract and the sensory discrimination possible with articulators like the tongue and lips (see Kent, Martin, & Sufit, 1990 for review), a more reasonable position is that taken by Weiss more than 50 years ago, when he suggested:

> Nobody in his senses would think of questioning the importance of sensory control of movement. But just what is the precise scope of that control? Weiss (p. 23, 1941)

That is, it is obvious that behavior is always sensorimotor in nature, and it is illogical to suggest that speech might be a special case in which sensory-mediated information is unimportant or simply ignored.

Control Processes

A related issue convolved with the role of sensory information in speech production has to do with the characteristics of the control process. Theoreticians focusing on the sensory-dependent vs. sensory-independent issue have also focused, simultaneously, on whether speech motor action is produced by a closed-loop or an open-loop system. In this context, "closed-loop" is often used synonymously with "feedback" and the term "open-loop" is often used synonymously with "no feedback." This simplistic dichotomy has had a long and unproductive history; furthermore, it is based on the inaccurate assumption that open-loop implies the lack of sensory influence (see discussion by Abbs & Cole, 1982).

Control systems are classified into two general categories: open-loop and closed-loop. The distinction is determined by the control action, which is the quantity responsible for activating the system to produce the output. For example, an open-loop control system is one in which the control action (the input) is independent of the output: a closed-loop control system is one in which the control action is somehow dependent on the output. In this regard, maintaining a limb in a specific position in the face of perturbing stimuli would be an example of the operation of closed-loop control: the control action is being adjusted by the deviations in the output of the system. A classic example of an open-loop control system is an automatic toaster in which the control action is controlled by a timer and once initiated, the action is completed without changes resulting from errors in the toasting quality. This simple and relatively straightforward example, however, assumes that the user, who can adjust the timer at anytime, is not part of the control system. More precisely, the user can only influence the control action in the future, not in the present, through changes in the timer. As such, a system can be open-loop but still be sensory dependent. One of the requirements of an open-loop system is accurate calibration, so that a precise input-output relation is established that allows the system to operate without direct feedback of any desired quantity.

Early motor control research emphasized the use of closed-loop control to regulate some physical variable (e.g., limb position, muscle length, muscle force, etc.). Servomechanistic models have been proposed in which the control of muscle length or limb position is regulated (Hammond, 1960; Merton, 1953) by comparing actual length or position with a model or reference, using the difference between the actual and model values as an error signal to generate a corrective movement. Similarly, speech theorists (Fairbanks, 1955; MacNeilage, 1970; Sussman, 1972) posited models of speech production employing closed-loop control of muscle length or spatial targets. Within the context of postural maintenance or slow tracking, sensory information from muscles or movement can provide the input necessary for the fine adjustments and error-correction capabilities characteristic of a closed-loop system. In the case of rapid coordinated movements, however, the inherent pathway delays and comparator-processing time, coupled with high loop gain, result in serious limitations for servocontrol (Rack, 1981).

An alternative proposal has been offered by Greene (1972), who proposed a style of control for purposeful movements that incorporates open-loop, or feedforward, control based on anticipatory or predictive adjustments that allow greater speech and flexibility required to regulate a complex dynamic system. Essentially, the input from one or several sensory systems is used to predict the requirements of an upcoming movement and produces a control signal that incorporates the predicted requirement. Before turning to a consideration of how best to describe the sensorimotor control system, some developmental considerations may be helpful in identifying the processes and objects that sensory mechanisms might influence.

Developmental Considerations

As in the acquisition of any other motor skill, the early stages of speech motor development involve practice and feedback. Of interest here are what role sensory influences have to play in that development. The period from perhaps one to six months of age, the period in which the infant engages in cooing, laughter, produces vowel sounds and reduplicated babbling, is the time when the most fundamental sensorimotor links and basic neuromotor patterns for speech are being established. Even at birth, however, the neonate exhibits coordinated activity of the physiological systems that underlie the speech production process. The earliest communicative behavior, crying, reflects the coordinated activity of the respiratory, laryngeal, and supralaryngeal systems exhibiting the same fundamental principles seen in the older child and adult. That is, vocalization involves manipulating discrete portions of the vocal tract, in the presence of a constant pressure source, to produce sound that is interpreted by the listener. What is elaborated during development is the ability to make finer and finer distinctions with the vocal apparatus. More precisely, speech motor development is a process in which the child learns how to make fractionated movements of the vocal tract to produce sounds corresponding to the adult phonological system.

Two conditions are required for these elaborations to occur: the child must possess the physiological capability to make configurational changes of the vocal tract and the ability to discriminate the acoustic properties of the sounds of the language. When these conditions are met, one generally finds that the acquisition of sounds progresses from rather gross movements of lips and jaw and tongue to more refined and fractionated action of the same articulators (Kent, 1992). For example, bilabial consonants are acquired earliest; labiodentals are acquired later. The production of labiodental sounds requires the ability to contract only a portion of orbicularis oris inferior and to use a different portion of mentalis to generate the correct movement vector for /f/ than that for /p/. Similarly, one can account for the development of other articulatorily complex sounds—for example, /ɝ/ is the last sound to be acquired, presumably because of its motor requirements (Shriberg & Kent, 1982). One of the hallmarks of the developing child is the ability to learn to contract a portion or a unique combination of muscles or groups of synergistic muscles and to produce a novel configuration. Motorically, the developing child is constantly modifying and refining its sensorimotor map to establish a degree of activation that can serve the needs of adult speech production—that is, the shaping and refining of vocal tract actions to the minimal functional structure necessary to produce the sounds of the language. This developmental process has a significant consequence: coarticulation, which can be viewed as the ultimate adaptation. By learning and refining motor patterns to their minimal functional structure, articulatory regions that are not necessary for a specific sound essentially become available for producing overlapping or contiguous sounds, and in so doing, maximize the speed at which sounds can be produced and information transferred.

From this perspective, speech development involves refining motor patterns and perceptual abilities associated with discriminating the phonetic segments of the language. During the early stages of development, the infant is developing an internal model of the vocal tract and establishing relations between movements of the vocal tract and their perceptual, auditory, and somatic sensory consequences. This learning phase is highly dependent on sensory stimulation from all possible sources: visual, auditory, and somesthetic. The different modalities provide different, but related, information to the child, establishing mappings between multimodal sensory information and the various neuromotor processes underlying adult speech production. Presumably, once development is complete, sensorimotor transformations that allow rapid and direct mappings between sensory modalities and levels of the motor output have been acquired.

Sensory Considerations

There are three sensory modalities that can influence the production of speech: the visual, auditory, and somatic modalities, each of which provides unique and somewhat redundant information to the speech motor control system. One of the characteristics that distinguishes the three modalities from one another is the information that can be extracted from the input modality. For example, visual input cannot provide direct information on the speaker's own production mechanism but can and does provide important environmental information. Visual information provides the speaker with cues regarding the successful transfer of information to the listener as well as information on the environment that may be relevant to successful information transfer. The visual modality may act to tune the output to the environment rather than have any specific effects on the output characteristics. The auditory system, in contrast, can provide more direct information on the state of the speaker's speech production mechanism including both the phonetic and prosodic properties of the signal (Svirsky, Lane, Perkell, & Wozniak, 1992). Inaccurate or distorted articulations, speech errors, and intonational information are carried very well and with a high degree of precision by the auditory modality. The predominant result of most of the studies using auditory distortions to evaluate the potential role of hearing in speech has been that changes are related, directly or indirectly, to loudness or intonation. Intuitively, this seems reasonable, since auditory information is only indirectly related to articulation and requires a complex mapping between acoustic characteristics and changes in the articulatory source. It is often the case that this mapping is not unique, and the ability of the nervous system to extract speech movement consequences directly can often be limited. The experiments by Kawahara (1993), mentioned above, also offer some evidence that the time scale for normal monitoring of the auditory stream for speaking may be longer than that required for discrete speech movement changes. Kawahara's experiments contain some of the best controlled auditory perturbation manipulations and have the most consistent experimental findings to date. One relevant finding is that the average latency of the response to the transformed auditory manipulation is 150 ms. Previous work by a number of investigators attempting to elicit audio-laryngeal reflexes with unnatural stimuli have found much shorter latencies.

The most direct source of sensory information for the control of movement is from the somatic sensory receptors located throughout the human body. The kind and density of receptors within various regions of the body is generally consistent with the kind of information coded by the receptor and the sensitivity of the region. The human vocal tract contains regions that are as tactilely sensitive as the fingers and face. The tongue, face, and fingers have a denser distribution of cutaneous sensory receptors than the upper arm or upper leg. As a consequence the finger tip, tongue, and face have very localized discrimination. Sensory receptors from the vocal tract that have the potential to contribute to the control of movement belong to a class of receptors called exteroceptors, whose information may principally serve either conscious sensation or regulation (Eyzaguire & Fidone, 1975). This class includes a variety of receptors, only two of which may have any direct affect on speech movement control: proprioceptors and cutaneous mechanoreceptors. Prioprioceptors are found in muscles and joints and provide information on the velocity and amount of muscle stretch and muscle tension. Cutaneous mechanoreceptors are found in the smooth and the hairy skin throughout the vocal tract and discharge in response to the state and changes in the state of the vocal tract. Somatic sensory receptors are functionally designated as having either static or dynamic properties depending on their discharge characteristics: static receptors discharge as long as a stimulus is present; dynamic receptors discharge only with the application or removal of an adequate stimulus. The static and dynamic receptor classification can be further delineated by consideration of the relative adaptive nature of the response to stimulation. (Detailed consideration of the specific receptor types within the vocal tract can be found in Dubner, Sessle, & Storey, 1978, and Kent, Martin, & Sufit, 1990). Generally, within the vocal tract somatic receptors are able to provide the relevant information to code the shape of the vocal tract as well as any change in vocal tract shape. Moreover, because of the low frequency nature of speech movements, the receptors can update the relevant information in real time.

As suggested above, whatever information is used must be used in a predictive manner, since once a movement is made it is too late to make an adjustment during running speech. Based on previously reported latencies (Abbs & Gracco, 1984) and the time necessary for using predictive estimates (coded velocity and/or acceleration) of the peripheral conditions to modify speech motor commands, it has been suggested that speech motor commands can be updated automatically from somatic sensory receptor information (Gracco, 1987). Even during rapid speech, time is available to make a simple coordinate transformation between somatic sensory input and somatic motor output. For the auditory modality, the same possibility is not as plausible, because of the additional transformation required and the potential ambiguity between the auditory signal and the state of the vocal tract. It is also possible that the auditory system may not operate with a sufficiently short time-constant to make adjustments that are required for movement-to-movement parameterization. While auditory-motor reflexes can be elicited with short latencies (15 ms), they are generally not functional and require stimulation that is not normally generated during ongoing speech. As mentioned above, the Kawahara (1993) results, using more natural and finer-grained changes in fundamental frequency, revealed average latencies on the order of 150 ms, latencies that would be too long to make adjustments on the order of individual vocal tract actions. These latencies, however, would be consistent with a suprasegmental role for the auditory system during ongoing speech. While speculative, it can be suggested that the auditory system may be contributing to the control process with a longer time constant than that of the somatic sensory system (see also Lane & Webster, 1991; Perkell, Lane, Svirsky, & Webster, 1992). Given the more direct relation between the somatic input and motor output than between the auditory input and motor output, this is a plausible position and one that is open to empirical evaluation.

The Sensorimotor Domain

In order to complete this limited overview of the potential role of sensorimotor mechanisms in speech production, it is critical to establish the context from which sensory-based adjustments are made. As suggested above, during speech motor development a child learns a set of coherent motor patterns that can be used to produce the unique sounds of the language, along with a set of coordinate transformations between the state of the vocal tract and the desired perceptual results. Once these mappings have been accomplished the child has acquired a set of characteristic motor commands that can be used as the input to the speech motor system and has also acquired a way of modifying the commands to adapt to changing peripheral conditions. These characteristic motor commands form the basis for speech production and act as the substrate upon which all modulating processes operate (Gracco, 1991; Perkell, Matthies, Svirsky, & Jordan, 1994). As such, speech movements are not generated *de novo* for each occurrence or production but rely on learned neuromotor commands that are stored in the nervous system. Based on arguments presented elsewhere it is suggested here that the learned motor patterns for speech reflect control structures or units that are organized according to the physiological (neuromuscular) components that generate the sounds of the language (see Gracco, 1990, 1991). Such a conceptualization is consistent with a model of speech production in which articulatory actions map onto the phonetic segments of the language, but this view is equally captured by the notion of gestural constellations (Browman & Goldstein, 1989, 1992), activation variables (Saltzman & Munhall, 1989), and coordinative structures (Easton, 1972; Fowler, 1980). Phoneme-based models have a long history, with many originating from Haskins Laboratories (Cooper Liberman, Harris, & Grubb, 1958; Halle, 1962; Halle, 1964; Halle & Stevens, 1964; Liberman, 1957; Liberman, Cooper, Harris, & MacNeilage, 1962; Liberman, Cooper, Harris, MacNeilage, & Studdert-Kennedy, 1967; Liberman, Cooper, Shankweiler, & Studdert-Kennedy, 1967; Lindblom, 1963; Moll & Shriner, 1967; Öhman, 1967; Stevens & Halle, 1967; Stevens & House, 1963). However, as noted above, a major problem with all theoretical positions positing units of any kind has been the failure to find invariant units at any peripheral level of observation such as the acoustic, kinematic or electromyographic. However, rather than reflecting a problem for models with invariant constructs, the lack of invariance should be considered the norm and a reflection of the adaptive nature of the speech motor control process. For example, the same set of motor commands issued without regard to changing peripheral conditions (phonetic context) would almost always result in errors. Contextual variations, which are ubiquitous in speech, necessitate a mechanism able to adjust to the output of a constantly changing peripheral environment. Combining the notion of a predictive control process, in which the current state of the vocal tract is used to update activated characteristic neuromotor patterns through a simple sensorimotor coordinate transformation, with invariant motor commands is one way in which flexibility is achieved for ongoing speech production. Two important aspects of speech motor control involve the notion of stored motor commands that activate vocal tract configurations and the sensorimotor updating of a future (not present) set of motor commands. It should be pointed out that characteristic motor patterns are only viewed as categorically invariant (see Gracco, 1991, for more detail) and that the motor commands and the sensory interactions are only two processes that modify the observable output. Other linguistic and cognitive influences interact to produce changes in speech motor output; that is, speech movements are an end-product reflecting a number of serial and parallel processes.

CONCLUSION

Speech production, as most skilled motor behavior, is viewed as a sensorimotor process in which central representations interact in complex ways with peripheral sensory signals to produce the resulting dynamic movement patterns (Abbs, Gracco, & Cole, 1984; Gentner, 1987; Gracco, 1987; Gracco, 1987). The issue dealt with here is when and how the various sensory signals modify speech motor output. The thesis developed is that the role of a sense modality in the speech production process is dependent upon the functional properties transduced by that modality, and that each modality (kinesthetic, auditory, and visual) provides different kinds of information to the nervous system that in turn influence the production process in unique but related ways. From a control standpoint, the control action is a set of motor commands for each unique sound of the language. These motor commands are modulated by the constantly changing peripheral environment in such a way that the positions of various articulators for one segment act to generate changes in future motor commands for another segment. The use of invariant motor commands, specified as equilibrium positions, integrated with explicit consideration of peripheral sensorimotor modulations, has been offered successfully in a model of jaw motion (Flanagan, Ostry, & Feldman, 1990; Laboissiere, Ostry, & Feldman, submitted). In spite of limitations of current models, it is clear that any model of speech production must have some explicit transformation of the output by afferent-dependent coding of vocal tract states, so that the control system will have the necessary adaptability. In conclusion, it appears that the concepts of variable articulations and invariant vocal tract targets can be integrated and understood once the potential contribution of sensory information is included. The empirical work done by Katherine Harris and her colleagues as well as the empirical work they motivated have facilitated this important theoretical synthesis.

REFERENCES

Abbs, J. H., & Cole, K. J. (1982). Consideration of bulbar and suprabulbar afferent influences upon speech motor coordination. In S. Grillner, B. Lindblom, J. Lubker, & A. Persson, (Eds.), *Speech motor control* (pp. 159-186). New York: Pergamon.

Abbs, J. H., & Gracco, V. L. (1984). Control of complex motor gestures: Orofacial muscle responses to load perturbations of the lip during speech. *Journal of Neurophysiology, 51(4),* 705-723.

Abbs, J. H., Gracco, V. L., & Cole, K. J. (1984). Control of multimovement coordination: Sensorimotor mechanisms in speech motor programming. *Journal of Motor Behavior, 16,* 195-232.

Borden, G. J. (1979). An interpretation of research on feedback interruption in speech. *Brain and Language, 7,* 307-319.

Browman, C. P., & Goldstein, L. (1989). Articulatory gestures as phonological units. *Phonology, 6,* 201-251.

Browman, C. P., & Goldstein, L. (1992). Articulatory phonology: An overview. *Phonetica, 49,* 222-234.

Cooper, F. S., Liberman, A. M., Harris, K. S., & Grubb, P. M. (1958). Some input-output relations observed in experiments on the perception of speech. In *Proceedings of the 2nd international congress of cybernetics* (pp. 930-941). Namur, Belgium.

Dubner, R., Sessle, B. J., & Storey, A. T. (1978). *The neural basis of oral and facial function.* New York: Plenum Press.

Easton, T. A. (1972). On the normal use of reflexes. *American Scientist, 60,* 591-599.

Elman, J. (1981). Effects of frequency-shifted feedback on the pitch of vocal productions. *Journal of the Acoustical Society of America, 70,* 45-50.

Eyzaguirre, C., & Fidone, S. J. (1975). *Physiology of the nervous system.* Chicago: Year Book Medical Publishers, Inc.

Fairbanks, G. (1955). Selective vocal effects of delayed auditory feedback. *Journal of Speech and Hearing Disorders, 20,* 333-346.

Fentress, J. C. (1973). Development of grooming in mice with amputated forelimbs. *Science*, *179*, 704-705.

Flanagan, J. R., Ostry, D. J., & Feldman, A. (1990). Control of human jaw and multi-joint arm movements. In G. Hammond (Ed.), *Cerebral control of speech and limb movements*. Amsterdam: Elsevier.

Folkins, J. W., & Abbs, J. H. (1975). Lip and jaw motor control during speech: Responses to resistive loading of the jaw. *Journal of Speech and Hearing Research*, *18*, 207-220.

Forrest, K., Abbas, P. J., & Zimmermann, G. N. (1986). Effects of white noise masking and low pass filtering on speech kinematics. *Journal of Speech and Hearing Research*, *29*, 549-562.

Fowler, C. A. (1980). Coarticulation and theories of extrinsic timing control. *Journal of Phonetics*, *8*, 113-133.

Gammon, S. A., Smith, P. J., Daniloff, R. G., & Kim, C. W. (1971). Articulation and stress/juncture production under oral anesthetization and masking. *Journal of Speech and Hearing Research*, *14*, 271-282.

Garber, S. R., & Moller, K. T. (1979). The effects of feedback filtering on nasalization in normal and hypernasal speakers. *Journal of Speech and Hearing Research*, *22*, 321-333.

Gentner, D. R. (1987). Timing of skilled motor performance: Tests of the proportional duration model. *Psychological Review*, *94*, 255-276.

Gracco, V. L. (1987). A multilevel control model for speech motor activity. In H. Peters & W. Hulstijn (Eds.), *Speech motor dynamics in stuttering* (pp. 57-76). Wien: Springer-Verlag.

Gracco, V. L. (1990). Characteristics of speech as a motor control system. In G. Hammond (Ed.), *Cerebral control of speech and limb movements* (pp. 1-28). Amsterdam: Elsevier.

Gracco, V. L. (1991). Sensorimotor mechanisms in speech motor control. In, H. Peters, W. Hulstijn, & C. W. Starkweather (Eds.), *Speech motor control and stuttering* (pp. 53-78). North Holland: Elsevier.

Gracco, V. L., & Abbs, J. H. (1985). Dynamic control of perioral system during speech: Kinematic analyses of autogenic and nonautogenic sensorimotor processes. *Journal of Neuroscience*, *54*, 418-432.

Gracco, V. L., & Abbs, J. H. (1988). Central patterning of speech movements. *Experimental Brain Research*, *71*, 515-526.

Greene, P. H. (1972). Problems of organization of motor systems. *Progress in Theoretical Biology*, 2303-338.

Halle, M. (1964). On the basis of Phonology. In J. A. Fodor & J. J. Katz (Eds.), *The structure of language*. Englewood Cliffs, N.J: Prentice-Hall.

Halle, M. (1962). Speech sound sequences. In, *Proceedings of the fourth international congress of phonetic sciences*. The Hague: Mouton.

Hammond, P. H. (1960). An experimental study of servo action in human muscular control. *Third International Conference of Medical Electronics*, 190-199.

Harris, K. S., Lysaught, G. F., & Schvey, M. M. (1965). Some aspects of the production of oral and nasal labial stops. *Language and Speech, 8*, 135-147.

Harris, K. S., Schvey, M. M., & Lysaught, G. F. (1962). Component gestures in the production of nasal labial stops. *Journal of the Acoustical Society of America, 34*, 743.

Howell, P., El-Yaniv, N., & Powell, D. J. (1987). Factors affecting fluency in stutterers. In H. F. M. Peters & W. Hulstijn (Eds.), *Speech motor dynamics in stuttering* (pp. 361-369). New York: Springer-Verlag.

Kawahara, H. (1993). Transformed auditory feedback: Effects of fundamental frequency perturbation. *Advanced Telecommunications Research Institute International Technical Report, TR-H-040.*

Kelso, J. A. S., & Stelmach, G. E. (1976). Central and peripheral mechanisms in motor control. In G. E. Stelmach (Ed.) *Motor control: Issues and trends*, New York: Academic Press.

Kelso, J. A. S., & Tuller, B. (1983). Compensatory articulation under conditions of reduced afferent information: A dynamic formulation. *Journal of Speech and Hearing Research, 26*, 217-223.

Kelso, J. A. S., Tuller, B., Vatikiotis-Bateson, E., & Fowler, C. A. (1984). Functionally specific articulatory cooperation following jaw perturbations during speech: Evidence for coordinative structures. *Journal of Experimental Psychology: Human Perception and Performance, 10*, 812-832.

Kent, R. D. (1992). The biology of phonological development. In C. A. Ferguson, L. Menn, & C. Stoel-Gammon (Eds.), *Phonological development* (pp. 65-90). Maryland: York Press.

Kent, R. D., & Moll, K. (1975). Articulatory timing in selected consonant sequences. *Brain and Language, 2*, 304-323.

Kent, R. D., Martin, R. E., & Sufit, R. L. (1990). Oral sensation: A review and clinical prospective. In H Winitz (Ed.), *Human communication and its disorders* (pp. 135-191). Norwood, NJ: Ablex Publishing.

Kollia, H. B., Gracco, V. L., & Harris, K. S. (1992). Functional organization of velar movements following jaw perturbation. *Journal of the Acoustical Society of America, 92*, 474.

Laboissiere, R., Ostry, D. J., & Feldman, A. (submitted). The control of multi-muscle systems: Human jaw and hyoid movements. *Journal of Neurophysiology.*

Ladefoged, P. J. (1967). *Three areas of experimental phonetics*. London: Oxford University Press.

Lane, H. L., & Tranel, B. (1971). The Lombard sign and the role of hearing in speech. *Journal of Speech and Hearing Research, 14*, 677-709.

Lane, H. L., & Webster, J. W. (1991). Speech deterioration in postlingually deafened adults. *Journal of the Acoustical Society of America, 89,* 859-866.

Lee, B. S. (1950). Effects of delayed speech feedback. *Journal of the Acoustical Society of America, 22,* 824-826.

Liberman, A. M. (1957). Some results of research on speech perception. *Journal of the Acoustical Society of America, 29,* 117-123.

Liberman, A. M., Cooper, F. S., Harris, K. S., & MacNeilage, P. F. (1962). A motor theory of speech perception. In *Proceedings of the speech communication seminar.* Stockholm: Speech Transmission Laboratory, Royal Institute of Technology.

Liberman, A. M., Cooper, F. S., Harris, K. S., MacNeilage, P. F., & Studdert-Kennedy, M. (1967). Some observations on a model for speech perception. In W. Wathen-Dunn (Ed.), *Models for the perception of speech and visual form* (pp. 68-87). Cambridge, MA: MIT Press.

Liberman, A. M., Cooper, F. S., Shankweiler, D. P., & Studdert-Kennedy, M. (1967). Perception of the speech code. *Psychological Review, 74,* 431-461.

Lindblom, B. (1963). Spectrographic study of vowel reduction. *Journal of the Acoustical Society of America, 35,* 1773-1781.

Löfqvist, A., & Gracco, V. L. (1991). Discrete and continuous modes in speech motor control. *PERILUS, XIV,* 27-34.

Lysaught, G. F., Rosov, R. J., & Harris, K. S. (1961). Electromyography as a speech research technique with an application to labial stops. *Journal of the Acoustical Society of America, 33,* 842.

MacNeilage, P. F. (1963). Electromyographic and acoustic study of the production of certain final clusters. *Journal of the Acoustical Society of America, 35,* 461-463.

MacNeilage, P. F. (1970). Motor control of serial ordering of speech. *Psychological Review, 77,* 182-196.

Merton, P. A. (1953). Speculations on the servo-control of movement. In G. W. Wolstenholme (Ed.), *The spinal cord.* London: Churchill.

Moll, K. L., & Shriner, T. H. (1967). Preliminary investigation of a new concept of velar activity during speech. *The Cleft Palate Journal, 4,* 58-69.

Munhall, K., Löfqvist, A., & Kelso, J. A. S. (1994). Lip-larynx coordination in speech: Effects of mechanical perturbations to the lower lip. *Journal of the Acoustical Society of America, 95,* 3605-3616.

Öhman, S. (1967). Peripheral motor commands in labial articulation. *Speech Transmission Laboratory: Quarterly Progress Report,* 430-63.

Perkell, J. S., Lane, H., Svirsky, M., & Webster, J. (1992). Speech of cochlear implant patients: A longitudinal study of vowel production. *Journal of the Acoustical Society of America, 91,* 2961-2978.

Perkell, J. S., Matthies, M., Svirsky, M., & Jordan, M. (1994). Goal-based speech motor control: A theoretical framework and some preliminary data. *Journal of Phonetics,* in press.

Polit, A., & Bizzi, E. (1979). Characteristics of motor programs underlying arm movements in monkeys. *Journal of Neurophysiology, 42,* 183-194.

Rack, P. M. H. (1981). Limitations of somatosensory feedback in control of posture and movement. In V. B. Brooks (Ed.), *Handbook of physiology. Section 1: The nervous system. Vol. II, Control, Part 1* (pp. 229-256). Bethesda, MD: American Physiological Society.

Ringel, R. L., & Steer, M. (1963). Some effects of tactile and auditory alterations on speech output. *Journal of Speech and Hearing Research, 6,* 369-378.

Rothwell, J., Traub, M., Day, B., Obeso, J., Thomas, P., & Marsden, C. (1982). Manual motor performance in a deafferented man. *Brain, 105,* 515-542.

Saltzman, E. L., & Munhall, K. G. (1989). A dynamical approach to gestural patterning in speech production. *Ecological Psychology, 1,* 333-382.

Sanes, J., & Evarts, E. V. (1983). Regulatory role of proprioceptive input in motor control of phasic or maintained voluntary contractions in man. In J. Desmedt (Ed.), *Motor control mechanisms in health and disease* (pp. 47-59). New York: Raven Press.

Scott, C. M., & Ringel, R. L. (1971). Articulation without oral sensory control. *Journal of Speech and Hearing Research, 14,* 804-818.

Shaiman, S. (1989). Kinematic and electromyographic responses to perturbation of the jaw. *Journal of the Acoustical Society of America, 86,* 78-88.

Shriberg, L. D., & Kent, R. D. (1982). *Clinical phonetics.* New York: Macmillan.

Siegel, G. M., & Pick, H. L. (1974). Auditory feedback in the regulation of the voice. *Journal of the Acoustical Society of America, 56,* 1618-1624.

Stevens, K. N., & Halle, M. (1967). Remarks on analysis by synthesis and distinctive features. In W. Wathen-Dunn (Ed.), *Models for the perception of speech and visual form.* Cambridge, Mass.: M. I. T. Press.

Stevens, K. N., & House, A. S. (1963). Perturbation of vowel articulations by consonantal context: An acoustical study. *Journal of Speech and Hearing Research, 6,* 111-128.

Sussman, H. M. (1972). What the tongue tells the brain. *Psychological Bulletin, 77,* 262-272.

Svirsky, M. A., Lane, H., Perkell, J. S., & Wozniak, J. (1992). Effects of short-term auditory deprivation on speech production in adult cochlear implant users. *Journal of the Acoustical Society of America, 92,* 1284-1300.

Taub, E., & Berman, A. J. (1968). Movement and learning in the absence of sensory feedback. In S. J. Freeman (Ed.), *The neuropsychology of spatially oriented behavior* (pp. 173-192). Homewood, IL: Dorsey Press.

Weiss, P. (1941). Self-differentiation of the basic patterns of coordination. *Comparative Psychology Monograph, 17,* 21-96.

29 Ephphatha[1]: Opening Inroads to Understanding Articulatory Organization in Persons with Hearing Impairment

Nancy S. McGarr
St. John's University
Haskins Laboratories

Melanie McNutt Campbell
The Graduate School and University Center of
The City University of New York
Haskins Laboratories

INTRODUCTION

In 1968, Dorothy Huntington, Katherine Harris, and George Sholes published a seminal paper entitled "An Electromyographic Study of Consonant Articulation in Hearing-Impaired and Normal Speakers" in the *Journal of Speech and Hearing Research*. This collaborative study was Harris' first in the area of speech production of hearing-impaired persons. It was also the very first to employ electromyographic (EMG) techniques to investigate articulatory

---

[1]"Ephphatha" is a transliteration of an Aramaic word meaning "Be opened." It appears in the New Testament book of Mark as the word used by Jesus when he performed a miracle by opening the ears of a man who was deaf and enabling him to speak.

movement in deaf speakers. Yet in ways other than technique it was also unlike many of the investigations of the speech of the hearing impaired that preceded it.

Some of the earliest known studies of the speech of the deaf were conducted in the 1930s, using kymographic recordings of air pressure inside and outside of the mouth or nose and movement tracings of the contacts of the lips, the tongue with the palate and the movements of the breathing muscles (Hudgins, 1934, 1937; Rawlings, 1935). Those investigations were largely limited to observations of speech rate, breath management (syllables per breath), and speech rhythm. One of the most famous studies, conducted in 1942 by Hudgins and Numbers, determined the relative effects of speech errors on the intelligibility of deaf students by obtaining phonographic recordings, presenting the recorded samples in intelligibility tests to listeners, and phonetically transcribing the productions. From pressure measures, movement tracings, and transcriptions, these early investigators made insightful speculations about the nature of the underlying articulatory behaviors, but had no means to directly observe such.

In the decades that followed, studies were conducted that used more direct measurement techniques. They included use of the oscillograph (Lindner, 1962), the sound spectrograph (Angelocci, Kopp, & Holbrook, 1964; Calvert, 1961), the glossal transducer (Brannon, 1964), cinefluorography (Crouter, 1963), and EMG (Huntington, Harris, & Sholes, 1968). Huntington, Harris, and Sholes (1968), however, employed a paradigm that has become a hallmark of work by Harris and her colleagues: collecting multiple channels of information simultaneously. In this particular study both audio and EMG signals were recorded from multiple repetitions of utterances by each subject. The audio recordings enabled the researchers to administer perceptual tests of intelligibility as well as to analyze of muscle activation. Observing multiple repetitions of tokens, allowed EMG patterns to emerge, and accorded the researchers an opportunity to examine the consistency of productions, rather than to rely on only one or two illustrative tokens from many speakers.

The paper was distinguished also by its perspective. As Harris wrote on another occasion, "studies of pathological phoneme production are generally directed towards treatment. We might be able to gain considerable insight into the function of various parts of the nervous system in controlling articulation, if we used clinical material on damage at different levels and sites in the sensorimotor system" (Harris, 1971, p. 208). The Huntington, Harris, and Sholes paper (1968) was just such a study: "Our immediate interest was to clarify, if possible, some of the factors which give deaf speech its perceived pathological quality; our more general goal was the better understanding of the organization of normal speech by considering the effects of various types of malfunction of the whole speech chain" (p. 147).

In that spirit, it is the aim of this paper to understand better the importance of auditory feedback to the organization of speech by adopting a specific perspective for a brief review of some literature about the speech of hearing-impaired persons and for a discussion of some research in progress. From this perspective, the intent of the literature overview will not be to list deaf speech errors, but to delineate broader parameters of speech motor control dependent on auditory feedback. This perspective incorporates a central assumption: that the speech production system of a hearing-impaired person is intact, save the benefit of complete and undistorted auditory feedback, and thus seeks and uses any available visual, vibrotactile, kinesthetic, proprioceptive, and residual auditory feedback to organize itself. The speech produced by such a system, while communicating with more variable success than one served by normal hearing, will still resemble it. The overview of the literature will show that the chances for successful, intelligible speech production will be higher when a speech target has inherent possibilities for visual and/or vibrotactile feedback, when the articulators involved have fewer degrees of freedom of movement, and when the timing of the movement of one articulator is not tightly interrelated to the movement of another. Such factors will affect all hearing-impaired talkers and thus contribute to similarities in their speech production. The availability and quality of auditory feedback and the ability to use that and other types of feedback will, however, vary from one hearing-impaired individual to another. Any two intact speech production systems seeking to organize themselves around different auditory information, or around secondary visual, vibrotactile, kinesthetic or proprioceptive cues in its absence, will vary in production strategies and thus contribute to differences among hearing-impaired speakers.

Huntington, Harris, and Sholes built the line of inquiry for their paper from Calvert's (1961) results. Calvert had shown that experienced listeners could not discriminate isolated vowels spoken by hearing speakers from those spoken by hearing-impaired talkers, but could discriminate them when they were in CV syllables. Huntington, Harris, and Sholes reasoned, then, that deaf speech quality was caused neither by simple abnormal control of vocalization nor by abnormal communication of suprasegmental phonemes. They considered that possible contributing factors could be assigned arbitrarily to either topological or dynamic categories. While the 1968 study addressed only questions about topology, this paper will begin with topology, discuss dynamics, and proceed to two additional areas that have emerged as relevant from later work mirroring the original Huntington, Harris, and Sholes methodology. A brief overview of some facts that are known about each will be provided and brought to bear on some current research concerning the articulatory organization of hearing-impaired talkers. While deaf speech quality most certainly includes aspects of respiration (reviewed in Metz & Schiavetti, this volume) and voice quality, the goal here is to focus on articulatory parameters, beginning with the question that Huntington, Harris, and Sholes addressed. When compared to hearing speakers, do hearing-impaired speakers:

1. use different articulatory configurations (Topology);
2. demonstrate movement patterns to and from consonants (Dynamics and Timing);
3. show less consistent speech patterns in multiple productions of a given target (Variability Among and Within Hearing-Impaired Talkers);
4. exhibit different interarticulatory coordination properties (Interarticulatory Organization)?

TOPOLOGY

In attempting to determine what factors "give deaf speech its pathological quality," Huntington, Harris, and Sholes (1968) speculated, "On the one hand, dynamic features may be distorted—that is, transitions from consonant articulatory configuration to vowel may be strange. On the other hand, purely topological features may be abnormal—that is, the 'wrong' articulatory configurations may be commonly produced by deaf talkers" (p. 148). They investigated the latter for both tongue and lip configurations. The lip data are presented first.

Surface EMG recordings were made of upper, lower, and corner lip muscle activity for 11 common consonants in a disyllabic frame (hə'CVk) spoken by two deaf and two hearing speakers. The 11 consonants included a set of visible segments, /p, b, m, w, f/. Upper lip and corner lip peak activity showed high positive correlations between the hearing and deaf subjects. The facial muscle postures of the deaf were generally correct, though usually exaggerated. Huntington, Harris, and Sholes concluded, "Deaf speakers are more likely to be like normal speakers where there is a possibility of getting visual feedback" (p. 157). Other evidence that visual information enhances accurate lip configuration was provided by Carr (1953). She recorded the spontaneous productions of 48 five-year-old children with congenital deafness and observed that the labial sounds /p,b,m,w,f,v/ occurred more frequently in the deaf than has been observed for hearing infants. However, as Locke (1980) points out in his review of Carr's study, it would be imprudent to conclude that deaf children display a different course of speech development based on visual information. They may use more labial sounds, but these consonants are also used frequently by hearing babies and are also found in many of the world's languages. The hearing-impaired child's pattern of use of such consonants then, still resembles the speech development pattern of a hearing child, but his successful and more frequent use of the labials is made more probable because visual cues are available.

Another visible aspect of articulation, jaw opening, appears to be correct in the speech of the hearing impaired, as shown in a cineflourographic study by Crouter (1963) of six hearing and six deaf talkers producing four vowels in target words in a carrier sentence. From film tracings of incisal opening (i.e., jaw lowering) she determined that the deaf used greater jaw lowering for the close vowels /u/ and /i/ than did the hearing, but maintained the jaw height distinction between high and low vowels. More recent studies using optical tracking (McGarr, Löfqvist, & Story, 1987), strain gauge (Tye-Murray & Folkins, 1990) and cinefluorography plus X-ray microbeam techniques (Tye-Murray, 1991) have suggested that deaf and hearing speakers use similar magnitudes of jaw displacement in vowel production (though ratios of tongue-to-jaw displacement may be different).

Conversely, the tongue, an articulator with not easily seen and one with many degrees of freedom for movement, appears to be less accurately shaped. Huntington, Harris, and Sholes (1968), also examined tongue EMG activity for production of /t,s,ʃ,r,l,k/. Electrodes were placed at tongue-tip, mid-tongue, and back-tongue positions. Peak values were correlated for various consonants for pairs of hearing and hearing-impaired subjects, one electrode at a time. All correlations were positive and only a few were not statistically significant, indicating that "the topology of deaf articulation is not totally uncorrelated to [the] normal pattern" (p. 157). Still, hearing speakers were more like each other than they were like the hearing-impaired speakers. The hearing-impaired subjects exhibited some configurations that were different from those of the hearing speaker. For example, one hearing-impaired talker used more tongue-tip activity for /s/ and /ʃ/ than for /t/, the reverse of the pattern for the hearing. More recently Waldstein and Baum (1991), in an acoustic study, have provided corroborative centroid and F_2 evidence that hearing-impaired children produce incorrect consonantal tongue configurations. Hearing-impaired children were shown to produce points of constriction for /ʃ,t,k/ farther back in the mouth than did hearing children.

Tongue configurations were also less accurate in vowels produced by hearing-impaired speakers. In addition to cinefluorographic tracings of incisal opening, described earlier, Crouter (1963) traced tongue shape at the middle frame of vowels. Despite great individual differences in tongue contour among all subjects, overall differences were evident between the hearing and the hearing impaired. Measurements were made of the minimal distance between the posterior pharyngeal wall and the base of the tongue. The airway was greater for high versus low vowels in both subject groups, but the hearing subjects had greater opening for high vowels. In other words, the hearing speakers moved their tongues farther forward and out of the pharyngeal cavity than did deaf speakers. Consistent retraction of the tongue root and a lowered tongue body in vowel production by the hearing impaired have also been reported by Subtelny, Li, Whitehead, and Subtelny (1989).

Angelocci et al. (1964) provided acoustic measures that further indicate that hearing-impaired subjects frequently employ vowel configurations that are topo-graphically incorrect. Examining the production of 10 vowels by hearing and hearing-impaired 11-to-14-year-old boys, they showed that F_1 and F_2 frequency means revealed a tendency for deaf speakers to follow the normal progression of a rising F_1 moving from close to open vowels and a lowering F_2 going from front to back vowels, but over a far more *limited* frequency range. On the other hand, drawing enclosures around plots of F_1 by F_2 values of tokens from deaf subjects for each vowel resulted in large areas entirely overlapping each other, indicating great variability across subjects and a high degree of inaccuracy in placement of articulators. Enclosures for the vowels of the hearing children were smaller, suggesting more accuracy. These findings have been supported by several acoustic (Monsen, 1976a; Monsen, 1976b; Shukla, 1989), cinefluorographic (Stein, 1980; Tye, Zimmermann, & Kelso, 1983), cinefluorographic plus X-ray microbeam (Tye-Murray, 1991), and intelligibility studies (Suonpää & Aaltonen, 1981). All indicate that hearing-impaired talkers exhibit a reduced articulatory space, resulting in vowel confusions for listeners. Monsen (1976a) suggests that the deaf may differentiate vowels primarily on the basis of F_1, which is greatly affected by height of the jaw, a visible articulator, and which carries acoustic information at lower frequencies, more often within the residual hearing of hearing-impaired listeners. F_2, on the other hand, is primarily affected by forward and backward movements of the tongue that are largely invisible and carry acoustic information at higher frequencies, often beyond residual hearing.

Do hearing-impaired speakers use different articulatory configurations than hearing speakers? The answer appears to be yes and no, depending upon the articulators involved and the inherent possibilities for visual, auditory or other feedback. The evidence that the configurations are not dissimilar in kind from those of hearing speakers, together with the great differences across deaf subjects, suggest that speech production systems of hearing-impaired talkers are sensitive to and are organized around any feedback available, including residual hearing.

DYNAMICS AND TIMING

A frequent observation about hearing-impaired speech is that it is slow (Boothroyd, Nickerson, & Stevens, 1974; Voelker, 1938). Closure durations of plosives and constriction durations of fricatives are extended (Calvert, 1961), vowel durations are longer (Zimmermann & Retaliata, 1981), repetitive syllables are uttered at a slower rate (Robb, 1985), and segmental shortening with increased length of utterance is not consistently exhibited (Tye-Murray & Woodworth, 1989). Rothman (1976), Stein (1980), and Zimmermann and Retaliata (1981) have observed slower consonant-to-vowel transitions. If speech production of hearing-impaired talkers is guided in part by vibrotactile cues, it is possible that articulator contacts, approximations and constrictions, and movement to and from such points, might be prolonged to maximize feedback, resulting in slower production overall. Conversely, with visible articulators, such as the lip and jaw, vision may augment and, in fact, reduce dependence on vibrotactile cues. Some evidence suggests that this is true. Open vowel posture durations (Tye-Murray, 1984) and opening and closing durations of the jaw (McGarr, Löfqvist, & Story, 1987; Tye-Murray & Folkins, 1990) and the lower lip (Tye-Murray & Folkins, 1990) have been found to be similar in hearing and hearing-impaired subjects. Jaw and lip peak velocities (McGarr et al., 1987; Tye-Murray & Folkins, 1990) have also been shown to be comparable for the two groups. Some have found these velocities to be faster in hearing-impaired subjects (Stein, 1980; Zimmermann & Retaliata, 1981). It is difficult to explain the findings of faster velocities of the lip and jaw other than to speculate that, because they are visible, their movements could serve to return the articulators from open postures with fewer sources of feedback to points of closure or constriction where vibrotactile feedback is abundant.

Consistent with the hypothesis that vibrotactile cues are a primary source of information for the hearing-impaired talker and that segments may be prolonged to maximize that feedback, is the evidence that hearing-impaired talkers exhibit less coarticulation than hearing speakers. Metz, Whitehead, and McGarr (1982) provided figures from high speed film data of laryngeal configurations by two women, one hearing and one hearing-impaired, during /h/ production of an /ihi/ syllable. The normal speaker placed the vocal folds in a semi-adducted posture in the medial glottal plane. The folds oscillated slightly during production of the /h/, but did not contact nor vibrate periodically. The hearing-impaired speaker, on the other hand, separated the vocal folds widely and did not position them in the medial glottal plane, making it difficult to make a smooth transition to the following vowel. Data from Whitehead and Metz (1980) had previously demonstrated that this subject had both extremely high air flow rates during the /h/ segment and perceptual discontinuities between individual segments of the /ihi/ syllable. At the acoustic level Waldstein and Baum (1991) and Baum and Waldstein (1991) examined both anticipatory and perseveratory coarticulation in hearing and hearing-impaired children. They measured duration and spectral characteristics of /ʃ,t,k/ when preceded or followed by the vowels /i/ and /u/. Both anticipatory and perseveratory lip rounding for /u/ affects the spectral characteristics of consonants and thus served as a measure of coarticulation. The highly intelligible hearing-impaired children in their studies showed evidence of both anticipatory and perseveratory liprounding, though the effects were not as robust as for the hearing children. Tye-Murray (1987) observed that hearing subjects lowered the jaw more during bilabial closure for "ba" and "pa" than for "bi" and "pi," but the deaf did not move in such an anticipatory fashion. The upper articulator studies just discussed are especially interesting because coarticulatory liprounding and jaw opening, while available visually, were not demonstrated at all or to the same degree as in hearing speakers. It may be that vibrotactile feedback for segment-by-segment production is used as a primary monitoring device even when visual information is available.

Acoustic studies have shown that formant transitions are more restricted in spectral range in hearing-impaired speech (Monsen, 1976b; Rothman, 1976). These studies offer still more data that different movement patterns are exhibited by hearing-impaired speakers. In the "Topography" section of this paper, evidence was presented that F_2 carries acoustic information often beyond the residual hearing of hearing-impaired talkers and that those speakers have a tendency to keep the tongue in a more retracted and low position during vowel production. Monsen (1976a) observed that F_2 "floats" around 1800 Hz regardless of the vowel produced. As Monsen pointed out, the lack of F_2 movement precludes the transmission of important consonant and vowel information. The restricted formant transitions presumably arise, then, from (1) movements to consonants from less accurate tongue configurations for vowels, and (2) from movements altered in order to exploit vibrotactile cues.

It was suggested in the "Introduction" of this paper that the probability for intelligible speech production in hearing-impaired speakers is diminished when rapid and precise temporal alignment of two or more articulatory events is required. McGarr and Harris (1983) studied the timing of the lips with the tongue for production of nonsense VCVCV utterances (e.g., /ə'pɑpip /). EMG techniques were employed to study orbicularis oris and genioglossus activity in one hearing and one deaf speaker. Lip muscle activity was similar for both, but the timing of tongue activity relative to lip muscle activity was different for each subject. The hearing speaker consistently displayed peak tongue muscle activity for /i/ coincident with the acoustic /p/ burst. The deaf speaker exhibited changing patterns of timing from token to token, and peak muscle activity for /i/ occurred frequently after lip release for /p/. Thus, the behavior of a visible articulator, the lips, appeared near-normal, but that of the invisible articulator, the tongue, with many degrees of freedom for possible movement, was shown to be variably timed with respect to the lips.

Another example of a particular problem for hearing-impaired speakers is the temporal alignment of activity of the larynx and upper articulators. Rothman (1977) showed with EMG and spectrographic data that deaf speakers frequently initiated or terminated voicing for a given segment incorrectly, either before or after tongue gestures were in place. Cinefluorographic data from Stein (1980) indicated that deaf speakers timed voice onset incorrectly in syllable-initial consonant production. Aerodynamic data from Whitehead and Barefoot (1980) showed that hearing and intelligible deaf speakers had greater average airflow during voiceless consonant production than voiced, but that unintelligible deaf speakers did not differentiate airflow for voiceless and voiced consonants. It was hypothesized that the unintelligible group did not employ normal glottal abductory gestures for voiceless consonants. Transillumination and transconductance data from McGarr and Löfqvist (1982) as well as fiberscopic evidence from Mahshie and Conture (1983) have provided further details. Both studies found that hearing-impaired speakers were variably successful in terms of voice onset timing. Some hearing-impaired talkers produced some coordinated gestures comparable to those of normals, whereas other productions differed in a variety of ways. A laryngeal gesture was sometimes used when none was required, or there was a failure to use an adductory or abductory gesture when expected, and patterns varied among the hearing-impaired speakers. Several studies have also shown that hearing-impaired speakers frequently time voice offset for vowels inappropriately late relative to the production of voiceless final consonants (Stein, 1980). Consequently they do not display normal, systematic differences in vowel duration as a function of the voiced or voiceless status of postvocalic consonants (Calvert, 1961; Monsen, 1974; Whitehead & Jones, 1977).

VARIABILITY AMONG AND WITHIN HEARING-IMPAIRED TALKERS

Hudgins and Numbers' classic study in 1942 of the speech intelligibility of the deaf revealed "recurrent" errors made by 192 deaf and partially deaf students. Seven categories of consonant errors and five of vowel errors were identified from the 5,701 total errors recorded. Clearly, speakers with hearing loss exhibit a certain commonality in speech errors. Furthermore, given a speech sample as small as a CV syllable (Calvert, 1961) or a vowel (Rubin, 1985), listeners can discriminate utterances spoken by hearing-impaired talkers from those spoken by hearing talkers. To the ear of the listener, then, hearing-impaired speakers make characteristically similar, if not stereotypical, articulation errors.

Nevertheless, many of the studies discussed in the preceding sections of this paper have revealed a great deal of variability *among* hearing-impaired talkers. Crouter (1963) obtained cineflourographic tongue shape tracings of vowels produced by hearing and hearing-impaired speakers. Great variability in tongue contour was seen in both groups, but the greatest differences were seen among deaf speakers. In their acoustic measures of vowels spoken by hearing and deaf subjects, Angelocci et al. (1964) plotted F_1 by F_2 values and found that the distributions of values for the vowels of the hearing were more limited, suggesting more consistency and accuracy in production. The greater dispersal of values for vowels spoken by the hearing-impaired children indicated a high degree of inconsistency and inaccuracy across speakers. Robb (1985) studied diadochokinetic syllable rates of 30 hearing-impaired high-school students. Results revealed large variability around durational means. Huntington, Harris, and Sholes (1968) found EMG patterns for tongue movement in consonant production by deaf subjects that differed both from those of the hearing controls and from each other. They concluded, "Deaf speakers are no more like each other than they are like normal speakers" (p. 157). Indeed, Monsen (1974) studied vowel duration preceding various final consonants in monosyllabic words spoken by hearing and hearing-impaired talkers. He concluded that deaf talkers were more different from the hearing than from other deaf talkers, yet stated that patterns of production varied so much from deaf speaker to deaf speaker as to be "idiolectal" in nature (p. 393).

Evidence suggests that reduced or absent auditory feedback results in both stereotypic productions common to most hearing-impaired talkers *and* differences from speaker to speaker. Most certainly, the kind and amount of residual hearing available to a talker influences production. Boothroyd's observations (1984) of varying speech perception abilities among hearing-impaired individuals with similar audiograms supports that conclusion. Though the following quotation is taken from Tye-Murray's (1992) observations about hearing-impaired children, it most certainly applies to hearing-impaired speakers of all ages: "The existence of idiosyncrasies suggests that latitude exists in how children compensate for deafness" (p. 255).

Production patterns by *any one* hearing-impaired speaker from speech sample-to-speech sample also appear to reflect the dichotomous form of stereotypicality and variability seen *among* speakers. Few early studies shed light on this question, since they sampled only one or two productions each of a variety of phonemes produced by multiple talkers (e.g., Crouter, 1963; Angelocci et al., 1964). Even when subjects produced several tokens of a particular phoneme, these phonemes were often spoken in different phonetic environments or only a few tokens were selected for analysis (e.g., Crouter, 1963; Hudgins and Numbers, 1942; Monsen, 1976a; Suonpää & Aaltonen, 1981). The Huntington, Harris, and Sholes study (1968) required that multiple tokens be collected in order to reveal any consistent EMG patterns and to "calibrate" the speech production of an individual. For clear patterns to emerge, movements had to be quite regular. Huntington, Harris, and Sholes commented that the results differed from observations they had made of repeated utterances by dysarthrics, which were extremely variable and displayed a tendency for all of the articulators to move at the same time. "In contrast, the deaf have well-organized habits, which may be 'correct' or 'incorrect' by comparison with normals, but are not dissimilar in kind" (pp. 157-158). They found that visible articulations were generally correct. They also found that tongue muscle patterns were "stereotyped," though frequently wrong, for production of the consonants /t,s,ʃ,r,l,k/. Vibrotactile cues might reasonably have allowed for the stability seen in the production of those consonants. Monsen's 1974 study of the durations of /i/ and /I/ before final consonants in monosyllables produced by hearing-impaired talkers revealed that the normal pattern of modification according to final consonant, seen in hearing speakers, was replaced by a different pattern in which the duration of /i/ and /I/ were almost mutually exclusive rather than more naturally overlapping as in the hearing. Monsen noted, "Although the variation is different from normal speech, in any one subject it appears systematic to a considerable extent" (pp. 393-394). He suggested that durational cues were more accessible to hearing-impaired talkers than spectral cues. Use and exploitation of durational differences may have competed with attention to and production of more subtle duration and timing differences of vowels with respect to final consonants. McGarr et al. (1987) used optical tracking to study the vertical dimension of jaw movement duration, displacement, and velocity for utterances such as "And Pa peals it" and "And Bea pops it." They found no good evidence of variability in any of the kinematic measures of the jaw, a visible articulator with relatively few degrees of freedom of movement.

On the other hand, several studies have provided evidence that there is also great variability in some articulatory gestures of deaf speakers. Bush (1981) examined the relations among fundamental frequency, formant frequency, and vowel intelligibility in monosyllabic words produced by deaf and hearing children and adolescents. Results revealed that the amount of change in fundamental frequency from vowel-to-vowel was greater in the deaf than in the hearing, but that these changes were a consequence of vowel height effects and not systematically used as a substitute for F_1 and F_2 movement to differentiate these vowels. McGarr and Löfqvist (1982, 1988) provided dramatic evidence of variability in laryngeal timing in relation to tongue and lip movements in hearing-impaired adults during obstruent and obstruent-cluster production. Even for phonemes judged by listeners as correct, considerable physiological variability from token-to-token was evident. As discussed previously, McGarr and Harris (1983) studied tongue-tip and lip coordination using bilabial-vowel-bilabial sequences such as /əˈpɑpip/. In both hearing and hearing-impaired speakers orbicularis oris peaks of activity for /p/ were well defined, though more prolonged in the hearing-impaired. However, the activity of the genioglossis for /i/ was variable, occurring too early, slightly late, or very late, relative to the release burst of the second /p/. Similar interarticulator variability has also been described by McGarr and Gelfer (1983), who found one deaf speaker with high token-to-token variability in onset and offset of genioglossus EMG activity relative to onset of voicing in six vowels in a [hVd] environment.

Rubin (1985) conducted a study to test directly whether deaf speakers produce vowels with the same variability as normal talkers. Her subjects were six orally trained, severely and profoundly hearing-impaired high school students and two age-matched normals. The subjects produced "You got me the /bVb/" with any of seven test vowels, and each of the seven types was said 15 times. Deaf talkers were significantly more variable than normals on measures of fundamental frequency, vowel duration, and F_1 and F_2 frequency. However, the pattern of variability varied from talker to talker. Some speakers showed small variability for point vowels but greater variability for half-close vowels like /ɛ/. Others showed overlap between front and back vowels. Still others showed a great deal of variability for all vowels.

There is much yet to be learned about the stability of articulatory movement in hearing-impaired populations. Observations and data presented by Metz, Schiavetti, Sitler, and Samar (1990) illustrate this well. First, they pointed out that Mahshie and Conture (1983) showed that lateral excursion of the arytenoid cartilages during abduction and adduction (observed using a fiberoptic nasolaryngoscope) and oral patterns coordinated with laryngeal movements (revealed from associated spectrograms) were more variable for two of four hearing-impaired subjects than for four hearing speakers. The other two hearing-impaired subjects exhibited consistent aberrant laryngeal articulatory patterns over multiple tokens. The two subjects whose productions were unstable had higher overall speech intelligibility ratings than the two whose productions were consistent, but wrong. Metz et al. noted that those findings are in contrast to those of McGarr and Löfqvist (1982) who also examined laryngeal articulatory behaviors and interarticulator timing patterns in obstruents produced by hearing and hearing-impaired adults. McGarr and Löfqvist reported that one subject who showed high production instability was their least intelligible speaker. Second, Metz et al. provided data in which they examined the relation between measures of the speech intelligibility of 40 hearing-impaired speakers and production stability as indicated by standard deviation calculations for eight acoustic measures including voice onset time, range of F_2 transitions, and difference in formant values between high and low vowels or between front and back vowels measured in a variety of utterances. Equal numbers of subjects were drawn from each of five groups representing five intelligibility ratings ranging from low to high. Results indicated that hearing-impaired subjects who were rated as having high intelligibility also exhibited high stability for the acoustic measures. Subjects rated as having low intelligibility, however, exhibited either of two overall patterns: highly stable, but aberrant patterns of acoustic values or highly variable patterns. The first may represent those who have learned a stable, yet wrong, set of production rules and consistently use them. The latter may represent those who did not learn a consistent set of speech production rules and who lack consistent underlying speech production motor strategies.

Harris, Rubin-Spitz, and McGarr (1985) have drawn comparisons between findings of greater variability in the speech of the hearing-impaired and that of the motorically immature young hearing speaker. It would appear that auditory feedback, just as speaking experience, constrains production variability. The accumulated data do not indicate that hearing-impaired talkers have a deviant phonology, reliably produced, but a partially correct phonology with varying patterns of variability. Correct and more stable production may occur when a speech target has inherent possibilities for visual, vibrotactile, and auditory feedback, when the articulators involved have fewer degrees of freedom, and when timing of the movement of one articulator is not highly dependent on another. Partially correct and less stable production may occur when these conditions are not met. From the perspective taken in this paper, examples of both stable and variable articulatory movement in hearing-impaired talkers are both explainable and expected.

INTERARTICULATORY ORGANIZATION

Monsen (1976a) proposed that the differences in speech between the hearing and the hearing-impaired can be explained by the characteristics of the feedback they receive. In his words, "the audiogram tends to imprint itself upon speech" (p. 197). Monsen supported this proposal with vowel data from hearing and hearing-impaired talkers. He examined five tokens each of the vowels /i/, /ɑ/, and /ɔ/ from words in sentences in order to map the rough dimensions of each individual's vowel triangle. Drastic reductions in articulatory space were seen in the productions of hearing-impaired subjects, yet differentiated regions of production in the vowel space were preserved. Monsen stressed that the hypothesis that hearing-impaired subjects do not have clearly defined vowel target areas was unsupported. To prove such "one would have to show that there is a nearly random relation between the formant frequency pattern and the intended vowels" (p. 197). Monsen suggested that hearing-impaired talkers distinguish vowels on the basis of the first formant, since a resonance peak in the lower frequency region is more often within the residual hearing of hearing-impaired speakers. (The second formant is in higher frequency regions, often beyond residual hearing, and has been shown to be more restricted in range in hearing-impaired talkers.) F_1 is most affected by tongue height. Jaw position greatly influences tongue height and, as a visible articulator, is observable by a hearing-impaired talker.

If vowel targets for the hearing impaired are formed in terms of visual and partial auditory information, the nature of the targets must be quite different than those for the hearing. These different targets might or might not marshal the neuromuscular articulatory system in the same way. In a study currently underway to explore some of these questions, Campbell (in preparation) is examining the coordination between the tongue and the jaw for vowel production in the hearing impaired, using a bite-block.

Previous studies have shown that hearing talkers speak normally in a variety of conditions in which normal movement has been disrupted. Thus, for example, articulatory compensation has been shown to occur when a speaker's jaw is artificially fixed in an open position by a bite-block placed between the teeth (e.g., Fowler & Turvey, 1980; Lindblom, Lubker, & Gay, 1979; Lindblom & Sundberg, 1971a, 1971b; Nooteboom & Slis, 1970). When the jaw is not free to move normally, the tongue displays greater movement to achieve vowel targets. Vowel formant frequencies in bite-block speech of hearing speakers have been shown to approximate those in normal speech. Because formant measures have usually been made at the onset of the vowel, many have assumed that auditory feedback plays no role in on-line speech motor control. Furthermore, it has also been speculated that while acoustic feedback may play some role in coordinating reciprocal movements of articulators, it may not be a substantial component. Testing for the presence or absence of articulatory compensation in hearing-impaired talkers probes the importance of acoustic feedback to compensation and the nature of neuromuscular articulatory coordination in speakers whose vowel targets have been developed around little or no reliance on audition. If hearing-impaired speakers use visual and kinesthetic feedback from jaw displacement as a primary means to differentiate one vowel from another and if they use little difference in tongue positioning (Stein, 1980; Stevens, Nickerson, & Rollins, 1983; Monsen, 1976a, 1976b), it may be that the jaw and tongue complex are not functionally linked in the same manner as in the hearing, and there may be no compensatory movements of the tongue. One cinefluorographic and perceptual study (Tye et al., 1983) has suggested that hearing-impaired speakers with acquired and congenital hearing losses exhibit articulatory compensation for a bite-block while an acoustic study (Osberger, Netsell, & Goldgar, 1986) indicated that prelingually deafened subjects do not.

Campbell's subjects were hearing-impaired adults with hearing loss onset before two years of age and an average hearing loss in either the severe or profound range. All were graduates of an oral school and had received consistent speech training throughout their school years. A group of hearing subjects, age and sex-matched to the hearing-impaired group, served as controls. Subjects produced repetitions of one of three vowels /i/, /ɪ/, or /æ/ in the carrier phrase "Say /hVt/ again." Each vowel was produced 16 times in random order in each of four, also randomized, conditions:

1. a normal condition;
2. a bite-block condition in which subjects wore bite-blocks custom-made to produce an interincisal distance of 20 mm. Hearing-impaired subjects wore their hearing aids.
3. a bite-block-plus-masking condition in which subjects wore the bite-block and were deprived of auditory feedback. Hearing-impaired subjects removed their hearing aids and both hearing and hearing-impaired subjects listened to masking noise delivered via headphones from a pink noise generator at a measured output of 81 dB SPL.
4. a masking noise condition in which hearing-impaired subjects removed their hearing aids and both hearing and hearing-impaired subjects listened to masking noise, as in #3, above.

Compensation was determined by measuring the formants of vowels and comparing the formant frequencies spoken in test the conditions with those obtained in the normal condition.

To discuss whether a subject compensates for a bite-block, one must first consider what would happen if he did not. That is, one must consider what would happen to formant values of vowels if they were produced with a lowered jaw, as for a bite-block, and if the rest of the vocal tract did not vary. Such measures are not possible to obtain with humans, so they are provided by computer simulation. Such simulations predict the direction and degree of change in F_1 and F_2, whose values go in opposite directions. A bite-block lowers the jaw and opens the mouth, thus raising F_1. The tongue rocks back in the mouth, thus lowering F_2. For the vowel /i/, the only vowel that will be discussed here, simulated values predict a rise in F_1 of approximately 250 to 275 Hz and a lowering of F_2 of approximately 375 to 400 Hz, when the jaw is lowered for a bite-block and there is no compensatory change in the vocal tract.

Returning to Campbell's study, preliminary results for one profoundly hearing-impaired subject and one subject with normal hearing, matched for age and gender, are illustrated in Figure 1. The hearing-impaired male subject (P3M) sustained a profound bilateral sensorineural hearing loss present since birth, possibly of genetic origin. His pure tone average threshold in the better ear was 107 dB HL.

Compared to the hearing speaker (N1M), the hearing-impaired subject (P3M) produced tokens of /i/ in a lower and more backed position in the vocal tract, reflected in higher first formant and lower second formant values. These values were also dramatically more variable.

In general, subject N1M showed a very small but systematic effect of test conditions in the direction, but not the degree, predicted by simulation. The most disrupting condition, bite-block plus masking, resulted in mean F_2 values that were decreased by 3.6% and F_1 values that were raised by 7.6%. Standard Deviations for F_2 doubled for the bite-block-plus-masking condition (SD=59.8) compared to the normal condition (SD=24.12).

The form of P3M's adjustments was quite different. In Figure 1 it can be seen that the tokens in the four conditions separate, revealing the effect of each condition in the direction, but not to the degree, predicted for F_1. The greatest change was seen in the bite-block-plus-masking condition in which F_1 increased 12.6% from a mean of 342 Hz in the normal condition to 385 Hz and the Standard Deviation went from 16.05 to 53.63. Clearly this subject made adjustments in upward movement of the tongue to adjust for a fixed open jaw. A surprising effect on F_1 was shown in the masking condition where mean F_1 (371 Hz) increased 8.5%. That effect was greater than the bite-block condition that produced a mean F_1 (359 Hz) 5% higher than the normal condition. Removal of the hearing aid and the application of masking noise affected this subject's productions in the F_1 dimension more than the bite-block. These results suggest that this profoundly hearing-impaired subject used residual hearing to monitor his production of this vowel and made upward adjustments of the tongue for a bite-block that were more successful with his hearing aid than without.

There was a surprising reversal from the predicted change in F_2. Mean F_2 was increased, rather than decreased, by 7.5%, 12.7%, and 14.8% in the masking, bite-block, and bite-block-plus-masking conditions, respectively, compared to the normal condition. It appears that the subject's response to each test condition was to move the tongue farther forward in the mouth. These adjustments may have been an attempt to maximize tactile feedback in the presence of a bite-block by bunching the tongue at the front of the vowel space.

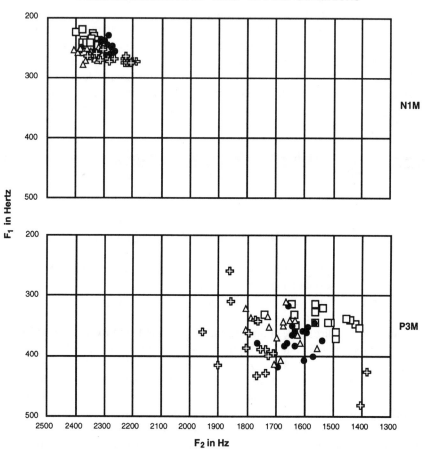

FIGURE 1. Formant measures for the vowel /i/ in the word "heat" spoken by one hearing subject, NIM, and one profoundly deaf subject, P3M, in four conditions:

- ☐ Normal Condition.
- ● Masking Condition (Hearing Aids Removed).
- △ Bite-Block Condition.
- ✛ Bite-Block-Plus Masking Condition (Hearing Aids Removed).

However, removal of the hearing aid and the application of masking noise with or without a bite-block also resulted in an increase in F_2. Furthermore, accuracy in the combined condition was greatly disrupted compared to the normal condition (SD=185.68 Bite-Block-plus-Masking; SD=63.25 Normal Condition) and was also more disrupted than in the bite-block alone condition (SD=55.99), again suggesting forward adjustments of the tongue for a bite-block that were more consistent with the hearing aid than without.

In response to a fixed jaw, this hearing-impaired speaker exhibited adjusted movements of the tongue for the vowel /i/ that resembled articulatory compensation in hearing speakers. Those adjustments appeared to be aided by residual hearing. However, the adjustments were less successful than, and different from, those of the hearing subject, perhaps, in part, as a result of attempts to maximize tactile feedback.

CONCLUSIONS

This brief overview of evidence from a variety of studies of the speech patterns of hearing-impaired persons and the discussion of research in progress suggest that the speech production system of a hearing-impaired speaker seeks and uses any available visual, vibrotactile, kinesthetic, proprioceptive, and residual auditory feedback to organize itself. The speech produced by such a system, while communicating with more variable success than that produced by a system served by normal hearing, still resembles it. Chances for successful, intelligible production are greater when a speech target has inherent possibilities for visual and/or vibrotactile feedback, when the articulators involved have fewer degrees of freedom of movement, and when the timing of the movement of one articulator is not highly constrained by the movement of another. These factors result in similarities among hearing-impaired talkers. However, the great contrasts among hearing-impaired subjects in many of these studies point to the possibility of large differences in availability, quality, and use of feedback by individuals within the categories of severe or profound deafness. Research addressing the organization and control of speech produced by hearing-impaired subjects clearly is warranted to increase knowledge and to provide the important data bases necessary to improve the efficacy of training.

EPILOG

As a Distinguished Professor at the Graduate School and University Center of the City University of New York, Katherine Safford Harris has served on the doctoral thesis committees of many of its students. Noted especially here are those students who studied aspects of the speech and language skills of persons with hearing impairments, students who profited from the wit and wisdom of Katherine Harris as a member of their thesis committees:

Clarissa R. Smith	(1972)	Residual Hearing and Speech Production in Deaf Children
Sr. Rosemary Helen Gaffney	(1977)	Assessing Receptive Language Skills of Five- to Seven-Year-Old Deaf Children
Nancy Steinmuller McGarr	(1978)	The Differences Between Experienced and Inexperienced Listeners in Understanding the Speech of the Deaf
Mary Joseph Osberger	(1978)	The Effect of Timing Errors on the Intelligibility of Deaf Children's Speech
Abigail Peterson Reilly	(1979)	Syllable Nucleus Duration in the Speech of Hearing and Deaf Children
Judith A. Rubin	(1983)	Static and Dynamic Information in Vowels Produced by the Hearing Impaired
Melanie McNutt Campbell	(in progress)	Articulatory Compensation in Hearing and Hearing-Impaired Speakers
Areti Okalidou	(in progress)	Coarticulation in Deaf Speakers

ACKNOWLEDGMENT

This manuscript and the research conducted as Campbell's thesis was supported by NIH Grant DC-00121 to Haskins Laboratories.

REFERENCES

Angelocci, A. A., Kopp, G. A., & Holbrook, A. (1964). The vowel formants of deaf and normal-hearing 11- to 14-year-old boys. *Journal of Speech and Hearing Disorders, 29,* 156-170.

Baum, S. R., & Waldstein, R. S. (1991). Perseveratory coarticulation in the speech of profoundly hearing-impaired and normally hearing children. *Journal of Speech and Hearing Research, 34,* 1286-1292.

Boothroyd, A. (1984). Auditory perception of speech contrasts by subjects with sensorineural hearing loss. *Journal of Speech and Hearing Research, 27,* 134-144.

Boothroyd, A., Nickerson, R., & Stevens, K. (1974). Temporal patterning in the speech of the deaf. *SARP Report #7.* Northampton, MA: Clarke School for the Deaf.

Brannon, J. B., Jr. (1964). *Visual feedback of glossal motions and its influence upon the speech of deaf children.* Unpublished doctoral dissertation, Northwestern University.

Bush, M. A. (1981). *Vowel articulation and laryngeal control in the speech of the deaf.* Unpublished doctoral dissertation, Massachusetts Institute of Technology.

Calvert, D. R. (1961). *Some acoustic characteristics of the speech of profoundly deaf individuals.* Unpublished doctoral dissertation, Stanford University.

Campbell, M. M. (in preparation). *Articulatory compensation in hearing and hearing-impaired speakers.* Unpublished doctoral dissertation, City University of New York.

Carr, J. (1953). An investigation of the spontaneous speech sounds of five-year-old deafborn children. *Journal of Speech and Hearing Disorders, 18,* 22-29.

Crouter, L. (1963). *A cinefluorographic comparison of selected vowels spoken by deaf and hearing subjects.* Master's thesis, University of Kansas.

Fowler, C. A., & Turvey, M. T. (1980). Immediate compensation in bite-block speech. *Phonetica, 37,* 306-326.

Harris, K. S. (1971). Children's language development and articulatory breakdown. In D. L. Horton & J. J. Jenkins (Eds.), *The perception of language* (pp. 207-215). Columbus, OH: Chas. E. Merrill.

Harris, K. S., Rubin-Spitz, J., & McGarr, N. S. (1985). The role of production variability in normal and in deviant developing speech. In J. Lauter (Ed.), *Proceedings of the conference on the planning and production of speech in normal and hearing-impaired individuals: A seminar in honor of S. Richard Silverman* (pp. 50-57). ASHA Reports 15. Rockville, MD: American Speech-Language-Hearing Association.

Hudgins, C. V. (1934). A comparative study of the speech co-ordination of deaf and normal subjects. *Journal of Genetic Psychology, 44,* 1-48.

Hudgins, C. V. (1937). Voice production and breath control in the speech of the deaf. *American Annals of the Deaf, 82,* 338-358.

Hudgins, C. V., & Numbers, F. C. (1942). An investigation of the intelligibility of the speech of the deaf. *Genetic Psychology Monographs, 25,* 289-392.

Huntington, D. A., Harris, K. S., & Sholes, G. N., (1968). An electromyographic study of consonant articulation in hearing-impaired and normal speakers. *Journal of Speech and Hearing Research, 11,* 147-158.

Lindblom, B. E. F., Lubker, J. F., & Gay, T. (1979). Formant frequencies of some fixed-mandible vowels and a model of speech motor programming by predictive simulation. *Journal of Phonetics, 7,* 147-161.

Lindblom, B. E. F., & Sundberg, J. (1971a). Acoustical consequences of lip, tongue, jaw, and larynx movement. *Journal of the Acoustical Society of America, 50*, 1166-1179.

Lindblom, B. E. F., & Sundberg, J. (1971b). Neurophysiological representation of speech sounds. *Papers from the Institute of Linguistics - University of Stockholm, 7*, 16-21.

Lindner, G. (1962). Über den zeitlichen Verlauf der Sprechweise bei Gehörlosen. *Folia Phoniatrica, 14*, 67-75.

Locke, J. L. (1980). The prediction of child speech errors: Implications for a theory of acquisition. In G. H. Yeni-Komshian, J. F. Kavanagh, & C. A. Ferguson (Eds.), *Child phonology, Volume 1: Production* (pp. 193-209). New York: Academic Press.

Mahshie, J. J., & Conture, E. G. (1983). Deaf speakers' laryngeal behavior. *Journal of Speech and Hearing Research, 26*, 550-559.

McGarr, N. S., & Gelfer, C. E. (1983). Simultaneous measurements of vowels produced by a hearing-impaired speaker. *Language and Speech, 26*, 233-246.

McGarr, N. S., & Harris, K. S. (1983). Articulatory control in a deaf speaker. In I. Hochberg, H. Levitt, & M. J. Osberger (Eds.), *Speech of the hearing impaired: Research, training and personnel preparation* (pp. 75-95). Baltimore, MD: University Park Press.

McGarr, N. S., & Löfqvist, A. (1982). Obstruent production by hearing-impaired speakers: Interarticulator timing and acoustics. *Journal of the Acoustical Society of America, 72*, 34-42.

McGarr, N. S., & Löfqvist, A. (1988). Laryngeal kinematics in voiceless obstruents produced by hearing-impaired speakers. *Journal of Speech and Hearing Research, 31*, 234-239.

McGarr, N. S., Löfqvist, A., & Story, R. S. (1987). Jaw kinematics in hearing-impaired speakers. *Proceedings XIth ICPhS: The eleventh international congress of phonetic sciences* (Vol. 4, pp. 173-176). Tallinn, Estonia, USSR: Academy of Sciences of the Estonian S.S.R.

Metz, D. E., Schiavetti, N., Sitler, R. W., & Samar, V. J. (1990). Speech production stability characteristics of hearing-impaired speakers. *The Volta Review, 92*, 223-235.

Metz, D. E., Whitehead, R. L., & McGarr, N. S. (1982). Physiological aspects of speech produced by deaf persons. *Audiology, 7*, 35-48.

Monsen, R. B. (1974). Durational aspects of vowel production in the speech of deaf children. *Journal of Speech and Hearing Research, 17*, 386-398.

Monsen, R. B. (1976a). Normal and reduced phonological space: The production of English vowels by deaf adolescents. *Journal of Phonetics, 4*, 189-198.

Monsen, R. B. (1976b). A taxonomic study of diphthong production in the speech of deaf children. In S. K. Hirsh, D. H. Eldredge, I. J. Hirsh, & S. R. Silverman (Eds.), *Hearing and Davis: Essays honoring Hallowell Davis* (pp. 281-290). St. Louis, MO: Washington University Press.

Nooteboom, S. G., & Slis, I. (1970). A note on the degree of opening and the duration of vowels in normal and "pipe" speech. *IPO Annual Progress Report*, *5*, 55-58.

Osberger, M. J., Netsell, R., & Goldgar, D. (1986). Articulatory reorganization in deaf talkers. *Journal of the Acoustical Society of America*, *79*, Supplement 1:S37. (A)

Rawlings, C. G. (1935). A comparative study of the movements of the breathing muscles in speech and in quiet breathing of deaf and normal subjects. *American Annals of the Deaf, 80*, 147-156.

Robb, M. P. (1985). Oral diadochokinesis in hearing-impaired adolescents. *Journal of Communication Disorders*, *18*, 79-89.

Rothman, H. (1976). A spectrographic investigation of consonant-vowel transitions in the speech of deaf adults. *Journal of Phonetics*, *4*, 129-136.

Rothman, H. (1977). An electromyographic investigation of articulation and phonetic patterns in the speech of deaf adults. *Journal of Phonetics*, *5*, 369-376.

Rubin, J. A. (1985). *Static and dynamic information in vowels produced by the hearing impaired*. Bloomington, IN: Indiana University Linguistics Club.

Shukla, R. S. (1989). Phonological space in the speech of the hearing impaired. *Journal of Communication Disorders*, *22*, 317-325.

Stein, D. M. (1980). *A study of articulatory characteristics of deaf talkers*. Unpublished doctoral dissertation, University of Iowa.

Stevens, K. N., Nickerson, R. S., & Rollins, A. M. (1983). Suprasegmental and postural aspects of speech production and their effect on articulatory skills and intelligibility. In I. Hochberg, H. Levitt, & M. J. Osberger (Eds.) , *Speech of the hearing impaired: Research, training, and personnel preparation* (pp. 35-51). Baltimore, MD: University Park Press.

Subtelny, J., Li, W., Whitehead, R., & Subtelny, J. D. (1989). Cephalometric and cineradiographic study of deviant resonance in hearing-impaired speakers. *Journal of Speech and Hearing Disorders*, *54*, 249-263.

Suonpää, J., & Aaltonen, O. (1981). Intelligibility of vowels in words uttered by profoundly hearing-impaired children. *Journal of Phonetics*, *9*, 445-450.

Tye, N., Zimmermann, G. N., & Kelso, J. A. S. (1983). "Compensatory articulation" in hearing impaired speakers: A cinefluorographic study. *Journal of Phonetics*, *11*, 101-115.

Tye-Murray, N. (1984). *Articulatory behavior of deaf and hearing speakers over changes in rate and stress: A cinefluorographic study*. Unpublished doctoral dissertation, University of Iowa.

Tye-Murray, N. (1987). Effects of vowel context on the articulatory closure postures of deaf speakers. *Journal of Speech and Hearing Research*, *30*, 90-104.

Tye-Murray, N. (1991). The establishment of open articulatory postures by deaf and hearing talkers. *Journal of Speech and Hearing Research*, *34*, 453-459.

Tye-Murray, N. (1992). Young cochlear implant users' response to delayed auditory feedback. *Journal of the Acoustical Society of America, 91,* 3483-3486.

Tye-Murray, N., & Folkins, J. W. (1990). Jaw and lip movements of deaf talkers producing utterances with known stress patterns. *Journal of the Acoustical Society of America, 87,* 2675-2683.

Tye-Murray, N., & Woodworth, G. (1989). The influence of final-syllable position on the vowel and word duration of deaf talkers. *Journal of the Acoustical Society of America, 75,* 629-632.

Voelker, C. H. (1938). An experimental study of the comparative rate of utterance of deaf and normal hearing speakers. *American Annals of the Deaf, 83,* 274.

Waldstein, R. S., & Baum, S. R. (1991). Anticipatory coarticulation in the speech of profoundly hearing-impaired and normally hearing children. *Journal of Speech and Hearing Research, 34,* 1276-1285.

Whitehead, R., & Barefoot, S. (1980). Some aerodynamic characteristics of plosive consonants produced by hearing-impaired speakers. *American Annals of the Deaf, 125,* 366-373.

Whitehead, R., L. & Jones, K. O. (1977). Air flow of plosive consonants in the speech of the hearing-impaired. Paper presented at the 94th Meeting of the Acoustical Society of America, Miami Beach, FL.

Whitehead, R., & Metz, D. (1980). Aberrant laryngeal devoicing gestures produced by deaf speakers: Evidence from acoustic, aerodynamic, and glottographic data. Paper presented at the 100th Meeting of the Acoustical Society of America, Los Angeles, CA.

Zimmermann, G., & Rettaliata, P. (1981). Articulatory patterns of an adventitiously deaf speaker: Implications for the role of auditory information in speech production. *Journal of Speech and Hearing Research, 24,* 169-178.

30 Auditory Feedback Delivered by Electrical Stimulation from a Cochlear Implant and Speech Production

Emily A. Tobey
Louisiana State University Medical Center

Nearly fifteen years ago, Borden (1979) concluded a review and interpretation of feedback research by quoting Paul Weiss (1941),

> Nobody in his senses would think of questioning the importance of sensory control on movement. But just what is the precise scope of that control? Is the sensory influx a constructive agent, instrumental in building up the motor patterns, or is it a regulative agent merely controlling the expression of autonomous patterns without contributing to their differentiation? (p. 406)

Many studies investigating the role of auditory feedback on the motor patterns associated with speech production have found these difficult questions to answer. Indeed, the more typical response to these questions is to presume that auditory feedback acts as both constructive and regulative agents, but that it may play these roles at different times developmentally. Several investigations suggest that auditory feedback is critical for the development of normal speech and language (see, for a review, Osberger & McGarr, 1982; Smith, 1975). Reduction or elimination of auditory feedback during the formative years of speech and language development may result in a variety of speech and language disorders, ranging from individuals who have fairly intelligible speech to individuals who must primarily rely on sign language for communication. Profound hearing impairments occurring after the development of speech and language, however, appear to have a less devastating impact on speech production (see, for example, Cowie & Douglas-Cowie, 1983; Waldstein, 1990). Indeed, some authors (Goehl & Kaufman, 1984) argue that speech production is minimally (if at all negatively) influenced by the elimination or reduction of auditory feedback (cf. Zimmermann & Collins, 1984).

Investigations of auditory feedback in normal hearing individuals have relied primarily on temporally separating air-and-bone conducted signals in delayed auditory feedback experiments, filtering signals, masking signals, or shifting the frequency of signals (see, for review, Borden, 1979). Speaker responses to such experimental conditions have led some investigators to argue that speech gestures are actively modified or controlled by auditory feedback. Other investigators, however, argue that auditory feedback cannot play an active role in the moment-to-moment control of speech, primarily because the temporal window involved in audition is long and the information tends to arrive after the end of movements it was to have corrected. Instead, these investigators propose that auditory feedback acts as a global calibrator for other sensory systems, such as proprioception and taction.

Recent technological advances in the management of profound hearing impairments provide an opportunity to explore further the role of auditory feedback on speech production. In the past ten years, great strides have been made in the prosthetic management of profound hearing impairments enabling investigators to scrutinize more closely how speakers use auditory feedback. Advances in technology now make it possible to distinguish between individuals who are deaf vs. individuals with profound hearing impairments. For individuals who receive no perceptual benefit from the sensory information provided by hearing aids, cochlear implants provide a prosthetic alternative.

Worldwide, cochlear implants have been surgically placed in more than 5000 adults and children with profound hearing impairments. Although the actual elements in cochlear implant systems differ from device to device, they essentially are composed of the same component parts. Cochlear implant systems have an external microphone designed to sense pressure variations in the sound field, processing strategies used to transform information from the microphone into electrical stimuli, and an electrode array surgically placed in or on the cochlea to excite any remaining intact auditory nerve fibers when electrically stimulated with signals derived from the processing strategies. Electrode arrays vary from a single channel of stimulation (used most commonly in individuals with ossified cochleas where only a single, short electrode may be placed) to multiple channels of stimulation through multiple electrodes (Wilson, 1993), thus taking advantage of the tonotopic nature of the cochlea for rate-place coding of frequency.

Auditory information provided by either a single or multiple channel cochlear implant differs substantially from the information received by the normal hearing ear (see, for review, Dorman, 1993). Single channel devices are limited to providing information about F_0 and duration. Neural refractory periods constrain the coding of frequency by single auditory nerve fibers to an upper limit of 1000 Hz, thus limiting single channel devices to a frequency range associated with F_0 or F_1 frequencies. Multichannel cochlear implants also supply limited information. Limitations are related to difficulties in implementing signal-processing strategies and the ability to insert an electrode array only into the first turn and a half of the cochlea. Additional limitations are imposed by the number of remaining auditory nerve fibers, which appears to contribute to the variability of performance across subjects.

The Nucleus cochlear implant developed in Australia is currently the only device approved by the Food and Drug Administration for implantation in children and adults in the United States. This device uses a speech processing strategy incorporating a speech-feature extraction algorithm where the pattern of stimulation to the 22-electrode array is based on an explicit analysis of the input signal (Blamey, Dowell, Brown, Clark, & Seligman, 1987). Initially, the speech processing strategy used by this system extracted F_0, intensity, duration, and spectral peaks at low and mid frequency ranges. Subsequently, spectral peaks of high frequency ranges has also been extracted. F_0 is coded on the electrode array by the rate of stimulation, intensity is coded by the amount of current delivered, and duration is coded by the length of stimulation. The spectral peaks associated with the first three formants are assigned to various electrodes: F_1 peaks are coded to more apical electrodes, F_2 peaks to less apical electrodes, and F_3 peaks are assigned to the three most basal electrodes. More recently, the Nucleus device has employed speech processing strategies that lean more toward a waveform representation than a speech-feature coding strategy, *per se* (McKay & McDermott, 1993). This strategy uses a scheme where energy is passed through sixteen bandpass filters and stimulates the cochlea based on the six channels containing the highest root mean square energy. Several additional devices and speech processing strategies are under development and may soon be commercially available to adults and children.

Perceptual evaluation of cochlear implants reveals considerable variation in performance across individuals (Dorman, 1993; Tyler, 1993). Some individuals perform with high degrees of accuracy while other individuals appear to receive no perceptual benefit. Most individuals, however, fall somewhere in between these two extremes. Variability among individuals is linked not only to the type of speech-processing strategy used to extract and code speech information, but also to differences among individuals in the survival of viable neural elements in the cochlea, the integrity of the central auditory nerve pathways, surgical placement of the electrode array, and cognitive/language skills. Although many individuals use the devices to communicate effectively, performance is typically enhanced when information from the cochlear implant is used in conjunction with visual cues associated with lip reading.

Cochlear implants provide a valuable research tool for exploring the role of auditory feedback on speech production since an investigator may manipulate the presence or absence of auditory feedback by turning the device on or off, as well as manipulating the amount and type of information provided by the device through variations to the speech processing strategies. The goal of a successful speech processing strategy is to mimic or restore perceptual performance to that of a normal hearing listener or an individual with a minimal hearing impairment. But if these devices are truly to aid communication, they must also aid speech production. Degree of hearing loss appears to play a significant role in overall speech intelligibility: the greater the amount of residual hearing, the more nearly normal speech appears to be. Boothroyd (1985) has stated that 5 dB worth of residual hearing may be equivalent to three years of speech therapy. Observations like these lead to corollaries like one that posits a speech-processing strategy that mimics a greater amount of residual hearing should facilitate more nearly normal speech production.

Investigations with adventitiously deafened adult users of cochlear implants support the notion that speech production will be influenced by feedback delivered via the electrical stimulation provided by implants (for a review, see Tobey, 1993). Lane and colleagues found significant changes in airflow during speech in three subjects following implantation (Lane, Perkell, Svirsky, & Webster, 1991). After implantation, two of the subjects demonstrated increases in airflow, which had been below normal before implantation, and one subject demonstrated decreased airflow, which had been above normal before implantation. Spectral changes in formant frequencies also occur after implantation. Perkell and his colleagues (Perkell, Lane, Svirsky, & Webster, 1992) report shifts in F_2 frequencies for three out of four subjects after implantation, shifts that were in directions more nearly approaching values reported by Peterson and Barney (1952). A study conducted by Tartter and colleagues (Tartter, Chute, & Hellman, 1989), however, reported formant frequency shifts in vowel production in one adolescent cochlear implant user that suggest a reduced vowel space. Alterations in mean durations, F_0, and sound pressure levels are also observed for speakers. The data suggest that information supplied by a cochlear implant may be used by speakers to adjust their speech, in some cases, to more nearly normal values and, in other cases, to less normal values. Although speech patterns do not always appear to change in similar ways across speakers, the data highlight the responsiveness of the speech production system to electrical stimulation of the auditory nerve.

Studies contrasting speech produced with an implant turned on vs. off suggest a dual role for auditory feedback (Svirsky & Tobey, 1991; Svirsky, Lane, Perkell, & Webster, 1992). Auditory feedback appears to be used for the long-term calibration of articulatory parameters, resulting in relatively gross adjustments, and also for short-term adjustments resulting in fine tuning. Evidence for a dual role arises from several sources. First, several acoustic parameters appear to change rapidly and consistently depending on whether an implant is turned on or off. These parameters include alterations in sound pressure level, F_0, duration, and some spectral features associated with certain vowels. Examination of utterances on a token-by-token basis reveals that spectral changes may occur as rapidly as in the first token or after only a few tokens (Svirsky & Tobey, 1991). Second, larger speech changes are observed when an implant is turned on after 24 hours of non-use than are observed when an implant is turned off after a half hour's use. Even a half hour's worth of feedback appears sufficient to reset or recalibrate speech production parameters that recur after the implant is turned off again (Svirsky et al., 1992). Short-term adjustments appearing within a few tokens reflect a temporal window whose duration may be ideally suited for prosodic feedback, whereas the longer temporal window associated with changes occurring over hours or days may be associated with a more global calibration of system parameters.

The amount and type of information provided by a cochlear implant also appears to influence speech production. Evidence for this observation arises from experimental studies manipulating the information provided by the signal processing strategies and from studies tracking the speech production performance of individuals who use different types of implant devices. Svirsky and Tobey (1991) contrasted the formant frequencies of a single token, "head," spoken under three different signal processing conditions. These conditions included a processor-off condition (no auditory feedback); a processor-on condition providing a full map of information (F_0, intensity, duration, F_1 and F_2 frequencies); and a single channel of information supplied by stimulating a single channel at a constant rate and amplitude. In this instance, the stimulation provided by the experimental single channel processing strategy signalled only the presence of a stimulus, but it failed to convey any direct information about the stimulus itself. Comparisons of F_2 frequencies produced under the three conditions revealed no significant differences between the processor off condition and the experimental single channel condition. However, significant increases in F_2 frequencies, to values more nearly approximating those of Peterson and Barney (1952), were observed in the full-map condition. Thus, this experimental manipulation of the speech-processing strategy suggested that electrical stimulation of the auditory nerve with a signal unrelated to the tokens being spoken was essentially similar to no stimulation at all.

Additional evidence suggesting that the content and type of information provided by a cochlear implant influences speech production is found in a longitudinal study contrasting the speech of a postlingually deafened preadolescent produced with different implants: a 3M single channel cochlear implant that failed, and its replacement, a multichannel Nucleus cochlear implant (Economou, Tartter, Chute, & Hellman, 1992). The 3M single channel device provides electrical stimulation via a modulated whole-wave signal that enables users to receive frequency information through rate-pitch mechanisms (up to 1000 Hz), as well as duration and amplitude information. The multichannel device, in contrast, used a speech processing strategy that provides information regarding F_0, intensity, duration, and the spectral peaks associated with F_1 and F_2 frequencies. Comparisons of F_0 during contrastive stress tasks revealed a significant increase of overall F_0 for both stressed and unstressed syllables as quickly as one day after the failure of the single channel device. F_0 decreased significantly after one year's use with the multichannel device, and contrastive stress patterns were appropriate for the first time. Vowel space, as inferred from measurements of F_1 and F_2 frequencies, also significantly expanded after a year's experience with the multichannel device. Indeed, formant frequency measurements suggest that a significant reduction in the vowel space occurred as soon as 18 days after the failure of the single channel device. These observations support a working hypothesis that the content and amount of information is an important part of auditory feedback. Moreover, the relative changes observed 18 days after implant failure, and the changes observed within a year after receiving a multichannel cochlear implant, provide evidence that auditory feedback may operate over a fairly long temporal window in setting system parameters.

Direct investigation of how the amount of information conveyed by a cochlear implant influences speech production also may be accomplished by systematic variations in signal processing strategies. A recent investigation by Tobey and her colleagues (Tobey et al., 1993) contrasted speech mediated by a newly developed processor: the spectral maxima sound processor (SMSP). The SMSP divides sound signals into 16 overlapping frequency bands using a bank of bandpass filters. The six bands with the greatest amplitude are used to stimulate the cochlea on a place basis at a constant rate. The sampling rate of the filterbank output is 250 Hz and six biphasic pulses are presented nonsimultaneously every 4 ms to the electrode array. In order to increase the amount of information presented to the array, the processing strategy was adjusted in two ways. First, the investigators increased the sampling rate of the frequency bands from 250 Hz to 450 Hz (High Rate), and second, they increased the number of current pulses presented to the array from 6 to 8 (8 pulse). Figure 1 shows the F_1 and F_2 frequencies for one subject under these conditions. As can be observed in the top panel, the vowels /æ/ and /u/ appear to be most clearly distinguished, with some formant frequency overlap for the vowels /i/, /ɪ/, and /ɛ/ in the SMSP condition. The addition of information by either increasing the sampling rate (middle panel) or number of current pulses (bottom panel) does not appear to assist the speaker further in distinguishing the high front vowels from each other. These data serve to illustrate the complexity of issues surrounding auditory feedback, since increasing the amount of information provided by an implant does not always result in improved performance in postlingually deafened adults, although the data do reinforce previous observations reporting the responsiveness of speech production to signal processing manipulations.

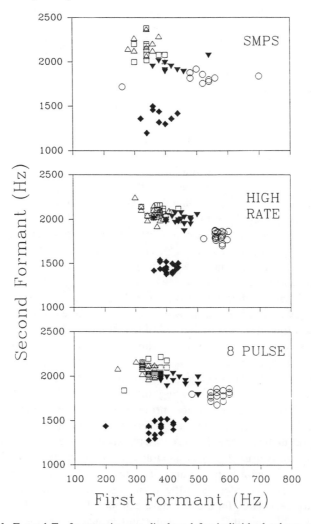

FIGURE 1. F_1 and F_2 frequencies are displayed for individual tokens of the words, "heed", "hid", "head", "had", and "who'ed", in the phrase, "Say _____again. Open circles reflect the values for "had", filled triangles represent "head", open squares are "hid", open triangles represent "heed", and the open diamonds reflect "whoed". The top panel represents the data collected with feedback delivered via a SMSP strategy, the middle panel represents the data collected with feedback delivered via a SMSP strategy sampling the filterbanks at 450 Hz, and the lower panel represents data collected with feedback delivered via a SMSP strategy presenting 8 pulses to the electrode array. See text for further explanation.

Another approach for investigating the role of auditory feedback in speech production is to examine the longitudinal performance of children who receive cochlear implants. Several studies contrast speech production before implantation with speech produced at various intervals after implantation (see for a review, Tobey, 1993). Significant increases in the ability to imitate segmental and nonsegmental speech features occur in children using single-channel devices and multichannel devices. Generally speaking, studies investigating the performance of children with cochlear implants report that the overall pattern of speech features present in samples before implantation is similar to those previously reported for profoundly hearing-impaired children (Smith, 1975). Following implantation, speech features appear to increase in accuracy and a wider range of features are observed (Osberger et al., 1991; Tobey & Hasenstab, 1991). These studies, however, are not without controversy. In part, the controversy focuses on how to define an appropriate control group for comparisons in order to tease apart performance changes due to maturation and aural habilitation from those directly associated with the sensory devices.

As a means of addressing these issues, Geers and Moog (1991) initiated a three-year longitudinal study examining the speech perception and production of children with profound hearing impairments using three types of sensory devices: cochlear implants, tactile aids, and hearing aids. Subjects were carefully matched for important variables such as chronological age, duration of hearing impairment, language ability, intelligence, and family support. Geers and Tobey (1992) recently reported on the speech production acquired under imitative and spontaneous speaking conditions for 18 of the children who had completed two years in the study: 72% of the speech features improved in the cochlear implant group by the second year, in contrast to a 51% increase in speech feature accuracy in the tactile aid group. Improved performance was noted for only 15% of the speech features in the hearing-aid group. In order to determine the proportion of variance in speech production features accounted for by chronological age and aided speech perception at pretest and after two years in the study, preliminary hierarchical regression analyses were conducted (Geers, Moog, Tobey, & Gustus, 1993). At the initiation of the study, chronological age accounted for the majority of variance in speech production performance, and perception skills accounted for little or no variance in the three groups of children. Two years later, for the hearing-aid group, chronological age continued to account for the majority of the variance and speech perception accounted for only 2% of the variance. For the tactile aid group, speech perception accounted for 32% of the variance in speech production at the two-year point. After the two years' experience with cochlear implants, the importance of chronological age to speech production was reduced from 40% to 11% and the importance of speech perception was increased from 1% to 36%. These data show greater changes in speech production and perception for children receiving cochlear implants than for children with tactile aids or hearing aids, suggesting that sensory information provided by electrical stimulation to the auditory nerve may be more useful than that supplied by an alternative sensory system or by traditional hearing aids for developing speech production skills.

Taken overall, the increasing use of cochlear implants in the prosthetic management of individuals with profound hearing impairments holds promise for further research investigating the role of auditory feedback on speech production. Investigations of speech production in users of cochlear implants may shed additional light on theoretical issues regarding auditory feedback. Studies experimentally manipulating the amount and type of information supplied by the speech processing strategies may provide additional information about how auditory feedback acts as a constructive or regulative agent. Comparisons of conditions contrasting no auditory feedback (cochlear implant off) to conditions with auditory feedback (cochlear implant on) also may provide a window for observing the time constants involved in such feedback. In addition, investigations of young children using such devices may provide insight into the developmental role of auditory feedback and perhaps open avenues for improved management of these children.

ACKNOWLEDGMENT

This work was supported, in part, from an Academic Research Enhancement Award from the National Institutes of Health.

REFERENCES

Blamey, P., Dowell, R., Brown, A., Clark, G., & Seligman, P. (1987). Vowel and consonant recognition of cochlear implant patients using formant-estimating speech processors. *Journal of Acoustical Society of America, 82,* 38-42.

Boothroyd, A. (1985). Auditory capacity and the generalization of speech skills. In J. Lauter (Ed.), *Speech planning and production in normal and hearing-impaired children.* (ASHA Reports, # 15), pp. 8-14. Rockville, MD: ASHA.

Borden, G. (1979). An interpretation of research on feedback interruption during speech. *Brain and Language, 7,* 302-319.

Cowie, R., & Douglas-Cowie, E. (1983). Speech production in profound postlingual deafness. In M. Lutman & M. Haggard (Eds.), *Hearing science and hearing disorders* (pp. 183-231). New York: Academic Press.

Dorman, M. (1993). Speech perception by adults. In R. Tyler (Ed.), *Cochlear implants: Audiological foundations* (pp. 145-190). San Diego: Singular Publishing Group, Inc.

Economou, A., Tartter, V., Chute, P., & Hellman, S. (1992). Speech changes following reimplantation from a single-channel to a multichannel cochlear implant. *Journal of the Acoustical Society of America, 92,* 1310-1323.

Geers, A., & Moog, J. (1991). Evaluating the benefits of cochlear implants in an education setting. *American Journal of Otology, 12 (Suppl)*. 116-125.

Geers, A., Moog, J., Tobey, E., & Gustus, C. (1993). Optimizing the benefits of cochlear implants for children. II: Speech production. *Asha, 35*, 164.

Geers, A., & Tobey, E. (1992). Effects of cochlear implants and tactile aids on the development of speech production skills in children with profound hearing impairments. *The Volta Review, 94, 135-164.*

Goehl, H., & Kaufman, D. (1984). Do the effects of adventitious deafness include disordered speech. *Journal of Speech and Hearing Disorders, 49*, 58-64.

Lane, H., Perkell, J., Svirsky, M., & Webster, J. (1991). Changes in speech breathing following cochlear implants in post-lingually deafened adults. *Journal of Speech and Hearing Research, 34*, 526-533.

McKay, C., & McDermott, H. (1993). Perceptual performance of subjects with cochlear implants using the spectral maxima sound processor (SMSP) and the mini speech processor (MSP). *Ear and Hearing, 14*, 350-367.

Osberger, M., & McGarr, N. (1982). Speech production characteristics of the hearing impaired. In N. Lass (Ed), *Speech and language: Advances in basic research and practice.* New York: Academic Press.

Osberger, M., Robbins, A., Berry, S., Todd, S., Hesketh, L., & Sedey, A. (1991). Analysis of spontaneous speech samples of children with a cochlear implant or tactile aid. *American Journal of Otology, 12 (Suppl)*, 173-181.

Perkell, J., Lane, H., Svirsky, M., & Webster, J. (1992). Speech of cochlear implant patients: A longitudinal study of vowel production. *Journal of Acoustical Society of America, 91*, 2961-2978.

Peterson, G., & Barney, F. (1952). Control methods used in a study of the vowels. *Journal of Acoustical Society of America, 24*, 175-184.

Smith, C. (1975). Residual hearing and speech production in deaf children. *Journal of Speech and Hearing Research, 18*, 795-811.

Svirsky, M., & Tobey, E. (1991). Effect of different types of auditory stimulation on vowel formant frequencies in multichannel cochlear implant users. *Journal of Acoustical Society of America, 89*, 2895-2904.

Svirsky, M., Lane, H., Perkell, J., & Webster, J. (1992). Effects of short-term auditory deprivation on speech production in adult cochlear implant users. *Journal of Acoustical Society of America, 92*, 1294-1300.

Tobey, E. (1993). Speech production. In R. Tyler (Ed.), *Cochlear implants: Audiological foundations* (pp. 257-316). San Diego: Singular Publishing Group, Inc.

Tobey, E., Blamey, P., Clark, G., McDermott, H., McKay, C., & Hogne, D. (in press). Manipulations of signal processing strategies and speech production in cochlear implant adult users. *Journal of Acoustical Society of America.*

Tobey, E., & Hasenstab, M. (1991). Effects of a Nucleus multichannel cochlear implant on speech production in children. *Ear and Hearing, 12 (Suppl).* 48S-54S.

Tartter, V., Chute, P., & Hellman, S. (1989). The speech of a postlingually deafened teenager during the first year of use of a multichannel cochlear implant. *Journal of Acoustical Society of America, 86,* 2113-2121.

Tyler, R. (1993). Speech perception by children. In R. Tyler (Ed.), *Cochlear implants: Audiological foundations* (pp. 191-256). San Diego: Singular Publishing Group, Inc.

Waldstein, R. (1990). Effects of postlingual deafness on speech production: Implications for the role of auditory feedback. *Journal of Acoustical Society of America, 88,* 2099-2114.

Weiss, P. (1941). Does sensory control play a constructive role in the development of motor coordination? *Schweizerische Medizinische Wochenschrift, NR12,* 406-407.

Wilson, B. (1993). Speech processing. In R. Tyler (Ed.), *Cochlear implants: Audiological foundations* (pp. 35-86). San Diego: Singular Publishing Group, Inc.

Zimmermann, G., & Collins, M. (1984). The speech of the adventitiously deaf and auditory information: A response to Goehl and Kaufman (1984). *Journal of Speech and Hearing Disorders, 6,* 220-221.

Producing Speech: The Segment (Reprise)

31

On the Dynamics of Temporal Patterning in Speech

Elliot Saltzman
Haskins Laboratories
Center for the Ecological Study of Perception and Action, University of Connecticut

Anders Löfqvist
Haskins Laboratories

Jeff Kinsella-Shaw
Haskins Laboratories
Center for the Ecological Study of Perception and Action, University of Connecticut

Bruce Kay
Brown University

Philip Rubin
Haskins Laboratories

With deep respect and affection for Katherine S. Harris, and her fascination with processes of durational change in the production of speech.

I. INTRODUCTION

What are the dynamics that underlie the temporal cohesion among the gestural or segmental components of a given speech utterance? The present chapter describes recent results from an ongoing series of studies at Haskins Laboratories that has begun to provide answers to this question (e.g., Kollia, 1994; Löfqvist & Gracco, 1991; Munhall, Löfqvist, & Kelso, 1994; Saltzman, 1992; Saltzman, Kay, Rubin, & Kinsella-Shaw, 1991; Saltzman, Löfqvist, Kinsella-Shaw, & Rubin, 1992; see also Gracco & Abbs, 1989, performed prior to the first author's move to Haskins). These studies employ a paradigm in which mechanical perturbations are applied to the articulatory periphery, typically the lower lip or jaw, during spoken utterances in order to probe the functioning of the speech production apparatus. The resultant changes in the temporal structure of the utterances are used to provide information about the dynamics of intergestural timing. For example, transient mechanical perturbations delivered to the speech articulators during repetitive speech sequences (Saltzman, 1992; Saltzman et al., 1991), or to the limbs during unimanual rhythmic tasks (Kay, 1986; Kay, Saltzman, & Kelso, 1991), can alter the underlying timing structure of the ongoing sequence and induce systematic shifts in the timing of subsequent movement elements. As will be elaborated shortly in greater detail, these data imply that the relative phasing of speech gestures is not rigidly specified over a given sequence. Rather, such results suggest that gestural patterning evolves fluidly and flexibly over the course of an ongoing sequence, governed by an intrinsic intergestural dynamics. Furthermore, these data suggest that the intergestural dynamical system functions as a sequence-specific central timing network that does not simply drive the articulatory periphery in a unidirectionally coupled manner. Rather, central and peripheral dynamics are coupled bidirectionally, so that feedback information from the articulatory periphery can influence the state of the central "clock."

There are two important methodological points to be made regarding this type of study. First, it is important to analyze extended, repetitive sequences, e.g., /...pæpæpæ.../, using steady-state, *phase-resetting* techniques. These methods were pioneered in studies of the effects of perturbations on the temporal structure of general biological rhythms (e.g., Glass & Mackey, 1988; Kawato, 1981; Winfree, 1980). In particular, such analyses are used to determine whether perturbations delivered during an ongoing rhythm have a permanent effect (i.e., phase shift) on the underlying temporal organization of the rhythm. Phase-resetting techniques have been used in many kinematic and neurophysiological studies of the control and coordination of rhythmic movements (e.g., Lennard & Hermanson, 1985; Lee & Stein, 1981). In such studies, what is measured is the amount of temporal shift introduced by the perturbation, relative to the timing pattern that existed prior to the perturbation. This phase shift is measured after the transient, perturbation-induced distortions to the rhythm have subsided, and the system has returned to its pre-perturbation, steady-state rhythm. The finding of a post-perturbation, steady-state phase shift using this method would support the hypothesis that there exists a central timing network that both drives the articulatory periphery and whose state is altered (phase-shifted) by feedback specific to events at the periphery. A further, crucial aspect of the phase-resetting methodology is that, across trials, perturbations are delivered so as to sample all phases of the utterance's repeated syllable "cycle," e.g., from maximum lip opening for one instance of /æ/ to the next in /...pæpæpæ.../. The complete analysis thus consists of testing for post-perturbation, steady-state phase shifts for all phases of perturbation delivery, in order to examine the variation, over the course of the syllable, in the sensitivity or receptivity of the central clock to peripheral, perturbation-induced events.

It is important to note that relatively lengthy, repetitive utterances are required for the steady-state, phase-resetting technique, in order to be able to distinguish temporal articulatory distortions that are attributable to central resetting processes from those that are attributable to the systematic yet transient behavior of the articulatory periphery. The utterances must be repetitive, since the units of analysis are cycles and, by definition, successive cycles must be approximately identical. The utterance must be relatively long since, even in the minimum cycle-period case, where each cycle is only one syllable long, one needs: a) a steady-state measure of pre-perturbation behavior that includes approximately 5–10 cycles; b) 1–2 (occasionally 3) more cycles during which the perturbation is applied, at least in our previous experiments (Saltzman, 1992; Saltzman et al., 1991) and in the experiment reported below; c) several more cycles (occasionally none) in order to settle back to within a criterion degree of closeness to the pre-perturbation behavior; and d) approximately 2 – 10 cycles to provide a steady-state measure of post-perturbation behavior.

The second methodological point hinges on the fact that normal speech does not consist of extended, rhythmic repetitions of a single syllable. Therefore, in order to be sure that central phase shifts identified using phase-resetting techniques actually reflect processes governing normal utterances, it is necessary to bridge the theoretical gap between phase-resetting results and those obtained from perturbing discrete, word-like sequences, e.g., /pəsæpæpl/. Because of the relatively short duration of such discrete sequences, the system cannot be relied upon to settle down and to "shake off" the effects of the perturbation in the time between the offset of the perturbation and the end of the utterance. In effect, one can reliably study only the transient responses to perturbations in such sequences. Thus, in order to relate steady-state, phase-resetting data meaningfully to transient data obtained by perturbing discrete utterances, it is necessary to study the transient responses of *repetitive* as well as discrete utterances, preferably using a within-subject experimental design. Once the relation between the steady-state and transient patterns is understood for the repetitive data, a conceptual link can be forged between the transient patterns of the repetitive and discrete data, and shared dynamical principles governing articulatory behavior can be identified.

Earlier work using this experimental paradigm has focused on intergestural timing changes that occur across syllables between successive bilabial closing gestures for /p/ during both the discrete utterance /pəsæpæpl/ and the repetitive utterance /...pæpæpæ.../ (Saltzman, 1992; Saltzman et al., 1991). The data described below represent preliminary results from a study (Saltzman, Löfqvist, Kinsella-Shaw, & Rubin, in preparation) that generalizes this approach to laryngeal devoicing gestures as well as to bilabial gestures. In addition, the study examines these same utterances for changes in intergestural timing that occur intrasegmentally between the bilabial closing and laryngeal devoicing gestures for /p/, and across syllables between successive laryngeal devoicing gestures for /p/. This study consists of data from two subjects, each of whom participated in two sessions. In each session, subjects produced alternating blocks of the repetitive sequence /...pæpæpæ.../ and the discrete sequence /pəsæpæpl/.

The preliminary data described below are of the first session's repetitive data for one subject (ES), and include 70 perturbed and 16 control trials. The data for the discrete sequence spoken by this subject (both sessions) have been described elsewhere (Saltzman et al., 1992), and are used here for comparisons with the repetitive data.

II. GENERAL EXPERIMENTAL METHODS: REPETITIVE DATA

A. Equipment and Data Processing

The subject sat in an adjustable dental chair, with his head restrained in an external frame. A small paddle connected to a torque motor was placed on the lower lip with a tracking force of 3 gm, in order to deliver step pulses of downward force (50 gm) at random times during the experimental trials. Timing of perturbation onset was controlled by a VAXstation II/GPX. Oral articulatory movements were measured optoelectronically using infrared light-emitting diodes mounted on the upper lip, lower lip, lip paddle, nose (the nose LED acted as a spatial reference), and a custom-made jaw splint. Laryngeal abduction and adduction movements were recorded using a transillumination technique in which a fiberoptic endoscope was introduced through the nose and placed in the pharynx in order to illuminate the larynx. The amount of light passing through the glottis, which depends of the degree of laryngeal opening, was detected by an optical sensor placed on a neck collar just below the cricoid cartilage. During the experimental session, the transilluminated larynx was displayed on a video monitor to ensure that the view of the larynx was unobstructed and that the endoscope's lens was not fogged. The acoustic speech signal and control voltage applied to the torque motor were recorded. All data were fed into a 16-track FM tape recorder for later digitization and signal processing.

B. Protocol

Twelve blocks of 25 trials were performed during each of two sessions. Each session lasted approximately 3 hours each. Blocks alternated between *repetitive* and *discrete* experimental conditions. In the discrete condition, each trial consisted of the sequence /pəsæpæpl/; in the repetitive condition, each trial consisted of a sequence of approximately 20-30 repetitions of the syllable /pæ/, spoken at a syllable rate comparable to that used in the discrete trials. Details of the protocol for the discrete-blocks condition are described in Saltzman et al. (1992); details of the protocol for the repetitive-blocks condition are described below.

For the repetitive blocks, perturbations were delivered during a random sampling of 80% of the trials; perturbation duration was preset in an external timing circuit to equal the subject's average syllable duration measured during pretest repetitive trials. On each perturbation trial, the perturbation was delivered during the *n*th syllable (n varied randomly from 8-11), and after *m*% of the predetermined syllable duration (m varied randomly from 1-100). Task instructions were to not actively resist the perturbation, and to return to a steady rhythm similar to that produced before the perturbation as quickly and easily as possible.

C. Definition of Cycle Types

Bilabial and laryngeal movements were analyzed using a lip-aperture (LA) trajectory that was defined by subtracting the upper lip signal (UL) from the lower lip signal (LL), i.e., LA = LL − UL (Figure 1, top panel), and a glottal opening trajectory that was defined using the transillumination signal (Figure 1, middle panel). Individual cycles were then defined between successive peak openings, and four cycle types were identified:

a) *pre-perturbation* cycles included the trial's first cycle through the last cycle before perturbation onset;
b) *perturbation* cycles included all cycles that overlapped the perturbation interval;
c) *transient* cycles were defined as those cycles following the perturbation during which cycle periods deviated from the average pre-perturbation cycle period by more than approximately 2.5 standard deviations; and
d) *post-return* cycles were defined from the last transient cycle to the end of the trial.

Analyses were limited to a maximum of 20 cycles for any given trial.

III. BILABIAL DATA

A. Phase-resetting Analyses

Cycle phase, ϕ, was defined to be zero at all peak bilabial openings. For all other points between peak openings, phase was defined as (t / T_i), where t is the time (in secs) from the most recent peak preceeding a given event of interest, and T_i is the period (in secs) of the cycle containing the event. Thus, phase values of events occurring in a given cycle were defined in a normalized range from zero to one. In the perturbation trials, the phase of perturbation delivery was defined with respect to the time of perturbation offset. This offset time served as a temporal anchoring point for "strobing" both backward and forward in time into the pre-perturbation and post-return cycle sequences, respectively, using the average pre-perturbation cycle period to define the strobe period. The within-cycle strobe phases from the pre-perturbation and post-return cycles were then averaged to define an *average old phase*, ϕ_{old}, and *average new phase*, ϕ_{new}, respectively. *Phase shift*, $\Delta\phi$, was then defined as $(\phi_{new} - \phi_{old})$ (modulo 1). Thus, $\Delta\phi$ is the amount that a given trial's post-return rhythm has been shifted relative to its pre-perturbation rhythm ($\Delta\phi > 0$ denotes phase advance; $\Delta\phi < 0$ denotes phase delay). The same measures were obtained for the control (no perturbation) trials, where calculations were anchored to the end of a randomly timed, but not delivered, "phantom perturbation."

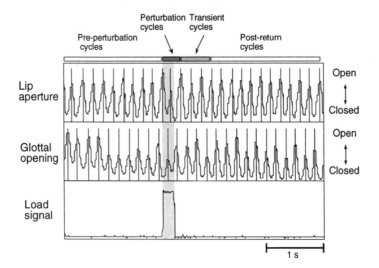

FIGURE 1. Data trajectories for a single repetitive trial: lip aperture (top panel), glottal opening (middle panel), and torque load signal (bottom panel). The sign of the loading signal is inverted in the figure to emphasize that the downward force on the lower lip acted to increase lip aperture. The boxes above the top panel mark the different cycle types for the trial.

Figure 2 illustrates the results of our analyses (one subject, one session), using $\Delta\phi$ data from the perturbation trials that were binned and averaged according to intervals of ϕ_{old}. For the control trials, $\Delta\phi$ data were simply pooled across all old phases of the "phantom perturbations." T-tests were computed for each bin to test whether the perturbation-induced phase shifts differed from control values. To adjust for an elevated Type I error rate due to multiple comparisons, α - levels were selected by dividing 0.01 and 0.05 by the number of comparisons made. As can be seen in the figure, the rhythm showed a phase advance in the 0.2–0.4 interval that was significantly different from the no-perturbation control trials ($p < 0.05$). This pattern replicates the bilabial phase-resetting results found in earlier studies (Saltzman, 1992; Saltzman et al., 1991), and lends further support to the hypothesis that central intergestural dynamics are sensitive to appropriately timed mechanical perturbations of the articulatory periphery, and that such events can permanently reset the rhythms of such central "clocks."

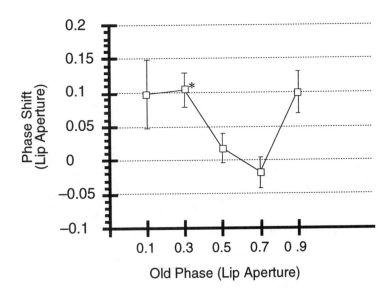

FIGURE 2. Phase shifts ($\Delta\phi$) of lip aperture trajectories (y-axis), binned and averaged according to average old phase (ϕ_{old}) values (x-axis). Bin labels represent the centers of these bins. Each point represents the average phase shift associated with a given bin. Error bars denote standard errors.

B. Transient Analyses

What are the origins of these steady-state phase shifts? One simple hypothesis is that most of these shifts are induced during the application of the perturbation. Thus, we focused next on the timing/phasing changes that occur during the first and second *perturbation cycles* (i.e., the first two cycles that overlapped the perturbation interval; see Figure 3), and their relation to steady-state phase-resetting. As with the phase-resetting analyses described earlier, control trial values were calculated for first and second "perturbation cycles" that were defined by randomly timed, but not delivered, "phantom" perturbations.

First perturbation cycle

For each trial, the duration of the bilabial first-perturbation cycles (dur_1) was normalized with respect to the session's average control "dur_1" values, using the formula (experimental − $\overline{control}$ / $\overline{control}$). Figure 4 (open squares) displays these duration change data in percentage form after binning and averaging according to the time of perturbation onset, which was normalized using the formula: (perton − t_{on_1}) / $\overline{prepert}$, where perton = onset time of perturbation, t_{on_1} = onset time of first-perturbation bilabial cycle, and $\overline{prepert}$ = the average pre-perturbation bilabial cycle duration for the trial. Protected t-tests indicated that cycle durations were significantly shortened relative to controls for the first ($p < 0.01$) and second ($p < 0.05$) bins, and significantly lengthened for the final bin ($p < 0.01$).

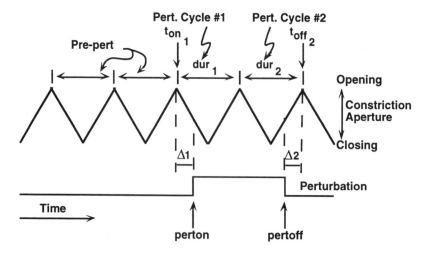

FIGURE 3. Schematic display of constriction (bilabial or glottal) aperture and perturbation trajectories for preperturbation cycles, and for perturbation cycles #1 and #2.

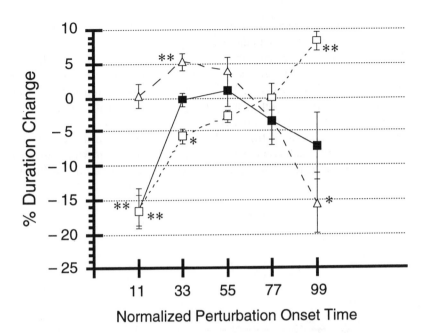

FIGURE 4. Percent durational change (100 × [experimental – $\overline{\text{control}}$] / $\overline{\text{control}}$) for the lip aperture trajectories' first perturbed cycles (open squares), second perturbed cycles (open triangles), and first-plus-second perturbed cycles (closed squares), binned and averaged according to normalized perturbation onset time (100 × [perton – t_{on_1}] / $\overline{\text{prepert}}$). Bin labels represent the centers of these bins. Error bars denote standard errors.

Second perturbation cycle

For each trial, the duration of the second-perturbation bilabial cycle (dur_2) was similarly normalized with respect to the session's average control "dur_2" values, using the formula (experimental – $\overline{\text{control}}$) / $\overline{\text{control}}$. Figure 4 (open triangles) displays these data after binning and averaging according to the same time base used for the first-perturbation bilabial cycle. Protected t-tests indicated that cycle durations were significantly lengthened relative to controls for the second ($p < 0.01$) and final ($p < 0.05$) bins, and significantly shortened for the final bin ($p < 0.01$).

C. Relationship between Steady-state and Transient Data

In order to test the hypothesis that the phase shifts observed in the steady-state were attributable to duration changes induced during the first two perturbation cycles, we first summed the normalized durations for each trial's first- and second-perturbed bilabial cycles, and binned these values using the same time base as in the previous separate analyses for these cycles (see Figure 4, filled squares). Protected *t*-tests indicated that the summed perturbation cycle durations were significantly different (shorter) than the corresponding nonperturbation controls only in the first bin ($p < 0.01$).

We then rebinned, reanalyzed, and replotted the steady-state phase shifts using this same normalized time base (as opposed to the $\overline{\phi_{old}}$ time base originally used earlier in Section III.A; see Figure 2). In Figure 5, we have compared the steady-state data curve (open triangles) with the summed perturbation cycles' curve (open squares); these curves are virtually mirror images of each other, with significant effects only in the first bin. Given the opposite sign conventions used to define phase shifts and duration changes, these curves are in essence identical, indicating that the steady-state phase shifts are indeed induced in the first two perturbation cycles.

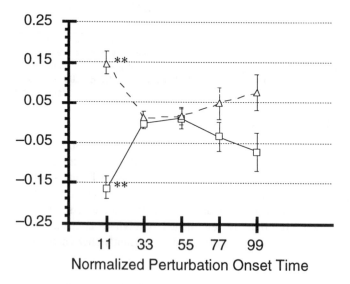

FIGURE 5. Lip aperture trajectory phase shifts ($\Delta\phi$: open triangles) and fractional duration changes ([experimental − control] / control: open squares) for the first-plus-second perturbed cycles (y-axis), binned and averaged according to normalized perturbation onset time ($100 \times$ [perton − t_{on_1}] / prepert) (x-axis). Bin labels represent the centers of these bins. Error bars denote standard errors.

IV. LARYNGEAL DATA

A. Phase-resetting Analyses

For the laryngeal trajectories, cycle type definition, phase-shift computation, binning and averaging of the $\Delta\phi$ data according to intervals of $\overline{\phi_{old}}$, as well as statistical testing, proceeded as with the lip aperture analyses. Figure 6 shows that the laryngeal rhythm showed a steady-state phase advance in the $0.6 - 0.8$ $\overline{\phi_{old}}$ interval that was significantly different from the no-perturbation control trials ($p < .01$).

   Note that the sensitive old phase intervals differed for the laryngeal (Figure 6) and the bilabial (Figure 2) data. This was simply because the laryngeal and bilabial cycles are themselves out of phase with one another (see the data trajectories in Figure 1). As will be shown below in Section V.A, when both phase shifts are plotted on a common time base, it becomes clear that the phase shift behaviors of lips and larynx are virtually identical.

B. Transient Data; Relationship between Steady-state and Transient Data

As with the bilabial data, we computed and summed the normalized duration changes for each trial's first- and second-perturbed laryngeal cycles, and binned these values using the normalized time of perturbation onset for the trial. Again, as with the bilabial data, time normalization used the formula (perton $-$ t_{on1}) / $\overline{prepert}$, where perton = onset time of perturbation, t_{on1} = onset time of the first-perturbation laryngeal cycle, and $\overline{prepert}$ = the average duration of the laryngeal pre-perturbation cycles for the trial. Protected t-tests indicated that the summed perturbation cycle durations were significantly different (shorter) than the corresponding nonperturbation controls only in the third bin ($p < 0.01$; see Figure 7, open squares).

   In order to test the hypothesis that the phase shifts observed in the steady-state were attributable to duration changes induced during the first two perturbation cycles, we rebinned, reanalyzed, and replotted the *steady-state* phase shifts using the same normalized time base (as opposed to the $\overline{\phi_{old}}$ time base originally used above in Section IV.A; see Figure 6), and compared the steady-state data curve (Figure 7, open triangles) with the summed perturbation cycles' curve (Figure 7, open squares). As with the earlier bilabial analyses, these laryngeal curves are virtually mirror images of each other with significant effects only in the third bin. Again, given the opposite sign conventions used to define phase shifts and duration changes, these curves are in essence identical, indicating that the steady-state phase shifts are indeed induced in the first two perturbation cycles.

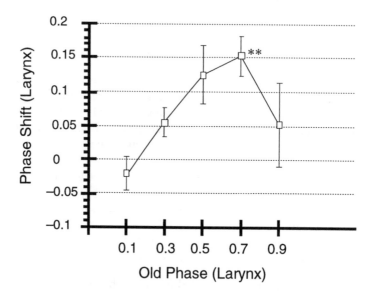

FIGURE 6. Phase shifts ($\Delta\phi$) for the laryngeal opening trajectories (y-axis) binned and averaged according to average old phase ($\overline{\phi_{old}}$) values (x-axis). Bin labels represent the centers of these bins. Error bars denote standard errors.

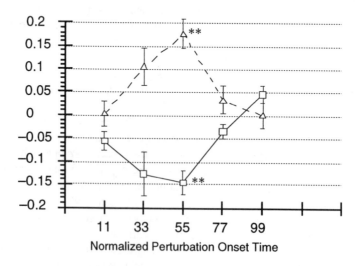

FIGURE 7. Laryngeal opening phase shifts ($\Delta\phi$: open triangles) and fractional duration changes ([experimental − $\overline{control}$] / $\overline{control}$: open squares) for the first-plus-second perturbed cycles (y-axis), binned and averaged according to normalized perturbation onset time ($100 \times$ [perton − t_{on_1}] / $\overline{prepert}$) (x-axis). Bin labels represent the centers of these bins. Error bars denote standard errors.

V. LIP-LARYNX ANALYSES

A. Steady-State Shifts in Relative Phase between Bilabial and Laryngeal Trajectories

The evolution of the relative phase of the bilabial and laryngeal trajectories during each trial was measured in the following manner. Successive peak laryngeal openings were used to define "strobe" events in each corresponding bilabial cycle. Relative phase for each laryngeally-strobed bilabial cycle could then be defined by:

$$\frac{\left[\begin{array}{c} \text{(time of the } i^{\text{th}} \text{ laryngeal event)} \\ - \text{ (time of the preceding bilabial peak)} \end{array} \right]}{\text{(period of the strobed bilabial cycle)}}$$

Average relative phase for each trial's strobed bilabial pre-perturbation cycles was defined computationally as [bilabial $\overline{\phi_{old}}$ – laryngeal $\overline{\phi_{old}}$]; average relative phase for the strobed bilabial postreturn cycles was defined computationally as [bilabial $\overline{\phi_{new}}$ – laryngeal $\overline{\phi_{new}}$]. Steady-state shifts in relative phase for each trial were then defined by the average postreturn relative phase minus the average pre-perturbation relative phase.

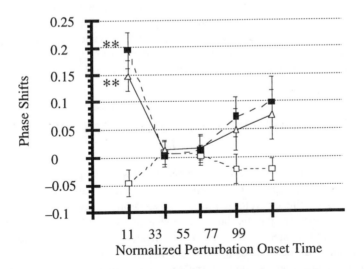

FIGURE 8. Phase shifts for laryngeal opening (closed squares), lip aperture (open triangles), and lip-larynx relative phase (open squares), binned and averaged according to normalized perturbation onset time ($100 \times [\text{perton} - t_{on_1}] / \overline{\text{prepert}}$). Bin labels represent bin centers. Error bars denote standard errors.

For ease of comparison with the results of the previous sections, we binned the data from each trial according to a normalized time base defined by the formula (perton $- t_{on1}$) / $\overline{prepert}$, where perton = onset time of perturbation, t_{on1} = onset time of the first-perturbation bilabial cycle, and $\overline{prepert}$ = the average duration of the bilabial pre-perturbation cycles for the trial. These binned data were averaged and tested against the corresponding mean value obtained in the control trials. As can be seen in Figure 8 (open squares), the perturbations did not induce any significant steady-state shifts in the relative phasing of lips and larynx. This makes sense since, as can also be seen in the figure, the individual phase-resetting behavior of the lips (open triangles) and larynx (filled squares) is virtually identical. When perturbations are delivered at the system's sensitive phase (i.e., the leftmost bin), the bilabial and laryngeal gestures appear to be phase-advanced as a relatively coherent unit, maintaining their relative phasing while they are advanced in absolute time.

B. Transient Data

The transient behavior of the relative phasing between bilabial and laryngeal trajectories was also examined. Transient shifts in relative phase for each trial were defined by the average relative phases computed for the first- and second-perturbed bilabial cycles, minus the average relative phases computed for the pre-perturbation bilabial cycles. We binned the data from each trial according to the same normalized time base used for the steady-state relative phase analyses. These binned data were averaged and tested against the corresponding mean value obtained in the control trials. There were no significant shifts in relative phase induced in the transient cycles (see Figure 9). These data indicate that even while a perturbation is being applied, the system acts so as to maintain the integrity of the bilabial-laryngeal intergestural unit.

VI. COMPARISONS OF REPETITIVE AND DISCRETE DATA

What are the relationships between the patterns described above for productions of the rhythmic, repetitive sequence /...pæpæpæ.../, and those found for productions of the discrete sequence /pəsæpæpl/ (Saltzman et al., 1992)? Because of their very nature, the discrete sequences cannot be counted upon to settle back to a steady-state after perturbation delivery. Consequently, comparisons must be made using transient methods of analysis.

The two measures studied in the discrete sequences that are the most comparable to those studied in the repetitive sequences are: a) the duration of the so-called *vowel cycle* defined between successive maximum bilabial openings for /æ/ in /pəsæpæpl/. Its analog in the repetitive analyses is the second-perturbation cycle; and b) the relative phase of the laryngeal peak inside the bilabial vowel cycle for devoicing the medial /p/. Its repetitive analog is the lip-larynx relative phase for the second-perturbation cycle. For historical reasons (e.g., Gracco & Abbs, 1989), we analyzed perturbation-induced changes in the duration of the discrete vowel cycle by binning the normalized duration change values according to a normalized time base defined by the formula (pertoff − t_{offd}) / $\overline{control_d}$, where pertoff = offset time of perturbation, t_{offd} = offset time of discrete bilabial cycle, and $\overline{control_d}$ = average duration of discrete bilabial cycles from the unperturbed control trials. The comparable time interval in the repetitive data is (pertoff − t_{offr}) / $\overline{control_r}$, where t_{offr} is the offset time of the second-perturbation cycle, and $\overline{control_r}$ is the average duration of unperturbed control second-perturbation cycles.

Comparisons between the discrete (open triangles) and repetitive (open squares) patterns are presented in the following two figures. The changes in bilabial cycle duration are comparable (Figure 10), showing significant lengthening and shortening, respectively, in the earlier and later time bins. The discrete relative phasing data shows significant increases in all bins, whereas the repetitive data show only a nonsignificant trend toward such increases (Figure 11).

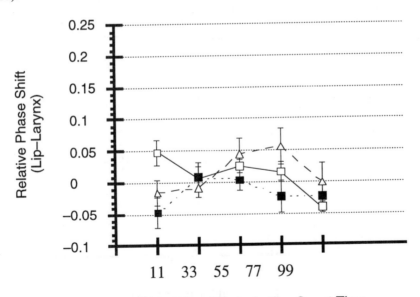

FIGURE 9. Shifts in lip-larynx relative phase for first-perturbed cycles (open squares), second-perturbed cycles (open triangles), and post-return cycles (closed squares), binned and averaged according to normalized perturbation onset time (100 × [perton − t_{on1}] / $\overline{prepert}$). Bin labels represent bin centers. Error bars denote standard errors.

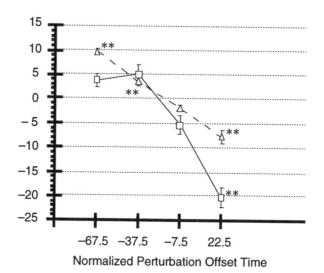

FIGURE 10. Normalized duration changes of lip aperture syllable (discrete data: open triangles; defined by [100 x {experimental − $\overline{control_d}$} / $\overline{control_d}$]) and second-perturbed cycle (repetitive data: open squares; defined by [100 x {experimental − $\overline{control_r}$} / $\overline{control_r}$]), binned and averaged according to their respective normalized time bases ([100 × {pertoff − t_{off_d}} / $\overline{control_d}$] for discrete data; [100 × {pertoff − t_{off_r}} / $\overline{control_r}$] for repetitive data). Bin labels represent bin centers. Error bars denote standard errors.

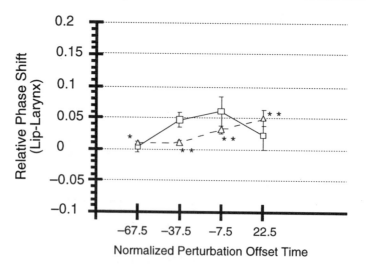

FIGURE 11. Shifts in lip-larynx relative phase for lip aperture syllable (discrete data: open triangles) and second perturbed cycle (repetitive data: open squares), binned and averaged according to their respective normalized time bases ([100 × {pertoff − t_{off_d}} / $\overline{control_d}$] for discrete data; [100 x {pertoff − t_{off_r}} / $\overline{control_r}$] for repetitive data). Bin labels represent bin centers. Error bars denote standard errors.

VII. DISCUSSION/CONCLUSIONS

The data described above provide several hints regarding the nature of the dynamics that underlie intergestural temporal cohesion, both within segments and between successive syllables. First, the steady-state phase shifts induced by the perturbations during the repetitive utterances (Sections III.A & IV.A) provide further support for the hypothesis that central intergestural dynamics can be "permanently" reset by peripheral articulatory events. These effects occur, however, only when the perturbation is delivered within a particularly "sensitive phase" of the cycle. During this period, the downwardly directed lower lip perturbation is opposing the just-initiated, actively controlled bilabial closing gesture for /p/. If one assumes that efferent commands to the periphery are strongest at gestural initiation, then these results imply that the timecourse of afferent sensitivity mirrors that of efferent strength.

The second hint is provided by the transient analyses of the repetitive utterances (Sections III.B and IV.B). These analyses indicate that, although systematic durational changes are induced in the first two perturbed cycles when perturbations are delivered during the utterances' nonsensitive phases, these changes are simply transient peripheral responses to the perturbation and do not indicate central phase-resetting. For example, Figure 4 shows how a perturbation-induced shortening of the first-perturbed cycle in bin 33 is cancelled by a lengthening of the following second-perturbed cycle; similarly, a lengthening of the first-perturbed cycle in bin 99 is cancelled by a shortening of the following second-perturbed cycle. These systematic but transient durational changes can be understood as cases in which the offset time of the second-perturbed cycle (t_{off_2} in Figure 3) is not altered by the perturbation, although the time of the preceding peak opening is either advanced (bin 33) or delayed (bin 99).

Relatedly, the perturbation-induced steady-state phase shifts are almost totally attributable to changes occurring during the first two perturbed cycles. For example, Figure 4 shows how a perturbation-induced shortening of the first-perturbed cycle in the utterance's sensitive phase bin (bin 11) is followed by no change in the second-perturbed cycle. Thus, the steady-state phase shift can be understood as a case in which the offset time of the second-perturbed cycle (t_{off_2} in Figure 3) and the times of the immediately preceding and all subsequent peak openings are advanced by the same amount. Methodologically, this result suggests that in order to distinguish centrally from peripherally induced durational changes in perturbed speech sequences, it is probably necessary to examine durational changes in at least two successive articulatory intervals (syllables?).

Third, steady-state and transient analyses of the repetitive data both support the hypothesis of greater intergestural temporal stability within segments/phonemes than between syllables. This conclusion is based on the existence of comparable steady-state phase resetting curves for the bilabial and laryngeal trajectories (Figure 8), and the absence of shifts in the relative phasing of lips and larynx in the both the steady-state and in the first- and second perturbed cycles (Figure 9). Thus, even when the bilabial and laryngeal gestures of the /p/s in successive syllables are shifted in time relative to one another by the perturbation, they are shifted as a coherent unit.

Finally, comparisons of the discrete and repetitive data suggest that intergestural temporal cohesion between successive syllables is comparable in the two types of utterance (Figure 10). However, intergestural cohesion within segments/phonemes may be greater for the repetitive sequences (Figure 11). This difference may be attributable to the sustained, rhythmic nature of the repetitive sequences and/or to differences in the within-syllable position of the bilabial stop /p/ in the repetitive (syllable initial position) and discrete (syllable final position) sequences. Differentiating among these possibilities is a task for future research.

ACKNOWLEDGMENTS

This work was supported by grant support from the following sources: NIH Grant #DC-00121 (Dynamics of Speech Articulation) and NSF Grant #DBS-9112198 (Phonetic Structure Using Articulatory Dynamics) to Haskins Laboratories.

REFERENCES

Gracco, V. L., & Abbs, J. H. (1989). Sensorimotor characteristics of speech motor sequences. *Experimental Brain Research, 75,* 586-598.

Glass, L. G., & Mackey, M. C. (1988). *From clocks to chaos: The rhythms of life.* Princeton, NJ: Princeton University Press.

Kawato, M. (1981). Transient and steady state phase response curves of limit cycle oscillators. *Journal of Mathematical Biology, 12,* 13-30.

Kay, B. A. (1986). *Dynamic modeling of rhythmic limb movements: Converging on a description of the component oscillators.* Unpublished doctoral dissertation, Department of Psychology, University of Connecticut, Storrs, CT.

Kay, B. A., Saltzman, E. L., & Kelso, J. A. S. (1991). Steady-state and perturbed rhythmical movements: A dynamical analysis. *Journal of Experimental Psychology: Human Perception and Performance, 17,* 183-197.

Kollia, H. (1994). *Functional organization of velar movements following jaw perturbation.* Unpublished doctoral dissertation, Department of Speech and Hearing Sciences, City University of New York.

Lee, R. G., & Stein, R. B. (1981). Resetting of tremor by mechanical perturbations: A comparison of essential tremor and Parkinsonian tremor. *Annals of Neurology, 10,* 523-531.

Lennard, P. R., & Hermanson, J. W. (1985). Central reflex modulation during locomotion. *Trends in Neuroscience, 8*, 483-486.

Löfqvist, A., & Gracco, V. L. (1991). Discrete and continuous modes in speech motor control. *Perilus XIV, Institute of Linguistics, University of Stockholm, Stockholm, Sweden*, 27-34.

Munhall, K. G., Löfqvist, A., & Kelso, J. A. S. (1994). Lip-larynx coordination in speech: Effects of mechanical perturbations to the lower lip. *Journal of the Acoustical Society of America, 95*, 3605-3616.

Saltzman, E. L. (1992). Biomechanical and haptic factors in the temporal patterning of limb and speech activity. *Human Movement Science, 11*, 239-251.

Saltzman, E., Kay, B., Rubin, P., & Kinsella-Shaw, J. (1991). Dynamics of intergestural timing. *Perilus XIV* (pp. 47-56). Institute of Linguistics, University of Stockholm, Stockholm, Sweden.

Saltzman, E., Löfqvist, A., Kinsella-Shaw, J., Rubin, P., & Kay, B. (1992). A perturbation study of lip-larynx coordination. In J. J. Ohala, T. M. Nearey, B. L. Derwing, M. M. Hodge, & G. E. Wiebe, (Eds.), *Proceedings of the International Conference on Spoken Language Processing (ICSLP '92): Addendum* (pp. 19-22). Edmonton, Canada: Priority Printing.

Winfree, A. T. (1980). *The geometry of biological time.* New York: Springer-Verlag.

32 Recovering Task Dynamics from Formant Frequency Trajectories: Results Using Computer "Babbling" to form an Indexed Data Base

Richard S. McGowan
Haskins Laboratories

INTRODUCTION

The recovery of vocal tract articulation from speech acoustics, the inverse problem, has been a long-standing problem. Many of the chapters in this book are concerned with the forward problem: How do humans move their articulators to produce the sounds heard as speech? The question here is: Can vocal tract articulation be inferred from the speech signal using machines?

The purpose of the project described in this chapter is to find the extent to which it is possible to recover human articulatory movement from the speech signal. First, it is important to examine the idea of recovering articulation from acoustics. An attempt could be made to recover the area of the vocal tube as a function of distance of a constriction from the lips, but this would tell little of articulation. An articulatory model is needed so that the relations between the area function and aerodynamics and human physiology are defined. What kind of articulatory model is to be used as the domain of the recovered articulation? Is the knowledge required that of the exact path of rigid articulators, such as the jaw, and the exact shape of soft structures, such as the tongue, at every instant? Is even more information sought, such as the muscle activity that causes motion and shape change? It seems unlikely that such details of articulation will be recovered by a general method in the near future because detailed articulatory models of the entire vocal system do not exist. Even if they did exist, there may be theoretically derivable limits to the amount of detail that can be recovered from the speech signal.

The articulatory models that currently exist contain a relatively low number of degrees of freedom (Coker, 1976; Maeda, 1982; Mermelstein, 1973). Because of the known limitations of these models, such as constraining motion to the midsaggital plane, they can only caricature articulatory movements. In the work described here, the recovery is attempted at a level of description even further removed from the movement of the individual articulators and the details of physiology. This level is the parameter space of task dynamics of the tract variables, known here, simply, as task dynamics (Saltzman, 1986; Saltzman & Kelso, 1987). Task dynamics is instantiated using a present-day articulatory model (Mermelstein, 1973), and it describes the coordinated activity of these model articulators in forming and breaking constrictions in the vocal tract during speech (Saltzman & Munhall, 1989).

It has been argued that a description of a vocal tract based on constrictions is closer to its speech acoustics than a description based on individual articulators (e.g., Boë, Perrier, & Bailly, 1992; McGowan, 1994). Acoustic features, such as resonance frequencies, are more sensitive to the placement and degree of the maximally constricted regions than to any other part of the area function. While task dynamic positions are specified in terms of constriction degree and location, this is not true of an articulatory position specification. Because the task-dynamic level is apparently closer to the resulting acoustics than any of the other articulatory levels discussed, it should be the most readily recoverable from the acoustics among these levels. Also, using task dynamics in the solution of the inverse problem is consistent with the historical trend of using as many constraints as possible to map from the acoustic domain to articulation. For instance, an articulatory model is often employed (e.g., Flanagan, Ishizaka, & Shipley, 1980) instead of an arbitrary area function (e.g., Wakita, 1973) to constrain the articulatory domain. Further, the constraint that articulatory movement be continuous can be imposed. Other constraints arise when the movement of the articulators is assumed to have a particular functional form, such as that of a damped sinusoid (e.g., Parthasarthy & Coker, 1992). By using task dynamics, the movements of the tract variables are constrained to a particular functional form, that of a damped sinusoid.

Recent work on recovering task dynamics from acoustics has shown some promise in model tests (McGowan, 1993, 1994). (It is recommended that the second reference be reviewed by the reader.) These model tests were performed running a program called ASYINV, using formant frequency data produced by an articulatory synthesizer (Rubin, Baer, & Mermelstein, 1981) that incorporates Mermelstein's model vocal tract (Mermelstein, 1973). The data were produced by writing gestural scores (Browman & Goldstein, 1990) that specify task-dynamic parameters, from which the movement patterns of the articulators in the model vocal tract could be computed. Because each tract variable is assumed to have second-order dynamics, the gestural scores specified, among other things, natural frequency, damping, target position, and the activation intervals over which the these second-order specifications were active. There were constraints imposed to reduce the number of parameters. All systems were assumed to be critically damped, and the duration of activation intervals were always greater than 100 ms and assumed to be equal to the period of oscillation based on the natural frequency. The tract variables used in these tests came in three groups of two, one of each group describing constriction location, and the other constriction degree (Saltzman & Munhall, 1989). One group comprised lip protrusion (LP) and lip aperture (LA); another group, tongue body constriction location (TBCL) and tongue body constriction degree (TBCD); and the third group, tongue tip constriction location (TTCL) and tongue tip constriction degree (TTCD). Each group had the same activation intervals. Thus, within each group only the target specifications for place and degree of constriction were independently specified. The same model vocal tract that produced the data was used in an analysis-by-synthesis procedure to find the optimum task-dynamic parameters--optimum in the sense of obtaining formant frequency trajectories that matched the data as closely as possible in a least-squares sense. The results of these tests encouraged some further research using this approach, and the procedure of testing by using the same models that produced the acoustic data to recover the articulation was retained here.

The recovery program, ASYINV, depends on using constrained models that have been derived from human motor control (task dynamics) and anatomy (Mermelstein's articulatory model). Beyond these biological models, biological analogies can be found in other parts of ASYINV. While the forward problem of speech synthesis is solved using physical law to relate tube shape and aerodynamics to sound, a nondeterministic search procedure is necessary to solve the inverse problem. The program ASYINV employs a genetic algorithm as such a procedure as an alternative for standard optimization procedures. This algorithm is described in a text by Goldberg (1989) and the practical reasons for using this procedure for the inverse problem can be found in McGowan (1994). In this algorithm, the individuals of a population are assigned randomly chosen task-dynamic parameter sets that are coded into binary strings called "chromosomes," and each is assigned a fitness. The fitness used in the present work was the inverse of the sum of square differences between the data formant frequency and the synthesized formant frequency. The sum was taken over the first three formant frequencies in 10 ms intervals for the duration of the utterance. Individuals are chosen to breed with others to form a new population of chromosomes, with the probability of being chosen made equal to each individual's fitness divided by the sum of the fitnesses of the other individuals.

When two individuals mate their chromosomes split at a randomly chosen location with each of two progeny obtaining one part of their chromosome from each parent. The children's fitnesses are evaluated based on their parameter sets as coded in their chromosomes. That is, the task-dynamic model is run based on the parameters specified by each child's chromosome, and the resulting fitness is computed for each child. A small probability of mutation is allowed. Most importantly, this procedure is not strictly an optimization procedure, but it is an adaptive procedure. Given the current population of proposed solutions, it leads to plausible regions in the parameter space that provide better adaptation, even if the fitness function is not stationary. Such an adaptive procedure should be useful if, as Studdert-Kennedy (1991) argues, language development can be "...cast fruitfully in an evolutionary and recapitulatory framework..." (p. 24) with attendant differentiation and assembly for adaptation.

The idea of using biological analogies and models appears reasonable for the inverse problem because it can be argued that human children, especially blind children, in learning to talk must recover articulation from speech. (It will be argued later that children, in some ways, have a less daunting task than recovering the detailed articulation of individual talkers from speech acoustics.) Children are equipped with a mechanism similar enough to other talkers so that they are able to do this (just as ASYINV has a model vocal tract), and in the terminology of learning systems they have an internal model (Holland, 1992; Jordan & Rumelhart, 1992). Not only does the child have a similar mechanism for sound production, but he has plenty of practice through babbling and interaction with others, and babbling sounds have commonalities with adult utterances (e.g., /CV/ constructions). In fact, learning to talk is delayed when a hearing child cannot babble (e.g., Bleile, Stark, & McGowan, 1993). The child also can use experience to help build a mapping between articulation and the resulting sound. While an internal model has been incorporated into the method of articulatory recovery in the previous work, the work to be described here continues the trend of using human-like constraints by allowing "babbling" and access to a data base of previous babbling activity. The quotes are used because the babbling performed here was a random search of the task-dynamic parameter space, while human babbling is more constrained. Also, the term data base is used instead of memory because the formation and access of the computer data base here has little resemblance to human memory (Rose, 1992).

Building a data base from known task dynamic-acoustic pairs was also warranted strictly on practical grounds, because the function evaluations can be computationally costly (say 2 seconds to run task dynamics simulating a 400 ms utterance on a DEC 3000), and a sufficient number of function evaluations needs to be performed to sample the parameter space properly (on the order of 2,000 to 16,000 evaluations in the previous work with a genetic algorithm). It appeared to be prudent to save task dynamic-acoustic pairs from function evaluations (i.e. syntheses) for future use. The task dynamic-acoustic pairs would facilitate something of a codebook lookup as used by Schroeter, Meyer, and Parthasarthy (1990). That is, the task-dynamic parameter recovery might be enhanced by accessing the data base of function evaluations to help in optimization/adaptation.

In these experiments it was hypothesized that providing the initial population with individuals whose acoustics closely matched that of the data would accelerate the optimization/adaptation. To test this hypothesis, a system for creating a reservoir of task dynamic-acoustic pairs, along with a method for recognizing similarities with the acoustic data needed to be devised. These procedures will now be described, as well as the results of some testing. The emphasis in the testing was to find out whether building a data base of task dynamic-acoustic pairs could help in recovering task dynamics. Measures of the efficiencies of various procedures were not emphasized.

Procedure

The recovery of task-dynamic parameters of four utterances was attempted under various conditions. These utterances were specified by gestural scores for /ədæ/, /əbæ/, /əbi/, and /ədi/ modified to meet constraints imposed by the author, and they are referred to in their unmodified phonetic transcription. The constraints were the same as those in previous work, as described above (see also McGowan, 1994). These scores were used as input to the task-dynamic simulation in order to create four different sets of formant trajectories that served as data in various recovery processes. The genetic algorithm as implemented for the recovery procedure was very close to the Simple Genetic Algorithm described by Goldberg (1989, pp. 59-70), as modified by McGowan (1994). There was only one change to the genetic algorithm from its previous implementation: the binary strings were decoded as Gray code (Forrest, 1993). Gray code ensures that nearest neighbors in an integer representation map to nearest neighbors in a binary representation.

The program used in the previous numerical experiments, ASYINV, was modified so that data structures containing task-dynamic chromosomes and formant frequency trajectories could be stored in an indexed file. These indexed files constituted the data bases in this work. An indexed file allows access to the data structures by means of character-coded keys, here known as acoustic keys. The number of formants tracked, in this case always 3, and the length of the formant tracks in tens of milliseconds, in this case always 39, were recorded in the first 6 characters of the key. The next three characters recorded the direction of change of each formant frequency from 90 ms to 390 ms: increasing, decreasing, or steady. The reason this time interval was chosen was that in all four test utterances the constriction was released at 90 ms and the following vowel extended through 390 ms. A formant frequency was declared steady if changes were less than 10%, and otherwise it was declared increasing or decreasing, depending on the direction of change. The next 6 characters of the acoustic key quantified the changes in 10% intervals. Thus, the key for /ədæ/ was 390300IDS201010, the key for /əbæ/ was 390300III202010, the key for /əbi/ was 390300III202020 and the key for /ədi/ was also 390300III202020.

Because four different utterances were recovered sharing common indexed files, it was necessary to make the chromosomes a constant length and the decoding of the parameters consistent. Otherwise, the chromosome length and decoding method would have provided extra knowledge as to the values of the task dynamic parameters. In all chromosomes, it was assumed that the lip tract variables LA and LP were activated at most once with the LP target treated as known, that the tongue body tract variables, TBCL and TBCD were activated at most once, and that the tongue tip tract variables TTCL and TTCD were activated at most once. There was a bit attached to each of the groups, lip, tongue body, and tongue tip, that coded whether each of these groups actually was activated. For instance, the gestural score for /əbæ/ did not activate the tongue tip tract variables, and it was possible to express this during recovery in a bit in the chromosome (McGowan, 1993). Also, there were instances where groups were activated more than once in the utterance creating the acoustic data. In these instances one of the activations was treated as known. For instance, the gestural score for /ədi/ had two activations of the tongue tip group, and in the recovery process one of these was assumed to be known. The ranges and resolutions of the parameter values that were covered by the chromosomes are shown in Table 1 (LP is omitted from Table 1 because it is predetermined).

The recovery procedure was run under five different conditions and under each condition, eight times for each utterance. In all of the conditions, the initial population size was 100, and the genetic algorithm was allowed to run for 60 generations with a probability of mating set to 0.6 and the probability of random mutation set to 0.001. The first condition was with the initial population produced randomly without an indexed file, just as in the previous work (McGowan, 1993, 1994), except that the initial population size was larger in this study. The other four conditions included individuals into the initial population from an indexed file derived from a previous session of random babbling.

TABLE 1. *Target value specifications.*

Tract Variable	Maximum/Minimum Target Value	Number of Bits in Chromosome	Resolution
LA	2.33/–0.383 cm	6	0.050 cm
TBCL	3.16/0.51 rad	6	0.042 rad
TBCD	1.63/–0.13 cm	6	0.028 cm
TTCL	1.17/0.40 rad	6	0.012 rad
TTCD	2.15/–0.65 cm	6	0.044 cm

These conditions were the result of combining two different factors: indexed file size and the detail of matching with the data acoustic key required to be taken from the indexed file and included into the initial population. There were two indexed files generated from random babbling. (The same constraints and ranges applied to the task dynamic parameters of the babbled utterances as to those of the original utterances producing the acoustic data.) The small file contained 11,156 individuals and the large file, about ten times as large, contained 111,596 individuals. The other factor that was varied was the number of characters in an individual's acoustic key needed to match the data acoustic key to be included into the initial population, either 9 or 15. With a match of 9 characters in the key, there needed to be agreement in the directions of all three formants, as well as length of utterance and number of formants recorded (and all individuals matched in the two latter parameters). This condition was called the unquantified formant trajectory matching condition. With a match of 15 characters there was the further requirement that the formant trajectories match in the designated 10% intervals. This condition was known as the quantified formant trajectory matching condition. If there were more than 100 individuals in memory that met the matching condition, the 100 individuals that had formants trajectories that matched those of the data best, in the sense of high fitness, were chosen for the initial population. In the cases where there were fewer than 100 individuals that produced the required match, the remaining initial population was filled with randomly generated individuals.

Results

The results of these babbling experiments are reported as the averages of maximum fitnesses found in the populations as a function of generation. The averages are over the eight trials of each utterance without the use of an indexed file, with the use of the small and large indexed file, both with unquantified and quantified formant trajectory matching.

Figures 1(a) and 1(b) show that both large and small indexed files helped increase maximum fitness during the recovery of utterance /ədæ/. In the case of the small indexed file the quantification of the formant trajectory change was not helpful, and in the case of the large indexed file it was actually a hindrance. The large indexed file condition outperformed the small indexed file condition in the final generations for unquantified formant matching. In the four conditions using an indexed file, all initial individuals were drawn from that file, even in the case of a small file requiring a match in quantified formant trajectories (see Table 2). This is unlike what happened for the other utterances.

The other utterances are interesting because their acoustic keys are identical in the first 9 characters (unquantified formant trajectory matching), and in the case of /əbi/ and /ədi/, there was a match in all 15 characters (quantified formant trajectory matching). Those utterances that shared the required number of key characters drew from the same pool of candidate individuals to be included into the initial population. If there were more than 100 such individuals matching the acoustic keys of two utterances to be recovered, then the initial populations for these two were not necessarily the same, because only the best 100 matches in terms of fitness with the data were taken in this case. However, when there were fewer than 100 matches, the same individuals were taken from the indexed file to be included in the different initial populations. Because there were fewer than 100 individuals taken from memory, /əbi/ and /ədi/ shared the same individuals taken from the small indexed file for both the unquantified and quantified formant trajectory conditions, and the utterances /əbæ/, /əbi/ and /ədi/ shared the same individuals taken from small memory in the unquantified formant trajectory condition.

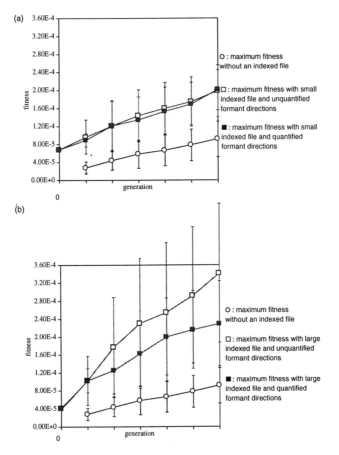

FIGURE 1. (a) Average maximum fitness as a function of generation number for /ədæ/ recovery with the small indexed file; (b) average maximum fitness as a function of generation number for /ədæ/ recovery with the large indexed file.

TABLE 2. *Number of individuals taken from memory.*

Utterance	number taken from the small indexed file for unquantified formant trajectories	number taken from the small indexed file for quantified formant trajectories
/ədæ/	100	100
/əbæ/	51	16
/ədi/	51	15
/ədi/	51	15

Utterance	number taken from the large indexed file for unquantified formant trajectories	number taken from the large indexed file for quantified formant trajectories
/ədæ/	100	100
/əbæ/	100	100
/ədi/	100	100
/ədi/	100	100

The results for /əbæ/ were interesting partly because of the sizable standard deviations (Figures 2(a) and 2(b)). The standard deviations in the fitnesses were on the same order of magnitude as the fitnesses themselves in the cases of no indexed file, and still very substantial for the small indexed file conditions. A detailed look at individual recoveries revealed that there was a tendency in these conditions to sometimes get trapped in local fitness maxima where the tongue tip was used to form the initial consonantal constriction instead of the lips. (For example, this happened 3 out of 8 times for the small indexed file, unquantified formant trajectory matching condition.) The cases where correct initial bilabial closure was attained after 60 generations attained a much higher fitness. Note that Table 2 shows that only 51 individuals were taken from the small indexed file in the unquantified formant trajectory matching condition and 16 in the case of the quantified formant trajectory matching condition. Based on the trials with the small indexed file, it was impossible to tell definitively whether an indexed file may have helped to alleviate the problem of attaining bad local maxima in the fitness function. However, it is obvious that this small indexed file did not expedite the growth of fitness for this utterance on the average. The large indexed file condition helped to sort things out. The unquantified formant trajectory matching condition actually did worse with this particular large indexed file than with the small indexed file, and still with large standard deviations through the generations. Again, 3 out of 8 times the tongue tip was used instead of the lips to form the initial constriction. The quantified formant trajectory matching condition with the large indexed file helped to improve performance above that of all other conditions for /əbæ/. The standard deviations about the average maximum fitness were reduced, and there were no instances where the initial bilabial closure was mistaken for a tongue-tip constriction. This was a case where more refined initial characterization of the acoustic signal (i.e., greater acoustic key matching) was helpful. Perhaps this is not surprising, given that /əbæ/ shares the same unquantified formant trajectory key with the utterance /ədi/.

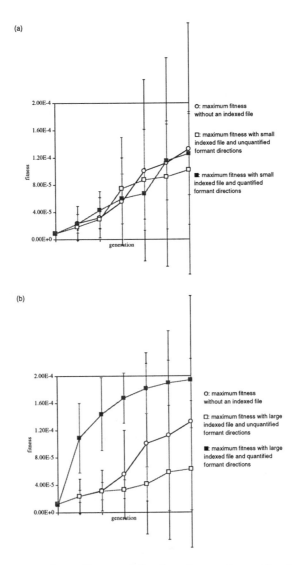

FIGURE 2. (a) Average maximum fitness as a function of generation number for /əbæ/ recovery with the small indexed file; (b) average maximum fitness as a function of generation number for /əbæ/ recovery with the large indexed file.

For the utterance /əbi/ there was clear improvement over the completely random initial population only in the condition of the large indexed file with unquantified formant trajectory matching (Figures 3(a) and 3(b)). This was an instance where the less restrictive, unquantified formant trajectory matching outperformed the quantified formant trajectory matching in the large indexed file condition. Also, despite the fact that this utterance shared the same initial population taken from the small indexed file in the unquantified formant trajectory matching condition with /əbæ/, the problem of substituting tongue-tip constriction for bilabial closure did not result in this condition for this utterance.

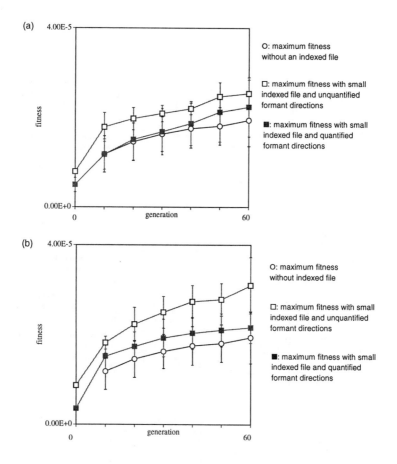

FIGURE 3. (a) Average maximum fitness as a function of generation number for /əbi/ recovery with the small indexed file; (b) average maximum fitness as a function of generation number for /əbi/ recovery with the large indexed file.

The trends for /ədi/ (Figures 4(a) and 4(b)) were very similar to those for /əbi/, with the only clear improvement being in the condition of the large indexed file with unquantified formant trajectory matching. For both cases the maximum fitness as a function of generation was not as great as that for /ədæ/ and /əbæ/, and the speculated reason is that finer control is needed for the tongue body and tip to produce the formants for /i/ than for /æ/. Other reasons were the instability of formant estimation during closure. Also, the lack of information on amplitude of sound and noise content meant that timing of closure release was difficult to estimate.

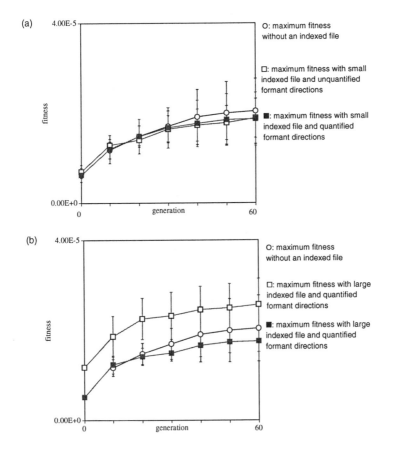

FIGURE 4. (a) Average maximum fitness as a function of generation number for /ədi/ recovery with the small indexed file; (b) average maximum fitness as a function of generation number for /ədi/ recovery with the large indexed file.

CONCLUSIONS

One of the conditions with an indexed file always did better than the condition without an indexed file for all four utterances in terms of average maximum fitness as a function of generation. In the case of /əbæ/, not only was the average of the maximum fitness raised, but the variance over the trials was decreased, so there was more consistency in the better result. The larger indexed file always seemed to outperform the smaller indexed file, but there was no consistency as to which key matching condition did the best. To generate that large indexed file took some computing effort, and may only be justified if a large number of utterances is to be recovered (at 16,000 function evaluations per utterance without memory, at least 7 recoveries with memory should be planned in order to break even). The inconsistency in the efficacy of different matching conditions remains to be explored. There appears to be a balance that needs to be struck between specificity and variety in the initial population. Obviously, if only individuals with correct behaviors are chosen with a highly specific matching condition, then variety is not an issue, but there is no guarantee that this will be the case. /ədi/ and /əbi/ had the same quantified formant trajectory keys despite the fact that the initial places of articulation were different. Perhaps because of this, and the fact that incorrect places of articulation can appear to provide a relatively good fit to the data in a new population, these utterances required an initial population with more variety than could be obtained from matching quantified formant trajectory keys.

There would be better results in all conditions if the number of unknown task dynamic parameters was reduced. There were 41 bits in each chromosome, which resulted in a search space of about 2.2×10^{12} possibilities. There may be ways of reducing the search space size that are more effective than others. For instance, for /əbæ/, to know that the lips formed the initial closure would greatly help the algorithm along. When measurements of articulatory movements are available they should be used to reduce the unknowns, so that other articulatory movement can be inferred directly. Aerodynamic information would be useful in obtaining timing of releases.

These model tests provide a means of determining how much detail can be recovered from an acoustic signal. For instance, it may be the case that task-dynamic parameters are recoverable, but the movements of individual articulators are not. The recovery of individual articulators was not addressed in this paper. However, if this is the goal in the future, the model vocal tract needs a sufficient number of degrees of freedom to move as an exact mimic of a person who creates speech. And, if the model is anatomically realistic, the number of degrees of freedom could grow to be very large. Also, to recover articulatory movement from speech acoustics using task dynamics requires that parameters called articulatory weightings be recovered (Saltzman & Munhall, 1989). These weights determine how much of each articulator is used in attaining a task-dynamic goal. For example, a bilabial closure can be attained with little jaw movement if the lips move sufficiently. The extra degrees of freedom, derived from a realistic articulatory model, and the extra unknown parameters provided by articulatory weights, may mean that there is no robust procedure for recovering the detailed movement of each articulator. This may be possible only when the acoustics is measured simultaneously with some articulatory movements.

An easier task than complete recovery is teaching a machine to repeat speech intelligibly, assuming that it has some internal model of vocal tract anatomy and motor coordination (e.g., task dynamics). A human could train a machine to utter intelligible utterances based on its own internal models. The human trainer would rate the machine on how well it thought it did, and this allows for a necessary subjective element in the fitness function. This fitness function will not be perfectly repeatable, so that the genetic algorithm as an adaptation, rather than as an optimization algorithm, would be best suited for the task: as the trainer changes his impressions, so do the utterances change. The fitness function evolves as the human-machine partnership evolves, thus resembling an endogenous fitness function (Holland, 1992, 1993). In this work, data bases have been used with a nonassociative reinforcement learning algorithm to expedite the inverse solution (see Barto, 1992, for associative and nonassociative learning). Instead of imposing acoustic keys to use to choose the initial population, an associative learning system should be employed so that the acoustic keys evolve with a human trainer along with the correct maps between acoustics and task dynamics. A classifier system could provide such a learning algorithm (Holland, 1992).

Is this closer to what children are required to do in learning speech, rather than recovering the detailed articulation of speakers? They perform a distal learning task with adult teachers judging the acoustical output as the children learn to speak (Jordan & Rumelhart, 1992). While the children have a physically instantiated internal forward model--a vocal tract with resultant speech--they do not use it to produce exact acoustic mimics of adult utterances. If children learn an articulatory-acoustic mapping for speech in general, they must have another internal model (an uninstantiated articulatory-acoustic mapping) that allows for differences in the anatomy and motor control biases of different speakers. It is for the adults to teach the children what in the acoustic signal is worth their attention. As stated by Jordan and Rumelhart (1992): "...the role of external 'teachers' is to help with... representational issues rather than to provide proximal targets directly to the learner" (p. 346). Thus, the adult may respond favorably to an utterance that was produced with a task dynamics that is appropriate and transcends the problem of anatomical scaling. This would indicate that such an adult knows the mapping between task dynamics and the speech acoustic signal. For the child, the physical vocal tract provides a tool in the construction of this more abstract acoustic mapping.

Could it be that children learn to talk by learning the mapping between task dynamics and the acoustics of speech? The task dynamic domain provides flexibility in relation to articulation, while its relation to the acoustic domain is sufficiently well determined. There must be sufficient flexibility in the articulatory domain, because vocal tracts vary anatomically among people. However, to recognize a particular gesture, or sound production as a particular meaningful utterance the acoustic output must be of a certain pattern. Task dynamics that map into particular acoustic features slowly mature as the child becomes an adult. As far as movement is concerned, the things that are targeted could be the task dynamics, because these parameters are acoustically the most salient. The articulation that achieves the task dynamics specification is assembled according to individual anatomy and motor control biases. These are plausible reasons for the articulatory-acoustic mapping to be learned through task dynamics, but this has yet to be shown to be the case.

ACKNOWLEDGMENTS

This work was supported by NIH Grant DC 01247 to Haskins Laboratories. I wish to thank Kathy Harris for helping me get started in speech research.

REFERENCES

Barto, A. G. (1992). Reinforcement learning and adaptive critic methods. In D. A. White & D. A. Sofge (Eds.), *Handbook of intelligent control* (pp. 469-91). New York: Van Nostrand Reinhold.

Bleile, K. M., Stark, R. E., & McGowan, J. S. (1993). Speech development in a child after decannulation: Further evidence that babbling facilitates later speech development. *Clinical Linguistics & Phonetics, 7,* 319-337.

Boë, L.-J., Perrier, P., & Bailly, G. (1992). The geometric vocal tract variables controlled for vowel production: proposals for constraining acoustic-to-articulatory inversion. *Journal of Phonetics, 20,* 27-38.

Browman, C. P., & Goldstein, L. (1990). Gestural specification using dynamically-defined articulatory structures. *Journal of Phonetics, 18,* 299-320.

Coker, C. H. (1976). A model of articulatory dynamics and control. *Proceedings of the IEEE, 64,* 452-460.

Flanagan, J. L., Ishizaka, K., & Shipley, K. L. (1980). Signal models for low bit-rate coding of speech. *Journal of the Acoustical Society of America, 68,* 780-791.

Forrest, S. (1993). Genetic algorithms: Principles of natural selection applied to computation. *Science, 261,* 872-878.

Goldberg, D. E. (1989). *Genetic algorithms in search, optimization, and machine learning.* Reading, MA: Addison-Wesley Publishing Company.

Holland, J. H. (1992). *Adaptation in natural and artificial systems.* Cambridge, MA: MIT Press.

Holland, J. H. (1993). *Echoing emergence: objectives, rough definitions, and speculations for echo-class models* (No. 93-04-023). Santa Fe Institute.

Jordan, M. I., & Rumelhart, D. E. (1992). Forward models: Supervised learning with a distal teacher. *Cognitive Science, 16,* 307-354.

Maeda, S. (1982). A digital simulation model of the vocal tract system. *Speech Communication, 1,* 199-229.

McGowan, R. S. (1993). Implementing a genetic algorithm to recover task-dynamic parameters of an articulatory synthesizer. *Haskins Laboratories Status Report on Speech Research, SR-113,* 95-106.

McGowan, R. S. (1994). Recovering articulatory movement from formant frequency trajectories using task dynamics and a genetic algorithm: Preliminary model tests. *Speech Communication, 14,* 19-48.

Mermelstein, P. (1973). Articulatory model for the study of speech production. *Journal of the Acoustical Society of America, 53,* 1070-1082.

Parthasarthy, S., & Coker, C. H. (1992). On automatic estimation of articulatory parameters in a text-to-speech system. *Computer speech and language, 6,* 37-75.

Rose, S. P. R. (1992). *The making of memory: from molecules to mind.* New York: Doubleday.

Rubin, P., Baer, T., & Mermelstein, P. (1981). An articulatory synthesizer for perceptual research. *Journal of the Acoustical Society of America, 70,* 1109-1121.

Saltzman, E. L. (1986). Task-dynamic coordination of speech articulators: A preliminary model. *Experimental Brain Research, 15,* 129-144.

Saltzman, E. L., & Kelso, J. A. S. (1987). Skilled action: A task-dynamic approach. *Psychological Review, 94.* 84-106.

Saltzman, E. L., & Munhall, K. G. (1989). A dynamic approach to gestural patterning in speech production. *Ecological Psychology, 14,* 333-382.

Schroeter, J., Meyer, P., & Parthasarthy, S. (1990). Evaluation of improved articulatory codebooks and codebook distance measures. *ICASSP '90.* Albuquerque.

Studdert-Kennedy, M. (1991). Language development from an evolutionary perspective. In N. A. Krasnegor, D. M. Rumbaugh, R. L. Schiefelbush, & M. Studdert-Kennedy (Eds.), *Biological and behavioral determinants of language development* (pp. 5-28). Hillsdale, NJ: Lawrence Erlbaum Associates, Publishers.

Wakita, H. (1973). Direct estimation of vocal tract shape by inverse filtering of acoustic speech wave forms. *IEEE Trans. Audio and Electroacoustics, 21,* 417-27.

33   Speech Dynamics

Betty Tuller and J. A. S. Kelso
Program in Complex Systems and Brain Sciences, Center for Complex Systems, Florida Atlantic University

Throughout her career, Katherine Harris refused to focus solely on one piece of the speech communication process, always remembering that communication involves both speakers and listeners. Early on, KSH was involved in seminal work illustrating processes intrinsic to the listener (e.g., Harris, 1958; Liberman, Harris, Hoffman, & Griffith, 1957). In the realm of speech production, she pioneered many of the techniques for articulatory monitoring and collaborated with the present authors (as well as many others) in exploring how articulatory maneuvers are coordinated (Kelso, Tuller, & Harris, 1983; Tuller, Kelso, & Harris, 1982) and modulated (Tuller, Harris, & Kelso, 1982). The language with which we describe the processes of speech production and speech perception are of course very different. Yet it seems quite likely to us that both speech production and speech perception involve self-organizing processes in the brain. Thus, the plan for an act constitutes a global organization generated by the cooperative activity of many interacting areas of the central nervous system, that unfolds in time. Likewise, the phenomenal percept is a global organization that arises due to the cooperative activity among many processing elements. The rules by which such self-organization occurs in the brain are not well understood. Nevertheless, here we try to establish that speech production and perception, like all complex systems in nature, are governed by *dynamics* and parameterized by intrinsic and extrinsic constraints. Such dynamics are a reflection of the coordinating activity of the nervous system, not only (and in some cases not at all) by the biophysical properties of the components themselves. That is, the dynamics are abstract, but measurable.

Is the underlying representation of speech intrinsically dynamic? In this chapter, we provide some recent evidence attesting to the central role of dynamics in both speech production and perception. First, we briefly review supporting evidence in speech production. Next, we provide evidence for *perceptual dynamics* in speech categorization and describe a dynamical model that captures the observed temporal evolution of perceptual states. Third, we investigate the neural underpinnings of perception by disambiguating the neural processes corresponding to acoustic aspects of a signal from those processes associated with the perception of speech. In these sections, the concepts of attractors, multistability, instability and so forth will play a central role. Last, we deal with spontaneous perceptual changes observed when a syllable is repeatedly presented. The results suggest an underlying dynamic that is intrinsically *metastable*, hence intermittent, allowing for both perceptual stability and flexibility.

I. Syllable Organization: Production Dynamics

In earlier work, we investigated the internal structure of syllables as a window into the dynamics of speech production (e.g., Kelso, Saltzman, & Tuller, 1986; Tuller & Kelso, 1990). Subjects were asked to produce the utterance /ip/ or /pi/ repetitively at a slow speaking rate. While they spoke, the experimenter signalled them to speed up their rate of production in a step-like manner. The relative phasing between glottal and lip movements was consistent, but different, for the two syllable types at slow speaking rates. At higher rates, /pi/ maintained the same phase relation as at slow rates. In contrast, as speaking rate increased the phase relation for /ip/ often changed to the phasing pattern for /pi/. Importantly, the change in articulatory coordination was always perceived by listeners as a change in the production's syllabic form (Tuller & Kelso, 1991). More recently, we asked subjects to produce the words "opt," "hopped," or "top" repetitively in time to a metronome that increased frequency in steps of 0.25 Hz while their tongue tip, tongue blade, and lip movements were monitored using an electromagnetometer. The results were similar to those observed for glottis-lip relations in that the relative phase of tongue and lip gestures differed for the three syllable types at slow speaking rates. At higher rates, the phase relations converged. These results suggest that the relative phase of articulatory gestures is an order parameter or collective variable whose dynamics are (minimally) bistable in one parameter regime but monostable in another. The shift in syllable organization is identified as a phase transition among articulatory gestures that shows characteristics of intrinsic (nonlinear) dynamics (multistability, loss of stability, switching, etc.). Moreover, pattern stability (within a broad range of speaking rates) and pattern change (at critical points) may arise from the same underlying nonlinear dynamics. In what follows, we propose a kindred conceptualization of the perception of speech categories.

II.A Speech Categorization: Perceptual Dynamics

The basic issue of how we sort a continuously changing signal into the appropriate category is still incompletely understood, whether the categories concern speech, objects, emotions, individuals, etc. (see, for example, Harnad, 1987). Our contention is that the perception of speech categories provides strong hints of what we term *perceptual dynamics*. Perceptual dynamics characterizes the time-dependent behavior of the speech system in terms of its (nonlinear) dynamics, that is, equations of motion describing the temporal evolution of the perceptual process, especially the stability and change of perceptual forms. We are exploring time-dependent behaviors explicitly (e.g., Tuller, Case, Ding, & Kelso, 1994), because we believe they reflect intrinsic properties of communicative processes.

   The stimulus continuum chosen, a 'say-stay' continuum with variation of the silent gap duration after the initial fricative, is one for which categorical perception has often been demonstrated (e.g., Best, Morongiello, & Robson, 1981). However, we varied the silent gap duration sequentially (between 0 ms and 76 ms in 4 ms steps) in order to explore the temporal evolution of phonemic categorization and change.[1] The subject's task was to identify each stimulus as either 'stay' or 'say' during a 2 second interstimulus interval. Comparing the switch location when gap duration increased from 0 ms with the switch location when gap duration decreased from 76 ms can produce three patterns of results, illustrated in Figure 1 for single runs selected from three different subjects. 1) *Critical boundary:* the boundary between responding "say" and responding "stay" remains fixed despite changes in the direction of gap change so that a given gap duration always gives rise to the same response (Figure 1(a)); 2) *hysteresis:* with increasing gap duration, a larger gap is necessary to induce a category change from "say" to "stay" than for the change from "stay" to "say" with decreasing gap duration (Figure 1(b)). That is, the initial perceptual state persists even when the current gap duration favors the alternative response when stimuli are randomized; 3) *enhanced contrast:* rather than "hold on" to the initial categorization, the subject switches from "say" to "stay" earlier when gap duration increases than when gap duration decreases (Figure 1(c)).

---

[1]Katherine Harris (personal communication) some years ago attempted sequential variation of an acoustic parameter but, without the theoretical tools for interpretation, abandoned the effort.

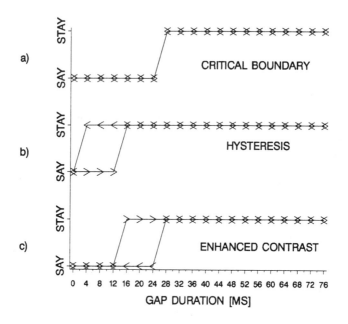

FIGURE 1. Single trials of sequential presentation of a "say-stay" continuum showing three distinct response patterns: (a) critical boundary; (b) hysteresis; (c) enhanced contrast.

The three patterns occurred 17%, 41%, and 42% of the time, respectively, across subjects and conditions.

The dependence of speech perception on the recent history and the direction of parameter change is strong evidence of nonlinearity and multistability. Next we describe a theoretical model that captures the observed patterns of category change within a unified dynamical account. The idea is that this theoretical description applies across particular acoustic cues or speech categories. Not surprisingly, the effects displayed in Figure 1 are not unique to perception of speech segments but are also observed in categorization of ambiguous sentences and visual patterns (Kelso et al., in press).

II.B Theoretical Modeling of Categorical Effects

Determination of a dynamical model is based on mapping stable perceptual categories onto attractors of a dynamical model. In this we follow essentially the same strategy as that employed by Haken, Kelso, and Bunz (1985) in studies of coordinated movement. For a single perceptual category, a local model containing a fixed-point is adequate. But hysteresis indicates that several stable percepts coexist, so a nonlinear dynamical model must be found. Task variables, here the changing acoustics, are presumed to act as parameters on the underlying dynamics.

Figure 2 shows the attractor layout for the following dynamical system:

$$V(x) = kx - x^2/2 + x^4/4, \tag{1}$$

where x is the perceptual form and k is the control parameter specifying the direction and degree of tilt for the potential $V(x)$. For ease of visualization, Figure 2 shows the potential for several values of k. With $k = -1$, only one stable point exists corresponding to a single percept (e.g., "say;" Figure 2(a)). As k increases the potential landscape tilts but otherwise remains unchanged in terms of the composition of attractor states (Figure 2(b)). However, when k reaches a critical point $k = -k_c$, an additional attractor (corresponding to "stay") appears by a saddle-node bifurcation in which a point attractor (with $x < 0$) and a point repeller ($x = 0$) are simultaneously created (Figure 2(c)). The co-existence of both percepts continues until $k = k_c$ where the attractor corresponding to "say" ceases to exist via a reverse saddle-node bifurcation, leaving only the stable fixed point corresponding to "stay" (Figure 2(d)). Further increases in k only serve to deepen the potential minimum corresponding to "stay" (Figure 2(e)).

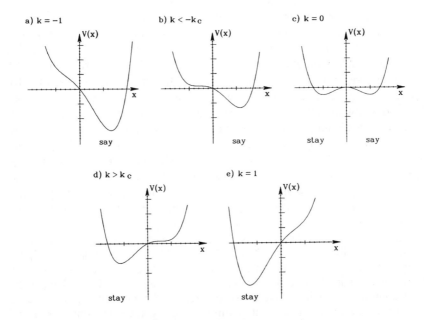

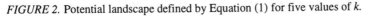

FIGURE 2. Potential landscape defined by Equation (1) for five values of k.

In order to account for the three response patterns observed (critical boundary, hysteresis, and enhanced contrast), it is necessary to examine in more detail the behavior of the "tilt" parameter k. Without going into too much detail, we propose the following equation describing the behavior of k as a function of the gap duration:

$$k(\lambda) = k_0 + \lambda + \epsilon/2 + \epsilon\,\theta(n-n_c)(\lambda-\lambda_f), \qquad (2)$$

where the value of k_0 specifies the percept at the beginning of a run, λ is linearly proportional to the gap duration, λ_f denotes the value of λ at the other extreme from its initial value, and n is the number of perceived stimulus repetitions in a run. Before a critical number of accumulated repetitions n_c is reached, $\theta(n-n_c) = 0$. When $n \geq n_c$ (during the second half of each trial run), $\theta(n-n_c)=1$. An additional parameter, ϵ, represents cognitive factors such as learning, linguistic experience and attention. Note that the introduction of criteria stemming from cognitive processes is not without precedent, for example, attention and previous experience play a large role in synergetic modeling of perception of ambiguous visual figures (Ditzinger & Haken, 1989, 1990; Haken, 1990) and contribute to factors that determine adaptation level in Helson's work (Helson, 1964).

When n and/or ϵ are sufficiently small, the tilt of the potential is only dependent on gap duration and the initial configuration. Figure 3 shows three regions corresponding to different states of the system in the $\epsilon - \lambda$ plane, in the first half of each run when n is small (Figure 3(a)) and in the second half of each run when n is large (Figure 3(b)). White regions indicate the set of parameter values for which a stimulus has but a single possible categorization in the represented portion of the run. Shaded regions indicate the set of parameter values for which a stimulus may be categorized as either one form or the other (the bistable region) and thus represent the condition from $-k_c$ (the lower border of each shaded region) to k_c (the upper border of each shaded region). Consider the condition with initial $k_0 = -1$ ("say") and the parameter λ increasing. As λ increases, the stimuli are categorized as "say" for any value of ϵ so long as the $\epsilon-\lambda$ coordinate remains below the shaded region. Within the shaded region, the stimuli are categorized as "say" despite the percept becoming progressively less stable. As λ continues to increase the percept switches from "say" to "stay" at the upper boundary of the shaded region (the heavy line) after which "say" is not a possible percept. Note that for different values of ϵ the switch to a new percept occurs at different durations of silent gap.

In the second half of the run, λ is decreasing and the resulting division of the $\epsilon -\lambda$ plane looks somewhat different (Figure 3(b)). This portion of the figure should be read from top to bottom, from large gap durations to small ones. As λ decreases, the stimuli are categorized as "stay" for any value of ϵ so long as the $\epsilon -\lambda$ coordinate remains above the shaded region. Again assuming the absence of a perturbing force, the subject continues to categorize the stimuli as "stay" within the bistable (shaded regions) despite the percept becoming less stable.

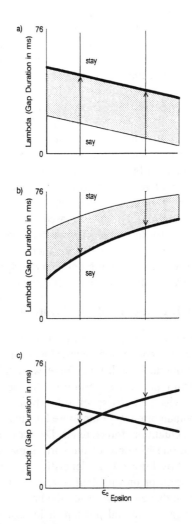

FIGURE 3. (a) Perceptual states in the ϵ –λ plane for increasing λ (small *n*); (b) perceptual states in the ϵ –λ plane for decreasing λ (large *n*); (c) superposition of reversed saddle node bifurcation points in *a* and *b*. (From Tuller et al., 1994.)

As λ continues to decrease (gap duration continues to shorten), the *lower* boundary of the shaded region (the heavy curve) marks the switching from "stay" back to "say." In Figure 3(c), we have simply superimposed the boundaries at which switching occurs in Figures 3(a) and 3(b). The line with negative slope represents the category switch in the first half of the run and the curve with positive slope represents the switch back to the initial category. Their intersection yields a critical value of ϵ, ϵ_c for which a critical boundary would be observed in the "say-stay" continuum. For $\epsilon > \epsilon_c$ (e.g., the vertical line to the right of ϵ_c), the system exhibits the enhanced contrast effect. In that region, λ is smaller for the negatively-sloped line than for the positively-sloped curve. For $\epsilon < \epsilon_c$ (e.g., the vertical line to the left of ϵ_c), the classical hysteresis phenomenon is obtained: λ is larger for the line than the curve. The main qualitative features of the observed data are thus captured in the model.

As stressed again and again by synergetics (e.g., Haken, 1983), any accurate portrayal of a real world problem must take into account the influence of random disturbances. In the current model, the parameter plane looks essentially the same when the effect of random fluctuations is explicitly considered. For a given point attractor, the degree of resistance to the influence of random noise is related to its *stability*, which, in general, depends on the depth and width of the potential well (basin of attraction). As k is increased successively, the stability of the attractor corresponding to the initial percept decreases (the potential well becoming shallower and flatter) leading to an increase in the likelihood of switching to the alternative percept as the boundary is approached rather than crossed. This prediction was confirmed in experiments (see Tuller et al., 1994).

III. Speech Categorization: Brain Dynamics

The presence of hysteresis and enhanced contrast in speech perception means that an individual listener categorizes the identical acoustic stimulus differently, depending on the experimental conditions. Here we exploit these nonlinear effects to dissociate brain processes pertaining to the physical (acoustic) stimulus, from brain activity patterns specific to phonetic perception. Neural activity patterns in the human brain were monitored using a multi-sensor SQUID array (Superconducting Quantum Interference Device) capable of detecting signals that are less than one billionth of the earth's magnetic field. SQUIDs pick up signals generated by intracellular dendritic currents without having to place electrodes in the brain or on the skull. Because the skull and scalp are transparent to magnetic fields generated inside the brain, and because the array is large enough to cover a substantial portion of human neocortex, this new research tool opens a (noninvasive) window into the brain's dynamic patterns and their relation to perception and cognition.

The SQUID array was placed over the left auditory evoked potential field (determined empirically for the individual before the start of the experiment). Details of the placement and recording set-up are reported in Kelso et al. (1992). Figure 4 (left) shows the location of the montage (the small circles) with respect to the subject's skull and cortex (reconstructed from MRI slices).

The speech categorization paradigm was similar to that described in the previous section, with gap duration of the "say-stay" continuum ranging from 4 ms to 48 ms. During the interstimulus interval of 1.5 seconds, the subject pressed one of a pair of microswitches to indicate which word he had just heard. Discussion will be limited to the stimulus with a gap duration of 32 ms because analysis of responses showed that it was perceived as "say" as often as it was perceived as "stay". For each of the 37 sensors, we averaged the raw signal across presentations with the same response, then calculated the root mean square (RMS) amplitude of the averaged waveforms. Differences in the topographic distributions of the RMS amplitude corresponding to the percepts "say" and "stay" are minimal throughout most of the vowel (the differences just after vowel onset are shown in the top right portion of Figure 4). However, a few milliseconds following vowel offset, a large difference in field amplitude between the two percepts is observed in the posterior part of the array (Figure 4, bottom right), despite the physical stimulus being identical for the two percepts. By the end of the epoch the topographic distributions are again identical across percepts.

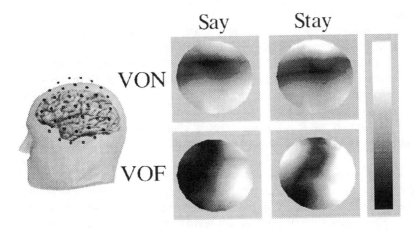

FIGURE 4. Left: A phantom view of the subject's head showing the SQUID sensor array's position relative to skull and cortex; Right: Topographic distributions of RMS amplitude just after vowel onset (VON, top) and just after vowel offset (VOF, bottom). (Adapted from Kelso et al., in press.)

These results are encouraging and open up a wide range of exciting possibilities for examining perceptual processing, not just in speech but in other modalities as well. The methodology allows a clear separation between the neural analogues of the acoustic (or any other) signal, and what the brain is doing when people perceive. Although preliminary, to our knowledge this is the first clear evidence of brain activity patterns that are specific to changes in phonetic percept.

IV. From Bistability to Multistability

When a person listens to a syllable (or a word or sentence) that is repeated over and over, with relatively short separations between neighboring repetitions, the person's perception of the syllable is initially veridical. After a variable time period, however, perception begins to shift intermittently among other syllables. These illusory changes, referred to as verbal transforms (Warren, 1961), received considerable attention for years after Warren and Gregory's (1958) original observations. In general, the literature has concentrated on discovering statistical regularities relating stimuli or subjects to the reported transforms. Here we treat the time-dependent perceptual changes as a dynamical process (Ding, Tuller, & Kelso, in press). To this end, we slow the rate of stimulus presentation and extend the total number of repetitions in order to examine temporal regularities in the evolving pattern of illusory changes.

The syllables [kɛ] and [gɛ] spoken by a male talker were edited digitally to a duration of 200 ms by truncating the vowel (by five pitch pulses or fewer) and ramping the amplitude of the final 50 ms for naturalness. Two audio tapes were constructed, each with a single syllable reproduced 1000 times with a 500 ms silent interval between successive tokens. Listeners were told that they would hear 12-minute segments of auditory material, that certain sounds might be repeated over and over, but that other sounds might also be heard. Their task was to listen to each sound and, during the short silence between sounds, to write down what they heard. All subjects reported hearing changes in the syllables presented, although the total number of variants in a single trial ranged from a low of 2 to a high of 26 across subjects and stimuli.

In this section, we focus on the temporal characteristics of perceptual change and ignore the phonetic identity of the perceived forms. Figure 5(a) shows a time series for a single trial of 1000 repetitions of [kɛ], illustrating the temporal pattern of perceptual switches. A change in value indicates that a switch took place, regardless of the sounds reported. One can see that bursts of switching are interspersed with more prolonged periods during which no perceptual change takes place. Moreover, there does not seem to be any consistent pattern in the time series data. For example, the duration of a percept does not appear to reach asymptote with listening time, a percept held a long time is not necessarily followed by briefly persisting states, and so forth.

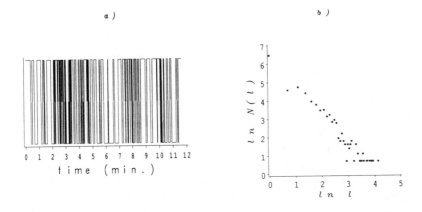

FIGURE 5. (a) Single trial time series of switching behavior (see text). (b) Distribution (across subjects) of the number of occurrences of phonetic strings that persist for length *l*.

These trends were examined further by determining the distribution of dwell times, defined as the length of time, *l*, that any given percept is maintained before a switch to another percept occurs. Figure 5(b) shows the log-log distribution of $N(l)$, the number of occurrences of phonetic strings that persist for length *l*. Note that since the stimulus presentation rate was constant, the percept length, *l*, may be converted to time by multiplying by 700 ms. The relationship between $N(l)$ and *l* can be fit rather well by a power law,

$$N(l) \sim l^{-\alpha}, \tag{3}$$

where α is the scaling exponent whose value is about 1.5.

The quantity $P(l)=N(l)/N$, with N the total number of stimulus presentations, approximates the probability that any percept persists for length *l*. If the strings have a finite average length, $< l >$, then listeners maintain a percept for a characteristic time interval, on average, before the perceived phonetic string switches to some other form. Switches among percepts could then be said to occur with an average rate of $1/< l >$. But with $\alpha < 2$ in Equation (3), $< l >$ is in fact infinite due to the divergent sum $\Sigma_1^{\infty} lP(l) \sim \Sigma_1^{\infty} = l^{1-d} = \infty$. This means that a characteristic time scale is not established for the shifting perceptions. On the other hand, since $\alpha > 0$ the temporal pattern of perceptual switches is not random.

One strong possibility is that the experimental methodology did not allow the perceptual system sufficient time to establish a consistent time scale. That is, 1000 repetitions of the stimulus may be insufficiently large to emerge from a region of perceptual transients. Four of the subjects listened to the stimulus tapes 9 additional times, for a total of 10,000 repetitions of each syllable. Despite this huge extension of the listening time, the relationship between $N(l)$ and l was still described well by a power law with a scaling exponent of $0<\alpha<2$.

What kind of dynamics might generate these temporal patterns? Clearly, they must be intermittent. Intermittency means that the temporal pattern of switching is neither perfectly predictable nor random, neither perfectly stable nor unstable. Rather, the perceptual system is intrinsically metastable, living at the edge of instability where it can switch spontaneously among perceptual states. Indeed, perceptual states themselves may be metastable, as opposed to corresponding to stable, fixed point attractors. In the intermittency regime, there is attractiveness but, strictly speaking, no attractors.

We note that so-called *type 1 intermittency*, which occurs near *tangent* or *saddle-node* bifurcations (Pomeau & Manneville, 1980), has been implicated in recent work on perceptual switching with visual ambiguous figures (Kelso et al., in press). Moreover, there is a formal resemblance between perceptual intermittency and previous analyses of relative coordination (von Holst, 1973) in terms of intermittent dynamics (e.g., Kelso & DeGuzman, 1991; Kelso, DeGuzman & Holroyd, 1991a, 1991b). The latter work has demonstrated, both empirically and theoretically, that the neural coordination dynamics is seldom (except under highly prepared conditions) frequency- and phase-locked in an asymptotic sense. Rather, to retain flexibility and stability the nervous system lives near, but not in mode-locked states.

CONCLUSIONS

In this chapter we have demonstrated that the basic concepts of self-organized pattern formation (synergetics) and the mathematical tools of nonlinear dynamics capture specific phenomena in speech production and perception. The existence of attractive states, multistability, instability, transitions, and hysteresis, along with theoretical modeling of these complex behaviors suggest that the underlying neural representation takes the form of a self-organized dynamical system. We have also shown that there is no characteristic (average) time for perceptual persistence and change under conditions of repetitive syllable presentation. Instead, the perceptual system is marginally stable or metastable operating *close* to stable fixed points. Rather than requiring active processes to destabilize and switch from one stable state to another (e.g., through change in parameter(s), increase of fluctuations), intermittency is built into the dynamics as a generic mechanism for flexibly entering and exiting coherent perceptual states. As a historical note, the approach described here resonates with early Gestalt theory that stressed the search for natural principles of autonomous order formation in perception and brain function (Köhler, 1940). Though not limited to Gestalt theory, the present dynamical modeling *cum* empirical approach may provide the tools for exploring the phenomenological aspects of behavior emphasized by Gestaltists.

ACKNOWLEDGMENTS

This work was supported by NIMH (Neurosciences Research Branch) Grant MH 42900, BRS Grant RR07258, and NIDCD Grant DC001411.

REFERENCES

Best, C., Morongiello, B., & Robson, R. (1981). Perceptual equivalence of acoustic cues in speech and nonspeech perception. *Perception and Psychophysics, 29,* 191-211.

Ding, M., Tuller, B., & Kelso, J. A. S. (in press). Characterizing the dynamics of auditory perception. *Chaos.*

Ditzinger, T., & Haken, H., (1989). Oscillations in the perception of ambiguous patterns: A model based on synergetics. *Biological Cybernetics, 61,* 279-287.

Ditzinger, T., & Haken, H., (1990). The impact of fluctuations on the recognition of ambiguous patterns. *Biological Cybernetics, 63,* 453-456.

Haken, H., (1983). *Advanced synergetics: Instability hierarchies of self-organizing systems and devices.* Berlin: Springer.

Haken, H. (1990). Synergetics as a tool for the conceptualization and mathematization of cognition and behavior—How far can we go? In H. Haken & M. Stadler (Eds.), *Synergetics of cognition* (pp. 2-31). Berlin: Springer.

Haken, H., Kelso, J. A. S., & Bunz, H. (1985). A theoretical model of phase transitions in human hand movements. *Biological Cybernetics, 51,* 347-356.

Harnad, S. (1987). *Categorical perception: The groundwork of cognition.* Cambridge: Cambridge University Press.

Harris, K. S. (1958). Cues for the discrimination of American English fricatives in spoken syllables. *Language and Speech, 1,* 1-7.

Helson, H. (1964). *Adaptation level theory: An experimental and systematic approach to behavior.* New York: Harper and Row.

Kelso, J. A. S., Bressler, S. L., Buchanan, S., DeGuzman, G. C., Ding, M., Fuchs, A., & Holroyd, T. (1992). A phase transition in human brain and behavior. *Physics Letters A, 169,* 134-144.

Kelso, J. A. S., Case, P. Holroyd, T., Horvath, E., Rączaszek, J., Tuller, B., & Ding, M., (in press). Multistability and metastability in perceptual and brain dynamics. In M. Stadler & P. Kruse (Ed.), *Multistability in cognition.* Berlin: Springer.

Kelso, J. A. S., & DeGuzman, G. C. (1991). An intermittency mechanism for coherent and flexible brain and behavioral function, In J. Requin & G. E. Stelmach (Eds.), *Tutorials in motor neuroscience* (pp. 305-310). Dordrecht: Kluwer.

Kelso, J. A. S., DeGuzman, G. C., & Holroyd, T. (1991a). The self-organized phase attractive dynamics of coordination. In A. Babloyantz (Ed.), *Self-organization, emerging properties and learning* (pp. 41-62). New York: Plenum.

Kelso, J. A. S., DeGuzman, G. C., & Holroyd, T. (1991b). Synergetic dynamics of biological coordination with special reference to phase attraction and intermittency. In H. Haken & H. P. Köepchen (Eds.), *Rhythms in physiological systems* (pp. 195-213). Berlin: Springer.

Kelso, J. A. S., Saltzman, E. L., & Tuller, B., (1986). The dynamical perspective in speech production: data and theory. *Journal of Phonetics, 14,* 29-60.

Kelso, J. A. S., Tuller, B. & Harris, K. S. (1983). A 'dynamic pattern' perspective on the control and coordination of movement. In P. MacNeilage (Ed.), *The production of speech* (pp. 137-173). New York: Springer-Verlag.

Köhler, W. (1940). *Dynamics in psychology.* New York: Liveright.

Liberman, A. M., Harris, K. S., Hoffman, H. S., & Griffith, B. C. (1957). The discrimination of speech sounds within and across phoneme boundaries. *Journal of Experimental Psychology, 54,* 358-368.

Pomeau, Y., & Manneville, P. (1980). Intermittent transitions to turbulence in dissipative dynamical systems. *Communications in Mathematical Physics, 74,* 189.

Tuller, B., Case, P., Ding, M., & Kelso, J. A. S. (1994). The nonlinear dynamics of speech categorization. *Journal of Experimental Psychology: Human Perception and Performance, 20,* 1-14.

Tuller, B., Harris, K. S., & Kelso, J. A. S., (1982). Stress and rate: Differential transformations of articulation. *Journal of the Acoustical Society of America, 71,* 1534-1543.

Tuller, B., & Kelso, J. A. S. (1990). Phase transitions in speech production and their perceptual consequences. In M. Jeannerod (Ed.), *Attention and performance XIII* (pp. 429-452). Hillsdale, NJ: Erlbaum.

Tuller, B., & Kelso, J. A. S. (1991). The production and perception of syllable structure. *Journal of Speech and Hearing Research, 34,* 501-508.

Tuller, B., Kelso, J. A. S., & Harris, K. S. (1982). Interarticulator phasing as an index of temporal regularity in speech. *Journal of Experimental Psychology: Human Perception and Performance, 8,* 460-472.

von Holst, E. (1973). Relative coordination as a phenomenon and as a method of analysis of central nervous system function (1939). In R. Martin (Ed.), *The collected papers of Erich von Holst* (pp. 33-135). Coral Gables, FL: University of Miami.

Warren, R. M. (1961). Illusory changes of distinct speech upon repetition—The verbal transformation effect. *British Journal of Psychology, 52,* 249-258.

Warren, R. M., & Gregory, R. L. (1958). An auditory analogue of the visual reversible figure. *American Journal of Psychology, 71,* 612-613.

34 The Spatial Control of Speech Movements

K. G. Munhall and J. A. Jones
Queen's University

INTRODUCTION

Speech production, like all movement, requires the control of an articulator's position in three spatial dimensions. While some articulators like the mandible may move primarily in two dimensions during speech (Vatikiotis-Bateson and Ostry, in press), the tongue, pharynx, and lips produce significant motion in three dimensions (Stone, 1990; Abry & Boë, 1986). How the nervous system can control accurate movement through space is a question that is not well understood. In this paper we will try to outline some of the obstacles that must be overcome by the nervous system in controlling the position of articulators. We can divide the kinds of problems that the nervous system faces in speech into those that have to do with the general geometry of motor control and those that have to do with unique aspects of sound production. It is our position that a good deal of the complexity of spatial planning in speech owes to the former, more general kinds of problems. As a result, we will discuss both speech and nonspeech research.

Throughout the chapter, we will distinguish between the *path* an articulator takes and its *trajectory* along that path. By path we mean the sequence of positions in space that is occupied by an object or an articulator. By trajectory we mean the timing of this sequence of positions. In principle, these are independent descriptions. The hand can draw a circle in the air (path) with many different timing functions (trajectory). The hand could move very slowly, very quickly, or in a series of accelerations and decelerations. In each case the circular path that is drawn could be identical.[1] While much recent work has addressed the trajectories in speech, we are concerned here with the paths that are taken in the oral cavity. Our rationale is that the spatial form of a movement can reveal much about the planning and control processes.

For a number of reasons, speech research has tended to focus on trajectories in one spatial dimension. One of the major reasons for this reduced spatial measurement of speech movements is a technical one: Measurement techniques are often limited to one or two spatial dimensions. For example, the pulsed-echo ultrasound system used in the past at McGill University allowed measurement of the movements of the surface of the tongue, pharynx, and larynx only along a single measurement axis. The traditional lateral X-ray images and the more recent electromagnetic articulometer and X-ray microbeam systems are restricted to 2-D views of the vocal tract.

A second reason is more conceptual in nature. It is difficult to visualize data that represent two or three spatial dimensions plus time. We characteristically examine our data using 2-D plots with one axis being defined by time and the other the amplitude variation in one spatial dimension. To construct the 2-D and, on rare occasions, 3-D spatial patterns we commonly compare the average patterns for individual spatial dimensions (e.g., Parush, Ostry, & Munhall, 1983). We thus get a somewhat fragmented view of the spatial relations in speech production.

From our available information, the control of the positioning of the speech articulators *appears* to be a daunting task. As can be seen in recent MRI studies of the static oral cavity (Baer, Gore, Gracco, & Nye, 1991), the vocal tract varies in shape in complex ways for different sounds. These changes in shape of the oral cavity during sound production are determined by a large number of articulators. In many cases, the articulators work in concert to control a specific area of the vocal tract. For example, the lips and jaw jointly determine the oral aperture. Each of these articulators is controlled by more than a single muscle and some of the muscles produce more than a single biomechanical action. The anterior digastric, for example, both lowers and retracts the jaw during jaw opening. The tongue has these complexities and more. The intrinsic tongue

---

[1] The path and trajectory of a movement are probably not always independent. The idea of reduction or undershoot in speech assumes that the path taken is influenced by the timing of the movement. In reaching, there is a well-known relation between the curvature of the path and instantaneous velocity (Abend, Bizzi, & Morasso, 1982). Our purpose here is to draw attention back to the geometry of speech movements without underestimating the importance of timing as a constraining force in coordination.

muscles must shape and support themselves and thus the control of the tongue is more similar to the control of hydrostatic appendages such as in cephalopods than to the control of joints such as the jaw (Kier & Smith, 1985). A network of external muscles also contributes to the tongue's shape and position. Through all of this complexity the nervous system must guide the various articulators to positions that produce distinct classes of sound. To understand how this is accomplished we consider data from other types of movements (e.g., arm movements, eye movements) since many of the same problems facing researchers in speech have been studied by researchers interested in general issues in motor control. It is useful to consider the progress that has been made in these areas before examining the specific problems of speech production.

SPATIAL PLANNING OF ARM MOVEMENTS

In reaching, the hand characteristically follows a relatively straight-line path in space from the hand's starting position to the object to be grasped (Morasso, 1981). This straight-line path of the hand is produced by a particular sequence of joint motions at the shoulder, elbow, and wrist. The motions at each of the joints are produced by forces generated by the muscles spanning those joints and by reaction forces from the motion and environment. For the muscles at a joint the direction of force generation will be determined by the anatomy of the origins and insertions of the muscles. However, the effective force generation at a joint by a particular muscle will depend on the task being carried out and the manner in which the movement is produced (see Zajac & Gordon, 1989). For example, Hollerbach and Flash (1982) have shown that even during the slow movement of a multiple segment limb there are a number of significant torques that arise due to interactions among the moving segments. Thus, a given amount of force generation by a muscle at the elbow may correspond to quite different amounts of force generated around the elbow joint because of the motion of the rest of the arm.

In this description of arm movement, we can see many of the issues that have been the focus of recent research on limb movements. These include the issue of which coordinates a movement is planned in, how paths are determined, how inverse problems are dealt with, and how quite different coordinate systems can be integrated. In the following pages we will deal with these issues for arm movements and then discuss those issues that relate to speech production.

1. The Levels of Planning Problem

If we examine a stylized, two-joint limb, we can see how two alternative coordinate systems can be used to describe the same behavior. In Figure 1(a) we see a limb with two hinge joints (θ_1, θ_2) that each produce only rotation. The two segments (L_1, L_2) can be positioned in a variety of orientations depending on the angular settings at the two joints, and the end of the limb (E) can thus be placed at many different positions in the workspace (the area that can be reached). In Figure 1(b) we see the sequential change in position of the limb as it moves from a high position to a lower position. Figure 1(c) shows the curved path followed by the endpoint of the of the limb (E) in Cartesian coordinates. Finally, Figure 1(d) shows the relative changes in the two joint angles plotted against each other. As can be seen in this last plot, a straight line is observed, indicating that the motions at the joints start and stop at the same time and have similar velocity functions at the two joints.

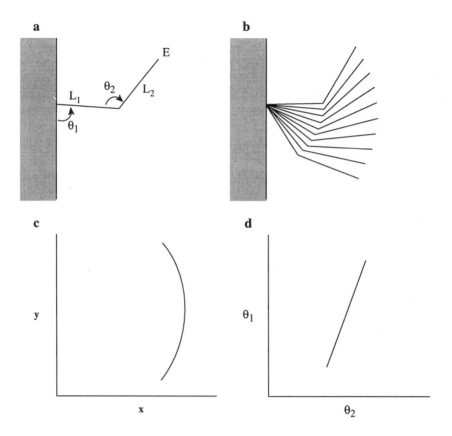

FIGURE 1. (a) A schematic two-joint limb is shown. Each of the joints are hinge joints. (b) The sequence of positions of the limb as the limb moves from the top position to the bottom configuration are shown. (c) The path traced by the endpoint of the limb (E) during the movement in 1(b) is shown. (d) The path of the lowering motion in 1(b) plotted in joint-angle coordinates.

For the movement from the upper position to the lower position in Figure 1(b) we have two alternative descriptions—the joint path shown in Figure 1(d) and endpoint path shown in Figure 1(c). For this hypothetical movement, there is no particular reason to choose between the two descriptions. For the nervous system, however, there may be good reasons to choose between the different frames of reference. Movements that optimize simplicity or energy efficiency or smoothness at one level of description may not simultaneously show the same optimization at another level. As a result, it is common (e.g., Soechting & Flanders, 1992) to ask if movements are planned from the perspective of any single coordinate system. In other words, does the nervous system try to use a coordinate system that is optimal in some sense?

In the example in Figure 1, we can see two candidate spatial coordinate systems—a joint level, and a hand or endpoint level. These two coordinate systems have featured prominently in a number of arm control schemes.[2] Soechting and Lacquaniti (1981), for example, proposed that reaches are planned with respect to joint variables (see also Flanagan & Ostry, 1990). Others have suggested that reaches are planned with respect to the hand's movement in Cartesian space (Georgopoulos, 1986; Morasso, 1981; Flash & Hogan, 1985). Both of these claims have been supported by the observance of kinematic invariances. Soechting and Lacquaniti (1981) reported that the angular velocities of shoulder and elbow movements were tightly coupled during the deceleration portion of a movement. This pattern was found to be independent of the load on the limb and the overall duration of the movement.[3]

A number of different kinds of evidence have been presented in favour of the idea that reaches are planned in the coordinates of the hand (Hogan, Bizzi, Mussa-Ivaldi, & Flash, 1987). Primary amongst this evidence is the observance of straight-line paths of the hand during reaching (Morasso, 1981). The inference is made that a straight-line path is optimal and that it is evidence that the planning is at a kinematic level for the hand. Other evidence in favor of hand-level planning is the observance of unimodal tangential velocity profiles for the hand. This is not always the case for the trajectories of the joints moving the arms. Again, the metric is simplicity and the unimodal velocity profiles are preferred. A final piece of evidence in favor of hand-level planning is the fact that the straight-line paths and unimodal velocity profiles for the motion of the hand are not influenced by the location of the reaches in the workspace. For the

---

[2]It is possible to model the same behaviors with dynamic coordinates. Wada and Kawato (1993) can account for many of the features of natural reaches with a minimum torque-change model.

[3]However, Hollerbach, Moore, and Atkeson (1987) have argued that the observed pattern results from working out at the extremes of the workspace. Soechting and Lacquaniti's subject performed reaches out from the body to the extremes of the space reachable by the arms. Under these conditions the joint angular velocities converge to a constant.

arm then, there is some indication that the nervous system coordinates the joints in the arm so as to produce a smooth and simple kinematic path for the endpoint of the arm.

In speech, similar issues can be addressed. Recently, we have addressed the issue of coordinate systems for mandibular movements. The mandible is a complex joint that has three rotational and three translational degrees of freedom. Its motion can be described in joint space or in terms of the position of points on the mandible in some vocal tract coordinate space. Using a number of different speech and nonspeech tasks we (Ostry & Munhall, in press) examined whether the paths of the mandible in joint space were linear. If so, this would suggest that the nervous system was controlling the degrees of freedom in a relatively simple, joint-based fashion, by tightly coupling their motion. We defined the joint space in terms of the rotation of the jaw in the sagittal plane as a function of translation along an axis parallel to the occlusal plane. Across different speaking rates, stress levels, and vowels and consonants, and for mastication, relatively straight-line motion was found in these joint space plots.

While the slopes of the functions varied with the consonants that were spoken, the same form was seen for a number of different speakers and conditions. It appears that the control of motion around the mandibular joint is simplified by controlling the trajectories for each degree of freedom in a similar fashion. Straight-line paths in joint space require that the motions in each degree of freedom start and stop at the same time and that the motions in each degree of freedom have similar velocity profiles. By doing this the nervous system may be optimizing control at the joint level rather than at a more global vocal tract level.

While such a proposal would simplify the control of the mandible, it does not fit with common assumptions about speech motor control: Sounds are specified in terms of articulatory targets or acoustic targets in which the region of the vocal tract and the degree of constriction can be specified. As Gay, Boë, Perrier, Feng, and Swayne (1991) point out, most acoustic modelling of the vocal tract works within this tradition. In this view, speech movements are planned with respect to a coordinate space defined by the vocal tract shape or its acoustic transfer function.

A number of pieces of evidence are consistent with this view. Hamlet's studies of adaptation to palatal prostheses are particularly informative (e.g., Hamlet & Stone, 1976, 1978; Hamlet, Stone, & McCarty, 1978). In her studies, speakers were fitted with prostheses that thickened the palate in the alveolar region and thus changed the shape and size of the oral cavity. Hamlet's subjects required a number of days to adapt to the prosthesis. When the subjects were fitted with new prostheses there was no shortening of the adaptation period. It appeared that each prosthesis had to be learned individually. The removal of the prosthesis produced a gradual return to the normal pattern of articulation.

In direct contrast to these data is evidence from static (Gay, Lindblom, & Lubker, 1981) and dynamic (Gracco & Abbs, 1985) perturbations of moving parts of the vocal tract, like the jaw and lips. In these latter studies, relatively rapid compensations are observed. In Hamlet's data the subjects slowly recalibrated their motion paths and in effect learned the dimensions of a new vocal tract. The data, as a whole, suggest that the position of the hard palate has a privileged position in the spatial planning of speech. The movable parts of the vocal tract appear to register their movements with respect to a consistent framework (Abbs, Gracco, & Cole, 1984).

2. The Shaping of Paths

While straight-line paths are frequently observed for unconstrained motions of the hand between a starting point and a target, motion of the tongue is quite complex in what appears to be similar circumstances (Houde, 1968). For example, in the production of a CVC, the tongue moves forward during the closure and thus the syllable is often produced with a looped path. Figure 2 shows 2-D data collected using the X-ray microbeam system at the University of Wisconsin. The speaker is producing the nonsense sequence /kæk/ under 2 speaking conditions. The top tracing is the mid-sagittal contour of the palate. The top three pairs of paths correspond to the movements of pellets placed on the tongue tip, blade, and dorsum during loud and normal speaking volumes. The bottom two pairs of paths correspond to pellets placed on the mandibular incisors and one of the mandibular molars. The tongue in this data is producing markedly curved paths. When more natural sequences are produced, the paths of individual tissue points on the tongue can trace even more complicated patterns. If speech were instead depicted as the transition between a series of targets in the vocal tract, we might expect a different pattern of paths; in particular, we might expect more straight-line paths, as are observed for limb movement.

While the observed movement patterns suggest a difference in control styles between limb movement and oral movement, two issues make this comparison difficult. First, the data that are presented in Figure 2, for example, are for arbitrary tissue points on the tongue surface. While they certainly reflect the general patterns of motion of the tongue they do not always perfectly reflect the goal-directed movements of the production system. Paths in limb movement can be defined because there is a well developed idea of the intentional focus of the movement. The hand is intimately tied to any goal or task description. In the oral cavity, we cannot define a single point on the tongue that is the goal-directed part of the articulator. The tongue's surface moves as a whole, and over a sequence one part is more important acoustically than another for a period of time. Bookstein (1992) has pointed out that this is one of the weaknesses of the idea of a Cartesian coordinate system for biological systems. That is, biological systems may not use the kind of orthogonal coordinates that we commonly use in experimental measurement. Furthermore, as we will point out below, sensory systems do not always provide such coordinates.

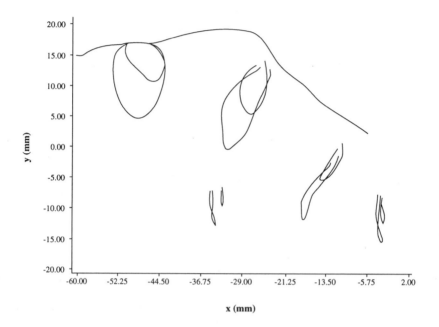

FIGURE 2. X-ray microbeam data showing tongue and jaw motion during the production of the nonsense utterance /kæk/ in normal and loud speaking conditions. The subject is facing right. The upper trace shows the mid-sagittal contour of the palate. The bottom two pairs of paths are from pellets on the mandibular incisors and a mandibular molar. The top three pairs of paths are from the motion of the tongue tip, blade, and dorsum. The larger path in each pair was recorded during the loud speaking condition.

The second issue is that, for the researcher, spatial targets cannot be well specified in speech. Classical limb movement effects like Fitts' law, cannot be tested in speech since target location and target width are unknown quantities. Clarity of speech certainly can be manipulated (Moon & Lindblom, 1989; Picheny, Durlach, & Braida, 1986) but this cannot be easily translated into a measure of how precise the articulation is.

While unconstrained reaches normally are produced with relatively straight-line paths, it is clear that we can guide our hands through space with curved or more complex paths. We do this naturally when reaching around obstacles and during drawing and handwriting. Two different kinds of explanations have been offered to explain this ability. The first class of explanation assumes that movements are planned in endpoint coordinates; that is, that curved paths are explicitly planned. Flash and Hogan (1985) have suggested that the nervous system may specify a small number of 'via points' along the planned path and the hand must pass through these via points during the movement. The via points thus provide accuracy constraints beyond the starting point and ending point of the movement. In the Flash and Hogan model, the presence of via points and a smoothness constraint account for the characteristics of observed curved movements.

A second approach to planning in endpoint coordinates is to construct curved paths from a combination of small path segments (Viviani & Terzuolo, 1982). Curved movements are constructed as parametric composite curves in which several sequential of line segments are joined smoothly. Formally, this approach has much in common with spline interpolation and it is implied that in curved movements the path can be segmented into component 'pieces.' Research has shown that the trajectories of curved movements can be used to segment the movements. The instantaneous curvature of the movement correlates with the instantaneous velocity of the hand movement. In a simple curved arc motion, more than a single velocity peak will be found and a velocity minimum will be observed at a peak in curvature (Abend, Bizzi, & Morasso, 1982).

In contrast to this explicit planning, curved paths may be produced 'unintentionally' as a by-product of other constraints on coordination. One example of this can be seen in Figure 1, in which a curved endpoint path results from attempts to simplify coordination in joint coordinates. The straight-line path in the joint coordinates (Figure 1(d) produces the curved path in the hand coordinates observed in Figure 1(c). In other cases, the path may be influenced by external forces. For example, Atkeson and Hollerbach (1985) have reported that in vertical reaches there is some curvature that is added to the movements by the effects of gravity. Wolpert, Ghahramani, and Jordan (in press) have argued that the slight curvature that is observed in unconstrained horizontal reaches may be due to a visual perceptual distortion of the curvature of the paths.

The question arises whether the paths of speech movements are specifically controlled or whether the observed paths are the end product of the process of coordination. There are many types of influences that could contribute to the observed paths in speech. Hoole and Munhall (1994) have explored the effects of the air pressure behind the oral closure on the paths. In speech with an ingressive airflow, the direction of the movement along the path changed. In loud speech, the curvature of the path increased. These observations suggest that the paths of the tongue might be influenced by small differences in the forces within the vocal tract during speech production. A second type of influence on the paths is the coarticulation of successive speech sounds (Perkell, 1969). Much evidence exists that suggests that the "targets" and the paths to these positions change in different contexts (e.g., Parush et al., 1983).

On the other hand, the paths in speech could be planned explicitly to be curved. Perkell (1969) suggested that moving the tongue continuously during speech may be more efficient. The forward motion during consonant closure may serve this purpose (cf. Munhall, Ostry, & Flanagan, 1991). Coker (1976) has suggested that the curved paths of the tongue may be necessary to create acoustic patterns that are important for distinguishing certain speech contrasts. For example, Coker (1976) noted that if the elliptical paths were not reproduced in articulatory synthesis, /g/ and /d/ were confused in some contexts. Currently, we cannot determine the relative contributions of the different explanations for the observed speech paths, but it is likely that a number of different influences shape the form of the path. This is consistent with the approach taken by researchers at ATR Laboratories (e.g., Hiroyama, Vatikiotis-Bateson, & Kawato, 1993; Vatikiotis-Bateson, Hiroyama, Wada, & Kawato, 1993). In their work, speech production is modeled computationally using artificial neural networks that learn the relationships between EMG signals, kinematic patterns and acoustic signals. The actual path of the articulators is determined jointly by the via points (targets), smoothness constraints, and the dynamics of the physiological system.

3. Inverse Problems

If curved reaches and speech movements were explicitly planned in kinematic coordinates with reference to the motion of the hand or vocal tract surface, the question would arise as to how joint and muscle activations might be determined to produce the planned paths correctly. This is known as an inverse kinematic and/or inverse dynamics problem. The nervous system must work backwards from the desired path of the hand to the forces and motions of component joints. There are two separate difficulties related to these inverse problems. One of the major difficulties in inverse kinematics is the many-to-one mapping that exists between the component articulators and the overall pattern of motion. In a natural reach, the arm is a redundant system. This means that the limb has more degrees of freedom spread across its joints than are needed to specify the position of the endpoint. A second difficulty is simply the computations required to solve the inverse kinematics and dynamics. It has been suggested that the nervous system does not have the computational resources to solve the inverse dynamics equations in real time (Loeb, 1983).

As noted above, speech goals are frequently specified in terms of vocal tract positions or a particular acoustic pattern. If the nervous system plans speech movements in this way, it must determine the relative contributions of the individual articulators to the achievement of this goal. In Saltzman's task-dynamics model (Saltzman, 1986; Saltzman & Munhall, 1989), a series of inverse mappings is carried out. Speech goals are defined in a task space that determines the dynamic characteristics of the motion; then they are mapped onto tract variables. Tract variables are local vocal tract constrictions. For example, oral aperture is a tract variable that is controlled by the upper lip, lower lip, and jaw. Finally, the tract variable is mapped onto the articulator motions.

Inherent in the idea of solving inverse equations is the idea of a precisely specified goal. Thus, in the task-dynamics model, tract variable constrictions are precisely specified in vocal tract coordinates. The jaw data reported by Ostry and Munhall (in press) suggest that vocal tract constrictions determined by lip-jaw and tongue-jaw coordination may be influenced by the strategy for jaw control. As has been said above, however, such joint-based control seems at odds with ideas about controlling the production of speech at the level of vocal tract shapes. There are two solutions to this contradiction. The first is that there may be little interference from the jaw's control style in achieving vocal tract goals, because the translation components of the jaw motion are relatively small and thus the path of the mandibular incisor in Cartesian coordinates is relatively linear (Vatikiotis-Bateson & Ostry, in press). The second solution is that precision in speech motor control may not have to be high. Acceptable bilabial closures, for example, can be produced with a wide range of contact forces between the lips. Or, although the vowel /i/ is produced with lateral tongue contact with the palate and teeth, Fujimura and Kakita (1977) have shown with a finite element model of the tongue, /i/ can be produced with a wide range of contractions of the genioglossus muscle. Finally, a similar "saturation" effect can be found for the hyoglossus muscle in the production of the vowel /ɑ/ (Perkell, 1990).

4. Coordinating the Coordinate Spaces

There are two parts to the many-to-one problem described above. The first is that some movement systems are redundant. The second part is that dimensions of the coordinate systems are different for each of the levels of description. Thus, the muscles can be described in terms of the coordinates of muscle length and tension, whereas the joints can be described kinematically by rotational coordinates and dynamically in terms of torques around a center of rotation, and the position of the hand can be described kinematically in terms of Cartesian coordinates and dynamically in terms of linear force vectors.

The spatial coordinates of sensory feedback systems like the kinesthetic system add another level of complexity (see Gracco, this volume). Sensory coordinates are determined by the orientation and position of the sensory transducers and they do not necessarily correspond to other coordinates used in planning and control. For example, in the vestibulo-ocular reflex there is a mismatch between the coordinates of the vestibular system and the muscles controlling eye movements. The vestibular system is composed of three semi-circular canals oriented in three nonorthogonal planes and it provides information in the coordinates defined by the canals' orientations regarding the head's position and movement through space. In the vestibulo-ocular reflex, the eyes produce compensatory movements that stabilize visual images on the retina during head or whole body movement. To produce this compensation by the eyes, the motor system must be provided with information about the extent and speed of the motion of the head. A coordinate transformation of the vestibular signals seems to be necessary because the planes of the semicircular canals are not perfectly aligned with the lines of action of the three extraocular muscle pairs used to rotate the eye (Baker, Banovetz, & Wickland, 1988).

In the oral cavity we have a rich supply of sensory information (Kent, Martin, & Sufit, 1990) as well as information provided by the auditory feedback from speaking. Each of these sources of information potentially defines a unique set of coordinates that must be integrated if feedback is to be used to control speech movements. For example, muscle spindles are a primary source of information about articulator positions, and spindles are distributed throughout the muscles in the oral cavity. The coordinates of the information provided by these transducers are determined by the number of spindles and the lines of action of the muscles they are embedded in. In both the oral cavity and the limbs, the muscle spindles are unevenly distributed across muscles. In jaw-opening muscles, like the digastric and lateral pterygoid, there have been few reports of muscle spindles, whereas in the jaw-closing muscles such as the masseter and temporalis there is a rich supply of spindles. This suggests that the opening motion of the jaw may be less tightly controlled than jaw closing. Recently, Scott and Loeb (1993) have examined the theoretical distribution of muscle spindles in the human arm. In their simulations, they estimated the relative accuracy of joint motions and the accuracy of the positioning of the limb endpoint based on the actual number of spindles in muscles controlling each joint. They found that the density of spindles across the various muscles predicted the psychophysical positioning data very well. Thus, muscle spindles may provide a spatial framework that is not a Cartesian representation of space and a spatial framework that does not have the same level of accuracy in all directions.

In speech, acoustic feedback poses an additional coordinates problem. The relation between the movements in the vocal tract and the resulting acoustics is nonlinear. More than a single tract shape can produce the same acoustic output (Atal, Chang, Mathews, & Tukey, 1978) and this increases the difficulty of the motor learning problem in acquiring speech. Jordan (1994) has shown that a connectionist model learning the articulatory-acoustic mapping has more difficulty learning the vowel /u/ than other vowels because there is more than a single vocal tract solution.

In spite of this nonlinearity, evidence consistent with planning in acoustic coordinates has been reported by Perkell and Nelson (1985). They examined the variance of tongue points during the production of vowels, and found that the spatial variation of the points was greater in directions that were acoustically less important for that part of the tongue. Thus, in the production of /i/, the height variance of the tongue tip was found to be less than the front/back variance. In contrast, the tongue dorsum varied less in the front/back plane for /ɑ/ than it did in the tongue height plane. In these examples, we see a strong relationship between the coordinates of vocal tract constrictions, the coordinates defined by the formant frequencies for the vowels, and the "saturation" effects in tongue control described above (Perkell, 1990). Unfortunately, there is nothing to suggest that the simple correspondence between coordinates observed by Perkell will be the general rule.

SPATIAL CONTROL OF SPEECH MOVEMENTS

Even from our limited views of the moving vocal tract, it is clear that the articulators produce a rapid and complicated spatial pattern during speech. In Houde's (1968) and Perkell's (1969) monographs, we can see 2-D patterns of motion for points on the tongue. The spatial path taken by the tongue is often elliptical in CVC productions, and, as can be seen in Figure 2, the tongue and jaw produce distinctly different spatial paths. To coordinate this behavior, the nervous system must work with many different coordinate systems during articulation. Whether some coordinates are more important during speech, however, is unclear. Flanders, Helms Tiller, and Soechting (1992) have presented data that suggest that the coordinate frame in which movements are planned may switch with task conditions. In their experiments, the origin of the coordinates seemed to be different during reaching movements with visual feedback than during reaching movements in the dark. This raises the possibility that no single coordinate system will suffice for all tasks.

The idea that speech production involves a spatial frame of reference is not a new idea. MacNeilage (1970) proposed that targets in speech were specified with respect to an "internalized space coordinate system." MacNeilage's proposal was intended to account for some of the complexities of serial ordering in speech and in his scheme a sequence was translated into a series of spatial target specifications. MacNeilage (1980) and others have pointed out some of the weaknesses of the overall model of serial order. These include the fact that not all goals in speech are positions (e.g., diphthongs) and that more than one vocal tract configuration can be used for a particular vowel. The fact that difficulties have been raised with the model, however, does not mean that the nervous system does not have to organize its movements with respect to a space coordinate system.

In this chapter we have tried to outline some of the key issues in the spatial control of speech. In many cases, these are 'in principle' issues that *could* plague a motor control system. In fact, the remarkable fluency of natural speech is evidence that these problems are usually circumvented by the production system. It is our goal to understand how the nervous system overcomes these potential difficulties.

ACKNOWLEDGMENTS

This work was supported by grants from NIH (DC-000594) and NSERC.

REFERENCES

Abbs, J., Gracco, V., & Cole, K. (1984). Control of multimovement coordination: Sensorimotor mechanisms in speech motor programming. *Journal of Motor Behavior, 16,* 195-232.

Abend, W., Bizzi, E., & Morasso, P. (1982). Human arm trajectory formation. *Brain, 105,* 331-348.

Abry, C., and Boë, L-J. (1986). "Laws" for lips. *Speech Communication, 5,* 97-104.

Atal, B., Chang, J., Mathews, M., & Tukey, J. (1978). Inversion of articulatory-to-acoustic transformation in the vocal tract by a computer-sorting technique. *Journal of the Acoustical Society of America, 63,* 1535-1555.

Atkeson, C. G., & Hollerbach, J. M. (1985). Kinematic features of unrestrained vertical arm movements. *The Journal of Neuroscience, 5,* 2318-2330.

Baer, T., Gore, J. C., Gracco, L. C., & Nye, P. W. (1991). Analysis of vocal tract shape and dimensions using magnetic resonance imaging: Vowels. *Journal of the Acoustical Society of America, 90,* 799-828.

Baker, J. F., Banovetz, J. M., & Wickland, C. R. (1988). Models of sensorimotor transformations and vestibular reflexes. *Canadian Journal of Physiological Pharmacology, 66,* 532-539.

Bookstein, F. L. (1992). Error analysis, regression, and coordinate systems. *Behavioral and Brain Sciences, 15,* 327-329.

Coker, C. (1976). A model of articulatory dynamics and control. *Proceedings of the IEEE, 64,* 452-460.

Flanagan, J. R., & Ostry, D. J. (1990). Trajectories of human multi-joint arm movements: Evidence of joint level planning. In V. Hayward (Ed.), *Experimental robotics.* New York: Springer-Verlag.

Flanders, M., Helms Tiller, S. I., & Soechting, J. F. (1992). Early stages in a sensorimotor transformation. *Behavioral and Brain Sciences, 15,* 309-362.

Flash, T., & Hogan, N. (1985). The coordination of arm movements: An experimentally confirmed mathematical model. *The Journal of Neuroscience, 5,* 1688-1703.

Fujimura, O., & Kakita, Y. (1979). Remarks on the quantitative description of lingual articulation. In B. Lindblom & S. Iman (Eds.), *Frontiers of speech communication research* (pp. 17-24). London: Academic Press.

Gay, T., Boë, L-J., Perrier, P., Feng, G., & Swayne, E. (1991). The acoustic sensitivity of vocal tract constrictions: A preliminary report. *Journal of Phonetics, 19,* 445-452.

Gay, T., Lindblom, B., & Lubker, J. (1981). Production of bite-block vowels: Acoustic equivalence by selective compensation. *Journal of the Acoustical Society of America, 69,* 802-810.

Georgopoulos, A. P. (1986). On reaching. *Annual Reviews of Neuroscience, 9,* 147-170.

Gracco, V., & Abbs, J. (1985). Dynamic control of the perioral system during speech: Kinematic analyses of autogenic and nonautogenic sensorimotor processes. *Journal of Neurophysiology, 54,* 418-432.

Hamlet, S., & Stone, M. (1976). Compensatory vowel characteristics resulting from the presence of different types of experimental dental prostheses. *Journal of Phonetics, 4,* 199-218.

Hamlet, S., & Stone, M. (1978). Compensatory alveolar consonant production induced by wearing a dental prosthesis. *Journal of Phonetics, 6,* 227-248.

Hamlet, S., Stone, M., & McCarty, T. (1978). Conditioning dentures viewed from the standpoint of speech adaptation. *Journal of Prosthetic Dentistry, 40,* 60-66.

Hiroyama, M., Vatikiotis-Bateson, E., & Kawato, M. (1993). Physiologically-based speech synthesis using neural networks. *IEICE Transactions on Fundamentals of Electronics, Communications, and Computer Sciences,* Vol. E76-A, 1898-1910.

Hogan, N., Bizzi, E., Mussa-Ivaldi, F. A., & Flash, T. (1987). Controlling multijoint motor behavior. *Exercise and Sports Science Reviews, 15,* 153-190.

Hollerbach, J. M., & Flash, T. (1982). Dynamic interactions between limb segments during planar arm movement. *Biological Cybernetics, 44,* 67-77.

Hollerbach, J. M., Moore, S. P., & Atkeson, C. G. (1986). Workspace effect in arm movement kinematics derived by joint interpolation. In G. Gantchev, P. Gatev., & B. Dimitrov (Eds.), *Motor control.* New York: Plenum.

Hoole, P., & Munhall, K. G. (1994). Do air-stream mechanisms influence tongue movement paths? *Journal of the Acoustical Society of America, 95,* 2821.

Houde, R. A. (1968). A study of tongue body motion during selected speech sounds. In SCRL Monograph (No. 2) Santa Barbara, CA: Speech Communications Research Laboratory.

Jordan, M. I. (1994). Unpublished data.

Kent, R. D., Martin, R., & Sufit, R. L. (1990). Oral sensation: A review and clinical prospective. In W. Winitz (Ed.), *Human communication and its disorders,* Vol. 3. Ablex: Norwood.

Kier, W. M., & Smith, K. K. (1985). Tongues, tentacles and trunks: the biomechanics of movement in muscular-hydrostats. *Zoological Journal of the Linnean Society, 83,* 307-324.

Loeb, G. (1983). Finding common ground between robotics and physiology. *Trends in NeuroScience, 6,* 203-204.

MacNeilage, P. F. (1970). Motor control of serial ordering of speech. *Psychological Review, 77,* 182-196.

MacNeilage, P. F. (1980). Distinctive properties of speech motor control. In G. E. Stelmach & J. Requin (Eds.), *Tutorials in motor behavior* (pp. 607-621).

Moon, S.-J., & Lindblom, B. (1989). Formant undershoot in clear and citation-form speech: A second progress report. 121-123 in STL-QPSR 1/1989 (Department of Speech Communication, RIT, Stockholm).

Morasso, P. (1981). Spatial control of arm movements. *Experimental Brain Research, 42,* 223-227.

Munhall, K. G., Ostry, D. J., & Flanagan, J. R. (1991). Coordinate spaces in speech planning. *Journal of Phonetics, 19,* 293-307.

Ostry, D. J., & Munhall, K. G. (1994). Control of jaw orientation and position in mastication and speech. *Journal of Neurophysiology, 71,* 1528-1545.

Parush, A., Ostry, D. J., & Munhall, K. G. (1983). A kinematic study of lingual coarticulation in VCV sequences. *Journal of the Acoustical Society of America, 77,* 640-648.

Perkell, J. S. (1969). *Physiology of speech production.* Cambridge, MA: The MIT Press.

Perkell, J. S. (1990). Testing theories of speech production: Implications of some detailed analyses of variable articulatory data. In W. J. Hardcastle & A. Marchal (Eds.), *Speech production and speech modelling* (pp. 263-288). Dordrecht: Kluwer.

Perkell, J. S., & Nelson, W. L. (1985). Variability in production of the vowels /i/ and /a/. *Journal of the Acoustical Society of America, 77,* 1889-1895.

Picheny, M. A., Durlach, N. I., & Braida, L. D. (1986). Speaking clearly for the hard of hearing II: Acoustic characteristics of clear and conversational speech. *Journal of Speech and Hearing Research, 29,* 434-446.

Saltzman, E. (1986). Task dynamic coordination of the speech articulators: A preliminary model. *Experimental Brain Research, series 15,* 129-144.

Saltzman, E. L., & Munhall, K. G. (1989). A dynamic approach to gestural patterning in speech production. *Ecological Psychology, 1(4),* 333-382.

Scott, S. H., & Loeb, G. E. (1993). The computation of position-sense from mono- and multiarticular muscle spindles. *Society for Neuroscience Abstracts, 19* (part 2), 993.

Soechting, J. F., & Flanders, M. (1991). Arm movements in three-dimensional space: Computation, theory, and observation. *Exercise and Sport Sciences Reviews, 19,* 389-418.

Soechting, J. F., & Flanders, M. (1992). Moving in three-dimensional space: Frames of reference, vectors, and coordinate systems. *Annual Reviews of Neuroscience, 15,* 167-191.

Soechting, J. F., & Lacquaniti, F. (1981). Invariant characteristics of a pointing movement in man. *The Journal of Neuroscience, 1,* 710-720.

Stone, M. (1990). A three-dimensional model of tongue movement based on ultrasound and X-ray microbeam data. *Journal of the Acoustical Society of America, 87,* 2207-2217.

Vatikiotis-Bateson, E., Hiroyama, M., Wada, Y., & Kawato, M. (1993). Generating articular motion from muscle activity using artificial neural networks. *Annual Bulletin Research Institute of Logopedics and Phoniatrics (RILP), 27,* 67-77.

Vatikiotis-Bateson, E., & Ostry, D. J. (in press). An analysis of the dimensionality of jaw motion in speech. *Journal of Phonetics.*

Viviani, P., & Terzuolo, C. (1982). Trajectory determines movement dynamics. *Neuroscience, 7,* 431-437.

Wada, Y., & Kawato, M. (1993). A neural network model for arm trajectory formation using forward and inverse dynamics models. *Neural Networks, 6,* 919-932.

Wolpert, D. M., Ghahramani, Z., & Jordan, M. I. (1994). Perceptual distortion contributes to the curvature of human reaching movements. *Experimental Brain Research, 98,* 153-156.

Zajac, F. E., & Gordon, M. E. (1989). Determining muscle's force and action in multi-articular movement. *Exercise and Sports Science Reviews, 17,* 187-230.

Author Index

Numbers in *italics* refer to reference pages.

Subject Index

Numbers in *italics* refer to reference pages.

557